Harmonic Analysis
on Homogeneous Spaces

PURE AND APPLIED MATHEMATICS

A Series of Monographs and Textbooks

1. K. Yano. Integral Formulas in Riemannian Geometry (1970)
2. S. Kobayashi. Hyperbolic Manifolds and Holomorphic Mappings (1970)
3. V. S. Vladimirov. Equations of Mathematical Physics (A. Jeffrey, editor; A. Littlewood, translator) (1970)
4. B. N. Pshenichnyi. Necessary Conditions for an Extremum (L. Neustadt, translation editor; K. Makowski, translator) (1971)
5. L. Narici, E. Beckenstein, and G. Bachman. Functional Analysis and Valuation Theory (1971)
6. D. S. Passman. Infinite Group Rings (1971)
7. L. Dornhoff. Group Representation Theory (in two parts). Part A: Ordinary Representation Theory. Part B: Modular Representation Theory (1971, 1972)
8. W. Boothby and G. L. Weiss (eds.). Symmetric Spaces: Short Courses Presented at Washington University (1972)
9. Y. Matsushima. Differentiable Manifolds (E. T. Kobayashi, translator) (1972)
10. L. E. Ward, Jr. Topology: An Outline for a First Course (1972)
11. A. Babakhanian. Cohomological Methods in Group Theory (1972)
12. R. Gilmer. Multiplicative Ideal Theory (1972)
13. J. Yeh. Stochastic Processes and the Wiener Integral (1973)
14. J. Barros-Neto. Introduction to the Theory of Distributions (1973)
15. R. Larsen. Functional Analysis: An Introduction (1973)
16. K. Yano and S. Ishihara, Tangent and Cotangent Bundles: Differential Geometry (1973)
17. C. Procesi. Rings with Polynomial Identities (1973)
18. R. Hermann. Geometry, Physics, and Systems (1973)
19. N. R. Wallach. Harmonic Analysis on Homogeneous Spaces (1973)

In Preparation:

B. Y. Chen. Geometry of Submanifolds
J. Dieudonné. Introduction to the Theory of Formal Groups
R. Larsen. Banach Algebras: An Introduction
M. Marcus. Finite Dimensional Multilinear Algebra (in two parts)
C. Procesi. Rings with Polynomial Identities
K. B. Stolarsky. Algebraic Numbers and Diophantine Approximation
I. Vaisman. Cohomology and Differential Forms

Harmonic Analysis on Homogeneous Spaces

NOLAN R. WALLACH

Department of Mathematics
Rutgers, The State University of New Jersey
New Brunswick, New Jersey

1973

MARCEL DEKKER, INC., New York

MARCEL DEKKER, INC.
95 Madison Avenue, New York, New York 10016

LIBRARY OF CONGRESS CATALOG CARD NUMBER: 72–91437
ISBN: 0–8247–6010–7

PRINTED IN THE UNITED STATES OF AMERICA

To

Barbara

Contents

Preface

The intent of this book is to give students of mathematics and mathematicians in diverse fields an entry into the subject of harmonic analysis on homogeneous spaces. It is hoped that the book could be used as a supplement to a standard one-year course in Lie groups and Lie algebras or as the main text in a more unorthodox course on the subject.

Pains have been taken to build the subject matter gradually from fairly easy material to the more advanced material. Exercises have been included at the end of each chapter to help the reader's comprehension of the material. We have also adopted a decimal system of labeling statements, lemmas, theorems, etc. (For example, 3.7.2(1) means the first displayed formula or assertion in the second paragraph of the seventh section of Chapter 3.) We felt after much soul-searching that this system of presentation would in the long-run make the book more accessible since it simplifies the cross-referencing. However, once one decides to consistently use a system of labeling such as ours, one is forced to write the material of the book in a "quantized" form. The writing also becomes dry and discussions of material are hard to fit into the text. It is hoped that the underlying mathematics will motivate the reader to tolerate the exposition. To somewhat temper the dryness of exposition, each chapter has an introduction which will, hopefully, help the reader to organize and motivate the material of the chapter.

We have also opted to label theorems generally only by a number. Thus, for example, the Peter-Weyl Theorem is *2.8.2. Theorem*. At the end of each chapter is a section labeled *Notes*. In this section we do our best to acknowledge the original source of a result and to credit proofs used to their discoverers. This being said, the reader should not assume that unattributed results in this book are due to the author. Much of the unattributed material has become part of the folklore of this subject or has at least become a part of the author's subconscious knowledge of the subject. We are, however, proud to say that we have taken the trouble to give new proofs of at least some of the material we thought was unnecessarily complicated and that some of the material in this book has not appeared elsewhere (to the author's knowledge) in published form.

At the end of this book there are eight appendices which give the reader a "guided tour" through the prerequisites of the book. Many proofs are included, and in fact appendices 3, 6, and 7 are essentially self-contained. Appendix 8 contains a technical result on asymptotics for certain eigenfunction problems (the main idea in that appendix is taken from Dunford and Schwartz [1]). In the other appendices we give the definitions of terms used and the statements of the precise forms of theorems we use. The unproven theorems are referenced to books that have proofs of almost precisely the same statements.

It can reasonably be said that this book is the study of group actions on cross-sections of vector bundles (or more simply induced representations). We develop in the first chapter the material on vector bundles and differential operators that will be used throughout the book. It is suggested that a reader use Chapter 1 basically for reference purposes.

In Chapter 2, we develop the elementary representation theory that will be used throughout the book. In Section 2.4 we give the definition of "induced representation" which will be studied in the book (the author learned this definition from R. Hermann [1]). Chapter 2 also contains the Peter-Weyl theorem (the idea of our proof is taken from Gelfand, Graev, Pyatetskii-Shapiro [1]) and elementary character theory.

In Chapter 3, we develop the basic structure theory of Weyl, Cartan for compact and semisimple Lie algebras, that is, maximal tori, roots, and the Weyl group. The material of this chapter is standard except for the inclusion of a result of Borel-Mostow on automorphisms of semisimple Lie algebras and a theorem of Samelson giving the Euler characteristics of a homogeneous space as an integral.

In Chapter 4, we give the basic structure of the universal enveloping algebra. Using the universal enveloping algebra we prove the theorem of the highest weight (following Cartier's simplification of the Harish-Chandra proof). We use the theorem of the highest weight to develop the topology of compact Lie groups (due to E. Cartan and Stiefel). We then prove the Weyl integral and character formulas.

In Chapter 5, we apply the Peter-Weyl theorem and Weyl character formula to the problem of decomposing induced representations of compact Lie groups. We develop the basic material on the Sobolev spaces attached to a homogeneous space in a manner completely analogous to the classical Sobolev spaces on a torus (the exposition of this section follows the exposition of F. Warner [1] on Sobolev spaces for the n-torus). We then use the

Sobolev theory to prove regularity theorems for homogeneous differential operators. We conclude the chapter with a formula due to Bott for the index of a Fredholm homogeneous differential operator on a homogeneous vector bundle.

In Chapter 6, the theory of Chapter 5 is used to give explicit realizations of the representations of compact Lie groups using "holomorphic induction." The pioneering work in this direction is due to Borel and Weil. We also use the index theorem of Chapter 5 to derive a formula of Borel-Hirzebruch [1]. The formula of Borel-Hirzebruch has been recently generalized to noncompact semisimple Lie groups by Okamoto and Narasimhan [1], W. Schmid [1], and J. Wolf [1].

In Chapter 7, we develop preliminary structural and analytic results to be used in the representation theory of Chapter 8. In this chapter the Cartan, Iwasawa, Bruhat, Gelfand-Naimark decompositions of semisimple Lie groups are proven. The integral formulas (due to Harish-Chandra) for these decompositions are then developed. We then prove integral formulas for the adjoint action of a semisimple Lie group on its Lie algebra (also due to Harish-Chandra). We then establish the structure theory of the class of Lie groups for which the deeper results of Chapters 7 and 8 will apply (the class contains, in particular, the complex semisimple Lie groups and "half the Lorentz groups"). We conclude the chapter by proving a formula which implies the Plancherel theorem for the class of groups alluded to above. The general form of this formula for semisimple Lie groups is due to Harish-Chandra [7]. We, however, generalize the more elementary techniques of Harish-Chandra's earlier paper [3].

In Chapter 8 representation theory for noncompact Lie groups is developed. We first introduce the principal series (induced representations in the sense of this book). We give three models of the principal series so that the reader will be able to read the extensive literature on these representations. We then develop the character theory of K-finite representations and in particular prove a character formula for the principal series. We then develop the theory of intertwining operators for principal series representations, using the formalism of Schiffman [1]. We do the analytic continuation of the intertwining operators by explicitly computing (using a lemma of B. Kostant and a technique of S. Helgason) the Fourier transforms of intertwining operators in terms of gamma functions. We relate the intertwining operators to the asymptotics of the matrix elements of the principal series for rank one groups. These asymptotics have been worked out in complete generality by Harish-Chandra [9]. The specific result in this book can be found in Knapp and Stein [1], however our proof is quite different. We then prove that "almost all" of the principal series representations are irreducible. We then give the Kunze-Stein [1] normalization of the intertwining operators. Finally the Plancherel formula, for the class of semisimple Lie groups described above, is given using the technique of Harish-Chandra [3].

As is apparent from this preface, the first half of this work owes a considerable debt to the pioneering work of H. Weyl and E. Cartan and the last two chapters develop part of the work of Harish-Chandra. It also makes free use of results and techniques of Gelfand-Naimark, Kunze-Stein, B. Kostant, S. Helgason, G. Schiffmann, and Knapp-Stein. To all those whose work we use in this book without proper acknowledgement, we express our deepest apologies.

No book on the subject of Lie groups can be written without expressing a debt to C. Chevalley's *Theory of Lie Groups* and S. Helgason's *Differential Geometry and Symmetric Spaces*. These two books allowed students access to the profound work of S. Lie, E. Cartan, H. Weyl, and Harish-Chandra in the theory of Lie groups. When the author was a student, it could accurately be said that these two books were the only prerequisites to research in the subject. Although the subject has grown tremendously since the publication of these remarkable books, they are the basic texts of the subject. We would be highly honored if our book would be added to the shelf of a student containing the books of Chevalley, Helgason, and the recent books of G. Warner [1] to name but a few.

We thank the Alfred P. Sloan Foundation for the fellowship that allowed us time to write this book and the National Science Foundation for partial support.

Finally, the author proudly expresses a different sort of debt to his wife Barbara, to whom this book is dedicated.

Suggestions to the Reader

Appendices 1–7 combined with Chapters 1, §1–5, 2, 3, 4, and 6 (assuming several results of Chapter 5), Chapter 7, §1–6, Chapter 8, §1–6, §9 could be a one-year course in Lie group theory.

A second course in Lie theory emphasizing the analysis could be Chapters 5, 7, and 8 (with Appendix 8).

It is suggested to the general reader that he peruse the book, read introductions, decide what is interesting to him and to "cross-reference" the necessary background material.

Harmonic Analysis on Homogeneous Spaces

CHAPTER 1

Vector Bundles

1.1 Introduction

The purpose of this chapter is to develop the basic material about vector bundles and differential operators. We actually develop more material than we will need in this book. If a reader is already fluent with the formalism of differential operators on vector bundles then he should by all means skip this chapter, using it only as a dictionary of notation. On the other hand, if the reader is not very fluent with vector bundles then there is enough material in this chapter to give him the "feel" of the subject. We also suggest the exercises at the end of this chapter to anyone wishing to "brush up" on his understanding of vector bundles and differential operators.

1.2 Preliminary Concepts

1.2.1 *Definition* Let X be a topological space. Let K denote either R or C (the real or complex numbers). A K-vector bundle over X is a topological space E and a continuous map p of E to X satisfying

(1) If x is in X then $E_x = p^{-1}(x)$ is a K-vector space.

(2) For each x in X there is a neighborhood U of x and a homeomorphism

F of $E|_U = p^{-1}(U)$ with $U \times K^n$ such that $F(v) = (p(v), f(v))$ and f is a linear map of the K-vector space $E_{p(v)}$ onto K^n.

1.2.2 *Examples* (1) Let $E = X \times K^n$ with $p(x, v) = x$.

(2) Let X be a differentiable manifold. A.1.2.6 says that $T(M)$ is a vector bundle over M.

1.2.3 *Definition* Let E and F be vector bundles over X. A homomorphism of the vector bundles E and F is a continuous map h of E to F so that $h(E_x)$ is contained in F_x and h is a linear map from E_x to F_x for each x in X. A vector bundle homomorphism is called an isomorphism if it is a homeomorphism.

1.2.4 Let (E, p) be a vector bundle over X. Let $\mathfrak{U}$ be an open covering of X so that for each U in $\mathfrak{U}$ there is a map F_U of $E|_U$ onto $U \times K^n$ satisfying 1.2.1(2). Such a covering is called a trivializing covering for E.

Consider for each U, V in $\mathfrak{U}$ the map $F_V \circ F_U^{-1}$ of $U \cap V \times K^n$ onto $U \cap V \times K^n$. Then for each y in X, v in K^n,

$$F_V \circ F_U^{-1}(y, v) = (y, g_{V,U}(y)v)$$

with $g_{U,V}(y)$ in $GL(n, K)$ (see A.2.1.2). Furthermore, since F_U and F_V are homeomorphisms, the map $g_{V,U}$ of $U \cap V$ into $GL(n, K)$ is continuous. We also note that if y is in $U \cap V \cap W$ then

(1) $$g_{W,V}(y) \cdot g_{V,U}(y) = g_{W,U}(y).$$

Equation (1) is called the cocycle condition. $g_{U,V}$ is called the cocycle with values in $GL(n, K)$ associated with the trivialization (U, F_U).

We now abstract this concept.

1.2.5 *Definition* Let X be a topological space and let $\mathfrak{U}$ be an open covering of X. Let g be an assignment of a continuous map $g_{U,V}$ of $U \cap V$ into $GL(n, K)$ for each pair U, V of elements of $\mathfrak{U}$, so that the cocycle condition 1.2.4(1) is satisfied. Then $\mathfrak{U}$ is called a $\mathfrak{U}$-cocycle with values in $GL(n, K)$.

1.2.6 *Theorem* Let $\mathfrak{U}$ be an open covering of the topological space X and let g be a $\mathfrak{U}$-cocycle with values in $GL(n, K)$. Then there is a unique (up to isomorphism) vector bundle E over X so that $\mathfrak{U}$ is a trivializing covering for E and g is a corresponding cocycle.

PROOF Let $Y = X \times K^n \times \mathfrak{U}$ where $\mathfrak{U}$ is given the discrete topology and Y is given the product topology. Let $T = \{(x, v, U)|x \text{ is in } U\}$. Give T the subspace topology in Y. We define an equivalence relation on T as follows: $(x, v, U) \equiv (y, w, V)$ if $x = y$ and $w = g_{V,U}(x)v$. We note that the cocycle condition is just the statement that $\equiv$ is an equivalence relation. Let E be the set of all equivalence classes of T. Let, for (x, v, U) in T, $q(x, v, U)$ be the equivalence class of (x, v, U). Give E the topology that makes q continuous and open (this is called the quotient topology). Let $\tilde{p}(x, v, U) = x$. Then $\tilde{p}$ defines a continuous map of T to X. Since $\tilde{p}(x, v, U)$ depends only on $q(x, v, U)$, $\tilde{p}$ induces a continuous map p of E onto X.

Let now x be in X, v, w be in $p^{-1}(x)$. Fix U in $\mathfrak{U}$ so that x is in U. Then $v = q(x, v', U)$ and $w = q(x, w', U)$. If a, b are in K we set $av + bw = q(x, av' + bw', U)$; the cocycle condition guarantees that $av + bw$ is well defined. Thus with this vector space structure on $p^{-1}(x)$, E satisfies 1.2.1(1).

By definition of E the map F_U^{-1} of $U \times K^n$ to E defined by $F_U^{-1}(x, v) = q(x, v, U)$ is a homeomorphism of $U \times K^n$ with $p^{-1}(U)$. Furthermore the vector space operations on $p^{-1}(x)$ are defined so that F_U satisfies 1.2.1(2). This proves the existence of E.

Suppose that E' is a vector bundle over X and that relative to $\mathfrak{U}$ there are mappings F'_U of E'_U to $U \times K^n$ satisfying 1.2.1(2) and such that the corresponding cocycle is g. We define a map h from T to E'

$$h(x, v, U) = F_U'^{-1}(x, v).$$

Then by the definition of the cocycle associated with a trivialization, h respects equivalence classes and thus induces a continuous map h of E to E'. It is clear that h is a vector bundle homomorphism. Defining for v in $E'|_U$, $t(v) = q(F_U(v), U)$. Then t is well defined and is a vector bundle homomorphism. Since t defines a right and a left inverse for h, h is a vector bundle isomorphism.

1.2.7 *Proposition* Let g and h be $\mathfrak{U}$-cocycles with values in $GL(n, K)$. Let E_1 and E_2 be, respectively, the corresponding vector bundles over X. Then E_1 and E_2 are isomorphic if and only if for each U in $\mathfrak{U}$ there is a continuous map s_U of U into $GL(n, K)$ so that

(1) $$g_{U,V}(x) = s_U(x)h_{U,V}(x)s_V(x)^{-1}$$

for x in $U \cap V$.

PROOF Suppose that (1) is satisfied. Let T be as in the proof of Theorem 1.2.6. Let $\underset{g}{\equiv}$, $\underset{h}{\equiv}$ be the equivalence relations on T corresponding to g and h, respectively. Define a map $\tilde{\rho}$ of T to T by $\tilde{\rho}(x, v, U) = (x, s_U(x)v, U)$.

Then $\tilde{\rho}$ is continuous. Suppose that $(x, v, U) \underset{h}{\equiv} (x, w, V)$. We assert that this implies that $\tilde{\rho}(x, v, U) \underset{g}{\equiv} \tilde{\rho}(x, w, V)$. Indeed, $w = h_{V,U}(x)v$. Now

$$g_{V,U}(x)s_U(x)v = s_V(x)h_{V,U}(x)s_U(x)^{-1}s_U(x)v = s_V(x)h_{V,U}(x)v = s_V(x)w.$$

This proves the assertion. $\tilde{\rho}$ therefore induces a continuous map ρ of E_1 to E_1, which is easily seen to be a vector bundle isomorphism.

Suppose now that E_1 and E_2 are isomorphic. Let ρ be the isomorphism. If for each U in $\mathfrak{U}$, F_U^1 and F_U^2 are the maps satisfying 1.2.1(2) so that g and h are the corresponding cocycles, then defining, for x in U, $s_U(x)$ by $F_U^1(\rho(F_U^{1^{-1}}(x, v))) = (x, s_U(x)v)$ it is easy to see that (1) is satisfied.

1.2.8 Let g and h be $\mathfrak{U}$-cocycles with values in $GL(n, K)$. Then g and h are said to be equivalent if there is for each U in $\mathfrak{U}$ a map s_U of U into $GL(n, K)$ which is continuous and is such that g and h satisfy 1.2.7(1). The set of all equivalence classes of $\mathfrak{U}$-cocycles with values in $GL(n, K)$ is denoted by $H^1(\mathfrak{U}; GL(n, K))$. 1.2.6 and 1.2.7 combine to prove that there is a bijection between $H^1(\mathfrak{U}; GL(n, K))$ and the isomorphism classes of K-vector bundles over X of fiber dimension n having $\mathfrak{U}$ as a trivializing covering.

1.3 Operations on Vector Bundles

1.3.1 Let X be a topological space and let E be a vector bundle over X. Let Y be a topological space and suppose that f is a continuous mapping of Y into X. Let $E \times Y$ be given the product topology. Set

$$f^*E = \{(v, y) | v \text{ in } E_{f(y)}\}.$$

Give f^*E the subspace topology. Taking the fiber over y to be $E_{f(y)} \times \{y\} = f^*E_y$ and the natural vector space structure on f^*E_y, it is easy to see that f^*E is a vector bundle over Y. f^*E is called the pull-back of E relative to f.

1.3.2 Let E_1 and E_2 be vector bundles over X. We define a vector bundle over $X \times X$ by giving $E_1 \times E_2$ the product topology and the obvious projection. Let $\text{diag}(x) = (x, x)$. Then diag maps X into $X \times X$. We set $E_1 \oplus E_2 = \text{diag}^*(E_1 \times E_2)$. $E_1 \oplus E_2$ is called the Whitney sum of E_1 and E_2.

1.3.3 Let E_1, E_2, and E_3 be vector bundles over X. A bilinear map of

$E_1 \oplus E_2$ to E_3 is defined to be a continuous map that maps fibers to fibers, that is bilinear on the fibers (here we make the natural identification of $(E_1 \oplus E_2)_x$ with $E_{1_x} \times E_{2_x}$). A tensor product of E_1 with E_2 is a pair (V, j) of a vector bundle V over X and a bilinear map j of $E_1 \oplus E_2$ into V so that if h is a bilinear map of $E_1 \oplus E_2$ into E_3 then there exists a unique vector bundle homomorphism $\tilde{h}$ of V to E_3 so that $\tilde{h} \circ j = h$. As in the case of the tensor product of vector spaces (see A.3.1.5), if a tensor product exists it is unique up to isomorphism. We now show that a tensor product of E_1 with E_2 exists. Let $\mathfrak{U}$ be a trivializing covering for E_1 and E_2. Let g and h be corresponding $\mathfrak{U}$-cocycles with values in $GL(n, K)$ and $GL(m, K)$, respectively. We assume that E_1 and E_2 have been constructed from g and h as in the proof of Theorem 1.2.6. Let, for U, W in $\mathfrak{U}$ and x in $U \cap W$,

$$k_{U,W}(x) = g_{U,W}(x) \otimes h_{U,W}(x).$$

Then k defines a $\mathfrak{U}$-cocycle with values in $GL(nm, K)$. Let V be the corresponding vector bundle over X. Let j be the map of $E_1 \oplus E_2$ to V induced by

$$((x, v, U), (x, w, U)) \mapsto (x, v \otimes w, U).$$

We leave it to the reader to check that j is well defined and bilinear and that (V, j) defines a tensor product of E_1 with E_2.

We use the notation $E_1 \otimes E_2$ for the tensor product of E_1 with E_2. The definition of the tensor product implies that $E_1 \otimes (E_2 \otimes E_3)$ is isomorphic with $(E_1 \otimes E_2) \otimes E_3$.

1.3.4 We leave it to the reader to formulate the universal problem to define $\Lambda^r E$, $r = 0, 1, \ldots,$ fiber dimension E (see A.3.1.16).

1.3.5 Let E be a real vector bundle over X. Let $\mathbf{C}$ be the trivial vector bundle $X \times C$. We note that the fiber of the bundle $E \otimes \mathbf{C}$ at x is $E_x \otimes C$. Thus $E \otimes \mathbf{C}$ is a complex vector bundle. $E \otimes \mathbf{C}$ is called the complexification of E and denoted $E \otimes C$ or E^C.

1.3.6 Let E be a vector bundle over X. Let $\mathfrak{U}$ be a trivializing covering for E and let g be a corresponding cocycle. Let, for U, V in $\mathfrak{U}$ and x in $U \cap V$,

$$g^*_{U,V}(x) = {}^t g_{U,V}(x)^{-1}$$

where ${}^tA$ denotes the transpose of the linear map A (that is, if z^* is an element of $(K^n)^*$ and if z is an element of K^n then ${}^tAz^*(z) = z^*(Az)$). Then g^*

is a $\mathfrak{U}$-cocycle with values in $GL(n, K)$. Let E^* denote the corresponding vector bundle over X. We note that up to isomorphism E^* is independent of all choices used to define it. Also if x is in X then E_x^* can be identified with $(E_x)^*$. E and E^* are called dual vector bundles.

Using the various universal properties we have

$$E^* \otimes F^* = (E \otimes F)^*$$

$$\Lambda^r E^* = (\Lambda^r E)^*.$$

1.3.7 If E and F are vector bundles over X let $\operatorname{Hom}(E, F)$ denote the vector bundle $E^* \otimes F$. Then there is a natural identification of $(E^* \otimes F)_x$ with $E_x^* \otimes F_x$ which is naturally identified with $L(E_x, F_x)$ (the space of all K-linear maps from E_x to F_x). Thus $\operatorname{Hom}(E, F)$ is the vector bundle over X with fiber over x the space of all linear maps of E_x to F_x.

1.3.8 Let M be a differentiable manifold (resp. complex manifold). Let E be a K-(resp. complex) vector bundle over M. Then E is called a C^∞ (resp. holomorphic) vector bundle if E has the structure of a C^∞ (resp. complex) manifold, the projection is C^∞ (resp. holomorphic), and there is a trivializing covering for E which has maps satisfying 1.2.1(2) that are C^∞ (resp. holomorphic).

The basic example of a C^∞ vector bundle is the tangent bundle of a C^∞ manifold. A basic example of a holomorphic vector bundle is the holomorphic tangent bundle of a complex manifold. (See Appendices 1 and 2 for pertinent definitions.)

1.3.9 *Definition* Let M be a C^∞ manifold (resp. complex manifold). Let $\mathfrak{U}$ be an open covering of M. Then a C^∞ $\mathfrak{U}$-cocycle (resp. holomorphic $\mathfrak{U}$-cocycle) with values in $GL(n, K)$ (resp. $K = C$) is a $\mathfrak{U}$-cocycle with values in $GL(n, K)$ so that if U, V are in then $g_{U,V}$ is a C^∞ (resp. holomorphic) map of $U \cap V$ into $GL(n, K)$.

1.3.10 *Theorem* Let E be a vector bundle over M. Then E is a C^∞ (resp. holomorphic) vector bundle over M if and only if there is a trivializing covering for E so that a corresponding cocycle is C^∞ (resp. holomorphic).

PROOF The necessity is clear. Let $\mathfrak{U}$ be an open covering of M which is a trivializing covering for M and so that if, for each U in $\mathfrak{U}$, F_U is a map satisfying 1.2.1(2), then the corresponding cocycle is C^∞ (resp. holomorphic). We now show how to make E into a C^∞ (resp. holomorphic) vector bundle.

We first note that by possibly refining the covering $\mathfrak{U}$ we may assume that for each U in $\mathfrak{U}$ there is a C^∞ (resp. holomorphic) mapping f_U of U into R^n (resp. C^n) so that (U, f_U) is a chart for M. We define an atlas for E by defining $G_U = (f_U \times 1) \circ F_U$, mapping $E|_U$ to $R^m \times K^n$ (resp. $C^m \times C^n$). Here $(f_U \times 1)(x, v) = (f_U(x), v)$. The cocycle condition guarantees that $G_V \circ G_U^{-1}$ is C^∞ (resp. holomorphic) on $G_U(E|_U \cap E|_V)$.

1.3.11 *Corollary* Let E and F be C^∞ (resp. holomorphic) vector bundles over M, a C^∞ (resp. complex) manifold.

(a) The vector bundles $E \oplus F$, $E \otimes F$, E^*, and $\Lambda^r E$ are C^∞ (resp. holomorphic) vector bundles over M.

(b) Let N be a C^∞ (resp. complex) manifold, and let f be a C^∞ (resp. holomorphic) mapping of N to M. Then f^*E is a C^∞ (resp. holomorphic) vector bundle over N.

PROOF Let $\mathfrak{U}$ be a trivializing covering for E and F. Let g and h be, respectively, C^∞ (resp. holomorphic) cocycles corresponding to $\mathfrak{U}$ for E and F. Then the $\mathfrak{U}$-cocycle corresponding to $E \oplus F$ is given for x in $U \cap V$ by $g_{U,V}(x) \oplus h_{U,V}(x)$. That of $E \otimes F$ is $g_{U,V}(x) \otimes h_{U,V}(x)$. The cocycle for E^* is given by ${}^t g_{U,V}(x)^{-1}$. The cocycle for $\Lambda^r E$ is given by $\Lambda^r g_{U,V}(x)$. Thus by Theorem 1.3.10 all the bundles of (a) are C^∞ (resp. holomorphic).

The proof of (b) follows from Theorem 1.3.10 with the observation that relative to the covering $\{f^{-1}(U)\}$ of N a corresponding cocycle for f^*E is given by $k_{f^{-1}(U), f^{-1}(V)}(x) = g_{U,V}(f(x))$.

1.4 Cross Sections

1.4.1 *Definition* Let E be a K-vector bundle over a topological space X. A cross section of E is a continuous map f of X to E such that, for each x in X, $f(x)$ is in E_x. Let ΓE denote the set of all cross sections of E. We make ΓE into a K-vector space by defining for f, g in ΓE, a, b in K, $(af + bg)(x) = af(x) + bg(x)$.

1.4.2 Suppose that M is a C^∞ (resp. complex) manifold. Let E be a C^∞ (resp. holomorphic) vector bundle over M. Let $\Gamma^\infty E$ (resp. $H^0(E)$) be the space of all cross sections of E which are C^∞ (resp. holomorphic) mappings.

It is not hard to see that the space $\Gamma^\infty E$ is quite large (cf. Proposition 1.4.4). This is not necessarily true of $H^0(E)$. For example, if M is a compact complex manifold then a holomorphic cross section of the trivial bundle $M \times C$ is just a holomorphic map of M into C. Liouville's theorem implies that this map must be constant on connected components. Thus if M is a connected, compact, complex manifold, dim $H^0(M \times C) = 1$. We will see in Chapter 6 that there are holomorphic vector bundles over compact, complex manifolds that have no holomorphic cross sections other than the zero section.

1.4.3 *Examples* (1) If $E = X \times K^n$ then ΓE may be identified with the space of all continuous maps of X into K^n.

(2) If M is a C^∞ manifold and if $E = T(M)$ then $\Gamma^\infty E$ is the space of all vector fields on M.

(3) Differential forms on M are elements of $\Gamma^\infty \Lambda^k T(M)^*$.

(4) Let E and F be vector bundles over X. Then a vector bundle homomorphism of E to F is an element of Γ Hom(E, F) (and conversely).

1.4.4 *Proposition* Let M be a C^∞ manifold. Let E be a C^∞ vector bundle over M. Suppose that Y is closed subset of M and that f' is a C^∞ cross section of $E|_Y$. Then there is a C^∞ cross section f of E so that $f|_Y = f'$. (A function g' from a subset X of M into a C^∞ manifold N is said to be C^∞ if for each x in X there is an open subset U_x of M and a C^∞ mapping g of U_x into N so that $g|_{U_x \cap X} = g'|_{U_x \cap X}$.)

PROOF Let, for each x in Y, U_x be an open subset of M containing x and so that $E|_{U_x}$ is trivial and there is a an F_{U_x} which is C^∞ and satisfies 1.2.1(2). Relative to F_{U_x}, $f'|_{U_x \cap Y}$ is a C^∞ map of $U_x \cap Y$ into K^n. Thus there is an open covering $\{V_x | x \text{ in } Y\}$ of Y and for each x in Y a C^∞ cross section f_x of E_{V_x} extending $f'|_{V_x \cap Y}$. Let $\{\varphi_x\} \cap \{\varphi\}$ be a partition of unity subordinate to $\{V_x\} \cup \{M - Y\}$ (supp $\varphi_x \subset V_x$). Let $f = \sum \varphi_x f_x$. Then f is the desired extension.

1.4.5 Using the Tietze extension theorem it is not hard to prove the continuous analogue of 1.4.4 (see 1.9.2). Using this result the results of 1.4.6, 7, 8 have obvious extensions to versions for locally compact, paracompact spaces.

1.4.6 *Corollary* Let M be a C^∞ manifold. Let E and F be C^∞ vector bundles over M. Let Y be a closed subset of M. If f' is a C^∞ vector bundle isomorphism of $E|_Y$ with $F|_Y$ then there is an open subset U containing Y and a C^∞ vector bundle isomorphism f of $E|_U$ onto $F|_U$ extending f'.

PROOF f' is a cross section of $\mathrm{Hom}(E, F)|_Y$. Proposition 1.4.4 implies that there is a C^∞ cross section of $\mathrm{Hom}(E, F)$ extending f'. Call it f. Let U be the set of all p in M so that f is a bijection of E_p with F_p. Clearly U is open in M and contains Y. Then $f|_U$ is the desired extension.

1.4.7 *Theorem* Let M and N be C^∞ manifolds and let f be a C^∞ map of $N \times [0, 1]$ into M. Set $f_t(y) = f(y, t)$. Let E be a C^∞ vector bundle over M. Then f_0^*E and f_1^*E are C^∞ vector bundle isomorphic.

PROOF Let U be an open subset of N so that $\bar{U}$ is compact. Let h be the restriction of f to $\bar{U} \times [0, 1]$. Let $q(y, t) = y$. Then q is a C^∞ mapping of $\bar{U} \times [0, 1]$ into $\bar{U}$. Now

$$h^*E = \{(v, (y, s))|v \text{ in } E_{h(y,s)}\},$$

$$q^*h_t^*E = \{(v, (y, s))|v \text{ in } E_{h(y,t)}\}.$$

Thus $h^*E|_{\bar{U}\times\{t\}}$ and $q^*f_t^*E|_{\bar{U}\times\{t\}}$ are vector bundle isomorphic (indeed equal!). 1.4.6 and the compactness of $\bar{U}$ imply that there is $\varepsilon > 0$ so that $h^*E|_{\bar{U}\times(t-\varepsilon,t+\varepsilon)}$ and $q^*h_t^*E|_{\bar{U}\times(t-\varepsilon,t+\varepsilon)}$ are C^∞ vector bundle isomorphic. Thus the set of all t in $[0, 1]$ so that h_t^*E is vector bundle isomorphic with h_0^*E is open and non-empty. Applying 1.4.6 we find that the set of all t so that h_t^*E is not C^∞ isomorphic with h_0^*E is open in $[0, 1]$. Hence, since $[0, 1]$ is connected, h_0^*E is C^∞ isomorphic with h_1^*E.

Let U_j be a sequence of open subsets of N so that U_j is compact, $\bar{U}_j$ is contained in U_{j+1}, and $\bigcup_j U_j = N$. By the above $f_0^*E|_{U_j}$ is C^∞ isomorphic with $f_1^*E|_{U_j}$. Let $\{V_{j,k}\}$ be an open cover of $\bar{U}_j - U_{j-1}$ so that $V_{j,k}$ is contained in U_{j+1}. If $V_{j,k} \cap V_{r,m}$ is not empty then $|j - r| \leqslant 2$. Suppose furthermore, that $f_0^*E|_{V_{j,k}}$ and $f_1^*E|_{V_{j,k}}$ are C^∞ trivial and that g^0 and g^1 are the corresponding C^∞-cocycles. Then applying the fact that $f_0^*E|_{U_j}$ and $f_1^*E|_{U_j}$ are C^∞ isomorphic we see that there are C^∞ maps

$$\lambda_{j,k}\colon V_{j,k} \to GL(n, K)$$

giving an equivalence between g^0 and g^1. This implies that f_0^*E and f_1^*E are C^∞ isomorphic.

1.4.8 *Corollary* Let M be a contractible C^∞ manifold (that is, there is a C^∞ map f of $M \times [0, 1]$ into M so that if $f_t(x) = f(x, t)$, then f_0 is the identity map and f_1 maps M to a single point). Then every C^∞ vector bundle over M is C^∞ isomorphic with a trivial bundle.

PROOF Let f be as in the statement of the corollary. Then 1.4.7 implies that f_0^*E is C^∞ isomorphic with f_1^*E. Let $f_1(M) = \{p\}$. Then since $f_0^*E = E$ and $f_1^*E = M \times E_p$ the result follows.

1.5 Unitary Structures

1.5.1 *Definition* Let E be a complex (resp. real) vector bundle over a topological space X. A unitary (resp. orthogonal) structure on E is a *real* bilinear map of $E \oplus E$ to $X \times C$ (see 1.3.3), denoted by $\langle\ ,\ \rangle$, such that

(1) If v, w are in E_x then $\langle v, w\rangle = \overline{\langle w, v\rangle}$ ($\bar{a}$ denotes the complex conjugate of the complex number a).

(2) If v is in E_x then $\langle v, v\rangle$ is nonnegative and $\langle v, v\rangle = 0$ if and only if $v = 0$.

(3) If v, w are in E_x and c is in C (resp. R) then $\langle cv, w\rangle = c\langle v, w\rangle$.

1.5.2 *Lemma* Let M be a C^∞ manifold and let E be a complex (resp. real) vector bundle over M. Then E has a unitary (resp. orthogonal) structure. If E is a C^∞ vector bundle then the structure may be taken to be C^∞.

PROOF Let $\mathfrak{U}$ be a trivializing covering for E. Let $\{\varphi_U\}$ be a partition of unity subordinate to $\mathfrak{U}$ (supp $\varphi_U \subset U$). Let for each U in $\mathfrak{U}$, F_U satisfy 1.2.1(2) and be C^∞ if E is. Let $F_U(v) = (p(v), f_U(v))$. On $E|_U$ we define a bilinear map $\langle\ ,\ \rangle_U$ by setting for v, w in E_x, $\langle v, w\rangle_U = \varphi_U(x)\,(f_U(v), f_U(w))$ where $(\ ,\ )$ is the standard Hermitian inner product on C^n (resp. R^n). Extend $\langle\ ,\ \rangle_U$ to be 0 outside of U. Set $\langle\ ,\ \rangle = \sum \langle\ ,\ \rangle_U$ the sum extended over $\mathfrak{U}$.

1.5.3 The proof of Lemma 1.5.2 (except for the C^∞ statement) used only the paracompactness of M. Thus vector bundles over paracompact spaces have unitary structures.

1.6 $K(X)$

1.6.1 Let S be a commutative semigroup written additively. We define $(K(S), j)$ to be the solution of the following universal problem:

$K(S)$ is a commutative group, j is a semigroup homomorphism of S into $K(S)$, and if A is a group and if α is a semigroup homomorphism of S into A, then there is a unique group homomorphism $\bar{\alpha}$ of $K(S)$ into A so that $\bar{\alpha} \circ j = \alpha$.

1.6.2 *Lemma* If S is a commutative semigroup then $(K(S), j)$ exists and is unique up to isomorphism. If S is a semiring (that is, S has another associative binary operation distributive over addition), then $K(S)$ has a unique ring structure so that if A is a ring and if α is a semiring homomorphism of S into A then $\bar{\alpha}$ is a ring homomorphism.

PROOF Let $S \times S$ be the product semigroup. Set $K(S) = (S \times S)/\Delta(S)$ where $\Delta(S) = \{(s, s) | s \text{ in } S\}$. Let $j(t) = (t + s, s) + \Delta(S)$ for t in S; s is an arbitrary element of S. Clearly j is independent of s and defines a semigroup homomorphism of S into $K(S)$. $K(S)$ is actually a group since

$$-((t, s) + \Delta(S)) = (s, t) + \Delta(S).$$

Suppose that α is a semigroup homomorphism of S into a group A. Since the subgroup of A generated by $\alpha(S)$ is Abelian, we may assume that A is an Abelian group and we write A additively. Set $\bar{\alpha}(s, t) = \alpha(s) - \alpha(t)$. Then $\bar{\alpha}$ defines a semigroup homomorphism of $S \times S$ into A with kernel containing $\Delta(S)$. Thus $\bar{\alpha}$ induces a group homomorphism of $K(S)$ into A. The homomorphism is unique since $j(S)$ generates $K(S)$. This proves the existence of $(K(S), j)$. The uniqueness of $(K(S), j)$ follows directly from the universal problem that it solves.

If S has a semiring structure then make $S \times S$ into a semiring by defining $(u, v) \cdot (s, t) = (us + vt, ut + vs)$. With this multiplication $\Delta(S)$ is an ideal in $S \times S$. This implies that $K(S)$ inherits a ring structure. The uniqueness statements for this part of the proof are direct and easy.

1.6.3 *Definition* Let X be a compact topological space and let $B(X)$ be the collection of all isomorphism classes of complex vector bundles over X. If E is a vector bundle over X then let $[E]$ denote the isomorphism class of E. $B(X)$ is a commutative semiring with sum $[E] + [F] = [E \oplus F]$, and multiplication $[E] \cdot [F] = [E \otimes F]$. We set $K(X) = K(B(X))$.

1.6.4 Let X and Y be compact topological spaces. Let f be a continuous mapping from X to Y. If E is in $E(Y)$ then set $f^*[E] = [f^*E]$ in $K(X)$. Then f^* defines a semiring homomorphism of $E(Y)$ into $K(X)$. Thus f^* extends to a ring homomorphism of $K(Y)$ into $K(X)$.

1.6.5 *Theorem* $K(X)$ is a homotopy invariant of X.

PROOF This theorem follows from the continuous versions of 1.4.7 and 1.4.8.

1.6.6 Further properties of $K(X)$ will be developed in the exercises.

1.7 Differential Operators

1.7.1 *Definition* Let M be a C^∞ manifold. Let E and F be C^∞ K-vector bundles over M where K is either R or C. A linear map $D\colon \Gamma^\infty E \to \Gamma^\infty F$ is called a differential operator of order m if for each x in M there is a neighborhood U, of x in M with local coordinates $x_1, \ldots x_n$ and C^∞ trivializations $E|_U \overset{\Psi}{\to} UxK^l$ and $F|_U \overset{\Phi}{\to} UxK^r$ such that if f is in $\Gamma^\infty E$, if y is in U, if $(\Psi f)(y) = (y, F(y))$, and if $(\Phi D f)(y) = (y, G(y))$, then

$$G(y) = \sum_{i_1 + \cdots + i_n \leqslant m} A_{i_1 \cdots i_n}(y) \frac{\partial^{i_1 + \cdots + i_n}}{\partial x_1^{i_1} \cdots \partial x_n^{i_n}} F$$

with $A_{i_1 \cdots i_n}$ a C^∞ map from U into $L(K^l, K^r)$.

1.7.2 *Example* Let $E = \Lambda^k T(M)^*$, $F = \Lambda^{k+1} T(M)^*$, and let d be a linear map from $\mathscr{D}^k(M)$ to $\mathscr{D}^{k+1}(M)$ satisfying:

(1) If $k = 0$ then $f_{*p}(v) = (f(p), df_p(v))$ for f in $C^\infty(M) = \mathscr{D}^0(M)$ and v in $T(M)_p$.

(2) $d^2 = 0$.

(3) $d(\omega \wedge \eta) = d\omega \wedge \eta + (-1)^k \omega \wedge d\eta$ if ω is in $\mathscr{D}^k(M)$, η is in $\mathscr{D}^r(M)$.

(4) If ψ is a C^∞ mapping from M to N then $d\psi^*\omega = \psi^* d\omega$ for ω in $\mathscr{D}^k(N)$.

Using (1)–(4) we find that if such a d exists and if U is an open subset of M with local coordinates $x_1, \ldots, x_n$ then if ω is in $\mathscr{D}^k(M)$ and

$$\omega|_U = \sum a_{i_1 \cdots i_k} dx_{i_1} \wedge \cdots \wedge dx_{i_k}$$

then

$$(*) \qquad d\omega|_U = \sum da_{i_1 \cdots i_k} \wedge dx_{i_1} \wedge \cdots \wedge dx_{i_k}.$$

Here we have used (4), (3), (1), (2) in that order. This shows that if d exists, d is unique and is a first-order differential operator. Using (*) it is an exercise to show that d exists.

1.7.3 In what follows we deal only with complex vector bundles. We leave it to the reader to work out the analogous concepts for real vector bundles.

1.7.4 Let M be an orientable C^∞ manifold and let ω be a volume element for M. Let E be a C^∞ vector bundle over M. Let $\Gamma_0^\infty(E)$ be the space of all C^∞ cross sections of E with compact support.

Let F be another C^∞ vector bundle over M and let $\langle\ ,\ \rangle$ be a C^∞ unitary structure on E or F.

Let D be a differential operator from E to F. We say that a differential operator D^* from F to E is a formal adjoint for D if for f in $\Gamma_0^\infty(E)$, g in $\Gamma_0^\infty(F)$,

$$\int_M \langle Df, g\rangle\omega = \int_M \langle f, D^*g\rangle\omega.$$

1.7.5 *Theorem* Let D be a differential operator from E to F. Then D has a unique formal adjoint D^* (depending only on the choices of unitary structures and volume element). D^* is of order m.

PROOF Suppose that D_1 and D_2 are formal adjoints for D. Then for each f in $\Gamma_0^\infty(E)$, g in $\Gamma_0^\infty(F)$, $\int_M\langle f, D_1 g - D_2 g\rangle\omega = 0$. If we take $f = D_1 g - D_2 g$ then $\int_M\langle f, f\rangle\omega = 0$. Hence $f = 0$. Thus by the local definition of a differential operator $D_1 = D_2$.

Let $\mathfrak{U} = \{U_\alpha\}$ be an open covering of M so that

(1) M has local coordinates $x_1^\alpha, \ldots, x_n^\alpha$ on U_α.

(2) $E_{U\alpha} \xrightarrow{\Psi} U_\alpha \times C^l$, $F_{U\alpha} \xrightarrow{\Phi} U_\alpha \times C^r$ are C^∞ trivializations of E and F over U_α.

(3) $\{U_\alpha\}$ is locally finite with a partition of unity $\{\varphi_\alpha\}$ subordinate to $\mathfrak{U}$.

Suppose that we have found a formal adjoint D^* for $D|_{E_{U\alpha}}$ for each α. Then by uniqueness $D^*_\alpha = D^*_\beta$ on $U_\alpha \cap U_\beta$. If f is in $\Gamma^\infty F$ then $f = \sum \varphi_\alpha f$. Define $D^*f = \sum D^*_\alpha \varphi_\alpha f$. Then D^* defines a formal adjoint for D. We may thus assume that

(a) $M = R^n$
(b) $E = R^n \times C^l$, $F = R^n \times C^r$
(c) $\omega = \rho\, dx_1 \wedge \cdots \wedge dx_n$ with ρ a positive C^∞ function on R^n.

Fix a Hermitian inner product $(\ , \ )_1$ and $(\ , \ )_2$ on C^l and C^r, respectively. Then the inner products on E and F are given by $\langle v, w\rangle_x = (A(x)v, w)_1$ for v, w in C^l and $\langle v, w\rangle_x = (B(x)v, w)_2$ for v, w in C^r where A and B are C^∞ mappings from R^n to the spaces of Hermitian positive definite $l \times l$ and $r \times r$ matrices.

Let for $\alpha = (\alpha_1, \ldots, \alpha_n)$ an n-tuple of nonnegative integers $D^\alpha = \partial^{|\alpha|}/\partial x_1^{\alpha_1} \cdots \partial x_n^{\alpha_n}$, $|\alpha| = \sum \alpha_i$.

Then $D = \sum c_\alpha D^\alpha$ the sum taken over $|\alpha| \leqslant m$ and c_α is a C^∞ function from R^n into $L(C^l, C^r)$. Let f and g be, respectively, C^∞ maps from R^n to C^l with compact supports. Set $h = (f, g)_1$ and set

$$\eta = h dx_1 \wedge \cdots \wedge dx_{k-1} \wedge dx_{k+1} \wedge \cdots \wedge dx_n.$$

Then

$$d\eta = (-1)^{k-1}\left(\left(\frac{\partial f}{\partial x_k}, g\right)_1 + \left(f, \frac{\partial q}{\partial x_k}\right)_1\right) dx_1 \wedge \cdots \wedge dx_n.$$

Stokes' theorem implies that $\int_{R^n} d\eta = 0$. This implies that

$$\int_{R^n}\left(\frac{\partial f}{\partial x_k}, g\right)_1 dx = -\int_{R^n}\left(f, \frac{\partial q}{\partial x_k}\right)_1 dx.$$

We therefore find by induction that

$$\int_{R^n} (D^\alpha f, g)_1 dx = (-1)^{|\alpha|}\int_{R^n} (f, D^\alpha g)_1\, dx.$$

If T is a linear map of C^l to C^r then let T' denote the adjoint of T relative to $(\ , \ )_1$ and $(\ , \ )_2$ (that is, $(Tu, v)_2 = (u, T'v)_1$).

Suppose that f is in $\Gamma_0^\infty E$ and that g is in $\Gamma_0^\infty F$. Then

$$\begin{aligned}\int_{R^n} \langle Df, g\rangle\omega &= \sum_{|\alpha|\leqslant m}\int_{R^n} (B(x)c_\alpha(x)(D^\alpha f)(x), g(x))_2\rho(x)dx\\ &= \sum_{|\alpha|\leqslant m}\int_{R^n} (D^\alpha f(x), \rho(x)c_\alpha(x)'B(x)'g(x))_1 dx\\ &= \int_{R^n}\sum_{|\alpha|\leqslant m} (-1)^{|\alpha|}(f, D^\alpha \rho c'_\alpha B' g)_1 dx.\end{aligned}$$

Setting

$$D^*g = \sum_{|\alpha| \leqslant m} (-1)^{|\alpha|}\rho(x)^{-1}A(x)^{-1}D^\alpha(\rho c'_\alpha B'g)$$

then D^* is a formal adjoint for D.

1.7.6 Let D be a differential operator from E to F of order m. Let x be in M and let ξ be an element of $T(M)^*_x$. Let $x_1, \ldots, x_n$ be local coordinates around x. Then $\xi = \sum \xi_i dx_i$. If $\alpha = (\alpha_1, \ldots, \alpha_n)$ is an n-tuple of nonnegative integers then set $\xi^\alpha = \xi_1^{\alpha_1} \cdots \xi_n^{\alpha_n}$. In a suitably small neighborhood of x, $D = \sum_{|\alpha| \leqslant m} A_\alpha D^\alpha$ (see 1.7.1). Set $\sigma(D, \xi) = \sum_{|\alpha| = m} A_\alpha \xi^\alpha$. Then $\sigma(D, \xi)$ is a linear map of E_x to F_x which a priori seems to depend on the choices made to define it. The next lemma shows that $\sigma(D, \xi)$ is well defined and gives a means to compute it. $\sigma(D, \;.\,)$ is called the symbol of D.

1.7.7 *Lemma* Let v be in E_x, ξ be in $T(M)^*_x$, and let f be in $\Gamma^\infty E$, φ in $C^\infty(M)$ be so that $f(x) = v$, $d\varphi_x = \xi$. Then

$$\sigma(D, \xi)v = \left(\frac{1}{m!}\right)\frac{d^m}{dt^m}(e^{-t\varphi}De^{t\varphi}f)(x)|_{t=0}.$$

PROOF In local coordinates around x and relative to trivializations of E and F we see that $Df = \sum_{|\alpha| \leqslant m} A_\alpha D^\alpha f$. Thus

$$D(e^{t\varphi}f)(x) = e^{t\varphi}\{t^m \sum_{|\alpha| = m} \xi^\alpha A_\alpha v + \text{lower order terms in } t\}.$$

The result follows from these observations.

1.7.8 *Proposition* (a) Let E, F, G be C^∞ vector bundles over M. Let D_1 be a differential operator from E to F and let D_2 be a differential operator from F to G. Then $\sigma(D_2D_1, \xi) = \sigma(D_2, \xi) \circ \sigma(D_1, \xi)$ for ξ in $T(M)^*_x$.

(b) Suppose that we have put unitary structures on E and on F and that ω is a volume element for M. Let D be a differential operator from E to F of degree m. Let D^* be the formal adjoint of D relative to these choices. Then $\sigma(D^*, \xi) = (-1)^m\sigma(D, \xi)^*$ for ξ in $T(M)^*_x$ where $\sigma(D, \xi)^*$ is the adjoint of $\sigma(D, \xi)$ relative to the inner products on E_x and F_x.

PROOF The proof of this proposition is straightforward and we leave it as an exercise for the reader.

1.7.9 *The symbol of d.* Let x be in M, ξ in $T(M)^*_x$. Let β be in $\Lambda^k T(M)^*_x$.

We wish to find $\sigma(d, \xi)\beta$. Let η be in $\mathscr{D}^k(M)$ so that $\eta_x = \beta$ and let φ be in $C^\infty(M)$ so that $d\varphi_x = \xi$. Then

$$\sigma(d, \xi)\beta = \frac{d}{dt}(e^{-t\varphi}de^{t\varphi}\eta)_x|_{t=0} = \frac{d}{dt}\{td\varphi_x \wedge \beta + d\eta_x\}|_{t=0} = \xi \wedge \beta.$$

Thus if we define, for ξ in $T(M)^*$, $\varepsilon(\xi)\beta = \xi \wedge \beta$ then $\sigma(d, \xi) = \varepsilon(\xi)$.

1.7.10 *The symbol of d^* and Δ* Suppose that M has a Riemannian structure $\langle\ ,\ \rangle$. We extend this Riemannian structure to a unitary structure on $\Lambda^k T(M)^* \otimes C$ for each k. Let $v \mapsto v^\#$ be given by $v^\#(w) = \langle v, w\rangle_x$ for v in $T(M)_x$, w in $T(M)_x$. We put on $T(M)^*$ the orthogonal structure that makes $v \mapsto v^\#$ an isometry for each x in M. We extend the orthogonal structure to a unitary structure on $T(M)^* \otimes C$ in the usual manner

$$(\langle v + (-1)^{\frac{1}{2}}w, u + (-1)^{\frac{1}{2}}z\rangle = \langle v, u\rangle + \langle w, z\rangle + (-1)^{\frac{1}{2}}(\langle w, u\rangle - \langle v, z\rangle)).$$

If $\omega_1, \ldots, \omega_k$ and $\eta_1, \ldots, \eta_k$ are in $T(M)^* \otimes C$ define

$$\langle\omega_1 \wedge \cdots \wedge \omega_k, \eta_1 \wedge \cdots \wedge \eta_k\rangle = \det(\langle\omega_i, \eta_j\rangle).$$

This defines a unitary structure on $\Lambda^k T(M)^* \otimes C$ for each k.

Let d^* be the formal adjoint to d. d^* is a first-order differential operator from $\Lambda^{k+1}T(M)^* \otimes C$ to $\Lambda^k T(M)^* \otimes C$.

1.7.8(b) says that $\sigma(d^*, \xi) = -\varepsilon(\xi)^*$ on $\Lambda^{k+1}T(M)_x^* \otimes C$ for ξ in $T(M)_x^*$. We compute $\varepsilon(\xi)^*$. Let $\omega_1, \ldots, \omega_n$ be an orthonormal basis of $T(M)_x^*$ so that $\xi = c\omega_1$. We assume that $c \neq 0$. Let $1 \leqslant i_1 < \cdots < i_k \leqslant n$, $1 \leqslant j_1 < \cdots < j_{k=1} \leqslant n$. Then

$$\begin{aligned}\langle\varepsilon(\xi)\omega_{i_1} \wedge \cdots \wedge \omega_{i_k}, \omega_{j_1} \wedge \cdots \wedge \omega_{j_{k+1}} \\ = c\langle\omega_1 \wedge \omega_{i_1} \wedge \cdots \wedge \omega_{i_k}, \omega_{j_1} \wedge \cdots \wedge \omega_{j_{k+1}}\rangle.\end{aligned}$$

Thus $\varepsilon(\xi)^*(\omega_{j_1} \wedge \cdots \wedge \omega_{j_{k+1}}) = 0$ if $j_1 \neq 1$, $\varepsilon(\xi)^*(\omega_{j_1} \wedge \cdots \wedge \omega_{j_{k+1}}) = c\omega_{j_2} \wedge \cdots \wedge \omega_{j_{k+1}}$ if $j_1 = 1$.

Let for ξ in $T(M)_x^*$, ξ^b be define by $(\xi^b)^\# = \xi$. If v is in $T(M)_x$ define $i(v)\colon \Lambda^{k+1}T(M)_x^* \times C \to \Lambda^k T(M)_x^* \times C$ by

$$(i(v)\omega)(v_1, \ldots, v_k) = \omega(v, v_1, \ldots, v_k).$$

We have thus shown that $\varepsilon(\xi)^* = i(\xi^b)$. Hence $\sigma(d^*, \xi) = -i(\xi^b)$.

We note that the uniqueness of formal adjoints implies that $(d^*)^2 = 0$. Set $\Delta = (d + d^*)^2$. Then $\Delta = dd^* + d^*d$ and Δ is a second-order differential operator from $\Lambda^k T(M)^* \otimes C$ to itself for each k.

$$\begin{aligned}\sigma(\Delta, \xi) &= \sigma(d, \xi)\sigma(d^*, \xi) + \sigma(d^*, \xi)\sigma(d, \xi) \\ &= -\varepsilon(\xi)i(\xi^b) - i(\xi^b)\varepsilon(\xi) = -\langle\xi, \xi\rangle I.\end{aligned}$$

i(ξ^b) is an anti-derivation

Δ is called the Hodge Laplacian of $(M, \langle\ ,\ \rangle)$. The operator $-d^*d$ on $C^\infty(M)$ is called the Laplace–Beltrami operator of the Riemannian structure $\langle\ ,\ \rangle$ on M.

1.8 The Complex Laplacian

1.8.1 Let M be a complex manifold. Then $T(M) \otimes C = T(M)^C = \mathscr{T}(M) + \bar{\mathscr{T}}(M)$ as in A.5.2.3. Let $\mathscr{T}(M)^*$ and $\bar{\mathscr{T}}(M)^*$ be, respectively, the dual bundles to $\mathscr{T}(M)$ and $\bar{\mathscr{T}}(M)$. Then $(T(M)^C)^* = \mathscr{T}(M)^* + \bar{\mathscr{T}}(M)^*$. Furthermore

$$\Lambda^r(T(M)^C)^* = \sum_{k+l=r} \Lambda^k \mathscr{T}(M)^* \otimes \Lambda^l \bar{\mathscr{T}}(M)^*.$$

An element of $\Gamma^\infty(\Lambda^k \mathscr{T}(M)^* \otimes \Lambda^l \bar{\mathscr{T}}(M)^*)$ is called a differential form of type (k, l) (a (k, l)-form for short).

1.8.2 Let U be an open subset of M with holomorphic local coordinates $z_1, \ldots, z_n$. $z_i = x_i + (-1)^{\frac{1}{2}} y_i$ and $x_1, y_1, \ldots, x_n, y_n$ form a system of local coordinates on U. The complex linear extension of d satisfies $dz_i = dx_i + (-1)^{\frac{1}{2}} dy_i$. Now by definition of $\mathscr{T}(M)$ the vectors $\partial/\partial z_i$ form a basis of $\mathscr{T}(M)$ at each point of U. Hence $dz_1, \ldots, dz_n$ forms a basis of $\mathscr{T}(M)^*$ for each point of U. Similarly $d\bar{z}_i = dx_i - (-1)^{\frac{1}{2}} dy_i$ and $d\bar{z}_1, \ldots, d\bar{z}_n$ forms a basis of $\bar{\mathscr{T}}(M)^*$ for each point of U.

Using the chain rule we see that if $u_1, \ldots, u_n$ is another system of holomorphic coordinates around p in U then $du_i = \sum (\partial u_i/\partial z_j) dz_j$. Thus 1.3.10 implies that $\mathscr{T}(M)^*$ is a holomorphic vector bundle over M.

1.8.3 If ω is a C^∞ cross section of $\Lambda^r(T(M)_E)^*|_U$ then

$$\omega = \sum_{k+l=r} \sum_{\substack{|I|=k \\ |J|=l}} a_{I,J} dz_j \wedge d\bar{z}_J.$$

Here we use the notation $I = (i_1, \ldots, i_k)$, $J = (j_1, \ldots, j_l)$, $1 \leqslant i_1 < \cdots < i_k \leqslant n$, $1 \leqslant j_1 < \cdots < j \leqslant n$ and $dz_I = dz_{i_1} \wedge \cdots \wedge dz_{i_k}$, $d\bar{z}_J = d\bar{z}_{j_1} \wedge \cdots \wedge d\bar{z}_{j_l}$.

Now $d\omega = \sum da_{I,J} \wedge dz_j \wedge d\bar{z}_J$. A direct computation using the definitions of dz_i, $d\bar{z}_j$, $\partial/\partial z_i$, $\partial/\partial \bar{z}_j$ gives

$$da_{I,J} = \sum \left(\frac{\partial a_{I,J}}{\partial z_i}\right) dz_i + \sum \left(\frac{\partial a_{I,J}}{\partial \bar{z}_i}\right) d\bar{z}_i.$$

Define

$$\partial a_{I,J} = \sum \left(\frac{\partial a_{I,J}}{\partial z_i}\right) dz_i,$$

$$\bar{\partial} a_{I,J} = \sum \left(\frac{\partial a_{I,J}}{\partial \bar{z}_i}\right) d\bar{z}_i.$$

Set $\partial\omega = \sum \partial a_{I,J} \wedge dz_I \wedge d\bar{z}_J$ and $\bar{\partial}\omega = \sum \bar{\partial} a_{I,J} \wedge dz_I \wedge d\bar{z}_J$. Then $\partial + \bar{\partial} = d$. ∂ and $\bar{\partial}$ define first-order differential operators, respectively, from the forms of type (p, q) to the $(p + 1, q)$ forms and from the (p, q) forms to the $(p, q + 1)$ forms. The definition of ∂ and $\bar{\partial}$ implies that $\partial^2 = 0$, $\bar{\partial}^2 = 0$. $d^2 = 0$ implies that $\partial\bar{\partial} + \bar{\partial}\partial = 0$.

1.8.4 Let E be a holomorphic vector bundle over M. We define in this paragraph a first-order differential operator $\bar{\partial}$ from $E \otimes \Lambda^r \bar{\mathscr{T}}(M)^*$ to $E \otimes \Lambda^{r+1} \bar{\mathscr{T}}(M)^*$.

Let U be an open subset of M so that $E|_U$ is trivial (holomorphically) and so that there are holomorphic coordinates $z_1, \ldots, z_n$ on U. Let $e_1, \ldots, e_k$ be holomorphic cross sections of $E|_U$ so that they form a basis of E_p for each p in U. If f is an element of $\Gamma^\infty(E \times \Lambda^r \bar{\mathscr{T}}(M)^*)$ then $f|_U = \sum_i e_i \otimes \sum_J f_J^i d\bar{z}_J = \sum e_i \otimes \omega_i$ with ω_i a $(0, r)$ form for each i. Set $\bar{\partial} f = \sum e_i \otimes \bar{\partial}\omega_i$. We show that $\bar{\partial} f$ is well defined. If $f_1, \ldots, f_k$ is another set of holomorphic cross sections of $E|_U$ which form a basis for E_p for each p in U then $f|_U = \sum f_i \otimes \eta_i$ for some collection of $(0, r)$ forms η_i on U. Now $e_i = \sum g_{ji} f_j$ with g_{ij} a holomorphic function on U. Hence $\omega_i = \sum g_{ji}\eta_j$. $\bar{\partial} g_{ji} = 0$ for each i, j. Hence $\sum e_i \otimes \bar{\partial}\omega_i = \sum f_i \otimes \bar{\partial}\eta_i$. Thus $\bar{\partial}$ is independent of the choices made in its definition. We note that the definition of $\bar{\partial}$ immediately implies that $\bar{\partial}^2 = 0$.

1.8.5 *Lemma* Let f be a C^∞ cross section of E. Then f is a holomorphic cross section of E if and only if $\bar{\partial} f = 0$.

PROOF Let p be in M and let U be an open neighborhood of p so that $E|_U$ is trivial. Then f induces a C^∞ map $\tilde{f}$ from U to C^r, $\tilde{f} = (f_1, \ldots, f_r)$, with $f(z) = (z, \tilde{f}(z))$.

$$f_{*_z} = (I, (df_{1_z}, \ldots, df_{r_z}) = (I, (\partial f_{1_z}, \ldots, \partial f_{r_z}))$$

if $\bar{\partial} f = 0$. Thus if $\bar{\partial} f = 0$ then f_{*_z} is complex linear for each p in M. The lemma now follows.

1.8.6 Let $\langle\ ,\ \rangle$ be a Riemannian structure on $T(M)$. Extend $\langle\ ,\ \rangle$ to a unitary structure on $T(M)^C$. Extend this unitary structure to one on $\Lambda^r\bar{\mathcal{T}}(M)^*$ as in 1.7.10. Let $\langle\ ,\ \rangle$ also denote a unitary structure on E. Put the tensor product unitary structure on $E \otimes \Lambda^r\bar{\mathcal{T}}(M)^*$. Let $\bar{\partial}^*$ be the formal adjoint to $\bar{\partial}$ relative to these inner products and the Riemannian volume element on M. We compute the symbols of $\bar{\partial}$ and $\bar{\partial}^*$ in exactly the same way that we computed the symbols of d and d^*.

We first define a vector bundle map of $T(M)^*$ into $\bar{\mathcal{T}}(M)^*$ (here we take $\bar{\mathcal{T}}(M)^*$ to be a real vector bundle). Let p be in M and let U be a neighborhood of p with local holomorphic coordinates $z_1, \ldots, z_n$. If $z_i = x_i + (-1)^{\frac{1}{2}}y_i$ then $x_1, y_1, \ldots, x_n, y_n$ form a system of local coordinates on U. If ξ is in $T(M)^*_p$ then $\xi = \sum a_i dx_i + \sum b_i dy_i$. Set $\xi^0 = \sum (a_i + (-1)^{\frac{1}{2}}b_i)d\bar{z}_i$. (Note that $\xi \mapsto \xi^0$ is just the projection of ξ onto $\bar{\mathcal{T}}(M)^*$ relative to the direct sum decomposition $(T(M)^C)^* = \mathcal{T}(M)^* \oplus \bar{\mathcal{T}}(M)^*$.) If φ is in $C^\infty(M)$ then $(d\varphi)^0 = \bar{\partial}\varphi$.

Let p be in M, ξ be in $T(M)^*_p$, and let φ be in $C^\infty(M)$ so that $d\varphi_p = \xi$. Let v be in E_p and let ω be in $\Lambda^r\bar{\mathcal{T}}(M)^*$. Let f be an element of $\Gamma^\infty(E \otimes \Lambda^r\bar{\mathcal{T}}(M)^*)$ so that $f(p) = v \otimes \omega$. Then

$$\sigma(\bar{\partial}, \xi)(v \otimes \omega) = \frac{d}{dt}(e^{-t\varphi}\bar{\partial}e^{t\varphi}f)(p)|_{t=0}$$

$$= \frac{d}{dt}(t(\bar{\partial}\varphi)_p \wedge f(p) + \bar{\partial}f(p))_{t=0} = v \otimes (\bar{\partial}\varphi)_p \wedge \omega$$

$$= v \otimes \xi^0 \wedge \omega = v \otimes \varepsilon(\xi^0)\omega.$$

Thus $\sigma(\bar{\partial}, \xi) = I \otimes \varepsilon(\xi^0)$.

Extending the map $\xi \mapsto \xi^\flat$ of 1.7.10, using the Hermitian structure on $(T(M)^C)$, to $(T(M)^C)^*$ we see that $\sigma(\bar{\partial}^*, \xi) = -I \otimes i((\xi^0)^\flat)$.

Setting, as in the real case, $\Box = (\bar{\partial}^* + \bar{\partial})^2$. We find that $\Box = \bar{\partial}\bar{\partial}^* + \bar{\partial}^*\bar{\partial}$ and that $\sigma(\Box, \xi) = -\langle\xi, \xi\rangle I$.

$\Box$ is called the complex Laplacian of E.

1.9 Exercises

1.9.1 Use 1.2.7 to give a necessary and sufficient condition for the triviality of a vector bundle.

1.9.2 Use the Tietze extension theorem (cf. Hocking and Young [1], p. 62, Theorem 2–33) to prove the continuous version of 1.4.4.

1.9.3 Use 1.9.2 to prove continuous analogs of 1.4.6, 7, and 8.

1.9.4 Let K be either R or C. Let $G_{k,n}$ denote the set of all k-dimensional subspaces of K^n. Let $G = GL(n, K)$ act on $G_{k,n}$ by $g \cdot V = g(V)$ for g in G and V in $G_{k,n}$. Let $V_0 = \{(z_1, \ldots, z_k, 0, \ldots, 0) | z_i \text{ in } K\}$. Show that $G \cdot V_0 = G_{k,n}$. Find the isotropy group of V_0, H, and note that it is closed in G. Topologize $G_{k,n}$ as G/H. ($G_{k,n}$ is known as the Grassmann manifold of k-planes in K^n.) Let $E_{k,n} = \{(V, v) | v \text{ in } V\} \subset G_{k,n} \times K^n$. Show that if $p(V, v) = V$ then $(E_{k,n}, p)$ is a vector bundle over $G_{n,k}$.

1.9.5 Let X be compact and let $p: E \to X$ be a vector bundle over X. Show that there is a surjective vector bundle homomorphism $h: X \times K^n \to E$ for n sufficiently large.

(Hint: $X = V_1 \cup \cdots \cup V_k$ with V_i open in X and $E|_{V_i}$ trivial. Let $f_1, \ldots, f_k$ be a partition of unity subordinate to $V_1, \ldots, V_k$. Let $h_1^i, \ldots, h_m^i$ be cross sections of $E|_{V_i}$ so that $h_1^i(x), \ldots, h_m^i(x)$ is a basis for E_x for each x in V_i. Set $f_{ij} = f_i h_j^i$. Define $h(x, a_{ij}) = \sum a_{ij} f_{ij}(x)$.)

1.9.6 Let E, X be as in 1.9.5. Show that there is an injective vector bundle homomorphism of E into $X \times K^n$ if n is sufficiently large. (Hint: Use 1.9.5 and a unitary structure on E.)

1.9.7 Let E, X be as in 1.9.5. Suppose that the fiber dimension of E is k. Show that there is n sufficiently large and $h: X \to G_{k,n}$ continuous so that E is vector bundle isomorphic with $h^* E_{k,n}$ ($G_{k,n}$ and $E_{k,n}$ are as in 1.9.4).

1.9.8 Let $E = \{(z, v) | \langle z, v \rangle = 0\} \subset S^n \times R^{n+1}$. Show that if $p(z, v) = z$ then E is a vector bundle over S^n. Show that E is vector bundle isomorphic with the tangent bundle of S^n. (Here S^n is made into a C^∞ manifold by letting $U_i^\pm = \{(z_1, \ldots, z_{n+1}) = z | z \text{ in } S^n, \pm z_i > 0\}$ and

$$\psi_i^\pm(z) = (z_1, \ldots, z_{i+1}, z_{i+1}, \ldots, z_{n+1}).)$$

1.9.9 Let G be the semigroup whose set is the positive integers and whose addition is given by $x, y \mapsto \min(x, y)$. Show that $K(G) = 1$.

1.9.10 Let M be an orientable C^∞ manifold with Riemannian structure $\langle\ ,\ \rangle$. A diffeomorphism $f\colon M \to M$ is called an isometry if f_{*p} is an isometry of $T(M)_p$ with $T(M)_{f(p)}$ for each p in M. Let the notation be as in 1.7.10. Let $f\colon M \to M$ be a diffeomorphism and suppose that there is $0 \leqslant k \leqslant \dim M$ so that $\Delta f^*\omega = f^*\Delta\omega$ for all ω in $\mathscr{D}^k(M)$. Show that f is an isometry of M.

1.9.11 Let the notation be as in 1.7.10. Suppose that f is in $C_0^\infty(M)$ and that $\Delta f \geqslant 0$. Show that f is constant on each connected component of M.

1.9.12 Let E and F be C^∞ vector bundles over the C^∞ manifold M. Show that the following is an alternate definition of a differential operator:

(1) A differential operator of degree 0 is a map $D\colon \Gamma^\infty E \to \Gamma^\infty F$ so that there is a C^∞ vector bundle homomorphism $h\colon E \to F$ such that $Df = h \circ f$.

(2) Suppose that we have defined differential operators of degree $\leqslant k$. Then D is a differential operator of degree $\leqslant k + 1$ if whenever f is in $C^\infty(M)$ and h is in $\Gamma^\infty E$, then the operator $D_f h = D(fh) - fDh$ defines a differential operator of degree $\leqslant k$.

1.10 Notes

1.10.1 Basic references for the material of 1.2 through 1.5 are Atiyah [1] and Steenrod [1].

1.10.2 1.6 is taken directly from Atiyah [1].

1.10.3 The material of 1.7 and 1.8 is essentially the same as the material on the same topics in Palais *et al.* [1].

1.10.4 The idea in 1.9.12 was taught to the author by Robert Hermann.

CHAPTER 2

Elementary Representation Theory

2.1 Introduction

In this chapter we introduce the concept of a representation of a Lie group. Our definition (2.2.1) is quite strong, since we insist that the representation space be a Hilbert space, and in many applications it has been convenient to allow the representation space to be a Banach or (even) a Frechet space. However, our definition gives the most widely used class of representations and is certainly sufficient for the purposes of this book.

Section 3 gives the basic theory of finite dimensional representations. In Section 3 the main result is 2.3.7. In the proof of 2.3.7 it is easily seen that the assumption of continuity was not used. That is, 2.3.7 is an algebraic result.

In Section 4 we give a technique for constructing representations of Lie groups. The notion of "induced representation" that we use is the basic construction of this book. Some of the deepest work in this book is the study of these induced representations.

In Section 7 we develop the idea of a completely continuous representation. In Section 8 we prove the Peter–Weyl theorem using 2.7.4. In Section 9 we prove the orthogonality relations for compact Lie groups and use them to show that the characters of irreducible representations of a compact Lie group form an orthonormal basis of the space of central functions on the group.

2.2 Representations

2.2.1 *Definition* Let G be a Lie group and let $(H, \langle\ ,\ \rangle)$ be a complex Hilbert space. A representation of G on H is a homomorphism π of G into $GL(H)$ (see A.6.1.11) so that the map of $G \times H$ into H given by $(g, v) \mapsto \pi(g)v$ is continuous.

(π, H) is said to be unitary if $\pi(g)$ is a unitary operator for each g in G.

2.2.2 The notion of representation is equally meaningful for H a real Hilbert space. We will however concentrate our attention on the complex case.

2.2.3 *Definition* Let G be a Lie group and let (π, H), (ρ, V) be representations of G. Then (π, H) and (ρ, V) are said to be equivalent if there is a continuous linear isomorphism of H onto V so that $A\pi(x) = \rho(x)A$ for all x in G.

Let $\mathrm{Hom}_G(H, V)$ denote the space of all continuous linear maps of H into V so that $\rho(x)A = A\pi(x)$ for all x in G. Then $\mathrm{Hom}_G(H, V)$ is called the space of all G-homomorphisms of H to V, or the space of operators intertwining π and ρ.

2.2.4 *Definition* Let (π, H) be a representation of the Lie group G. Let W be a subspace of H. Then W is said to be invariant if for each g in G, $\pi(g)W \subset W$. (π, H) is said to be irreducible if the only closed invariant subspaces of H are H and (0).

2.2.5 *Definition* Let (π_i, H_i) be a countable collection of representations of G. Then a representation (π, H) of G is said to be a direct sum of the (π_i, H_i) if for each i there is an injective element A_i in $\mathrm{Hom}_G(H_i, H)$ so that the sum $V = \sum_i A_i(H_i)$ is direct and V is a dense subspace of H. We write $H = \sum_i H_i$, a direct sum.

2.2.6 These notions have unitary counterparts. If (π, H) and (ρ, V) are unitary representations then they are said to be unitarily equivalent if there is a unitary bijective intertwining operator from H to V. (π, H) is a unitary direct sum of (π_i, H_i) if $H = \sum_i H_i$ and the H_i are mutually orthogonal (here we identify H_i with its image in H).

2.3 Finite Dimensional Representations

2.3.1 *The contragradient representation* Let (π, H) be a finite dimensional representation of G. Let H^* be the dual space of H. Equip H^* with the Hilbert space structure that makes the dual basis to an orthonormal basis of H an orthonormal basis of H^*. We define a representation of G on H^* by setting $\pi^*(g)\lambda = \lambda \circ \pi(g)^{-1}$. To see that π^* is actually a representation we need only check the continuity. But if v is in H then the map

$$g, \lambda \mapsto (\pi^*(g)\lambda)(v) = \lambda(\pi(g^{-1})v)$$

is continuous. We note that (π^*, H^*) is unitary if and only if (π, H) is.

The functions $g \mapsto \lambda(\pi(g)v)$, λ in H^*, v in H are called matrix elements of the representation (π, H).

2.3.2 *Tensor product* Let (π_1, H_1) and (π_2, H_2) be finite dimensional representations of G. Let $H = H_1 \otimes H_2$. Give H the Hilbert space structure that makes the basis $\{v_i \otimes w_j\}$ orthonormal if $\{v_1, \ldots, v_n\}$ and $\{w_1, \ldots, w_m\}$ are orthonormal bases of H_1 and H_2, respectively. If g is in G we define $\pi(g) = \pi_1(g) \otimes \pi_2(g)$. Since

$$\lambda \otimes v(\pi(g)(v \otimes w)) = \lambda(\pi_1(g)v) \cdot v(\pi_2(g)w)$$

we see that (π, H) defines a representation of G, called the tensor product representation of H_1 and H_2.

2.3.3 *Grassman product* Let (π, H) be a finite dimensional representation of G. Equip $\Lambda^r H$ with the Hilbert space structure that makes

$$\{v_{i_1} \wedge v_{i_2} \wedge \cdots \wedge v_{i_r} | i_1 < i_2 < \cdots < i_r\}$$

an orthonormal basis of $\Lambda^r H$ if $\{v_1, \ldots, v_n\}$ is an orthonormal basis of H. Let G act on $\Lambda^r H$ by $\Lambda^r \pi(g) = \Lambda^r(\pi(g))$. Computing matrix elements one finds that $(\Lambda^r \pi, \Lambda^r H)$ defines a representation of G.

2.3.4 *Lemma* Let (π, H) be a finite dimensional irreducible representation of G. Then $\mathrm{Hom}_G(H, H) = CI$.

PROOF Let A be in $\mathrm{Hom}_G(H, H)$. Since H is finite dimensional, A has an eigenvector, $v \neq 0$, with eigenvalue λ. Let $H_\lambda = \{w \text{ in } H | Aw = \lambda w\}$.

Then H_λ is invariant and nonzero and thus by irreducibility $H_\lambda = H$. This implies that $A = \lambda I$.

2.3.5 *Lemma* Let (π, H) be a finite dimensional unitary representation of G. Then (π, H) is a unitary direct sum of irreducible representations.

PROOF By induction on the dimension of H. If H is one dimensional then (π, H) is irreducible and the result is trivially true. Suppose that the result is true for dimensions less than that of H. If H is irreducible we are done. If H is not irreducible then there is a nonzero invariant subspace V. Let V_1 be the orthogonal complement of V in H. Then $H = V + V_1$ unitary direct sum of representations. The result now follows by applying the induction hypothesis to V and to V_1.

2.3.6 *Exterior tensor product* Let (π_1, H_1) and (π_2, H_2) be finite dimensional representations of G. Let, for (g, h) in $G \times G$, $\pi_1 \hat{\otimes} \pi_2(g, h) = \pi_1(g) \otimes \pi_2(h)$. Then $(\pi_1 \hat{\otimes} \pi_2, H_1 \otimes H_2)$ is a representation of $G \times G$ which is unitary if (π_1, H_1) and (π_2, H_2) are unitary (here we use the Hilbert space structure on $H_1 \otimes H_2$ of 2.3.2); this representation is called the exterior tensor product representation.

2.3.7 *Proposition* Let (π, H) be a unitary irreducible representation of $G \times G$. Then (π, H) is equivalent to the exterior tensor product of two irreducible unitary representations of G. Conversely, the exterior tensor product of two irreducible unitary representations of G is an irreducible unitary representation of $G \times G$.

PROOF Let $\tilde{\pi}(g) = \pi(g, e)$ (e is the identity of G). Then $(\tilde{\pi}, H)$ is a unitary representation of G. Lemma 2.3.5 implies that H is a unitary direct sum $\sum_i H_i$ of irreducible unitary representations of G. We assert that all of the H_i are equivalent. Let W be the sum of all invariant subspaces U of H so that $\mathrm{Hom}_G(H_1, U) = (0)$. We assert that W is $G \times G$ invariant. Indeed, if g is in G then $\tilde{\pi}(x)\pi(e, g) = \pi(e, g)\tilde{\pi}(x)$ for all x in G. Thus if U is contained in W then $\pi(e, g)U$ is contained in W for all g in G. This proves the $G \times G$ invariance of W. Since $W \neq H$ and H is irreducible this implies that $W = (0)$. This proves that under the action of $\tilde{\pi}$, H is the unitary direct sum of $H_1, \ldots, H_n$ with H_i unitary, irreducible, and H_i is equivalent with H_j for all i, j.

Let P_i be the projection of H onto H_i corresponding to the above direct sum decomposition of H. 2.3.4 implies that $\dim \mathrm{Hom}_G(H_i, H_j) = 1$ for all i, j. Indeed if B intertwines H_i and H_j and if $B \neq 0$ then B is an equivalence by irreducibility. Thus if A, B are in $\mathrm{Hom}_G(H_i, H_j)$ and, say, $A \neq 0$ then $A^{-1}B$ is in $\mathrm{Hom}_G(H_i, H_i)$; hence $B = cA$ with c a complex scalar. Fix now for each i, j, $E_{i,j}$ a nonzero element of $\mathrm{Hom}_G(H_i, H_j)$. If A is in $\mathrm{Hom}_G(H, H)$ then $A = \sum P_i A P_j = \sum a_{i,j} E_{i,j}$ with $a_{i,j}$ in C.

Now let $V = \mathrm{Hom}_G(H_1, H)$. Then $\dim V = n$. If g is in G and A in V define $\pi_2(g)A = \pi(e, g) \circ A$. Then (π_2, V) is a unitary representation of G (here we define an inner product (,) on V by $(A, B) = \mathrm{tr}\, B^*A$, where B^* is the adjoint of B). Let π_1 be the restriction of $\tilde{\pi}$ to H_1. We define a map B of $H_1 \otimes V$ into H by $B(v \otimes A) = Av$. Then

$$B(\pi_1(g)v \otimes \pi_2(h)A) = \pi(e, h)A\pi(g, e)v = \pi(g, h)Av.$$

This says that B is an element of $\mathrm{Hom}_{G \times G}(H_1 \otimes V, H)$. Since $B \neq 0$ and H is irreducible B is surjective. Now, $\dim H = \dim H_1 \otimes V$, hence B is an equivalence. Finally, the irreducibility of H implies that (π_2, V) is an irreducible representation of G.

Suppose that (π_1, H_1) and (π_2, H_2) are irreducible, unitary, finite dimensional representations of G. We show that their exterior tensor product is irreducible. Let V be a nonzero, $(G \times G)$-invariant, irreducible subspace of $H_1 \otimes H_2$. Retracing the argument above we find that as a $G \times \{e\}$ representation V is isomorphic with the unitary direct sum of k copies of H_1. If v is in H_2 define for w in H_1, $A(v)w = w \otimes v$. Then using the above arguments it is easy to see that A is a bijection between H_2 and $\mathrm{Hom}_G(H_1, H_1 \otimes H_2)$. We have also shown that V is equivalent with $H_1 \otimes \mathrm{Hom}_G(H_1, V)$. Now $A^{-1}(\mathrm{Hom}_G(H_1, V))$ is an invariant subspace of H_2. Since V is nonzero and H_2 is irreducible this implies that $\dim \mathrm{Hom}_G(H_1, V) = \dim H_2$. But then $\dim V = \dim H_1 \otimes H_2$. Hence $V = H_1 \otimes H_2$.

2.3.8 *Lemma* Let (π, H) be an irreducible, finite dimensional, unitary representation of G. Then any invariant Hermitian form on H is a real multiple of the Hilbert space structure of H.

PROOF Let $\langle\ ,\ \rangle$ be the Hilbert space structure of H. If h is an invariant Hermitian form on H then $h(\pi(g)v, \pi(g)w) = h(v, w)$. Furthermore, there is a Hermitian matrix A so that $h(v, w) = \langle Av, w\rangle$. Now, the above condition of invariance says that A is in $\mathrm{Hom}_G(H, H)$. Lemma 2.3.4 implies that $A = cI$ with c a complex number. Since A is Hermitian c must be real.

2.3.9 *Corollary* If two unitary, irreducible, finite dimensional representations are equivalent they are unitarily equivalent.

PROOF Let (π_i, H_i), $i = 1, 2$, be unitary, irreducible, finite dimensional representations of G. Let A define an equivalence between H_1 and H_2. If $\langle\ ,\ \rangle_i$, $i = 1, 2$, are, respectively, the Hilbert space structures on H_1 and H_2 then 2.3.8 implies that $\langle Av, Aw\rangle_2 = c\langle v, w\rangle_1$ with c positive for all v, w in H_1. Set $B = c^{-\frac{1}{2}}A$. Then B defines the unitary equivalence.

2.3.10 *Lemma* Let (π, H) be an irreducible, finite dimensional, unitary representation of $G \times G$ such that there is a nonzero vector v in H so that $\pi(g, g)v = v$ for all g in G. Then there is an irreducible, unitary finite dimensional representation (π_1, H_1) of G so that (π, H) is unitarily equivalent with $(\pi_1 \hat{\otimes} \pi_1^*, H_1 \otimes H_1^*)$.

PROOF Proposition 2.3.7 says that we may assume that $(\pi, H) = (\pi_1 \hat{\otimes} \pi_2, H_1 \otimes H_2)$ with (π_i, H_i) an irreducible, finite dimensional, unitary representation of G for $i = 1, 2$. Let $V = L(H_1^*, H_2)$ and let $\rho(g, h)A = \pi_2(h) \circ A \circ \pi_1^*(g)^{-1}$, for g, h in G and A in V. We make V into a Hilbert space by defining $(A, B) = \operatorname{tr} B^*A$ for A, B in V. Then (ρ, V) is a unitary representation of $G \times G$. We define a linear map, B from $H_1 \otimes H_2$ to V via $B(v \otimes w)(z^*) = z^*(v)w$ for v in H_1, w in H_2, and z^* in H_1^*. A computation shows that B is in $\operatorname{Hom}_{G \times G}(H_1 \otimes H_2, V)$. Since $B \neq 0$ and $H_1 \otimes H_2$ is irreducible, B is injective. Since $\dim V = \dim H_1 \otimes H_2$, B is surjective. Hence B is an equivalence. By hypothesis, there is thus a nonzero element U of V so that $\pi_2(g)U\pi_1^*(g)^{-1} = U$ for all g in G. But then by irreducibility of H_2, (π_2, H_2) is unitarily equivalent with (π_1^*, H_1^*).

2.4 Induced Representations

2.4.1 *The modular function of an action* Let M be an orientable C^∞ manifold (see A.4.1.6) with volume form ω. Let G be a Lie group acting on M (see A.2.4.6 and A.4.1.8 for notation and definitions). Then $(g^*\omega)_x = c(g, x)\omega_x$, for each g in G and x in M. Furthermore c is a C^∞ real valued function on $G \times M$.

If g and h are in G then

$$((gh)^*\omega)_x = (h^*g^*\omega)_x = (h^*c(g, \cdot)\omega)_x = c(h, x)c(g, hx)\omega_x.$$

We therefore see that c satisfies the cocycle condition

$$c(gh, x) = c(g, hx)c(h, x), \text{ for } g, h \text{ in } G, x \text{ in } M. \tag{1}$$

If η is another volume form for M then $\eta = f\omega$, where f is a nowhere zero C^∞ function on M. If d is the cocycle corresponding to η then a computation shows that

(2) $$d(g, x) = f(gx)c(g, x)f(x)^{-1}.$$

This expresses the dependence of c on ω.

The change of variables formula (see A.4.2.10) says that if for h in $C_0(M)$ we set

$$\int_M h(x)dx = \int_M h\omega$$

then

(3) $$\int_M h(gx)|c(g, x)|dx = \int_M h(x)dx.$$

2.4.2 Let E be a complex vector bundle over M. Let G be a Lie group acting on M. Then E is called a G-vector bundle if G acts on E such that if g is in G, x is in M, then $gE_x \subset E_{gx}$ and the map of E_x to E_{gx} given by g is linear.

2.4.3 *Lemma* Let E be a G-vector bundle over M. Let $\langle\ ,\ \rangle$ be a unitary structure on E. Then there is a continuous function h on $G \times M$ so that $\langle gv, gv\rangle_{gx} \leqslant h(g, x)\langle v, v\rangle_x$.

PROOF For each g in G there is a continuous cross section A_g of $\mathrm{Hom}(E, E)$ so that if v is in E_x then $\langle gv, gv\rangle_{gx} = \langle A_g(x)v, A_g(x)v\rangle_x$. We may take $h(g, x) = \|A_g(x)\|^2$ (see A.6.1.9).

2.4.4 Let E be a vector bundle over M with unitary structure $\langle\ ,\ \rangle$. Then $(E, \langle\ ,\ \rangle)$ is called a unitary G-vector bundle if E is a G-vector bundle such that the map of E_x to E_{gx} given by g is unitary for each g in G and x in M. $(E, \langle\ ,\ \rangle)$ is called G-admissible if E is a G-vector bundle and if the function h of 2.4.3 may be chosen so that for each compact subset C of G there is a constant L_C so that $h(g^{-1}, gx) \leqslant L_C$ for all g in C, x in M.

2.4.5 Let E be a vector bundle over M with unitary structure $\langle\ ,\ \rangle$. Let $\Gamma_0 E$ be the space of all compactly supported cross sections of E. We define a pre-Hilbert space structure on $\Gamma_0 E$ by setting

(1) $$(f_1, f_2) = \int_M \langle f_1(x), f_2(x)\rangle_x\, dx.$$

We denote by $L^2(E, \omega)$ the corresponding Hilbert space. $L^2(E, \omega)$ is separable (see exercise 2.10.12).

2.4.6 *Theorem* Let $(E, \langle\ ,\ \rangle)$ be a G-admissible vector bundle over M. Let, for f in $\Gamma_0 E$, g in G, x in M

$$(\pi(g)f)(x) = |c(g^{-1}, x)|^{\frac{1}{2}} g f(g^{-1}x). \tag{1}$$

Then for each g in G, $\pi(g)$ extends to a continuous operator on $L^2(E, \omega)$ and $(\pi, L^2(E, \omega))$ defines a representation of G. If $(E, \langle\ ,\ \rangle)$ is a unitary G-vector bundle then $(\pi, L^2(E, \omega))$ is a unitary representation of G.

PROOF Let g be in G, f in $\Gamma_0 E$. Then

$$\begin{aligned}(\pi(g)f, \pi(g)f) &= \int_M \langle gf(g^{-1}x), gf(g^{-1}x)\rangle_x |c(g^{-1}, x)|\,dx \\ &\leqslant \int_M h(g^{-1}, x)\,\langle f(g^{-1}x), f(g^{-1}x)\rangle_{g^{-1}x} |c(g^{-1}, x)|\,dx \\ &= \int_M h(g^{-1}, gx)\langle f(x), f(x)\rangle_x\,dx \text{ (by 2.4.1(3))} \leqslant L_g(f, f),\end{aligned} \tag{2}$$

where h and L are as in 2.4.4. Thus (using A.6.1.8) $\pi(g)$ extends to a continuous operator on $L^2(E, \omega)$. Furthermore, if $|\cdots|$ is the norm on $H = L^2(E, \omega)$ and if $\|\cdots\|$ is the corresponding norm on $L(H, J)$ (see A.6.1.9) then if C is a compact subset of G and if $B_C = L_C^{\frac{1}{2}}$ then (2) implies that if g is in C then

$$\|\pi(g)\| \leqslant B_C. \tag{3}$$

Furthermore, the computation of (2) shows that if E is a G-unitary vector bundle then $\pi(g)$ is a unitary operator on H for each g in G. It is a simple computation using 2.4.1(1) to see that if g, h are in G then $\pi(gh) = \pi(g)\pi(h)$. Thus to prove that π is a representation we need only prove that the map $g, v \mapsto \pi(g)v$ is continuous. We first prove

$$\lim_{g \to e} \pi(g)f = f \text{ for } f \text{ in } H. \tag{i}$$

To prove (i) we first assume that f is in $\Gamma_0 E$. Then by uniform continuity of f and admissibility of E there is for each positive integer n a neighborhood V_n of e in G so that $V_{n+1} \subset V_n$, $\overline{V}_1$ is compact and

$$\langle \pi(x)f(u) - f(u), \pi(x)f(u) - f(u)\rangle < 1/n \tag{4}$$

for all u in M and x in V_n. Now if x is in V_n then

$$(5)\qquad \|\pi(x)f - f\|^2 = \int_M \langle \pi(x)f(u) - f(u), \pi(x)f(u) - f(u)\rangle\, du < C\left(\frac{1}{n}\right),$$

where C is the volume of $V_1\mathrm{supp}(f)$.

This proves (i) for f in $\Gamma_0 E$. Suppose that f is in H. Let $\varepsilon > 0$ be given. Let W be an open neighborhood of e so that $\overline{W}$ is compact. Let $D > B_{\overline{W}} + 1$ (see (3)). Let f_1 be in $\Gamma_0 E$ be such that $\|f_1 - f\| < \varepsilon/4D$. Let V be a neighborhood of e so that $V \subset W$ and if x in V then $\|\pi(x)f_1 - f_1\| < \varepsilon/4$. If x is in V then

$$\|\pi(x)f - f\| \leqslant \|\pi(x)f - \pi(x)f_1\| + \|\pi(x)f_1 - f_1\| + \|f_1 - f\|$$

$$< B_W\frac{\varepsilon}{4D} + \frac{\varepsilon}{4} + \frac{\varepsilon}{4D} = (B_W + 1)\frac{\varepsilon}{4D} + \frac{\varepsilon}{4} < \frac{\varepsilon}{2}.$$

This completes the proof of (i).

Suppose now g is in G, f is in H, and $\varepsilon > 0$ is given. We must find a neighborhood U of g in G and $\delta > 0$ so that if h is in U and $\|f' - f\| < \delta$ then $\|\pi(h)f' - \pi(g)f\| < \varepsilon$. First of all let W be a neighborhood of g in G so that $\overline{W}$ is compact. Then $\|\pi(h)f' - \pi(g)f\| \leqslant B_{\overline{W}}\|f' - \pi(h^{-1}g)f\|$. Thus we need only show that given $\varepsilon > 0$ there is a neighborhood U of e in G and $\delta > 0$ so that if $\|f - f'\| < \delta$ and h is in U then $\|\pi(h)f - f'\| < \varepsilon$. But $\|\pi(g)f - f'\| \leqslant \|\pi(g)f - f\| + \|f - f'\|$. Hence (i) guarantees that U and δ can be found.

2.5 Invariant Measures on Lie Groups

2.5.1 *Haar measure* Let G be a Lie group. Let $\mathfrak{g}$ be the Lie algebra of G looked upon as the space of all left invariant (resp. right invariant) vector fields on G. Let ω be an alternating n-form on $\mathfrak{g}$ ($n = \dim G$). Then ω defines an n-form on G by

$$\omega_g((X_1)_g, \ldots, (X_n)_g) = \omega(X_1, \ldots, X_n)$$

for g in G, $X_1, \ldots, X_n$ in $\mathfrak{g}$. If g is in G then $L_g^*\omega = \omega$ (resp. $R_g^*\omega = \omega$). Furthermore if ω_1 is an n-form on G so that $L_g^*\omega_1 = \omega_1$ for all g in G (resp. $R_g^*\omega_1 = \omega_1$) then $\omega_1 = c\omega$ with c a real constant. Let ω be denoted ω^L if $\mathfrak{g}$ is looked upon as the space of left invariant vector fields on G and ω^R if $\mathfrak{g}$ is looked upon as the space of all right invariant vector fields on G.

2.5.2 *Lemma* Let η be defined by $\eta(g) = g^{-1}$. Then $\eta^*\omega^L = (-1)^n\omega^R$.

PROOF Clearly $L_g \circ \eta = \eta \circ R_{g^{-1}}$. This formula implies that $R_g^*(\eta^*\omega^L) = \eta^*\omega^L$. Hence by the remarks of 2.5.1, $\eta^*\omega^L = c\omega^R$ with c a real constant. Now $\eta(e) = e$ and $\eta_{*e}(X_e) = -X_e$ for X in $\mathfrak{g}$. Hence $c = (-1)^n$.

2.5.3 *The modular function* Since ω^L is an everywhere nonvanishing n-form on G, $\omega^R = \delta \cdot \omega^L$, with δ a nowhere vanishing C^∞ real valued function on G. 2.5.2 states that

(1) $$\eta^*\omega^L = (-1)^n\delta \cdot \omega^L.$$

Using the formula $R_g L_x = L_x R_g$ for all x, g in G we see that

(2) $$R_g^*\omega^L = c(g)\omega^L,$$

where $c(g)$ defines a function on G. Using the definition of δ we see that $c(g) = \delta(x)\delta(xg)^{-1}$, for all x in G. Since $\delta(e) = 1$ we see that

(3) $$c(g) = \delta(g)^{-1}$$

for all g in G. Equation (3) implies that

(4) $$\delta(gh) = \delta(g)\delta(h), \text{ for all } g, h \text{ in } G.$$

If f is in $C_0(G)$ set

(5) $$\int_G f\omega^L = \int_G f(x)dx.$$

Then the change of variables formula implies that

(6) $$\int_G f(xg)dx = \int_G f(x)\,|\delta(g)|\,dx$$

and that

(7) $$\int_G f(x^{-1})dx = \int_G f(x)|\delta(x)|^{-1}dx.$$

2.5.4 *Definition* G is called unimodular if $|\delta(g)| = 1$ for all g in G. We note that if G is connected and unimodular then $\omega^L = \omega^R$.

2.5.5 *Proposition* Let $D^1(G)$ be the closure of the subgroup of G generated by all commutators of G. (A commutator in G is a product $ghg^{-1}h^{-1}$ for g, h in G.) If $G/D^1(G)$ is compact then G is unimodular.

PROOF 2.5.3.(4) implies that δ is identically 1 on $D^1(G)$. Thus δ defines a representation of $H = G/D^1(G)$ on C. If for some h in H, $|\delta(h)| \neq 1$

then we may assume $|\delta(h)| > 1$ (if $|\delta(h)| < 1$ then $|\delta(h^{-1})| > 1$). Since $|\delta(h^n)| = |\delta(h)|^n$, we see that the sequence $|\delta(h^n)|$ is unbounded. But δ is continuous and H is assumed to be compact. This contradiction proves the proposition.

2.5.6 *Corollary* A compact Lie group is unimodular.

2.5.7 *Proposition* Let G be a Lie group. Let, for g in G, $\mathrm{Ad}(g)$ be the automorphism of $\mathfrak{g}$ that is the differential of $L_g \circ R_{g^{-1}}$. If $|\det \mathrm{Ad}(g)| = 1$ for all g in G then G is unimodular.

PROOF Let $\mathfrak{g}$ be looked upon as the space of all left invariant vector fields on G. If X is in $\mathfrak{g}$ then $(R_{g*}X)_e = (R_{g*} \circ L_{g^{-1}}X)_e = (\mathrm{Ad}(g^{-1})X)_e$. Thus $(R^*\omega^L)_e = \det \mathrm{Ad}(g^{-1})\omega^L$. Hence $\delta(g) = \det \mathrm{Ad}(g^{-1})$ for g in G.

2.5.8 *Lemma* Let G be a compact Lie group. Let (π, H) be a finite dimensional representation of G. Then there is an inner product $(\ ,\ )$ on H so that $(\pi, (H, (\ ,\ )))$ is a unitary representation of G.

PROOF Let $\langle\ ,\ \rangle$ be any inner product on H. Set

$$(u, v) = \int_G \langle \pi(g)v, \pi(g)w \rangle dg.$$

The invariance of dg implies that $(\pi(g)v, \pi(g)w) = (v, w)$.

2.6 The Regular Representation

2.6.1 Let G be a Lie group and let dx be Haar measure on G. Let $L^2(G)$ denote the completion of $C_0(G)$ relative to the inner product

$$(f_1, f_2) = \int_G f_1(x)\overline{f_2(x)}dx.$$

2.6.2 *Proposition* Let for f in $C_0(G)$, g in G,

(1) $$(\pi(g)f)(x) = f(g^{-1}x).$$

Then $\pi(g)$ extends to a unitary operator on $L^2(G)$ for each g in G. $(\pi, L^2(G))$ is a unitary representation of G.

If G is unimodular define for f in $C_0(G)$, g, h in G

$$(\tau(g, h)f)(x) = f(g^{-1}xh). \tag{2}$$

Then $\tau(g, h)$ extends to a unitary operator on $L^2(G)$ and $(\tau, L^2(G))$ is a unitary representation of $G \times G$.

PROOF This result is a special case of Theorem 2.4.6. Indeed, $C_0(G) = \Gamma_0(G \times C)$.

2.6.3 *Proposition* Let G be a compact Lie group. Let H be a closed, invariant subspace of $L^2(G)$ so that as a representation of G, H is irreducible. Then H is finite dimensional.

PROOF If $H = (0)$ there is nothing to prove. Otherwise, let f be a unit vector in H. Let, for v in $L^2(G)$, $Pv = (v, f)f$. Define T by the formula

$$(Tv, w) = \int_G (P\pi(g)v, \pi(g)w)dg,$$

for v, w in $C(G)$. Now

$$\begin{aligned}(Tv, w) &= \int_G (\pi(g)v, f)(f, \pi(g)w)dg \\ &= \int_G \int_G v(g^{-1}x)\overline{f(x)}dx \int_G f(y)\overline{w(g^{-1}y)}dydg \\ &= \int_G \int_G v(x)\overline{f(gx)}dx \int_G f(gy)\overline{w(y)}dydg \\ &= \int_G \left(\int_G \overline{f(gx)}f(gy)dg\right)v(x)\overline{w(y)}dxdy.\end{aligned}$$

Set

$$k(x, y) = \int_G \overline{f(gx)}f(gy)dg.$$

Then k is a continuous function on $G \times G$. The above computation shows that

$$Tv(x) = \int_G k(x, y)v(y)dy,$$

for v in $C(G)$. Hence T is completely continuous (see A.7.2.3). Clearly $k(x,y) = \overline{k(y,x)}$. Thus T is self-adjoint. $TH \subset H$ and $T|_H \neq 0$ since $(Tf,f) > 0$. Clearly $\pi(g)T = T\pi(g)$ for all g in G. Furthermore A.6.2.15 implies that T has an eigenvalue, $\lambda \neq 0$. The irreducibility of H implies that $T|_H = \lambda I$. This implies that the identity map of H to H is completely continuous. Hence H is finite dimensional.

2.7 Completely Continuous Representations

2.7.1 Let G be a Lie group and let (π, H) be representation of G. If f is in $C_0(G)$ define $\pi(f)$ by the formula

$$\langle \pi(f)v, w\rangle = \int_G \langle f(g)\pi(g)v, w\rangle \, dg. \tag{1}$$

We write (1) symbolically as

$$\pi(f) = \int_G f(g)\pi(g)\, dg. \tag{2}$$

2.7.2 *Lemma* If f is in $C_0(G)$ then $\pi(f)$ is a continuous operator on H.

PROOF Let v be a unit vector in H. Then

$$\langle \pi(f)v, \pi(f)v\rangle = \int_G \int_G f(g)\overline{f(h)}\langle \pi(g)v, \pi(h)v\rangle \, dg dh.$$

Set $C = \sup\{|f(x)| \mid x \text{ in } G\}$. Let Ω be the support of f and let $K = \int_G \varphi(x)\, dx$ where φ is a nonnegative function on $C_0(G)$ which is identically one on Ω. Set $D = \sup\{\|\pi(x)\| \mid x \text{ in } \Omega\}$. Then $\langle \pi(f)v, \pi(f)v\rangle \leqslant C^2KD$.

2.7.3 *Definition* A representation (π, H) of G is said to be completely continuous if for each f in $C_0^\infty(G)$, $\pi(f)$ is completely continuous.

2.7.4 *Theorem* If (π, H) is a completely continuous, unitary representation of G and if G is unimodular then $H = \sum H_i$, a unitary direct sum of irreducible representations of G. Furthermore if i is fixed then there are only a finite number of j so that H_j is equivalent with H_i.

PROOF Let U_i be a countable basis for neighborhoods of e in G (that is, if U is open in G and if e is in U then there is U_i so that $U_i \subset U$) so that if g is in U_i then g^{-1} is in U_i. Let, for each i, h_i be an element of $C_0^\infty(G)$ which is nonnegative, has support in U_i, and is such that $h_i(e) > 0$. Set $f_i(x) = h_i(x)h_i(x^{-1})$. By multiplying f_i by a positive constant we may assume that for each i, $\int_G f_i(x)dx = 1$.

Since $\pi(f_i)$ is completely continuous $H = H_{0,i} + \sum_{j=1} H_{j,i}$, a unitary direct sum of subspaces, where $H_{0,i} = \ker(\pi(f_i))$ and $H_{j,i}$ is an eigenspace of $\pi(f_i)$ with nonzero eigenvalue. Furthermore dim $H_{j,i}$ is finite, for $j \neq 0$. Let V be the closure of the subspace $\sum_{j=1}\sum_{i=1} H_{i,j}$ of H. We assert that $V = H$.

Indeed if v is in H and if v is perpendicular to V then $\pi(f_i)v = 0$ for all i. Now, given $\varepsilon > 0$ there is i so that if x is in U_i then $\|\pi(x)v - v\| < \varepsilon$. If u is in H then

$$\begin{aligned} |\langle \pi(f_i)v - v, u\rangle| &= \left|\int_G f_i(x)\langle \pi(x)v - v, u\rangle\, dx\right| \\ &\leqslant \int_G f_i(x)|\langle \pi(x)v - v, u\rangle| dx \\ &\leqslant \int_G f_i(x)\|\pi(x)v - v\|\ \|u\|\ dx \\ &\leqslant \varepsilon\|u\| \text{ (since supp } f_i \text{ is contained in } U_i). \end{aligned}$$

But we have assumed that $\pi(f_i)v = 0$. Hence $|\langle v, u\rangle| \leqslant \varepsilon\|u\|$ for any $\varepsilon > 0$. This implies that $\langle v, u\rangle = 0$. Since u is arbitrary $v = 0$.

Let, for each j, i, $V_{j,i}$ be the intersection of all closed invariant subspaces of H containing $H_{j,i}$. Clearly $V_{j,i}$ is a closed invariant subspace of $H_{j,i}$.

(1) If $U \subset V_{j,i}$ is aclosed invariant subspace and if $U \cap H_{j,i} = (0)$ then $U = (0)$.

In fact, if $U \cap H_{j,i} = (0)$ then $V_{j,i} \cap U^\perp \supset H_{j,i}$. Thus $V_{j,i} \cap U^\perp = V_{j,i}$. Hence $U \subset V_{j,i}^\perp \cap V_{j,i} = (0)$.

Let now $H_{j,i}^1$ be a subspace of $H_{j,i}$ of lowest positive dimension such that there is $U \subset V_{j,i}$ closed and invariant such that $U \cap H_{j,i} = H_{j,i}^1$. Let $V_{j,i}^1$ be the intersection of all closed invariant subspaces of $V_{j,i}$ containing $H_{j,i}^1$. If $W \subset V_{j,i}^1$ is closed and invariant then $V_{j,i}^1 = W \oplus W^\perp \cap V_{j,i}^1$. Now $H_{j,i}^1 = W \cap H_{j,i} \oplus (W^\perp \cap V_{j,i}^1) \cap H_{j,i}$. If $W \cap H_{j,i} \neq (0)$ and $(W^\perp \cap V_{j,i}^1) \cap H_{j,i} \neq (0)$ we have contradicted the definition of $V_{j,i}^1$. Hence (1) implies $W = V_{j,i}^1$ or $W = (0)$. That is, $V_{j,i}^1$ is irreducible.

Let now $H_{j,i}^2$ be defined in the same way as $H_{j,i}^1$ relative to $V_{j,i} \cap (V_{j,i}^1)^\perp$ and $H_{j,i} \cap (H_{j,i}^1)^\perp$. Let $V_{j,i}^2$ be the intersection of all closed invariant subspaces of $V_{j,i} \cap (V_{j,i}^1)^\perp$ containing $H_{j,i}^2$. As above $V_{j,i}^2$ is irreducible. Continue this procedure to find that $V_{j,i} = V_{j,i}^1 \oplus V_{j,i}^2 \oplus \cdots \oplus V_{j,i}^{n_{ji}}$, an orthogonal direct sum with $n_{ji} \leqslant \dim H_{j,i}$.

(2) If $\langle V_{j,i}^k, V_{r,s}^l \rangle \neq (0)$ then $V_{r,s}^l \subset V_{j,i}$.

To prove (2) we define $P_{j,i}^k: H \to V_{j,i}^k$ to be the orthogonal projection onto $V_{j,i}^k$. If $\langle V_{k,i}^j, V_{r,s}^l \rangle \neq (0)$ then $P_{j,i}^k(V_{r,s}^l) \neq (0)$. Since $V_{r,s}^l$ and $V_{j,i}^k$ are irreducible, we see that $P_{j,i}^k|_{V_{r,s}^l}$ defines an equivalence between $V_{r,s}^l$ and $V_{j,i}^k$. But then $V_{r,s}^l \cap H_{j,i} \neq (0)$. Hence $V_{r,s}^l \cap V_{j,i} \neq (0)$. Since $V_{r,s}^l$ is irreducible this implies that $V_{r,s}^l \subset V_{j,i}$.

By relabeling, we may replace the $V_{j,i}$ by $V_1, V_2, \ldots$, Set $U_1 = V_1$. Suppose that we have defined U_j. Define. $U_{j+1} = V_j \cap (U_1 + \cdots + U_j)^\perp$. (2) implies $V_1 + \cdots + V_{j+1} = U_1 \oplus \cdots \oplus U_{j+1}$ orthogonal direct sum. Also (2) combined with the direct sum decomposition of the V_j implies that U_j is a unitary direct sum of at most $m_j < \infty$ closed invariant subspaces. Suppose that f is in H and $\langle f, U_j \rangle = 0$ for all j. Then $\langle f, V_j \rangle = 0$ for all j. Hence $\langle f, H_{j,i} \rangle = 0$ for all j, $i \geqslant 1$. Hence by the beginning of this proof $f = 0$. Thus $H = \sum U_j$ unitary direct sum. Since U_j is a unitary direct sum of a finite number of closed irreducible subspaces, $H = \sum H_j$ unitary direct sum of closed, irreducible subspaces.

Suppose that $H_{j_1}, \ldots, H_{j_n}$ are all equivalent to H_k. Let i be such that $\pi(f_i)|_{H_k} \neq 0$. By the definition of $\pi(f_i)$, if λ is a nonzero eigenvalue for $\pi(f_i)$ on H_k it must be an eigenvalue for $\pi(f_i)$ on H_{j_r} for $r = 1, \ldots, k$. Since the λ eigenspace for $\pi(f_i)$ has dimension $m < \infty$ we see that $n \leqslant m$. The proof of the theorem is now complete.

2.8 The Peter–Weyl Theorem

2.8.1 Let G be a compact Lie group. Let, for each finite dimensional, irreducible, unitary representation (π, H) of G, $\{(\pi, H)\}$ denote its equivalence class. Let $\hat{G}$ be the set of all equivalence classes of irreducible, finite dimensional, unitary representations of G. Let for each γ in $\hat{G}$, (π_γ, V_γ) be a fixed representative. For each γ in $\hat{G}$ let A_γ be the map from $V_\gamma^* \otimes V_\gamma$ to $C(G)$ defined by $A_\gamma(\lambda \otimes v)(g) = \lambda(\pi_\gamma(g)v)$ for g in G, v in V_γ, λ in V_γ^*. We note that $A_\gamma(V_\gamma^* \otimes V_\gamma)$ is the space spanned by the matrix elements of V_γ. An easy computation shows that A_γ is an element of

$$\mathrm{Hom}_{G\times G}((\pi_\gamma^* \hat{\otimes} \pi_\gamma, V_\gamma^* \otimes V_\gamma), (\tau, L^2(G))).$$

2.8.2 *Theorem* $(\tau, L^2(G)) = \sum V_\gamma^* \otimes V_\gamma$, a unitary direct sum over $\hat{G}$ of representations of $G \times G$. Furthermore $\hat{G}$ is countable.

PROOF We first show that if H is a compact Lie group and if π is the usual action of H on $L^2(H)$ then $(\pi, L^2(H))$ is completely continuous.

Let f, f_1 be elements of $C(H)$. Then

$$\pi(f)f_1(x) = \int_H f(g)f_1(g^{-1}x)\,dg = \int_H f(xg)f_1(g^{-1})\,dg$$

$$= \int_H f(xg^{-1})f_1(g)\,dg$$

since H is unimodular. Hence $\pi(f)$ has kernel $k(x, y) = f(xy^{-1})$. This proves the assertion.

We now realize $(\tau, L^2(G))$ as a subrepresentation of the regular representation of $G \times G$. Let $B(f)(x, y) = f(xy^{-1})$ for x, y in G and f in $C(G)$. Using the unimodularity of G it is a computation to show that B is unitary. A simple computation shows that B intertwines τ with the regular representation of $G \times G$. Using the above result for $H = G \times G$, we see that $(\tau, L^2(G))$ is a completely continuous representation of $G \times G$. Hence Theorem 2.7.4 and Proposition 2.6.3 apply and we see that $(\tau, L^2(G)) = \sum H_i$, unitary direct sum of irreducible representations (hence finite dimensional) of $G \times G$. We now show that the H_i are mutually inequivalent. Indeed, suppose that V and W are equivalent $G \times G$ subrepresentations of $L^2(G)$ that are irreducible and finite dimensional. We show that $V = W$. Let $e_1, \ldots, e_n$ and $f_1, \ldots, f_n$ be, respectively, orthonormal bases of V and W so that the corresponding matrix representations of $G \times G$ are the same. Set $F(x, y) = \sum_i e_i(x)\overline{f_i(y)}$ for x, y in G. If g, h are in G then using the unitarity of the action of $G \times G$ on $L^2(G)$ we see that $F(gxh^{-1}, gyh^{-1}) = F(x, y)$. In particular $F(e, g^{-1}) = F(g, e)$. Let $Sf(g) = f(g^{-1})$ for f in $C(G)$. Since G is unimodular S extends to a unitary operator on $L^2(G)$. It is clear that if H is an invariant subspace of $L^2(G)$ then SH is invariant; also, if $\overline{H}\{f \mid f \text{ in } H\}$ then H is invariant. Now

$$\textstyle\sum_i \overline{f_i(e)}e_i(g) = F(g, e) = F(e, g^{-1}) = \sum_i e_i(e)\overline{f_i(g^{-1})} = \sum_i e_i(e)\overline{SF_i(g)}.$$

Thus $V \cap \overline{SW} \neq (0)$. Since V is irreducible $V = \overline{SW}$. But $\dim V = \dim \overline{SW}$, hence $V = \overline{SW}$. The same argument applies if we take $V = W$ to see that $W = \overline{SW}$. Hence $V = \overline{SW} = W$ as was to be shown.

To complete the proof of the theorem we need only show that if W is an irreducible finite dimensional $G \times G$ subrepresentation of $L^2(G)$ then there is a unit vector v in W so that $\tau(g, g)v = v$ for all g in G (here we use 2.3.10). Let $f_1, \ldots, f_n$ be an orthonormal basis of W. Then this basis defines a map of W to C^n, A, given by $A(c_1 f_1 + \cdots + c_n f_n) = (c_1, \ldots, c_n)$. Let for each (g, h) in $G \times G$, $\rho(g, h)$ be the linear operator on C^n defined by $A(\tau(g, h)f) = \rho(g, h)Af$. Let $T(x, y) = (f_1(xy^{-1}), \ldots, f_n(xy^{-1}))$, then T maps $G \times G$ into C^n. A simple computation shows that $T(gx, hx) = \overline{\rho(g, h)}T(x, y)$. Hence the unit vector in the direction $A^{-1}(T(e, e))$ is the desired v.

2.8.3 *Corollary* Let V and W be irreducible finite dimensional representations of G. Let f be a matrix element of V and let h be a matrix element of W. Then $(f, h) = 0$ if V and W are not equivalent.

2.8.4 *Corollary* Let G be a compact Lie group, then G is Lie isomorphic with a closed subgroup of $GL(n, C)$ for n sufficiently large.

PROOF Since G is countable we may write $\hat{G} = \{\gamma_j\}$. Let $\pi_j = \pi_{\gamma_1} \oplus \cdots \oplus \pi_{\gamma_j}$. Let $G_j = \ker \pi_j$. Then $G_j \supset G_{j+1}$. 2.8.2 implies that if j is sufficiently large then $\dim G_j = 0$. Hence since G is compact this implies that G_j finite. This implies that if we take some k larger than j (large enough that none of the elements of G_j go to the identity under π_k) then $G_k = (e)$. This proves the result.

2.8.5 *Theorem* The algebraic direct sum $\sum V_\gamma^* \otimes V_\gamma$ is dense in $C(G)$ relative to the uniform norm ($\|f\|_\infty = \sup_{x \text{ in } G} |f(x)|$).

PROOF Let f be in $C(G)$. Let U be an open subset of G containing e and so that $|f(y^{-1}x) - f(x)| < \varepsilon/2$ for all y in U, x in G. Let h be in $C(G)$ be a nonnegative function with support in U and so that $\int_G h(x)\,dx = 1$. We note that if p is in $H = \sum V_\gamma^* \otimes V_\gamma$ (algebraic direct sum) then $\pi(h)p$ is in H. Choose p in H so that $\|f - p\| < \varepsilon/\|h\|2$. Now

$$|\pi(h)f(x) - f(x)| = \left|\int_G h(y)(f(y^{-1}x) - f(x))dy\right|$$
$$\leqslant \int_G h(y)|f(y^{-1}x) - f(x)|\,dy < \frac{\varepsilon}{2}\int_G h(x)\,dx = \frac{\varepsilon}{2}.$$

Also if u is in $C(G)$ then

$$|\pi(h)u(x)| = \left|\int_G h(y)u(y^{-1}x)\,dy\right| \leqslant \|h\|\,\|u\|.$$

Now

$$\|f - \pi(h)p\|_\infty \leqslant \|f - \pi(h)f\|_\infty + \|\pi(h)f - \pi(h)p\|_\infty < \frac{\varepsilon}{2}$$
$$+ \|h\|\,\|f - p\| < \frac{\varepsilon}{2} + \frac{\varepsilon}{2} = \varepsilon.$$

This proves the theorem.

2.9 Characters and Orthogonality Relations

2.9.1 *Definition* Let (ρ, V) be a finite dimensional representation of G, a Lie group. If g is in G set $\chi_V(g) = \operatorname{tr} \rho(g)$. χ_V is called the character of (ρ, V).

2.9.2 Using the formula $\operatorname{tr} AB = \operatorname{tr} BA$ one sees that χ_V depends only on the equivalence class of (ρ, V). If G is compact let $\hat{G}$ be the set of all equivalence classes of irreducible, finite dimensional, unitary representations of G. Let, for each γ in $\hat{G}$, χ_γ be the character of any representative of γ. Then χ_γ is well defined, and is called a simple character of G.

2.9.3 *Lemma* Let (ρ, V) be an irreducible, finite dimensional, unitary representation of G, a compact Lie group. Let v and w be in V and let v^* and w^* be in V^*. Then

(1) $$\int_G v^*(\rho(g)v)\overline{w^*(\rho(g)w)}dg = \left(\frac{1}{n}\right)\langle v, w\rangle\langle v^*, w^*\rangle$$

where $n = \dim V$ and $\langle\ ,\ \rangle$ is the inner product of V and the dual inner product of V^*.

PROOF Since G is unimodular the usual action of $G \times G$ on $L^2(G)$ (see 2.6.2(2)) is unitary. Hence the right-hand side of (1) defines a $G \times G$ invariant inner product on $V^* \otimes V$ under the action $\rho^* \otimes \rho$. The usual tensor product inner product is given by $(v^* \otimes v, w^* \otimes w) = \langle v, w\rangle \langle v^*, w^*\rangle$. Applying 2.3.8 we find that the left-hand side of (1) is $c\langle v, w\rangle\langle v^*, w^*\rangle$. To complete the proof we need only compute c. Let $v_1, \ldots, v_n$ be an orthonormal basis of V and let $v_1^*, \ldots, v_n^*$ be the dual basis. Let $f_{ij}(g) = v_i^*(\rho(g)v_j)$. By the above f_{ij} defines up to a constant factor an orthonormal basis of the matrix elements of (ρ, V). Hence $\sum_{ij} f_{ij}(g)\overline{f_{ij}(g)} = r$ a constant. But

$$\textstyle\sum_{ij} f_{ij}(e)\overline{f_{ij}(e)} = \sum_{ij} |f_{ij}(e)|^2 = n.$$

But then

$$n = \sum_{ij} \int_G f_{ij}(g)\overline{f_{ij}(g)}\, dg = cn^2.$$

This shows that $c = 1/n$ and proves the lemma.

2.9.4 *Corollary* Let V and W be irreducible unitary representations of G, a compact Lie group. Then

$$\int_G \chi_V(g)\overline{\chi_W(g)}\,dg = \begin{cases} 0 \text{ if } V \text{ and } W \text{ are not equivalent.} \\ 1 \text{ if } V \text{ and } W \text{ are equivalent.} \end{cases}$$

PROOF The first assertion is an immediate consequence of 2.8.3. If V and W are equivalent then $\chi_V = \chi_W$. In the notation of the proof of 2.9.3, $\chi_V = \sum f_{ii}$. 2.9.3 implies that $\int_G f_{ii}(g)\overline{f_{jj}(g)}\,dg = \delta_{i,j}1/n$ where $n = \dim V$. Summing on i, j gives the second assertion.

2.9.5 *Corollary* Let $P_\gamma\colon L^2(G) \to V_\gamma \otimes V_\gamma^*$ (here $B_\gamma\colon V_\gamma \otimes V_\gamma^* \to C(G)$ is given by $B_\gamma(v \otimes v^*)(g) = v^*(\pi_\gamma(g)^{-1}v))$ be the orthogonal projection corresponding to $L^2(G) = \sum_{\gamma\in\hat{G}} V_\gamma \otimes V_\gamma^*$, unitary direct sum. Then $P_\gamma(f)(x) = d(\gamma)\int_G \overline{\chi_\gamma(g)}f(g^{-1}x)\,dg$ for f in $C(G)$ (here $d(\gamma) = \dim V_\gamma$).

PROOF As an element of $L^2(G)$, $f = \sum_{\gamma\in\hat{G}} P_\gamma f$. Let μ be in $\hat{G}$, and let $h = B_\mu(v \otimes v^*)$ with v in V_μ, v^* in V_μ^*. Then

$$h(g^{-1}x) = v^*(\pi_\mu(x)^{-1}\pi_\mu(g)v) = B_\mu(\pi_\mu(g)v \otimes v^*)(x).$$

Hence for fixed x, $g \mapsto h(g^{-1}x)$ is a matrix entry of π_μ. 2.9.3 therefore implies that if $\mu \neq \gamma$ then $\int_G \overline{\chi_\gamma(g)}P_\mu f(g^{-1}x)\,dg = 0$. It is therefore enough to show that if f is in $V_\gamma \otimes V_\gamma^*$ then

$$d(\gamma)\int_G \overline{\chi_\gamma(g)}f(g^{-1}x)\,dg = f(x).$$

Let $v_1, \ldots, v_d$ ($d = d(\gamma)$) be an orthonormal basis of V_γ and let $v_1^*, \ldots, v_d^*$ be the dual basis of V_γ^*. Then

$$\chi_\gamma(g) = \sum_{i=1}^{d} v_i^*(\pi_\gamma(g)v_i).$$

Suppose that $f(x) = v_i^*(\pi_\gamma(x)^{-1}v_j)$. Then

$$f(g^{-1}x) = \sum_k v_i^*(\pi_\gamma(x)^{-1}v_k)v_k^*(\pi_\gamma(g)v_j).$$

Hence

$$\begin{aligned} d&\int_G \overline{\chi_\gamma(g)}f(g^{-1}x)dg \\ &= \sum_{k,r} v_i^*(\pi_\gamma(x)^{-1}v_k)d\int_G \overline{v_r^*(\pi_\gamma(g)v_r)}v_k^*(\pi_\gamma(g)v_j)dg \\ &= \sum_{k,r} v_i^*(\pi_\gamma(x)^{-1}v_k)\delta_{kr}\delta_{rj} \text{ (by 2.9.3)} = f(x). \end{aligned}$$

2.9.6 *Corollary* Let f be in $L^2(G)$ and suppose that $f(xgx^{-1}) = f(g)$ for all x, g in G. Then $f = \sum_{\gamma\in\hat{G}} \langle f, \chi_\gamma\rangle\chi_\gamma$.

PROOF We identify $V_\gamma \otimes V_\gamma^*$ with $L(V_\gamma, V_\gamma)$ via $(v \otimes v^*)(w) = v^*(w)v$. Then under this identification $P_\gamma f(g^{-1}xg) = \pi_\gamma(g)P_\gamma f(x)\pi_\gamma(g)^{-1}$. But $f(gxg^{-1}) = f(x)$. 2.3.4 now implies that $P_\gamma f = c\bar{\chi}_\gamma$ (note that $\bar{\chi}_\gamma$ is in $V_\gamma \times V_\gamma^*$ and that $\bar{\chi}_\gamma \leftrightarrow I$ relative to the above identification). Now $\bar{\chi}_\gamma = \chi_{V_\gamma^*}$, hence 2.9.4 implies that $c = \langle P_\gamma f, \bar{\chi}_\gamma\rangle = \langle f, \bar{\chi}_\gamma\rangle$. Letting γ^* be the equivalence class of π_γ^* we see that the map $\gamma \to \gamma^*$ is a bijective map of $\hat{G}$ to $\hat{G}$ (indeed $(\gamma^*)^* = \gamma$). Hence

$$f = \sum_{\gamma\in\hat{G}} \langle f, \chi_{\gamma^*}\rangle\, \chi_{\gamma^*} = \sum_{\gamma\in\hat{G}} \langle f, \chi_\gamma\rangle\, \chi_\gamma.$$

2.10 Exercises

2.10.1 Let G and H be Lie groups. Let (π, V) and (ρ, W) be finite dimensional representations of G and H, respectively. Give a definition of $\pi \hat{\otimes} \rho$ on $V \otimes W$ analogous to 2.3.6. Show that every irreducible, finite dimensional, unitary representation of $G \times H$ is of the form $(\pi \hat{\otimes} \rho, V \otimes W)$ with (π, V) and (ρ, W) irreducible, finite dimensional, unitary representations of G and H, respectively.

2.10.2 (Burnside's theorem) Let G be a Lie group. Let (π, H) be an irreducible, finite dimensional, unitary representation of G. Show that if A is in $L(H, H)$ then $A = \sum c_i\pi(g_i)$ with g_i in G and c_i in C. (Hint: use 2.3.7.)

2.10.3 Let, for $z = (z_1, z_2)$, $w = (w_1, w_2)$ in C^2, $Q(z, w) = z_1\bar{w}_1 - z_2\bar{w}_2$. Let $U(1, 1)$ denote the group of all g in $GL(2, C)$ so that $Q(gz, gw) = Q(z, w)$. If g is in $U(1,1)$ and z is in $S^1 = \{z \text{ in } C \,|\, |z| = 1\}$ define $g \cdot z = (az + b)(cz + d)^{-1}$ where

$$g = \begin{bmatrix} a & b \\ c & d \end{bmatrix}.$$

Show that this defines an action of $U(1, 1)$ on S^1. Compute $c(g, e^{i\theta})$ relative to the volume element $d\theta$ (see 2.4.1).

2.10.4 Let G be a finite group. If f: $G \to C$ then define

$$\int_G f(g)dg = ({}^{\#}G)^{-1} \sum f(g).$$

Show that $L^2(G)$ is, as a representation of $G \times G$, equal to $\sum_{\gamma \in \hat{G}} V_\gamma \otimes V_\gamma^*$ (see 2.8.2 for notation). Prove the analog of 2.9.3. If f and h are in $L^2(G)$ define $f * h(x) = \int_G f(y)g(y^{-1}x)dy$. Show that, under the identification of $V_\gamma \otimes V_\gamma^*$ with $L(V_\gamma, V_\gamma)$ (see the proof of 2.9.6) the $*$ multiplication corresponds to matrix multiplication.

2.10.5 Let G be a Lie group. A function $\varphi: G \to C$ is said to be positive definite if φ is continuous and if for $c_1, \ldots, c_n$ in C, $g_1, \ldots, g_n$ in G, $\sum c_i \bar{c}_j \varphi(g_j^{-1} g_i) \geqslant 0$. Show that if $(\pi, (H, \langle\ ,\ \rangle))$ is a unitary representation of G and if v is in H then $\varphi(g) = \langle \pi(g)v, v\rangle$ is positive definite. Suppose that φ is positive definite. Let V be the linear span of all functions of the form $g \mapsto \varphi(gx)$ for x in G. If f, h are in V, $f(g) = \sum c_i \varphi(gx_i)$, $h(g) = \sum d_j \varphi(gy_j)$, then define $(f, h) = \sum c_i \bar{d}_j \varphi(y_j^{-1} x_i)$. Show that $(\ ,\ )$ is well defined. If f, h are in V let us say that $f \equiv h$ if $(f - h, u) = 0$ for all u in V. Let $V_0 = V/\equiv$. If v, u are in V_0 define $(u, v) = (f, h)$ for f in u, h in v. Show that $(\ ,\ )$ is well defined on V_0 and positive definite. If f is in V define $\pi(g)f(x) = f(xg)$. Show that if $f \equiv h$ then $\pi(g)f \equiv \pi(g)h$. Show that $(\pi, (V_0, (\ ,\ )))$ extends to a unitary representation of G.

2.10.6 Use an argument analogous to the proof of 2.9.3 to prove 2.8.3.

2.10.7 Let G be a compact Lie group (or a finite group as in 2.10.5). Let (π, H) be an irreducible finite dimensional representation of G. Let A be in $L(H, H)$. Show that $\int_G \pi(g)A\pi(g)^{-1}dg = (\dim H)^{-1}(\operatorname{tr} A)I$.

2.10.8 Let the notation be as in 2.10.7. Let χ be the character of π. Show that $\int_G \chi(xgyg^{-1})dg = (\dim H)^{-1}\chi(x)\chi(y)$. Conversely, show that if f is a continuous function on G so that

$$(*) \int_G f(xgyg^{-1})dg = f(x)f(y),$$

then $f = d(\gamma)^{-1}\chi_\gamma$ for some γ in $\hat{G}$. (Hint: Compute for μ, τ in $\hat{G}$,

$$\int_{G \times G \times G} \overline{\chi_\mu(x)\chi_\tau(y)} f(xgyg^{-1})dx\, dy\, dg$$

in two ways, using (*) and using the invariance of dx.)

2.10.9 Show how to prove 2.8.5 using 2.8.4 combined with the Stone–Weierstrauss theorem (see Lang [1] for the Stone–Weierstrauss theorem).

2.10.10 Let P^k be the space of all complex valued polynomials on R^n that are homogeneous of degree k. Let for f in P^k, $\Delta f = \sum (\partial^2/\partial x_i^2)f$. Let $H^k = \{f \text{ in } P^k | \ \Delta f = 0\}$. Let $r(x) = \sum x_i^2$. Show that $P^k = H^k \oplus rP^{k-2}$ $(P^{-j} = 0$ if $j > 0)$. Let $G = SO(n) = \{n \times n \text{ matrices } A \text{ over } R |\, {}^tAA = I, \det A = I\}$. Let for f in P^k, g in G, $\pi(g)f(x) = f(g^{-1}x)$. Show that H^k is an invariant subspace of P^k and that (π, H^k) is an irreducible representation of G if $n \geqslant 3$. Let G act on $S^{n-1} = \{x \text{ in } R^n | \langle x, x\rangle = 1\}$ by the restriction of its action on R^n. Let ω be a G-invariant volume element on S^{n-1}. Let ρ be the action of G on $L^2(S^{n-1}, \omega) = H$. Show that as a representation of G, $H = \sum H^k|_{S^{n-1}}$, a unitary direct sum.

2.10.11 Let G be a unimodular Lie group and let K be a closed unimodular subgroup of G so that G/K is compact. Let $M = G/K$ and ω be a volume form on M. Let $(\pi, L^2(M \times C, \omega))$ be the unitary representation defined in 2.4.6. Show that π is completely continuous.

2.10.12 Show that (in the notation of 2.4.5) $L^2(E, \omega)$ is separable. (Hint: Argue as in the proof of A.7.1.7 or reduce to A.7.1.7 using a partition of unity argument.)

2.11 Notes

2.11.1 2.3.7 is essentially Burnsides' theorem (see 2.10.2).

2.11.2 The notion of induced representation goes back to Frobenius [1]. In the context which we use 2.4.6 in this book our notion of induced representation is equivalent with that of Mackey [1] (at least for unitary representations). The author learned the basic idea of 2.4.6 from Hermann [1].

2.11.3 2.7.4 is taken from Gelfand, Graev, Pyatetskii-Shapiro [1], p. 23. our proof is a corrected version of theirs.

2.11.4 2.8.2 and 2.8.5 constitute the Peter–Weyl theorem (Peter and Weyl [1]).

2.11.5 The orthogonality relations were first discovered by Schur [1].

CHAPTER 3

Basic Structure Theory of Compact Lie Groups and Semisimple Lie Algebras

3.1 Introduction

In this chapter we develop the Cartan–Weyl structure theory of compact Lie groups, that is, the conjugacy of maximal tori, the root theory, and the Weyl group. We do this structure theory by means of the complexified Lie algebra. Doing a little extra work, we develop along the way the basic theory of complex semisimple Lie algebras.

In Section 3.2 we develop enough linear algebra to do the structure theory. In Section 3.3 we prove the basic result on finite dimensional representations of nilpotent Lie algebras (an amalgam of Lie's and Engel's theorems). In Sections 3.4 and 3.5 the root theory of complex semisimple Lie algebras is developed. The main results of Section 3.6 are 3.6.6 and 3.6.11; the former is known as Weyl's theorem. In Section 3.7 we study real forms of complex semisimple Lie algebras; in particular, we prove the existence of Cartan decompositions and Weyl's theorem on the existence of a compact form (the basis of the celebrated "unitarian trick"). In Section 3.8 we prove some lovely results of H. Samelson on the Euler characteristic of a homogeneous space. In Section 3.9 we prove a result of Borel and Mostow on automorphisms of compact Lie algebras. We use this result in Section 3.10 to show that every element of a compact connected Lie group is contained in some maximal torus. This result is, of course, an easy consequence of the Hopf–Rinow theorem. Rather than develop the necessary Riemannian geometry we have chosen to prove 3.9.1. We will use 3.9.1 quite often in the remainder of this book.

3.2 Some Linear Algebra

3.2.1 Let $M_n(C)$ be the C-algebra of all $n \times n$ complex matrices. Let Z be an element of $M_n(C)$. Then Z is said to be semisimple if there is a basis of C^n relative to which Z is diagonal. Z is said to be nilpotent if there is a positive integer k such that $Z^k = 0$.

3.2.2 *Proposition* Let Z be in $M_n(C)$. Then Z can be written uniquely in the form $Z = S + N$, where S is semisimple, N is nilpotent, and $SN = NS$. Furthermore, there is a polynomial without constant term $q(x)$ so that $q(Z) = S$.

PROOF Since $M_n(C)$ is finite dimensional there is a smallest integer k so that $I, Z, \ldots, Z^k$ are linearly dependent. Hence $\sum_{j=0}^{k} b^j Z^j = 0$. $b^k = 1$. Set $d(x) = \sum_{j=0}^{k} b_j x^j$. Then $d(x)$ can be factored

$$d(x) = \prod_{i=1}^{s} (x - c_i)^{t_i}.$$

Set

$$W_j = \prod_{\substack{i=1 \\ i \neq j}}^{s} (Z - c_i)^{t_i}.$$

Let $V_j = W_j C^n$. Then on V_j, $(Z - c_j)^{t_i} = 0$, but no smaller power is 0. Let v be an element of V_j so that $(Z - c_j)^{t_j - 1} v \neq 0$. Set $v_1 = v$, $v_{r+1} = (Z - c_j I)v_r$. Then $v_1, \ldots, v_{t_j}$ are linearly independent and $Zv_r = c_j v_r + v_{r+1}$. Thus $Z|_{V_j}$ has matrix

woefully INC

$$\begin{bmatrix} c_j & & 0 \\ & \ddots & \\ * & & c_j \end{bmatrix}.$$

Let U_j be a complement for V_j in C^n. Extending the v_i to a basis of C^n using elements of U_j, then Z has matrix

$$\left(\begin{array}{cc|c} c_j & 0 & 0 \\ & \ddots & \\ * & c_j & \\ \hline & * & Z' \end{array}\right).$$

Hence $p(x) = \det(xI - Z) = (x - c_j)^{t_j} \det(xI' - Z')$. Thus $(x - c_j)^{t_j}$ divides $p(x)$. This implies that the polynomial $d(x)$ divides the polynomial $p(x)$ (this assertion is the Cayley–Hamilton theorem). Hence $p(Z) = 0$.

$p(x)$ factors as

$$\prod_{i=1}^{k} (x - m_i)^{r_i}.$$

Let $M_i = \{v \mid (Z - m_i)^q v = 0 \text{ for some } q\}$. Clearly $ZM_i \subset M_i$. Suppose that w is in $M_i \cap M_j$ with $i \neq j$ and $w \neq 0$. If p, q are such that $(Z - m_i)^{p-1}w \neq 0$, $(Z - m_i)^p w = 0$, $(Z - m_j)^{q-1}w \neq 0$, and $(Z - m_j)^q w = 0$, then set $w_1 = (Z - m_i)^{p-1}(Z - m_i)^{q-1}w$. Then $Zw_1 = m_i w_1$ and $Zw_1 = m_j w_1$. Thus $w_1 = 0$. Set $w_2 = (Z - m_i)^{p-2}(Z - m_j)^{q-1}w$. Then by the above argument $w_2 = 0$. Continuing in this way we find $(Z - m_j)^{q-1}w = 0$. This contradiction implies that $w = 0$. Hence $M_i \cap M_j = (0)$.

Suppose that v is not in $M_1 + M_2 + \cdots + M_k$. Let $p_v(x)$ be the polynomial of minimal degree so that $p_v(Z)v = 0$. Let among all v not in $M_1 + M_2 + \cdots + M_k$, v_0 be so that $p_{v_0}(x)$ has minimal degree. Now the Cayley–Hamilton theorem implies that

$$P_{v_0}(x) = \prod_{i=1}^{k} (x - m_i)^{h_i}.$$

Suppose that h_i and h_j are not 0 with $i \neq j$. Then by definition of v_0, $Zv_0 = m_i v_0 \bmod M_1 + \cdots + M_k$ and $Zv_0 = m_j v_0 \bmod M_1 + \cdots + M_k$. But $m_i \neq m_j$ implies that v_0 is in $M_1 + \cdots + M_k$ which is contrary to the definition of v_0. Hence $p_{v_0}(x) = (x - m_i)^h$ for some i and h. But this again implies that v_0 is in $M_1 + \cdots + M_k$. Hence $C^n = M_1 + \cdots + M_k$.

Let S be defined by $S|_{M_i} = m_i I$. Then $(S - Z)^n = 0$. Hence $N = Z - S$ gives the decomposition $Z = S + N$. Clearly $SZ = ZS$, hence $NS = SN$. The uniqueness assertion is also clear from the proof of the existence of S and N. We leave it to the reader to write down the polynomial $q(x)$.

3.2.3 Let (,) be the standard Hermitian inner product on C^n. If Z is in $M_n(C)$ define Z^* by the formula $(Zv, w) = (v, Z^*w)$ for all v, w in C^n. As a matrix $Z^* = {}^t\bar{Z}$, that is, the complex conjugate of the transposed matrix.

Since C^n is finite dimensional every Z in $M_n(C)$ is completely continuous. The spectral theorem for completely continuous self-adjoint operators (see A.6.2.15) applies and we have the following proposition.

3.2.4 *Proposition* If Z is in $M_n(C)$ and if $Z^* = Z$ then there is a unitary operator on C^n so that UZU^{-1} is diagonal with real entries.

3.2.5 A self-adjoint matrix Z is said to be positive definite if for each v in C^n, $v \neq 0$, $(Zv, v) > 0$. Proposition 3.2.4. says that Z is positive definite if and only if there is a unitary matrix U so that UZU^{-1} is diagonal with real positive entries.

3.2.6 *Lemma* If Z is a self-adjoint element of $M_n(C)$ then e^Z is positive definite.

PROOF Let U be a unitary matrix so that UZU^{-1} is diagonal with real entries $a_1, \ldots, a_n$. Then $Ue^ZU^{-1} = e^{UZU^{-1}}$ is diagonal with entries $e^{a_1}, \ldots, e^{a_n}$. This combined with the observation of 3.2.5 proves the lemma.

3.2.7 *Lemma* Let F be a real or complex valued polynomial functions on $M_n(C)$. Suppose that Z is a Hermitian matrix so that $F(e^{mZ}) = 0$ for all m a nonnegative integer. Then $F(e^{tZ}) = 0$ for all t in R.

PROOF Let U be a unitary matrix so that UZU^{-1} is diagonal. Replacing F by $\tilde{F}(X) = F(U^{-1}XU)$ we may assume that Z is diagonal and $F = \tilde{F}$. We restrict the polynomials F to the diagonal matrices and assuming that Z is diagonal with entries $a_1, \ldots, a_n$ on the diagonal we see that $F(e^{ma_1}, \ldots, e^{ma_n}) = 0$ for all m a nonnegative integer. set $f(t) = F(e^{ta_1}, \ldots, e^{ta_n})$. Then $f(m) = 0$ for each nonnegative integer m. Now if f is not identically zero then $f(t) = \sum_{m=1} b_m e^{tA_m}$ with $A_1 > A_2 > \cdots > A_r$, and $b_m \neq 0$ for $m = 1, \ldots, r$. But if s is sufficiently large, and s is an integer then we see that

$$|b_1 e^{sA_1}| > |\sum_{m>1} b_m e^{sA_m}|$$

which contradicts the assumption $f(s) = 0$. Hence f is identically zero. This proves the lemma.

3.2.8 *Lemma* (1) If Z, W are self-adjoint and $[Z, W] = 0$ then Z and W can be simultaneously diagonalized.

(2) If A is positive definite then there is a unique self-adjoint matrix Z so that $e^Z = A$.

PROOF (1) Let for each c in R, $V_c = \{v | Zv = cv\}$. Then $C^n = \sum V_c$

and if $c \neq d$ then V_c is orthogonal to V_n. Furthermore if v is in V_c then $ZWv = WZv = cWv$. Hence $WV_c \subset V_c$. Diagonalizing W on each V_c gives the result.

(2) Let A be positive definite. There is a unitary operator U so that UAU^{-1} is diagonal with real positive entries $a_1, \ldots, a_n$. Let W be the diagonal matrix with diagonal entries $\log a_1, \ldots, \log a_n$. Then $e^{U^{-1}WU} = U^{-1}e^W U = U^{-1}(UAU^{-1})U = A$.

Let now V be self-adjoint and suppose that $e^V = A$. Let $F_{ij}(Z)$ be defined to be the i, j matrix element of $[A, Z]$. Then $F_{ij}(e^{mV}) = 0$ for all nonnegative integers. Thus $F_{ij}(e^{tV}) = 0$ for all t in R. Hence $[A, e^{tV}] = 0$ for all t in R. Differentiating we find that A and V commute. Hence A and V are simultaneously diagonalizable and on each eigenspace of A with eigenvalue c, V has eigenvalue $\log c$. Hence $V = Z$.

3.2.9 *Corollary* If A is positive definite then there is a unique positive definite matrix B so that $B^2 = A$. B is denoted $A^{\frac{1}{2}}$.

PROOF Let $A = e^Z$, with Z self-adjoint. If $B = e^{\frac{1}{2}Z}$ then $B^2 = A$. If D is positive definite and $D^2 = A$ then $D = e^W$ with W self-adjoint. $e^{2W} = A$ hence by 3.2.8, $2W = Z$.

3.2.10 *Proposition* Let $U(n)$ be the group of all $n \times n$ unitary matrices. Let $\mathfrak{p}_n$ be the real vector space of self-adjoint $n \times n$ matrices. Then the map $U, Z \mapsto Ue^Z$ is a diffeomorphism of $U(n) \times \mathfrak{p}_n$ onto $GL(n, C)$.

PROOF Let A be in $GL(n, C)$. Then A^*A is positive definite. Set $P = (A^*A)^{\frac{1}{2}}$. We assert that AP^{-1} is unitary. Indeed,

$$(AP^{-1}v, AP^{-1}v) = (v, P^{-1}A^*AP^{-1}v) = (v, P^{-1}P^2P^{-1}v) = (v, v)$$

for all v in C^n. This in light of 3.2.8 implies that the map above is surjective. If $A = UP = U'P'$ then $A^*A = P'^2$, hence 3.2.9 implies that $P = P'$. Hence the map is bijective. The map is clearly C^∞. Thus to complete the proof we need only show that the map is everywhere regular. A computation shows that if $f(U, Z) = Ue^Z$ for U in $U(n)$, Z in $\mathfrak{p}_n$, then if X is in the Lie algebra of $U(n)$, $\mathfrak{U}(n) = \{X | X^* = -X\}$ and if Y is in $\mathfrak{p}_n$, then $f_{*(U,1)}(X, Y) = UXe^Z + U de^{(Z+tY)}/dt|_{t=0}$. Now $de^{(Z+tY)}/dt|_{t=0}$ is in $\mathfrak{p}_n$. Thus if $f_{*(U,Z)}(X, Y) = 0$ then Xe^Z is in $\mathfrak{p}_n$. But then $Xe^Z = -e^Z X$. This says that $e^{-Z}Xe^Z = -X$. Now Z is diagonalizable. Let $e_1, \ldots, e_n$ be an orthonormal basis for C^n so that $Ze_i = \lambda_i e_i$ with λ_i in R. If $E_{ij}e_k = \delta_{jk}E_i$ then $\{E_{ij}\}$ forms a basis of $M_n(C)$. Clearly, $e^{-Z}E_{ij}e^Z = e^{(\lambda_j - \lambda_i)}E_{ij}$. Thus all the eigenvalues of $X \mapsto e^{-Z}Xe^Z$ are positive. This implies that $X = 0$. Using A.2.2.8 one finds

$$\frac{d}{dt} e^{(Z+tY)}\bigg|_{t=0} = e^Z\left(\frac{I - e^{-\operatorname{ad} Z}}{\operatorname{ad} Z}\right)\cdot Y.$$

Hence

$$\sum_{j=0}^{\infty} \frac{(\operatorname{ad} Z)^n(-1)^n}{(n+1)!} Y = 0.$$

But $(\operatorname{ad} Z)^{2k+1}\cdot Y$ is in $\mathfrak{U}(n)$ and $(\operatorname{ad} Z)^{2k}\cdot Y$ is in $\mathfrak{p}_n$. Hence,

$$\sum_{j=0}^{\infty} \frac{(\operatorname{ad} Z)^{2j}}{(2j+1)!}\cdot Y = 0.$$

But on $M_n(C)$ we find (using the E_{ij} again) that

$$\sum_{j=0}^{\infty} \frac{(\operatorname{ad} Z)^{2j}}{(2j+1)!}$$

has eigenvalues

$$\sum_{j=0}^{\infty} \frac{(\lambda_i - \lambda_k)^{2j}}{(2j+1)!} > 0.$$

Hence $Y = 0$. This proves that f_* is nonsingular at each point of $U(n) \times \mathfrak{p}_n$.

3.2.11 *Proposition* Let A be a subspace of $M_n(C)$ and let T be the set of all X in $M_n(C)$ satisfying $[X, A] \subset A$. If Z is in T and if $\operatorname{tr} ZU = 0$ for all U in T then Z is nilpotent.

PROOF Let $Z = S + N$, as in 3.2.2. Let $v_1, \ldots, v_n$ be a basis of C^n satisfying $Sv_i = a_i v_i$, $i = 1, \ldots, n$, a_i in C. Let V be the vector space over R spanned by $a_1, \ldots, a_n$ (dim $V = 0$, 1, or 2). We show that $V = (0)$ by showing that if f is a real linear form on V then $f = 0$. Let W be the endomorphism of C^n defined by $Wv_i = f(a_i)v_i$.

Let E_{ij} be the endomorphism of C^n defined by $E_{ij}v_k = \delta_{ik}v_i$. Then $\operatorname{ad}(S)E_{ij} = (a_i - a_j)E_{ij}$ and $\operatorname{ad} N$ is nilpotent (see 3.11.1). Since $\operatorname{ad} S$ and $\operatorname{ad} N$ commute, 3.2.2 implies that $\operatorname{ad} S$ is a polynomial without constant term in $\operatorname{ad} Z$, hence S is in T. Now $\operatorname{ad}(W)E_{ij} = (f(a_i) - f(a_j))E_{ij}$. Let $p(x)$ be a polynomial without constant term so that $p(a_i - a_j) = f(a_i) - f(a_j)$. Then $\operatorname{ad} W = p(\operatorname{ad} S)$. Hence W is in T. Hence $\operatorname{tr} ZW = 0$ by our hypothesis. Thus $0 = \operatorname{tr} ZW = \sum a_i f(a_i)$. This implies $0 = f(\sum a_i f(a_i)) = \sum f(a_i)^2$. Since f is real valued, $f = 0$.

3.3 Nilpotent Lie Algebras

3.3.1 *Definition* Let $\mathfrak{g}$ be a Lie algebra over a field K. Then $\mathfrak{g}$ is said to be nilpotent if for each X in $\mathfrak{g}$ there is a positive integer k so that $(\text{ad } X)^n = 0$.

3.3.2 *Theorem* Let $\mathfrak{g}$ be a nilpotent Lie subalgebra of $M_n(C)$. Then there is a basis of C^n so that relative to this basis the matrix of each element of $\mathfrak{g}$ is upper triangular.

PROOF We note that if we can show that there is v in C^n, $v \neq 0$ and a linear form f on $\mathfrak{g}$ so that $Xv = f(x)v$ for each X in $\mathfrak{g}$ then the result will follow. We prove this equivalent assertion by induction on dim $\mathfrak{g}$. If dim $\mathfrak{g} = 1$ then the result follows from the fact that every linear transformation of C^n has an eigenvector. Suppose that dim $\mathfrak{g} = r + 1$ and that the result is true for all nilpotent algebras of dimension $k \leqslant r$. Let $\mathfrak{g}_1$ be a proper subalgebra of $\mathfrak{g}$ of maximal dimension. If X is in $\mathfrak{g}_1$ then X induces a linear map X' of $\mathfrak{g}/\mathfrak{g}_1$ into $\mathfrak{g}/\mathfrak{g}_1$ given by $X'(Z + \mathfrak{g}_1) = \text{ad } X \cdot Z + \mathfrak{g}_1$. The map $X \to X'$ is a Lie algebra homomorphism of $\mathfrak{g}_1$ into $L(\mathfrak{g}/\mathfrak{g}_1, \mathfrak{g}/\mathfrak{g}_1)$. Hence by the inductive hypothesis there is a nonzero element $\tilde{Y}$ of $\mathfrak{g}/\mathfrak{g}_1$ so that $X' \tilde{Y} = c\tilde{Y}$ (c depending on X) for each X in $\mathfrak{g}_1$. Let $Y + \mathfrak{g}_1 = \tilde{Y}$. Then if X is in $\mathfrak{g}_1$, $(\text{ad } X)^k \cdot Y = c^k Y (\text{mod } \mathfrak{g}_1)$. Hence since $(\text{ad } X)^k = 0$ for some k, $c = 0$. This implies that $[\mathfrak{g}_1, Y] = 0$. Thus $CY + \mathfrak{g}_1$ is a subalgebra of $\mathfrak{g}$. The definition of $\mathfrak{g}_1$ now implies that $\mathfrak{g} = CY + \mathfrak{g}_1$.

By the inductive hypothesis there is a linear form f, on $\mathfrak{g}_1$ and a nonzero element v of C^n so that $Xv = f(X)v$ for all X in $\mathfrak{g}_1$. Now $XY \cdot v = YXv = f(X)v$. Thus if $V_f = \{W \text{ in } C^n | XW = f(X)W\}$ for all X in $\mathfrak{g}_1$ then $YV_f \subset V_f$. $V_f \neq (0)$. There is therefore V_0 in V_f, $V_0 \neq 0$ so that $Yv_0 = \lambda v_0$ for some λ in C. Clearly $\mathfrak{g} \cdot v_0 \subset Cv_0$. This completes the proof.

3.4 Semisimple Lie Algebras

3.4.1 *Definition* Let $\mathfrak{g}$ be a Lie algebra over R or C. Then $\mathfrak{g}$ is said to be simple if $\mathfrak{g}$ is not Abelian and $\mathfrak{g}$ has no ideals except for (0) and $\mathfrak{g}$. $\mathfrak{g}$ is said to be semisimple if $\mathfrak{g}$ can be written as a direct sum of simple ideals.

3.4.2 *Lemma* If $\mathfrak{g}$ is semisimple then the decomposition $\mathfrak{g} = \mathfrak{g}_1 \oplus \cdots \oplus \mathfrak{g}_n$ into simple ideals is unique up to order.

PROOF Let $\mathfrak{h}$ be a simple ideal of $\mathfrak{g}$. If $[\mathfrak{g}_i, \mathfrak{h}] \neq 0$ then $\mathfrak{g}_i \cap \mathfrak{h} \neq 0$, which implies that $\mathfrak{h} = \mathfrak{g}_i$. But now $[\mathfrak{h}, \mathfrak{h}] \neq 0$, hence by the above $\mathfrak{h} = \mathfrak{g}_i$ for some i.

3.4.3 *Definition* Let $\mathfrak{g}$ be a Lie algebra over R or C. Define for X, Y in $\mathfrak{g}$, $B_\mathfrak{g}(X, Y) = \operatorname{tr} \operatorname{ad} X \operatorname{ad} Y$. Then $B_\mathfrak{g}$ is a symmetric bilinear form on $\mathfrak{g} \times \mathfrak{g}$ called the Killing form of $\mathfrak{g}$.

3.4.4 *Lemma* Let $\mathfrak{g}$ be a Lie algebra over R or C and let D be a derivation of $\mathfrak{g}$ ($D[X, Y] = [DX, Y] + [X, DY]$). Then $B_\mathfrak{g}(DX, Y) = -B_\mathfrak{g}(X, DY)$. If A is an automorphism of $\mathfrak{g}$ then $B_\mathfrak{g}(AX, AY) = B_\mathfrak{g}(X, Y)$.

PROOF If D is a derivation of $\mathfrak{g}$ then $\operatorname{ad} DX = D \operatorname{ad} X - \operatorname{ad} XD$. This ✓ proves the first assertion. If A is an automorphism of $\mathfrak{g}$ then $\operatorname{ad} AX = A \operatorname{ad} XA^{-1}$. This proves the second assertion.

3.4.5 *Theorem* Let $\mathfrak{g}$ be a Lie algebra over $K = R$ or C. Then $\mathfrak{g}$ is semisimple if and only if $B_\mathfrak{g}$ is nondegenerate.

PROOF Suppose that $B_\mathfrak{g}$ is nondegenerate. Let $\mathfrak{g}_1$ be an ideal in $\mathfrak{g}$. We assert that $\mathfrak{g}_1^\perp = \{X | B_\mathfrak{g}(X, \mathfrak{g}_1) = (0)\}$ is an ideal in $\mathfrak{g}$ and that $\mathfrak{g}_1^\perp \cap \mathfrak{g}_1 = (0)$. If Y is in $\mathfrak{g}_1^\perp$ and if X is in $\mathfrak{g}$, Z is in $\mathfrak{g}_1$, then $B(\operatorname{ad} X \cdot Y, Z) = -B(Y, \operatorname{ad} X \cdot Z) = 0$. Hence $\mathfrak{g}_1^\perp$ is an ideal in $\mathfrak{g}$. If X, Y are in $\mathfrak{g}_1 \cap \mathfrak{g}_1^\perp$ and if Z is in $\mathfrak{g}$, then $B(Z, [X, Y]) = -B([X, Z], Y) = 0$, since $[X, Z]$ is in $\mathfrak{g}_1$, Y is in $\mathfrak{g}_1^\perp$. Hence since B is nondegenerate, $\mathfrak{g}_1 \cap \mathfrak{g}_1^\perp$ is Abelian. Let $X_1, \ldots, X_r$ be a basis for $\mathfrak{g}_1 \cap \mathfrak{g}_1^\perp$ and let $X_1, \ldots, X_n$ be an extension to a basis for $\mathfrak{g}$. If X is in $\mathfrak{g}_1 \cap \mathfrak{g}_1^\perp$ and if Y is in $\mathfrak{g}$ then $\operatorname{ad} X \operatorname{ad} Y \cdot X_i = 0$, $i = 1, \ldots, r$($\mathfrak{g}_1 \cap \mathfrak{g}_1^\perp$ is an ideal), $\operatorname{ad} X \operatorname{ad} Y \cdot X_j$ is in $\mathfrak{g}_1 \cap \mathfrak{g}_1^\perp$ for $j = r + 1, \ldots, n$. Hence $B(X, Y) = 0$ for all Y in $\mathfrak{g}$. This implies that $X = 0$. Thus $\mathfrak{g} = \mathfrak{g}_1 \oplus \mathfrak{g}_1^\perp$, a direct sum of ideals orthogonal relative to B. Continuing this process we find that $\mathfrak{g}$ is a direct sum of ideals that are either simple or abelian. The above argument shows a $\mathfrak{g}$ has no non-zero abelian ideals, and is thus semi simple.

Suppose that $\mathfrak{g}$ is semisimple. Then $\mathfrak{g} = \mathfrak{g}_1 \oplus \cdots \oplus \mathfrak{g}_n$, a direct sum of simple ideals. If $i \neq j$ then the argument above shows that $\mathfrak{g}_i$ is orthogonal to $\mathfrak{g}_j$ relative to B. We may thus assume that $\mathfrak{g}$ is simple. Since ker ad is an ideal in $\mathfrak{g}$ and since $\mathfrak{g}$ is (now assumed) simple, ker ad $= (0)$. We may thus identify $\mathfrak{g}$ with its image under ad in $L(\mathfrak{g}, \mathfrak{g})$. Let $\mathfrak{n} = \{X \text{ in } \mathfrak{g} | B(X, \mathfrak{g}) = (0)\}$. $\mathfrak{n}$ is an ideal in $\mathfrak{g}$. If $\mathfrak{n} = (0)$ the result is proved. We may thus assume that $\mathfrak{n} = \mathfrak{g}$. Choosing a basis of $\mathfrak{g}$, we may assume that $\mathfrak{g}$ is a subspace of $M_p(C)(p = \dim \mathfrak{g})$. Let A be the linear span of $\mathfrak{g}$ in $M_p(C)$. Let $T = \{Z \text{ in } M_p(C) | [Z, A] \subset A\}$. If Z is in A then since $[\mathfrak{g}, \mathfrak{g}] = \mathfrak{g}$ we see that $Z = \sum [X_i, Y_i]$, X_i and Y_i in A. If W is in T then $\operatorname{tr} WZ = \operatorname{tr} W \sum [X_i, Y_i] = \sum \operatorname{tr} W[X_i, Y_i] = \sum \operatorname{tr}[W, X_i]Y_i = 0$ since we have assumed that $B = 0$. Now 3.2.11 implies that if X is in $\mathfrak{g}$ then ad X is nilpotent. The argument of 3.3.2 implies that $\mathfrak{g}$ has a one dimensional ideal. This is a contradiction.

3.4.6 *Lemma* Let $\mathfrak{g}$ be a semisimple Lie algebra over R or C. Let D be a derivation of $\mathfrak{g}$. Then $D = \operatorname{ad} X$ for some X in $\mathfrak{g}$.

PROOF Let $f(Z) = \operatorname{tr} D \operatorname{ad} Z$. Since $B_{\mathfrak{g}} = B$ is nondegenerate, $f(Z) = B(X, Z)$ for some X in $\mathfrak{g}$. Now

$$\begin{aligned} B(DY, Z) &= \operatorname{tr} \operatorname{ad} DY \operatorname{ad} Z = \operatorname{tr} D \operatorname{ad} Y \operatorname{ad} Z - \operatorname{tr} \operatorname{ad} YD \operatorname{ad} Z \\ &= \operatorname{tr} D(\operatorname{ad} Y \operatorname{ad} Z - \operatorname{ad} Z \operatorname{ad} Y) = \operatorname{tr} D \operatorname{ad}[Y, Z] \\ &= f([Y, Z]) = B(X, [Y, Z]) = B([X, Y], Z). \end{aligned}$$

Thus, since Z is arbitrary, $DY = \operatorname{ad} X \cdot Y$ as was to be shown.

3.4.7 *Proposition* Let $\mathfrak{g}$ be a semisimple Lie algebra over C. If X is in $\mathfrak{g}$, let $\operatorname{ad} X = S + N$ be as in 3.2.2. Then there are elements X_s and X_n in $\mathfrak{g}$ so that $\operatorname{ad} X_s = S$ and $\operatorname{ad} X_n = N$.

PROOF Let for c in C, $\mathfrak{g}_c = \{Z \text{ in } \mathfrak{g} | (\operatorname{ad} X - cI)^k Z = 0 \text{ for some } k\}$. Then on $\mathfrak{g}_c$, $S = cI$. We note that $[\mathfrak{g}_c, \mathfrak{g}_d] \subset \mathfrak{g}_{c+d}$. This follows from the following formula (which can be proved by induction): if Z, W are in $\mathfrak{g}$, then

$$(\operatorname{ad} X - (c + d)I)^k Z, W = \sum \binom{k}{r} [(\operatorname{ad} X - cI)^r Z, (\operatorname{ad} X - dI)^{k-r} W].$$

We have thus proved that S is a derivation of $\mathfrak{g}$. Hence, by 3.4.5, $S = \operatorname{ad} X_s$ for some X_s in $\mathfrak{g}$.

3.5 Cartan Subalgebras

3.5.1 *Definition* Let $\mathfrak{g}$ be a semisimple Lie algebra over C. A Cartan subalgebra of $\mathfrak{g}$ is a maximal Abelian subalgebra of $\mathfrak{g}$, $\mathfrak{h}$, so that if X is in $\mathfrak{h}$ then ad X is semisimple.

3.5.2 If $\mathfrak{g}$ is a Lie algebra over C, let $\det(\operatorname{ad} X - tI) = \sum t^k D_k(X)$. Then D_k is a polynomial function on $\mathfrak{g}$. Let m be the smallest integer so that D_m is not identically zero. Set $D(X) = D_m(X)$. If X is in $\mathfrak{g}$ and if $D(X) \neq 0$ then X is called a regular element of $\mathfrak{g}$.

If X is in $\mathfrak{g}$ set $\mathfrak{g}^X = \{Y|(\operatorname{ad} X)^k Y = 0 \text{ for some } k\}$.

3.5.3 *Lemma* Let $\mathfrak{g}$ be a complex semisimple Lie algebra. If X is a regular element of $\mathfrak{g}$ then $\mathfrak{g}^X$ is a nilpotent subalgebra of $\mathfrak{g}$.

PROOF Let $X = X_s + X_n$ as in 3.4.6. Then $\mathfrak{g} = \mathfrak{g}_0 + \sum \mathfrak{g}_{a_i}$, where $\mathfrak{g}_0$ is the zero eigenspace for ad X_s and $\mathfrak{g}_{a_i}$ is the a_i eigenspace for ad X_s, $a_i \neq 0$. Clearly $\mathfrak{g}_0 = \mathfrak{g}^X$. Let $\mathfrak{g}' = \sum \mathfrak{g}_{a_i}$. Then ad X_s is nonsingular on $\mathfrak{g}'$. If Z is in $\mathfrak{g}^X$ then ad$Z$ commutes with ad X_s, hence ad $Z \cdot \mathfrak{g}' \subset \mathfrak{g}'$. The set of all Z in $\mathfrak{g}^X$ so that ad $Z|_{\mathfrak{g}'}$ is nonsingular is nonempty, hence dense in $\mathfrak{g}^X$. Since X (hence X_s) is regular we must have for a dense subset of Z in $\mathfrak{g}^X$, $(\operatorname{ad} Z)^m U = 0$ for U in $\mathfrak{g}^X$. Thus $(\operatorname{ad} Y)^m U = 0$ for all Y, U in $\mathfrak{g}^X$.

3.5.4 *Proposition* If X is a regular element of $\mathfrak{g}$ then $\mathfrak{g}^X$ is a Cartan subalgebra of $\mathfrak{g}$.

PROOF Let $\mathfrak{g} = \mathfrak{g}^X + \sum \mathfrak{g}_{a_i}$ as in 3.5.3. Set $\mathfrak{g}^X = \mathfrak{g}_{a_0}$. We assert that if $a_i \neq -a_j$ then $B(\mathfrak{g}_{a_i}, \mathfrak{g}_{a_j}) = (0)$. Indeed, if Z is in $\mathfrak{g}_{a_i}$, U is in $\mathfrak{g}_{a_j}$, then ad Z ad $U\, \mathfrak{g}_{a_k} \subset \mathfrak{g}_{a_i + a_j + a_k}$ for all k. Thus tr ad Z ad $U = 0$ as was to be shown. We therefore see that B is nondegenerate on $\mathfrak{g}^X$. Applying 3.5.3 and 3.3.2 we see that there is a basis of $\mathfrak{g}$ so that relative to this basis every element of ad $\mathfrak{g}^X$ is in upper triangular form. Thus if Z is in $[\mathfrak{g}^X, \mathfrak{g}^X]$ then relative to the above basis Z is upper triangular with zeros on the diagonal. Hence $B(Z, W) = 0$ for all W in $\mathfrak{g}^X$. Thus $Z = 0$. Also if Z is in $\mathfrak{g}^X$ then $[Z, X_s] = 0$, hence Z_s is in $\mathfrak{g}^X$; thus Z_n is in $\mathfrak{g}^X$. Now relative to the above basis ad Z_n is upper triangular with zeros on the diagonal. Hence as above $B(Z_n, W) = 0$ for all

W in $\mathfrak{g}^X$. We have thus shown that $\mathfrak{g}^X$ is Abelian and that if Z is in $\mathfrak{g}^X$ then ad Z is semisimple. Clearly $\mathfrak{g}^X$ is maximal Abelian in $\mathfrak{g}$.

3.5.5 3.5.4 implies that if $\mathfrak{g}$ is a semisimple Lie algebra over C then $\mathfrak{g}$ has a Cartan subalgebra. Let $\mathfrak{h}$ be a Cartan subalgebra of $\mathfrak{g}$. By the definition of Cartan subalgebra the elements ad$X$, X in $\mathfrak{h}$, can be simultaneously diagonalized. Let for λ in $\mathfrak{h}^*$, $\mathfrak{g}_\lambda = \{x \text{ in } \mathfrak{g} | \text{ad} H \cdot X = \lambda(H)X \text{ for all } H \text{ in } \mathfrak{h}\}$. Let $\Delta = \{\alpha \text{ in } \mathfrak{h}^* | \alpha \neq 0 \text{ and } \mathfrak{g}_\alpha \neq 0\}$. Then Δ is called the root system of $\mathfrak{g}$ relative to $\mathfrak{h}$. Clearly, $\mathfrak{g} = \mathfrak{h} + \sum \mathfrak{g}_\alpha$, the sum is direct over α in Δ. This is called the root space decomposition of $\mathfrak{g}$ relative to $\mathfrak{h}$.

3.5.6 *Lemma* $B(\mathfrak{g}_\alpha, \mathfrak{g}_\beta) = (0)$ if $\alpha \neq -\beta$. $B(\mathfrak{h}, \mathfrak{g}_\alpha) = (0)$ for all α in Δ. In particular, B restricted to $\mathfrak{h} \times \mathfrak{h}$ is nondegenerate; if α is in Δ then $-\alpha$ is in Δ and B is a nondegenerate pairing of $\mathfrak{g}_\alpha$ with $\mathfrak{g}_{-\alpha}$.

PROOF Set $\mathfrak{h} = \mathfrak{g}_0$. If X is in $\mathfrak{g}_\alpha$ and if Y is in $\mathfrak{g}_\beta$ then for each H in $\mathfrak{h}$, $B(\text{ad} HX, Y) = \alpha(H)B(X, Y)$ and $B(X, \text{ad} HY) = \beta(H)B(X, Y)$. Hence $\alpha(H)B(X, Y) = -\beta(H)B(X, Y)$ for all H in $\mathfrak{h}$. This implies that $B(X, Y) \neq 0$ only if $\alpha = -\beta$.

3.5.7 B is nondegenerate on $\mathfrak{h} \times \mathfrak{h}$. Thus for each α in Δ there is a unique element H_α of $\mathfrak{h}$ so that $B(H_\alpha, H) = \alpha(H)$ for all H in $\mathfrak{h}$.

3.5.8 *Lemma* Let X be in $\mathfrak{g}_\alpha$, Y in $\mathfrak{g}_{-\alpha}$. Then $[X, Y] = B(X, Y)H_\alpha$.

PROOF Using the notation of 3.5.6 it is easy to see that $[\mathfrak{g}_\alpha, \mathfrak{g}_\beta] \subset \mathfrak{g}_{\alpha+\beta}$. Hence $[X, Y]$ is in $\mathfrak{h}$. If H is in $\mathfrak{h}$ then $B([X, Y], H) = B(Y, [H, X]) = \alpha(H)B(X, Y)$. This proves the lemma.

3.5.9 *Proposition* (1) If α is in Δ then $\dim \mathfrak{g}_\alpha = 1$.

(2) Let $\mathfrak{h}_R = \sum RH_\alpha$, the sum over all α in Δ. Then $\mathfrak{h}_R$ spans $\mathfrak{h}$ and B is positive definite on $\mathfrak{h}_R$.

(3) If α and β are in Δ then $2\beta(H_\alpha)/\alpha(H_\alpha) = -s - r$, where s, r are integers so that $\beta + (r-1)\alpha$ and $\beta + (s+1)\alpha$ are not in Δ but $\beta + t\alpha$ is in Δ for $r \leqslant t \leqslant s$, t an integer.

PROOF $\mathfrak{g} = [\mathfrak{g}, \mathfrak{g}] = \sum [\mathfrak{h}, \mathfrak{g}_\alpha] + \sum [\mathfrak{g}_\alpha, \mathfrak{g}_\beta]$. Thus since $[\mathfrak{h}, \mathfrak{g}_\alpha] \subset \mathfrak{g}_\alpha$ and $[\mathfrak{g}_\alpha, \mathfrak{g}_\beta] \subset \mathfrak{g}_{\alpha+\beta}$ we see that $\mathfrak{h} = \sum [\mathfrak{g}_\alpha, \mathfrak{g}_{-\alpha}] = \sum CH_\alpha$ by 3.5.8. Hence $\mathfrak{h}_R$ does indeed span $\mathfrak{h}$.

Let α, β be in Δ. Set $\tilde{\mathfrak{g}} = \sum_{n=r}^{s} \mathfrak{g}_{\beta+n\alpha}$, where r and s are as in (3). Then $[\mathfrak{g}_\alpha, \tilde{\mathfrak{g}}] \subset \tilde{\mathfrak{g}}$, $[\mathfrak{g}_{-\alpha}, \tilde{\mathfrak{g}}] \subset \tilde{\mathfrak{g}}$, and $[\mathfrak{h}, \tilde{\mathfrak{g}}] \subset \tilde{\mathfrak{g}}$. Let X in $\mathfrak{g}_\alpha$, Y in $\mathfrak{g}_{-\alpha}$ be so that $[X, Y] = H_\alpha$ (i.e., $B(X, Y) = 1$). Then $\operatorname{ad} H_\alpha|_{\tilde{\mathfrak{g}}} = [\operatorname{ad} X|_{\tilde{\mathfrak{g}}}, \operatorname{ad} Y|_{\tilde{\mathfrak{g}}}]$. Thus $\operatorname{tr}(\operatorname{ad} H_\alpha|_{\tilde{\mathfrak{g}}}) = 0$. But $\operatorname{tr}(\operatorname{ad} H_\alpha|_{\tilde{\mathfrak{g}}}) = \sum_{n=r}^{s} (\beta + n\alpha)(H_\alpha) \dim \mathfrak{g}_{\beta+n\alpha}$. Hence

(i) $$\beta(H_\alpha) \sum_{n=r}^{s} \dim \mathfrak{g}_{\beta+n\alpha} = -\alpha(H_\alpha) \sum_{n=r}^{s} n \dim \mathfrak{g}_{\beta+n\alpha}.$$

Thus if $\alpha(H_\alpha) = 0$ then $\beta(H_\alpha) = 0$ for all β in Δ. But then $B(H_\beta, H_\alpha) = 0$ for all β in Δ. This implies that $B(\mathfrak{h}, H_\alpha) = 0$. But then $H_\alpha = 0$, contrary to the assumption $\alpha \neq 0$. Hence $\alpha(H_\alpha) \neq 0$.

Suppose that $\dim \mathfrak{g}_\alpha > 1$. Let X be in $\mathfrak{g}_\alpha$, Y in $\mathfrak{g}_{-\alpha}$ be so that $B(X, Y) = 1$. Let Z in $\mathfrak{g}_\alpha$ be so that $Z \neq 0$ and $B(Z, Y) = 0$. An induction shows that

(ii) $\operatorname{ad} Y \cdot (\operatorname{ad} X)^n Z = -(n(n+1)/2)\alpha(H_\alpha)(\operatorname{ad} X)^{n-1} Z$, n a positive integer. But then, if $(\operatorname{ad} X)^n Z \neq 0$ then $(\operatorname{ad} X)^{n+1} Z \neq 0$. Since Z is presumed to be nonzero and $(\operatorname{ad} X)^n Z$ is in $\mathfrak{g}_{(n+1)\alpha}$, we must have $\dim \mathfrak{g} = \infty$. This contradiction implies that $\dim \mathfrak{g}_\alpha = 1$. Hence (1) is true.

(i) now implies (3). (3) implies that if α is in Δ then $\alpha(H)$ is real for H in $\mathfrak{h}_R$. If H is in $\mathfrak{h}_R$, $H \neq 0$, then $B(H, H) = \operatorname{tr}(\operatorname{ad} H)^2 = \sum \alpha(H)^2$. Thus $B(H, H) \geqslant 0$. If $B(H, H) = 0$, then $\alpha(H) = 0$ for all α in Δ, hence $H = 0$. This proves (2).

3.5.10 *Lemma* If α, B are in Δ and if $\alpha + \beta \neq 0$ then $[\mathfrak{g}_\alpha, \mathfrak{g}_\beta] = \mathfrak{g}_{\alpha+\beta}$.

PROOF If $\alpha + \beta$ is not in Δ the result is clear. We may thus assume that $\alpha + \beta$ is a root. Let r and s be as in 3.5.9(3). Then $s \geqslant 1$. However, if $[\mathfrak{g}_\alpha, \mathfrak{g}_\beta] \neq \mathfrak{g}_{\alpha+\beta}$ then by 3.5.9(1), $[\mathfrak{g}_\alpha, \mathfrak{g}_\beta] = 0$. Thus if $\tilde{\mathfrak{g}} = \sum_{n=r}^{0} \mathfrak{g}_{\beta+n\alpha}$ we see that $[\mathfrak{g}_\alpha, \tilde{\mathfrak{g}}] \subset \tilde{\mathfrak{g}}$, $[\mathfrak{g}_{-\alpha}, \tilde{\mathfrak{g}}] \subset \tilde{\mathfrak{g}}$, and $[\mathfrak{h}, \tilde{\mathfrak{g}}] \subset \tilde{\mathfrak{g}}$. Thus $\operatorname{tr}(\operatorname{ad} H_\alpha|_{\mathfrak{g}}) = 0$. This implies that $\sum_{n=r}^{0} (\beta + n\alpha)(H_\alpha) = 0$. But this implies that $2\beta(H_\alpha)/\alpha(H_\alpha) = -r$. In light of 3.5.9(3) this is a contradiction.

3.5.11 *Lemma* If α is in Δ and $c\alpha$ is in Δ with c in C then $c = \pm 1$.

PROOF 3.5.9(3) implies that $2c$ and $2/c$ are integers. Setting $m/2 = c$ we know that m is an integer. Since $2/c = 4/m$ we know that m divides 4. Hence $m = \pm 1, \pm 2$, or ± 4. If $m = \pm 2$ then $c = \pm 1$ as was to be proved. If $m = 1$, then 3.5.10 implies that $\mathfrak{g}_\alpha = [\mathfrak{g}_{\frac{1}{2}\alpha}, \mathfrak{g}_{\frac{1}{2}\alpha}] = 0$. Hence $m = 1$ is impossible. Similarly $m = -1$ is impossible. If $m = 4$ then $c = 2$ and $\mathfrak{g}_{2\alpha} = [\mathfrak{g}_\alpha, \mathfrak{g}_\alpha] = 0$, thus $m = 4$ is impossible. Similarly $m = -4$ is impossible. We have thus proved the lemma.

3.5.12 *Lemma* Let α, β be in Δ and let s, r be as in 3.5.9(3). If X is in $\mathfrak{g}_\alpha$, Y is in $\mathfrak{g}_{-\alpha}$, and Z is in $\mathfrak{g}_\beta$ then

$$[Y, [X, Z]] = (s(1 - r)/2)\, \alpha(H_\alpha)B(X, Y)Z.$$

PROOF We first make some observations.

(1) If U, V are in $\mathfrak{g}$ then

$$\operatorname{ad} U(\operatorname{ad} V)^t = \sum_{j=1}^{t} (\operatorname{ad} V)^{t-n} \operatorname{ad}[U, V](\operatorname{ad} V)^{n-1} + (\operatorname{ad} V)^t \operatorname{ad} U.$$

From (1) we see that

(2) $\operatorname{ad} Y(\operatorname{ad} X)^t = -B(X, Y)\sum_{u=1}^{t} \operatorname{ad} H_\alpha(\operatorname{ad} X)^{u-1} + (\operatorname{ad} X)^t \operatorname{ad} Y.$

Using (1) on the first term of (2) and combining terms we see

(3) $\operatorname{ad} Y(\operatorname{ad} X)^t = -B(X, Y)t(\operatorname{ad} X)^{t-1}\operatorname{ad} H_\alpha$

$$-B(X, Y)\frac{t(t - 1)}{2}\alpha(H_\alpha)(\operatorname{ad} X)^{t-1} + (\operatorname{ad} X)^t \operatorname{ad} Y.$$

We now prove the lemma. Let W be in $\mathfrak{g}_{\beta + r\alpha}$ and so that $(\operatorname{ad} X)^{-r}W = Z$. Then

$$[Y, [X, Z]] = \operatorname{ad} Y(\operatorname{ad} X)^{1-r}W = -B(X, Y)(1 - r)(\alpha + r\beta)(H_\alpha)Z$$
$$-B(X, Y)\alpha(H_\alpha)((1 - r)(-r)/2)Z,$$

since $\operatorname{ad} Y \cdot W = 0$. The result now follows from 3.5.9(3).

3.5.13 *Lemma* Let for α in Δ, E_α be chosen in $\mathfrak{g}_\alpha$ so that $B(E_\alpha, E_{-\alpha}) = 1$. Define $N_{\alpha,\beta}$ to be 0 if $\alpha + \beta \neq 0$ and is not in Δ. If $\alpha + \beta$ is in Δ define $N_{\alpha,\beta}$ by $[E_\alpha, E_\beta] = N_{\alpha,\beta}E_{\alpha+\beta}$.

(1) If α, β, γ are in Δ and if $\alpha + \beta + \gamma = 0$ then $N_{\alpha,\beta} = N_{\beta,\gamma} = N_{\gamma,\alpha}$.

(2) If α, β, γ, δ are in Δ and if $\alpha + \beta + \gamma + \delta = 0$ but no two sum to zero then $N_{\alpha,\beta}N_{\gamma,\delta} + N_{\beta,\gamma}N_{\alpha,\delta} + N_{\gamma,\alpha}N_{\beta,\delta} = 0$.

The proof of this lemma is an exercise in the use of the Killing form. We leave it to the reader.

3.5.14 *Proposition* There is an automorphism A of $\mathfrak{g}$ so that $A|_{\mathfrak{h}} = -I$.

PROOF Let $H_1, \ldots, H_m$ be a basis of $\mathfrak{h}_R$. We order Δ lexicographically relative to this basis. That is, $\alpha > \beta$ if $\alpha(H_j) = \beta(H_j)$ for $j = 1, \ldots, k$ and $\alpha(H_{k+1}) > \beta(H_{k+1})$. That is, $\alpha > 0$ if $\alpha(H_j) > 0$ for j the smallest integer such that $\alpha(H_j) \neq 0$. If α is in Δ set $|\alpha| = \alpha$ if $\alpha > 0$, $|\alpha| = -\alpha$ if $\alpha < 0$. Let Δ^+ be the set of positive roots.

Let for α in Δ, E_α in $\mathfrak{g}_\alpha$ be chosen so that $B(E_\alpha, E_{-\alpha}) = 1$. Suppose that A exists. If H is in $\mathfrak{h}$ then $[H, AE_\alpha] = A[A^{-1}H, E_\alpha] = -A[H, E_\alpha] = -\alpha(H)AE_\alpha$. Thus $AE_\alpha = c_\alpha E_{-\alpha}$. 3.4.4 implies also that $c_\alpha c_{-\alpha} = 1$.

Let $N_{\alpha,\beta}$ be as in 3.5.13. Continuing with the assumption that A exists, $[AE_\alpha, AE_\beta] = c_\alpha c_\beta N_{-\alpha,-\beta} E_{-\alpha-\beta}$. $A[E_\alpha, E_\beta] = c_{\alpha+\beta} N_{\alpha,\beta} E_{-\alpha-\beta}$. We thus see (in light of 3.5.8) that in order to prove the proposition we need only find for each α in Δ, c_α in C so that

(1) $c_{-\alpha} = c_\alpha^{-1}$.

(2) If α, β and $\alpha + \beta$ are in Δ then $c_\alpha c_\beta N_{-\alpha,-\beta} = c_{\alpha+\beta} N_{\alpha,\beta}$.

Let for β in Δ^+, $\Delta_\beta = \{\alpha \| \alpha| < \beta\}$. If α_0 is the smallest positive root set $c_{\alpha_0} = 1$, $c_{-\alpha_0} = 1$. Suppose that we have found for each γ in $\Delta_\beta(\beta$ in $\Delta^+)$, c_γ in C satisfying (1) and

$(2)_\beta$ If γ, δ and $\gamma + \delta$ are in Δ_β then $c_\gamma c_\delta N_{-\gamma,-\delta} = c_{\gamma+\delta} N_{\gamma,\delta}$.

Let β' be the smallest positive root bigger than β. To complete the proof of the proposition we need only show that c_β and $c_{-\beta}$ may be found so that (1) and $(2)_{\beta'}$, are satisfied.

CASE 1 β is not of the form $\gamma + \delta$ for γ, δ in Δ_β. Set $c_\beta = c_{-\beta} = 1$. If γ, δ are in $\Delta_{\beta'}$ and $\gamma + \delta$ is in $\Delta_{\beta'}$ then it is easy to see (in this case) that γ, δ and $\gamma + \delta$ are in Δ_β. Hence $(2)_{\beta'}$ follows from $(2)_\beta$ in this case.

CASE 2 $\beta = \alpha + \gamma$, α, γ in Δ_β. Then in order for $(2)_{\beta'}$ to be satisfied we must set $c_\beta = c_\alpha c_\gamma N_{-\alpha,-\gamma}/N_{\alpha,\gamma}$ and $c_{-\beta} = c_{-\alpha} c_{-\gamma} N_{\alpha,\gamma}/N_{-\alpha,-\gamma}$. Clearly with this definition $c_\beta c_{-\beta} = 1$.

(i) If ρ, δ are in Δ_β and $\rho + \delta$ is in Δ_β then $(2)_{\beta'}$ follows from $(2)_\beta$ in this case.

(ii) Suppose that ρ, δ are in Δ_β and $\rho + \delta = \beta$. We may assume that $\{\rho, \delta\} \neq \{\alpha, \gamma\}$. We must show that $c_\rho c_\delta N_{-\rho,-\delta}/N_{\rho,\delta} = c_\alpha c_\gamma N_{-\alpha,-\gamma}/N_{\alpha,\gamma}$. That is, we must show that $c_\rho c_\delta N_{-\rho,-\delta} N_{\alpha,\gamma} = c_\alpha c_\gamma N_{-\alpha,-\gamma} N_{\rho,\delta}$. Now $\alpha + \gamma + (-\rho) + (-\delta) = 0$ and no two sum to zero. 3.5.13(2) implies

$$N_{\alpha,\gamma} N_{-\rho,-\delta} + N_{\gamma,-\rho} N_{\alpha,-\delta} + N_{-\rho,\alpha} N_{\gamma,-\delta} = 0$$

and

$$N_{\rho,\delta} N_{-\alpha,-\gamma} + N_{\delta,-\alpha} N_{\rho,-\gamma} + N_{-\alpha,\rho} N_{\delta,-\gamma} = 0.$$

We must thus show that

$$c_\rho c_\delta (N_{\gamma,-\rho} N_{\alpha,-\delta} + N_{-\rho,\alpha} N_{\gamma,-\delta}) = c_\alpha c_\gamma (N_{\delta,-\alpha} N_{\rho,-\gamma} + N_{-\alpha,\rho} N_{\delta,-\gamma}).$$

We can apply $(2)_\beta$ to the above terms.

$$\begin{aligned} c_\rho c_\delta N_{\gamma,-\rho} N_{\alpha,-\delta} &= c_\gamma c_\delta c_{\rho-\gamma} N_{-\gamma,\rho} N_{\alpha,-\delta} = c_\alpha c_\gamma c_{\delta-\alpha} c_{\rho-\gamma} N_{-\gamma,\rho} N_{-\alpha,\delta} \\ &= c_\alpha c_\gamma N_{\delta,-\alpha} N_{\rho,-\gamma} \end{aligned}$$

since N is skew-symmetric and $\delta - \alpha + \rho - \gamma = 0$ and $c_{\delta-\alpha}c_{\rho-\gamma} = 1$. The second term is handled similarly.

(iii) If ρ, δ are in Δ_β and $\rho + \delta = -\beta$ then the argument of (ii) gives $(2)_{\beta'}$ in this case.

(iv) ρ is in Δ_β, $\beta + \rho$ is in $\Delta_{\beta'}$ (hence in Δ_β). We must show that $c_\beta c_\rho N_{-\beta,-\rho} = c_{\beta+\rho}N_{\beta,\rho}$. Now, $\beta + \rho + (-\beta - \rho) = 0$, hence $N_{\beta,\rho} = N_{\rho,-\beta-\rho}$, $N_{-\beta,-\rho} = N_{-\rho,\beta+\rho}$. Now $\beta = (\beta + \rho) - \rho$. Thus (ii) implies that

$$c_\beta = c_{\beta+\rho}c_{-\rho}N_{-\beta-\rho,\beta}/N_{\beta+\rho,-\rho}.$$

We therefore see that

$$c_\beta c_\rho N_{-\beta,-\rho} = c_{\beta+\rho}N_{-\beta-\rho,\rho}N_{-\beta,-\rho}/N_{\beta+\rho,-\rho} = -c_{\beta+\rho}N_{-\beta-\rho,\rho} = c_{\beta+\rho}N_{\beta,\rho},$$

which was to be shown.

(v) ρ is in Δ_β and $\rho - \beta$ is in $\Delta_{\beta'}$. This case follows from the argument of (iv).

We have exhausted all possibilities, hence have shown $(2)_{\beta'}$ is satisfied. The proposition now follows.

3.6 Compact Lie Groups

3.6.1 *Lemma* Let G be a compact Lie group with Lie algebra $\mathfrak{g}$. Then $\mathfrak{g}_1 = [\mathfrak{g}, \mathfrak{g}] = D^1(\mathfrak{g})$ is semisimple and $\mathfrak{g} = \mathfrak{c} \oplus \mathfrak{g}_1$ with $\mathfrak{c}$ the center of $\mathfrak{g}$.

PROOF Let $\langle\ ,\ \rangle$ be any inner product on $\mathfrak{g}$. Set

$$(X, Y) = \int_G \langle \mathrm{Ad}(g)X, \mathrm{Ad}(g)Y\rangle\, dg.$$

Then, if X, Y, Z are in $\mathfrak{g}$, $([X, Y], Z) = -(Y, [X, Z])$. If $\mathfrak{h}$ is an ideal in $\mathfrak{g}$ then setting $\mathfrak{h}^\perp = \{Z|(Z, \mathfrak{h}) = (0)\}$ we see that $\mathfrak{h}^\perp$ is an ideal in $\mathfrak{g}$ and $\mathfrak{g} = \mathfrak{h} \oplus \mathfrak{h}^\perp$. Continuing this procedure we find that $\mathfrak{g} = \mathfrak{a}_1 \oplus \cdots \oplus \mathfrak{a}_r$ with $\mathfrak{a}_i$ an ideal which is either simple or one-dimensional. This proves the result.

3.6.2 *Lemma* Let $\mathfrak{g}$ be the Lie algebra of a compact Lie group. Then $B_\mathfrak{g}$ is negative semidefinite. If $\mathfrak{g}_1 = [\mathfrak{g}, \mathfrak{g}]$ then $B_\mathfrak{g}|_{\mathfrak{g}_1} = B_{\mathfrak{g}_1}$ and $B_{\mathfrak{g}_1}$ is negative definite.

PROOF Let $(\ ,\ )$ be the inner product on $\mathfrak{g}$ of 3.6.1. If X is in $\mathfrak{g}$ then

$\operatorname{ad} X$ is skew-symmetric relative to (,). Thus $(\operatorname{ad} X)^2$ is symmetric with nonpositive eigenvalues. This implies that $\operatorname{tr}(\operatorname{ad} X)^2 \leqslant 0$. This proves the first assertion. If $\operatorname{tr}(\operatorname{ad} X)^2 = 0$ then the above argument implies that $\operatorname{ad} X = 0$. The second assertion now follows from 3.6.1.

3.6.3 *Lemma* Let G be a Lie group and let H be a closed subgroup of G. Let f be in $C_0(G/H)$. Then there is f_1 in $C_0(G)$ so that

$$f(gH) = \int_H f_1(gh)dh, \qquad \text{for all } g \text{ in } G.$$

PROOF Let C be the support of f. Let p be the natural projection of G on G/H. If x is in C let U_x be an open subset of G so that $\bar{U}_x$ is compact and $p(U_x)$ is a neighborhood of x in G/H. Since C is compact, a finite number $p(U_{x_1}), \ldots, p(U_{x_r})$ covers C. Set $C' = \bar{U}_{x_1} \cup \cdots \cup \bar{U}_{x_r}$. Then C' is compact and $p(C') \cup C$. Let P be a nonempty open subset of H so that $\bar{P}$ is compact. Set $C'' = C'\bar{P}$. Then C'' is compact. Let h in $C_0(G)$ be so that $h|_{C''} = 1$ and h is nonnegative on G. Set $h_1(gH) = \int_H h(gh)dh$. h_1 is a continuous function on G/H and is positive on C. Set $f_1(g) = h(g)f(gH)h_1(gH)^{-1}$ for g in $p^{-1}(C)$, $f_1(g) = 0$ if g is not in $p^{-1}(C)$. Then f_1 is the desired function.

3.6.4 *Proposition* Let G be a connected Lie group. Let H be a closed central subgroup of G so that G/H is compact and so that Haar measure on G/H is G-invariant. Let φ be a continuous homomorphism of H to R (the additive group of real numbers). Then φ extends to a continuous homomorphism of G to R.

PROOF Since G/H is compact there is by 3.6.3, u in $C_0(G)$ so that $\int_H u(gh)dh = 1$ for all g in G. Set $f_1(x) = \int_H \varphi(h)u(xh)dh$. Then f_1 is a continuous function from G to R. Furthermore, if h_0 is in H then

$$\begin{aligned} f_1(xh_0) &= \int_H \varphi(h)u(xh_0h)dh = \int_H \varphi(hh_0^{-1})u(xh)dh \\ &= \int_H (\varphi(h) - \varphi(h_0))u(xh)dh = f_1(x) - \varphi(h_0). \end{aligned}$$

We therefore have

$$f_1(x) - f_1(xh) = \varphi(h), \text{ for } g \text{ in } G,\ h \text{ in } H. \tag{1}$$

Set for x in G, $g_x(z) = f_1(z) - f_1(xz)$. Then the map $x, z \to g_x(z)$ is a continuous map from $G \times G$ to R. Equation (1) implies that $g_x(zh) = g_x(z)$

for all h in H. Thus g_x induces a continuous function on G/H. Let dt be Haar measure on G/H. Set $\tilde{\varphi}(x) = \int_{G/H} g_x(t)dt$. We assert that $\tilde{\varphi}$ is the desired extension.

If h is in H then $\tilde{\varphi}(h) = \int_{G/H} g_h(t)dt$. But $g_h(t) = \varphi(h)$ for h in H by (1). We therefore see that $\tilde{\varphi}(h) = \varphi(h)$ for h in H.

Let x, t be in G. Then

$$\begin{aligned} g_{xt}(u) &= f_1(u) - f_1(xtu) \\ &= f_1(u) - f_1(tu) + f_1(tu) - f_1(xtu) \\ &= g_t(u) + g_x(tu). \end{aligned}$$

Thus

$$\begin{aligned} \tilde{\varphi}(xt) &= \int_{G/H} g_{xt}(u)du = \int_{G/H} g_t(u)du + \int_{G/H} g_x(tu)du \\ &= \int_{G/H} g_t(u)du + \int_{G/H} g_x(u)du = \tilde{\varphi}(t) + \tilde{\varphi}(x). \end{aligned}$$

3.6.5 *Definition* A Lie algebra $\mathfrak{g}$ over R is said to be compact if $B_{\mathfrak{g}}$ is negative definite.

3.6.6 *Theorem* Let G be a connected Lie group with compact Lie algebra. Then G is compact.

PROOF Let Z be the center of G. We assert that Z is discrete. Indeed, since G is connected $Z = \ker \mathrm{Ad}$. Now $\mathrm{Ad}(\exp(tX)) = e^{t\,\mathrm{ad}X}$ thus the Lie algebra of Z is (0). This clearly implies that Z is discrete. We also assert that $\mathrm{Ad}(G)$ is a compact Lie group. To see this, let G_0 be the identity component of the group of automorphisms of $\mathfrak{g}$. Then 3.4.6 implies that G_0 has Lie algebra $\mathrm{ad}\,\mathfrak{g}$. Thus $\mathrm{Ad}(G) = G_0$. Now G_0 is a closed subgroup of the group of all linear maps of $\mathfrak{g}$ to $\mathfrak{g}$ leaving invariant $B_{\mathfrak{g}}$. Hence G_0 is compact.

As in the proof of 3.6.3 there is a compact subset G_1 of G so that e is in the interior of G_1 and $\mathrm{Ad}(G_1) = \mathrm{Ad}(G)$. Thus $G_1 Z = G$. Now $G_1 G_1 G_1^{-1}$ is a compact subset of G. Hence $G_1 G_1 G_1^{-1} \subset G_1 Z_0$ with Z_0 a finite subset of Z. Let D be the subgroup of Z generated by Z_0. Let p be the canonical projection of G onto G/D. $p(G_1 Z_0)$ contains a neighborhood of $p(e)$ and is closed under multiplication and inverses. Hence $p(G_1 Z_0) = G/D$. This implies that G/D is compact. To prove that G is compact we need only prove that D is finite.

Suppose that D is infinite. Then the fundamental theorem of finitely generated Abelian groups (cf. Jacobson [1]) implies that there is a nontrivial homomorphism of D into the integers φ. 3.6.4 implies that φ extends to a Lie homomorphism of G to R. But $[\mathfrak{g}, \mathfrak{g}] = \mathfrak{g}$. This leads to the contradiction that φ is trivial. We have thus proved that D is finite.

3.6.7 *Definition* A compact, Abelian, connected, Lie group is called a torus.

3.6.8 *Lemma* Let T be a torus of dimension n. Then T is Lie isomorphic with R^n/Z^n.

PROOF Let $\mathfrak{h}$ be the Lie algebra of T. Since $\mathfrak{h}$ is Abelian the map exp is a Lie homomorphism of the additive group of $\mathfrak{h}$ to T. Since T is connected exp is onto. Let $\Gamma = \ker(\exp)$. We assert that Γ is discrete. If it were not then there would be an infinite sequence $\{z_n\}$ in Γ so that $\lim_{n\to\infty} z_n = 0$. But exp is one-to-one in a neighborhood of 0. We conclude that Γ is discrete. Since $\mathfrak{h}/\Gamma$ is compact, Γ must be of maximal rank. This implies that Γ is a lattice in $\mathfrak{h}$. The result now follows.

3.6.9 *Lemma* Let T be a torus. Let $\mathfrak{h}$ be the Lie algebra of T. Then there is an element x in $\mathfrak{h}$ so that if $\gamma = \exp(x)$ then the set $\{\gamma^n | n \text{ in } Z\}$ is dense in T.

PROOF By 3.6.8 we may assume that $T = R^n/Z^n$, that $\mathfrak{h} = R^n$, and that $\exp(z) = z + Z^n$. Let $x = (x_1, \ldots, x_n)$ be in R^n and be such that if $m_1x_1 + \cdots + m_nx_n = m_0$ with m_i in Z for $i = 0, \ldots, n$, then $m_i = 0$ for all i (e.g., $x_i = \pi^i$). We assert that x has the property of the conclusion of the lemma.

Indeed, we show that if Γ_1 is the subgroup of R^n generated by x and Z^n then $\overline{\Gamma}_1 = R^n$. Suppose not. Then $\overline{\Gamma}_1$ is a closed subgroup of R^n, hence $R^n/\overline{\Gamma}_1$ is a Lie group. Since $\overline{\Gamma}_1 \supset Z^n$, $R^n/\overline{\Gamma}_1$ is compact. Since R^n is Abelian and connected, $R^n/\overline{\Gamma}_1$ is a torus. If $R^n/\overline{\Gamma}_1$ is of dimension greater than zero then there is a continuous homomorphism f mapping $R^n/\overline{\Gamma}_1$ onto R/Z. Let ξ be the map of T to $R^n/\overline{\Gamma}_1$ defined by $\xi(z + Z^n) = z + \overline{\Gamma}_1$. Then ξ is a Lie homomorphism of T onto $R^n/\overline{\Gamma}_1$. Hence $f \circ \xi$ is a Lie homomorphism of T onto R/Z. Let μ be the differential of $f \circ \xi$. Then μ is a linear map of R^n to R. By the above $\mu(\overline{\Gamma}_1) \subset Z$. But if $e_1, \ldots, e_n$ is the standard basis of R^n then e_i is in Z^n for each i and thus $\mu(e_i)$ is in Z. Thus $\mu(x) = \sum \mu(e_i)x_i = m_0$ in Z. The initial assumption on x now implies that $\mu(e_i) = 0$ for all i. This implies the contradiction $\mu = 0$.

3.6.10 *Definition* A maximal, connected, Abelian subgroup of a compact connected Lie group G is called a maximal torus of G.

3.6.11 *Theorem* Let G be a connected compact Lie group. Then G has a maximal torus.

(1) A maximal torus of G is a torus.

(2) If T is a maximal torus of G and if $\mathfrak{g}$ and $\mathfrak{h}$ are, respectively, the Lie algebras of G and T then $\mathfrak{h}$ is a maximal Abelian subalgebra of $\mathfrak{g}$. If $\mathfrak{h}$ is a maximal Abelian subalgebra of $\mathfrak{g}$ then the connected subgroup of G corresponding to $\mathfrak{h}$ is a maximal torus of G.

(3) If $\mathfrak{h}_1$ and $\mathfrak{h}_2$ are maximal Abelian subalgebras of $\mathfrak{g}$ then there is an element g in G so that $\text{Ad}(g)\mathfrak{h}_1 = \mathfrak{h}_2$.

(4) If T_1 and T_2 are maximal tori of G then there is a g in G so that $gT_1g^{-1} = T_2$.

PROOF Let $\mathfrak{h}$ be a maximal Abelian subalgebra of $\mathfrak{g}$. Let T be the connected subgroup of G corresponding to $\mathfrak{h}$. $\overline{T}$ is connected and Abelian. Thus, by the definition of $\mathfrak{h}$, $\overline{T}$ has Lie algebra $\mathfrak{h}$. This implies that $T = \overline{T}$. Suppose that T' is a connected Abelian subgroup of G containing T. The argument above implies that T' has Lie algebra $\mathfrak{h}$. Hence $T' = T$. This shows that G has a maximal torus and proves the second assertion of (2).

Let T be a maximal torus of G. Let $\mathfrak{h}$ be the Lie algebra of T. If $\mathfrak{h}$ were not a maximal Abelian subalgebra of $\mathfrak{g}$ then T could not be a maximal Abelian connected subgroup of G. This proves the first assertion of (2) and (1).

Using (2) it is clear that (4) is an immediate corollary of (3). We now prove (3). Let x_1, x_2 be elements of $\mathfrak{h}_1$ and $\mathfrak{h}_2$, respectively, satisfying the condition of 3.6.9 for T_1 and T_2, the connected subgroups corresponding to $\mathfrak{h}_1$ and $\mathfrak{h}_2$. We assert that $\mathfrak{h}_i = \{x \mid [x, x_i] = 0\}$. Indeed if z is so that $[z, x_i] = 0$ then $(\exp tz)(\exp sx_i) = (\exp sx_i)(\exp tz)$ for all s, t in R. But the set $\{\exp sx_i \mid s \text{ in } R\}$ is dense in T_i. Hence $\exp tz$ commutes with every element of T_i for every t in R. This implies that $[z, x] = 0$ for all x in $\mathfrak{h}_i$. Hence z is in $\mathfrak{h}_i$.

Let $(\ ,\ )$ be an $\text{Ad}(G)$ invariant inner product on $\mathfrak{g}$. Set $f(g) = (\text{Ad}(g)x_1, x_2)$. Since G is compact, f takes a minimum, say, at g_0. If x is in $\mathfrak{g}$ then

$$\frac{d}{dt}(\text{Ad}(\exp tx\, g_0)x_1, x_2)\bigg|_{t=0} = 0.$$

This says that $([x, \text{Ad}(g_0)x_1], x_2) = 0$ for all x in $\mathfrak{g}$. But $e^{t\,\text{ad}\,x} = \text{Ad}(\exp tx)$ for all t in R, x in $\mathfrak{g}$. Hence $\text{ad}\,x$ is skew-symmetric relative to $(\ ,\ )$ for each x in $\mathfrak{g}$. This implies that $(x, [\text{Ad}(g_0)x_1, x_2]) = 0$ for all x in $\mathfrak{g}$. Thus

$$[\text{Ad}(g_0)x_1, x_2] = 0.$$

But then $\mathrm{Ad}(g_0)x_1$ is in $\mathfrak{h}_2$. This implies that $\mathrm{Ad}(g_0)\mathfrak{h}_1 \subset \mathfrak{h}_2$. Since $\mathfrak{h}_1$ is a maximal Abelian subalgebra of $\mathfrak{g}$, we see that $\mathrm{Ad}(g_0)\mathfrak{h}_1 = \mathfrak{h}_2$. This proves the theorem.

3.7 Real Forms

3.7.1 *Definition* Let $\mathfrak{g}$ be a Lie algebra over C. Then a real subalgebra $\mathfrak{g}_0$ of $\mathfrak{g}$ is called a real form of $\mathfrak{g}$ if $\mathfrak{g}_0 + (-1)^{\frac{1}{2}}\mathfrak{g}_0 = \mathfrak{g}$ and if $\mathfrak{g}_0 \cap (-1)^{\frac{1}{2}}\mathfrak{g}_0 = (0)$.

3.7.2 *Lemma* Let $\mathfrak{g}$ be a Lie algebra over C and let $\mathfrak{g}_0$ be a real form of $\mathfrak{g}$. Define for X, Y in $\mathfrak{g}_0$, $\sigma(X + (-1)^{\frac{1}{2}}Y) = X - (-1)^{\frac{1}{2}}Y$. Then σ defines an automorphism of $\mathfrak{g}$ as a real Lie algebra and $\sigma(cX) = \bar{c}\sigma(X)$ for X in $\mathfrak{g}$, c in C.

PROOF A direct consequence of the definitions.

3.7.3 *Definition* Let $\mathfrak{g}$ be a Lie algebra over C. Let σ be an automorphism of $\mathfrak{g}$ as a real Lie algebra. Then σ is called a conjugation of $\mathfrak{g}$ if $\sigma(cZ) = \bar{c}\sigma(Z)$ for c in C, Z in $\mathfrak{g}$, and if $\sigma^2 = I$. If $\mathfrak{g}_0$ is a real form of $\mathfrak{g}$ and if σ is the conjugation of $\mathfrak{g}$ defined in 3.7.2 then σ is called the conjugation of $\mathfrak{g}$ relative to $\mathfrak{g}_0$.

3.7.4 *Lemma* If $\mathfrak{g}$ is a Lie algebra over C and if σ is a conjugation of $\mathfrak{g}$ then $\mathfrak{g}_\sigma = \{Z | \sigma Z = Z\}$ is a real form of $\mathfrak{g}$ and the conjugation of $\mathfrak{g}$ relative to $\mathfrak{g}_\sigma$ is σ.

PROOF If X and Y are in $\mathfrak{g}_\sigma$ then $\sigma[X, Y] = [\sigma X, \sigma Y] = [X, Y]$. Thus $\mathfrak{g}_\sigma$ is a real subalgebra of $\mathfrak{g}$. If X is in $\mathfrak{g}$ then $X = \frac{1}{2}(X + \sigma X) + \frac{1}{2}(X - \sigma X)$. Since the eigenspace for the eigenvalue -1 for σ is $(-1)^{\frac{1}{2}}\mathfrak{g}_\sigma$ we see that $\mathfrak{g} = \mathfrak{g}_\sigma + (-1)^{\frac{1}{2}}\mathfrak{g}_\sigma$. Clearly $\mathfrak{g}_\sigma \cap (-1)^{\frac{1}{2}}\mathfrak{g}_\sigma = (0)$. Thus $\mathfrak{g}_\sigma$ is indeed a real form of $\mathfrak{g}$. The last assertion of the lemma is equally easy.

3.7.5 *Theorem* Let $\mathfrak{g}$ be a complex semisimple Lie algebra. Let $\mathfrak{h}$ be a Cartan subalgebra of $\mathfrak{g}$. Then there is a compact real form $\mathfrak{g}_u$ of $\mathfrak{g}$ so that $(-1)^{\frac{1}{2}}\mathfrak{h}_R$ (see 3.5.9(2) for notation) is a maximal Abelian subalgebra of $\mathfrak{g}_u$.

PROOF Let A be an automorphism of $\mathfrak{g}$ so that $A|_{\mathfrak{h}} = -I$ (the existence of A is guaranteed by 3.5.12). Let Δ be the root system of $\mathfrak{g}$ relative to $\mathfrak{h}$. Let for each α in Δ, E_α be chosen so that E_α is in $\mathfrak{g}_\alpha$ and $B(E_\alpha, E_{-\alpha}) = 1$ (B is defined as in 3.4.3). Then $AE_\alpha = c_\alpha E_{-\alpha}$. Let for each α in Δ, a_α be so that $a_\alpha^2 = c_{-\alpha}$ and $a_\alpha a_{-\alpha} = -1$. Set $X_\alpha = a_\alpha E_\alpha$. Define, for α, β in Δ, $\alpha + \beta \neq 0$, $N_{\alpha,\beta} = 0$ if $\alpha + \beta$ is not in Δ; if $\alpha + \beta$ is in Δ define $N_{\alpha,\beta}$ by $[X_\alpha, X_\beta] = N_{\alpha,\beta}X_{\alpha+\beta}$. A computation shows that $AX_\alpha = -X_{-\alpha}$.

Now $A[X_\alpha, X_\beta] = -N_{\alpha,\beta}X_{-(\alpha+\beta)}$ and $[AX_\alpha, AX_\beta] = [X_{-\alpha}, X_{-\beta}]$. Hence $N_{\alpha,\beta} = -N_{-\alpha,-\beta}$. By definition of the X_α we see that $B(X_\alpha, X_{-\alpha}) = -1$. Thus

$$\begin{aligned} B([X_\alpha, X_\beta], [X_{-\alpha}, X_{-\beta}]) &= N_{\alpha,\beta}N_{-\alpha,-\beta}B(X_{\alpha+\beta}, X_{-\alpha-\beta}) \\ &= -N_{\alpha,\beta}N_{-\alpha,-\beta} = N_{\alpha,\beta}^2. \end{aligned}$$

Thus to prove that $N_{\alpha,\beta}$ is real we need only show that $B([X_\alpha, X_\beta], [X_{-\alpha}, X_{-\beta}]) \geqslant 0$. But

$$\begin{aligned} B([X_\alpha, X_\beta], [X_{-\alpha}, X_{-\beta}]) &= -B([X_{-\alpha}, [X_\alpha, X_\beta], X_{-\beta}) \\ &= -\tfrac{1}{2}s(1 - r)\alpha(H_\alpha)B(X_\beta, X_{-\beta}) \end{aligned}$$

in the notation of 3.5.12. Thus $N_{\alpha,\beta}^2 = \frac{1}{2}\alpha(H_\alpha)s(1 - r)$ which is nonnegative. Hence $N_{\alpha,\beta}$ is real. Using this it is not hard to show that

$$\mathfrak{g}_u = (-1)^{\frac{1}{2}}\mathfrak{h}_R + \sum R(X_\alpha + X_{-\alpha}) + \sum R(-1)^{\frac{1}{2}}(X_\alpha - X_{-\alpha})$$

is the desired compact real form of $\mathfrak{g}$.

3.7.6 *Theorem* Let $\mathfrak{g}$ be a complex semisimple Lie algebra. Let $\mathfrak{g}_u$ be a compact real form of $\mathfrak{g}$, and let τ be the conjugation of $\mathfrak{g}$ relative to $\mathfrak{g}_u$. If σ is a conjugation of $\mathfrak{g}$ then there is a one-parameter group of automorphisms, $A(t)$, of $\mathfrak{g}$ so that $A(1)\tau A(1)^{-1}$ and σ commute.

PROOF Let for X, Y in $\mathfrak{g}$, $(X, Y) = -B(X, \tau Y)$. Then $(\, , \,)$ is a Hermitian form on $\mathfrak{g}$. We assert that $(\, , \,)$ is positive definite. Indeed, if X is in $\mathfrak{g}$ then $X = Y + (-1)^{\frac{1}{2}}Z$, Y, Z in $\mathfrak{g}_u$.

$$(X, X) = -B(Y + (-1)^{\frac{1}{2}}Z, Y - (-1)^{\frac{1}{2}}Z) = -B(Y, Y) - B(Z, Z).$$

Since $B_{\mathfrak{g}_u} = B|_{\mathfrak{g}_u}$ is negative definite we see that $(\, , \,)$ is positive definite.

Let $N = \sigma\tau$. Then N is an automorphism of $\mathfrak{g}$ as a complex Lie algebra. Now $(NX, Y) = -B(\sigma\tau X, \tau Y) = -B(X, \tau\sigma\tau Y) = -B(X, \tau NY) = (X, NY)$. Thus N is Hermitian relative to $(\, , \,)$. Since N is an automorphism of $\mathfrak{g}$, N^2 is positive definite. 3.2.8(2) implies that $N^2 = e^W$ with W a Hermitian endomorphism of $\mathfrak{g}$. Let $X_1, \ldots, X_n$ be a basis of $\mathfrak{g}$. If U is an endomorphism of $\mathfrak{g}$ then denote by $F_{ijk}(U)$ the kth coordinate of $U[X_i, X_j] - [UX_i, UX_j]$. If

$F_{ijk}(U) = 0$ for all i, j, k then U is a homomorphism of $\mathfrak{g}$ to $\mathfrak{g}$. Now $F_{ijk}(e^{mW}) = 0$ for all m an integer and all i, j, k. Hence 3.2.7 implies that $F_{ijk}(e^{tW}) = 0$ for all t in R and i, j, k. Thus e^{tW} is an automorphism of $\mathfrak{g}$ for all t in R.

Set $\tau_t = e^{tW}\tau e^{-tW}$. $\tau N \tau = \tau\sigma\tau\tau = \tau\sigma = N^{-1}$. Hence $N\tau N = \tau$. This implies that $e^{mW}\tau\, e^{mW} = \tau$ for all integers m. Applying 3.2.7 again we find that $e^{tW}\tau\, e^{tW} = \tau$ for all t in R. Thus $\sigma\tau_t = \sigma e^{tW}\tau\, e^{-tW} = \sigma\tau e^{-2tW} = Ne^{-2tW}$. Also, $\tau_t\sigma = e^{tW}\tau\, e^{-tW}\sigma = e^{2tW}\tau\sigma = e^{2tW}N^{-1}$. Using 3.2.7 again we see that $Ne^{tW} = e^{tW}N$ for all t in R. Therefore $\tau_t\sigma = N^{-1}e^{2tW}$. Hence $\tau_t\sigma = \sigma\tau_t$ if and only if $Ne^{-2tW} = N^{-1}e^{2tW}$. That is, if and only if $N^2 = e^{4tW}$. Thus $\tau_{\frac{1}{4}}\sigma = \sigma\tau_{\frac{1}{4}}$. This implies that $A(t) = e^{\frac{1}{4}tW}$ is the desired one-parameter group of automorphisms of $\mathfrak{g}$.

3.7.7 *Corollary* Let $\mathfrak{g}_1$ and $\mathfrak{g}_2$ be compact real forms of the complex semisimple Lie algebra $\mathfrak{g}$. Then there is a one-parameter group of automorphisms, $A(t)$, of $\mathfrak{g}$ so that $A(1)\mathfrak{g}_1 = \mathfrak{g}_2$.

PROOF Let τ_1, τ_2 be, respectively, the conjugations of $\mathfrak{g}$ relative to $\mathfrak{g}_1$, $\mathfrak{g}_2$. Let $A(t)$ be a one-parameter group of automorphisms of $\mathfrak{g}$ so that $A(1)\tau_1 A(1)^{-1}$ commutes with τ_2. Set $\mathfrak{g}_1' = A(1)\mathfrak{g}_1$. Then $\mathfrak{g}_1$ is a compact form of $\mathfrak{g}$ with conjugation $\tau_1' = A(1)\tau_1 A(1)^{-1}$. Since τ_1' commutes with τ_2 we see that $\mathfrak{g}_2 = (\mathfrak{g}_2 \cap \mathfrak{g}_1') + (\mathfrak{g}_2 \cap (-1)^{\frac{1}{2}}\mathfrak{g}_1')$. But B is positive definite on $\mathfrak{g}_2 \cap (-1)^{\frac{1}{2}}\mathfrak{g}_1'$ and is negative definite on $\mathfrak{g}_2$. Hence $\mathfrak{g}_2 \cap (-1)^{\frac{1}{2}}\mathfrak{g}_1' = (0)$. This implies that $A(1)\mathfrak{g}_1 = \mathfrak{g}_2$.

3.7.8 *Definition* Let $\mathfrak{g}$ be a semisimple Lie algebra over R. Then a Cartan involution of $\mathfrak{g}$ is an involutive automorphism σ of $\mathfrak{g}$ so that if $\mathfrak{k} = \{X | \sigma X = X\}$, $\mathfrak{p} = \{X | \sigma X = -X\}$ then $B|_{\mathfrak{k}\times\mathfrak{k}}$ is negative definite and $B|_{\mathfrak{p}\times\mathfrak{p}}$ is positive definite ($B = B_{\mathfrak{g}}$). The decomposition $\mathfrak{g} = \mathfrak{k} \oplus \mathfrak{p}$ is called a Cartan decomposition of $\mathfrak{g}$.

3.7.9 *Theorem* Let $\mathfrak{g}$ be a semisimple Lie algebra over R. Let $\mathfrak{g}_C$ be the Lie algebra gotten by complexifying $\mathfrak{g}$.

(1) $\mathfrak{g}_C$ is a complex semisimple Lie algebra.

(2) $\mathfrak{g}$ has a Cartan involution.

(3) If σ_1 and σ_2 are Cartan involutions of $\mathfrak{g}$ then there is a one-parameter group, $A(t)$, of automorphisms of $\mathfrak{g}$ so that $A(1)\sigma_1 A(1)^{-1} = \sigma_2$.

PROOF It is easy to see that $B_{\mathfrak{g}_C}$ is the complex bilinear extension of $B_{\mathfrak{g}}$. Since the complex bilinear extension of a nondegenerate symmetric bilinear form is nondegenerate, $B_{\mathfrak{g}C}$ is nondegenerate. (1) now follows from 3.4.5.

Let $\mathfrak{g}_u$ be a compact real form of $\mathfrak{g}_C$ and let σ, τ be, respectively, the conjugations of $\mathfrak{g}_C$ relative to $\mathfrak{g}$ and to $\mathfrak{g}_u$. Let A be an automorphism of $\mathfrak{g}_C$ so that $A\tau A^{-1}$ and σ commute. Then $A\tau A^{-1}(\mathfrak{g}) = (\mathfrak{g})$. Let $\sigma = A\tau A^{-1}|_{\mathfrak{g}}$. Then σ is a Cartan involution by the definition of compact real form. This proves (2).

If σ_1 and σ_2 are Cartan involutions of $\mathfrak{g}$ then let $= \mathfrak{k}_1 + \mathfrak{p}_1 = \mathfrak{k}_2 + \mathfrak{p}_2$ be the corresponding Cartan decompositions of $\mathfrak{g}$. Set in $\mathfrak{g}_C$, $\mathfrak{g}_i = \mathfrak{k}_i + (-1)^{\frac{1}{2}}\mathfrak{p}_i$, $i = 1, 2$. Then $\mathfrak{g}_i$ is a compact real form of $\mathfrak{g}_C$. Let τ_1 and τ_2 be, respectively, the conjugations of $\mathfrak{g}_C$ relative to $\mathfrak{g}_1$ and $\mathfrak{g}_2$. Then $\tau_1\tau_2(\mathfrak{g}) = \mathfrak{g}$. Now, in the proof of 3.7.6 if we add the real polynomial condition that says that an endomorphism of $\mathfrak{g}_C$ leaves $\mathfrak{g}$ invariant, we find that the one-parameter group of automorphisms of $\mathfrak{g}_C$, $B(t)$, so that $B(1)\tau_1 B(1)^{-1} = \tau_2$ can be chosen so that each $B(t)$ leaves $\mathfrak{g}$ invariant. Thus if $A(t) = B(t)|_{\mathfrak{g}}$ then $A(t)$ is a one-parameter group of automorphisms of $\mathfrak{g}$ and $A(1)\sigma_1 A(1)^{-1} = \sigma_2$.

3.7.10 *Theorem* Let $\mathfrak{g}$ be a complex semisimple Lie algebra. Let $\mathfrak{h}_1$ and $\mathfrak{h}_2$ be Cartan subalgebras of $\mathfrak{g}$. Then there is an element A in the connected open subgroup of the automorphism group of $\mathfrak{g}$ so that $A\mathfrak{h}_1 = \mathfrak{h}_2$.

PROOF Let $\mathfrak{g}_1$ and $\mathfrak{g}_2$ be compact forms of $\mathfrak{g}$ so that $(-1)^{\frac{1}{2}}\mathfrak{h}_{i_R} \subset \mathfrak{g}_i$ for $i = 1, 2$. Let $B(t)$ be a one-parameter group of automorphisms of $\mathfrak{g}$ so that $B(1)\ \mathfrak{g}_1 = \mathfrak{g}_2$. Let K be any connected Lie group with Lie algebra $\mathfrak{g}_2$. Then 3.6.5 implies that K is compact. Hence there is an element k in K so that $\mathrm{Ad}(k)B(1)((-1)^{\frac{1}{2}}\mathfrak{h}_{1_R}) = (-1)^{\frac{1}{2}}\mathfrak{h}_{2_R}$. Thus $A = \mathrm{Ad}(k)B(1)$ is the desired element of the connected open subgroup of the automorphism group of $\mathfrak{g}$.

3.8 The Euler Characteristic of a Compact Homogeneous Space

3.8.1 Let G be a compact, connected Lie group. Let K be a closed subgroup of G. Let ρ be the isotropy representation of K on $G/K = M$ (that is, $\rho(k) = k_{*eK}$ acting on $T(M)_{eK}$). Let $\chi(M)$ be the Euler characteristic of M (See, e.g., Greenberg [1]).

3.8.2 *Theorem*

(1) $\chi(G/K) = \int_K \det(I - \rho(k))dk.$

(2) Suppose that K is connected. Let G' be a compact, connected Lie group with Lie algebra isomorphic to that of G. Let K' be the connected subgroup of G' whose Lie algebra corresponds to the Lie algebra of K. If K' is a closed subgroup of G' then $\chi(G'/K') = \chi(G/K)$.

To prove this result we recall some results related to the de Rham theorem (see F. Warner [1], 4.17, 5.34).

de Rham's theorem says that $H^p(M, R) = Z_d^p/B_d^p$, where $H^p(M, R)$ is the pth singular cohomology group of M with real coefficients, Z_d^p is the space of all C^∞ p-forms, ω, on M so that $d\omega = 0$, B_d^p is the set of all $d\omega$ with ω a $(p-1)$-form on M. The isomorphism is given explicitly via Stokes' theorem. We recall the pertinent definitions.

A C^∞ p-simplex is a C^∞ map, σ of Δ_p into M, where $\Delta_p = \{(t_1, \ldots, t_p \mid \sum t_i \leqslant 1, t_i \geqslant 0\}$. We define for $0 \leqslant i \leqslant p+1$ a C^∞ mapping k_i^p of Δ_p to Δ_{p+1} as follows:

If $p = 0$ set $k_0^0(0) = 1$, $k_1^0(0) = 0$.

If $p > 0$ set $k_0^p(t_1, \ldots, t_p) = (1 - \sum t_i, t_1, \ldots, t_p)$ and $k_i^p(t_1, \ldots, t_p) = (t_1, \ldots, t_{i-1}, 0, t_i, \ldots, t_p)$.

Let C_p be the free vector space over R generated by the singular p-simplices. If σ is a singular p-simplex define $\partial\sigma = \sum (-1)^i \sigma \circ k_i^{p-1}$. Then ∂ extends to a linear map of C_p to C_{p-1}. If ω is a p-form and if σ is a p-simplex then define $\omega(\sigma) = \int_{\Delta_p} \sigma^*\omega$. Then each p-form induces a linear map of C_p to R. Now, Stokes' theorem says that if ω is a $(p-1)$-form and if σ is a p-simplex then $d\omega(\sigma) = \omega(\partial\sigma)$.

Define Z_p to be $\ker(\partial\colon C_p \to C_{p-1})$ and $B_p = \partial C_{p+1}$. Then $H_p(M, R) = Z_p/B_p$ is the pth singular homology group of M with real coefficients (this makes sense since $\partial^2 = 0$). Now if ω is in Z_d^p and if σ is in B_p then by the above $\omega(\sigma) = 0$. Hence each element of Z_d^p defines a linear form on $H_p(M, R)$. Let for ω in Z_d^p, $u(\omega)$ be the element of $H_p(M, R)^*$. We can now state de Rham's theorem.

$\ker(u|_{Z_p^d}) = B_d^p$ and the induced mapping of $H_p(M)$ to $H_\sigma(M, R)^*$ is bijective.

With this review in mind we prove the theorem. If ω is in Z_d^p define $\omega^\#$ by the formula $\omega_x^\#(X_1, \ldots, X_p) = \int_G (g^*\omega)_x(X_1, \ldots, X_p)dg$ for x in M, $X_1, \ldots, X_p$ in $T(M)_x$. Now since each g in G can be joined to e by a C^∞ curve in G we see that $\omega - g^*\omega$ is in B_d^p (this follows from a combination of Stokes' theorem and de Rham's theorem). Thus for each element σ of Z_p we have $\omega(\sigma) = g^*\omega(\sigma)$. But now $\omega^\#(\sigma) = \int_G g^*\omega(\sigma)dg = \omega(\sigma)$. Noting that if η is a p-form and if $\eta(B_p) = 0$ then $d\eta = 0$, we see that $d\omega^\# = 0$ and $\omega - \omega^\#$ is in B_d^p.

Let $I^p(M)$ be the space of all G-invariant p-forms on M. The above arguments imply that $H^p(M, R) = \ker(d|_{I^p(M)})/dI^{p-1}(M)$. Let (ρ^*, V) be the contragradient representation to ρ. If ω is in $I_p(M)$ then ω_{eK} in $\Lambda^p V$ completely determines ω. Furthermore if $(\Lambda^p V)^K = \{v \text{ in } \Lambda^p V | \Lambda^p \rho^*(k)v = v \text{ for all } k$ in $K\}$ then the map $\omega \mapsto \omega_{eK}$ defines a bijection of $I^p(M)$ with $(\Lambda^p V)^K$. Thus d induces a linear map $\tilde{d}$ of $(\Lambda^p V)^K$ into $(\Lambda^{p+1} V)^K$ so that $\tilde{d}^2 = 0$ and this complex gives the cohomology groups of M. Thus we see that $\chi(M) = \sum (-1)^p \dim((\Lambda^p V)^K)$. We note that if K is connected this formula for $\chi(M)$ depends only on how the Lie algebra of K sits in the Lie algebra of G. Hence (2) is true.

Let now $\rho_i^* = \Lambda^i \rho^*$. 2.9.4 implies that $\int_K \operatorname{tr}(\rho_i^*(k))dk = \dim(\Lambda^p V)^K$. Now $\det(I - \rho(k)) = \sum (-1)^p \operatorname{tr}(\rho_i^*(k))$. We have thus proved that $\int_K \det(I - \rho(k))dk = \chi(M)$.

3.8.3 *Corollary* Let G be a compact connected Lie group. Let T be a maximal torus of G.

(1) $\chi(G/T) > 0$.

(2) The center of G is contained in T.

PROOF Let $\mathfrak{g}$ be the Lie algebra of G. Let $\mathfrak{p}$ be an Ad(T)-inviarant complement for the Lie algebra $\mathfrak{h}$ of T. Then as a representation of T, the complexification $\mathfrak{p}_C$ of $\mathfrak{p}$ splits into a direct sum of subspaces $\mathfrak{p}_\lambda$ with $\operatorname{Ad}(t)v = \lambda(t)v$ for t in T (we leave this assertion to the reader, the spaces $\mathfrak{p}_\lambda$ are just the root spaces of $[\mathfrak{g}, \mathfrak{g}]_C$) and if $\mathfrak{p}_\lambda \neq 0$ then $\mathfrak{p}_{\bar{\lambda}} \neq 0$. Let $\lambda_1, \ldots, \lambda_r, \bar{\lambda}_1, \ldots, \bar{\lambda}_r$ be the distinct characters λ of T so that $\mathfrak{p}_\lambda \neq 0$. Now, in the notation of the proof of 3.8.2, ρ is equivalent with the representation of T on $\mathfrak{p}$. Thus

$$\det(I - \rho(t)) = \Pi((1 - \lambda_i(t))(1 - \bar{\lambda}_i(t)))^{\dim \mathfrak{p}_{\lambda_i}} = \Pi(2 - (\lambda_i(t) + \bar{\lambda}_i(t)))^{\dim \mathfrak{p}_{\lambda_i}}.$$

If t is in T we therefore see that $\det(I - \rho(t)) \geqslant 0$ since $\lambda_i \neq 1$ for any i. If t is such that the cyclic group generated by t is dense in T then $\det(I - \rho(t)) > 0$. This proves the first assertion.

$\mathfrak{g} = \mathfrak{z} + \mathfrak{g}_1$ with $\mathfrak{z}$ the center of $\mathfrak{g}$ and $\mathfrak{g}_1$ a compact ideal. Let Z be the connected subgroup of G corresponding to $\mathfrak{z}$ and let G_1 be the connected subgroup of G corresponding to $\mathfrak{g}_1$. 3.6.6 implies that G_1 is compact. Now the set of all elements zg_1 with z in Z and g_1 in G_1 is a subgroup of G which contains a neighborhood of the identity of G. Hence every element of g is of the form zg_1 with z in Z, g_1 in G_1. If zg_1 is central in G then clearly g_1 is central in G_1. Let T_1 be the maximal torus of G_1 so that T_1 is contained in T. Then $T = ZT_1$. Let Z_1 be the center of G_1. By the above, (2) will follow if we can show that Z_1 is contained in T_1. Since $\mathfrak{g}_1$ is compact the center of G_1 is discrete. Hence Z_1 is finite. Let $G_0 = G_1/Z_1$ and let p be the canonical map of G_1 onto G_0. Let $T_0 = p(T_1)$. 3.8.2(2) implies that $\chi(G_1/T_1) = \chi(G_0/T_0)$. But G_1/Z_1T_1 is diffeomorphic to G_0/T_0 and G_1/T_1 is an m-fold covering space of G_1/Z_1T_1 where m is the order of the group $Z_1/Z_1 \cap T_1$.

But then $\chi(G_1/T_1) = m\chi(G_1/Z_1T_1) = m\chi(G_0/T_0) = m\chi(G_1/T_1)$ (see Greenberg [1]). Now, the first part of this theorem implies that $\chi(G_1/T_1)$ is positive. Hence $m = 1$. Thus Z_1 is indeed contained in T_1.

3.9 Automorphisms of Compact Lie Algebras

3.9.1 *Theorem* Let $\mathfrak{g}$ be a compact Lie algebra. Let A be an automorphism of $\mathfrak{g}$. Then A has a nonzero fixed vector.

PROOF Suppose $\mathfrak{g}$ and A are such that the theorem is false. We derive some properties of $\mathfrak{g}$ and A. Let $\mathfrak{g}_C$ be the complexification of $\mathfrak{g}$.

Let for each complex number λ, $\mathfrak{g}_\lambda = \{X \text{ in } \mathfrak{g}_C | AX = \lambda X\}$. Then $\mathfrak{g}_C = \sum \mathfrak{g}_\lambda$.

(1) $[\mathfrak{g}_\lambda, \mathfrak{g}_\mu] \subset \mathfrak{g}_{\lambda\mu}$. Thus, since we are assuming that $\mathfrak{g}_1 = 0$, $[\mathfrak{g}_\lambda, \mathfrak{g}_{\lambda^{-1}}] = 0$.

(2) Let B be the Killing form of $\mathfrak{g}_C$. Then $B(\mathfrak{g}_\lambda, \mathfrak{g}_\mu) \neq 0$ only if $\lambda = \mu^{-1}$. Hence $B|_{\mathfrak{g}_\lambda + \mathfrak{g}_{\lambda^{-1}}}$ is nondegenerate.

The above observations are easily proved. The first is just a restatement of the assumption that A is an automorphism of $\mathfrak{g}$. The second is proved in exactly the same way as 3.5.6.

(3) If X is in $\mathfrak{g}_\lambda$ and if ad X is nilpotent then $X = 0$. In fact, if X is in $\mathfrak{g}_\lambda$ and ad X is nilpotent and if Y is in $\mathfrak{g}_{\lambda^{-1}}$ then since $[\mathfrak{g}_\lambda, \mathfrak{g}_{\lambda^{-1}}] = 0$ we see that $(\text{ad}\, X \,\text{ad}\, Y)^k = (\text{ad}\, X)^k(\text{ad}\, Y)^k = 0$ for k sufficiently large. Thus $B(X, Y) = 0$ for all Y in $\mathfrak{g}_{\lambda^{-1}}$. But then $X = 0$ as (3) asserts.

Now suppose that X is in $\mathfrak{g}_\lambda$. Then $X = X_s + X_n$ with ad X_s semisimple, ad X_n nilpotent, $[X_s, X_n] = 0$, and this decomposition of X is unique (see 3.4.7). Now $AX = \lambda X = \lambda X_s + \lambda X_n$. Thus the uniqueness of the above splitting for X implies that $AX_n = \lambda X_n$. Now (3) implies that $X_n = 0$. This proves the following.

(4) If X is in $\mathfrak{g}_\lambda$ then ad X is semisimple.

Suppose that $A^k \neq I$ for any positive integer k. Then there is $\mathfrak{g}_\lambda \neq 0$ so that $\lambda^k \neq 1$ for all positive integers k. But then if x is in $\mathfrak{g}_\lambda$, y is in $\mathfrak{g}_\mu$, $(\text{ad}\, x)^k y$ is in $\mathfrak{g}_{\lambda^k\mu}$. Since there can be only a finite number of eigenvalues of A, ad x is nilpotent for all x in $\mathfrak{g}_\lambda$. But then $\mathfrak{g}_\lambda = 0$. This contradiction implies

(5) $A^k = I$ for some positive integer k.

(6) If x is in $\mathfrak{g}_\lambda$ then $\text{ad} x|_{\Sigma_s \mathfrak{g}_{\lambda^s}} = 0$.

In fact, set $\tilde{\mathfrak{g}} = \sum \mathfrak{g}_{\lambda^s}$. Then $\text{ad}\, x(\tilde{\mathfrak{g}}) \subset \tilde{\mathfrak{g}}$. Furthermore if k is given then there is s so that $\lambda^s\lambda^k = 1$. Thus $\text{ad}\, x|_\lambda$ is nilpotent. But ad x is also semisimple, hence (6) follows.

We now prove the theorem with a double induction. We assume that the result is true for all $\mathfrak{g}$ and all automorphisms of order less than m the order of A. We also assume the result is true for all compact Lie algebras of dimension less than $n = \dim \mathfrak{g}$. We continue to assume that the theorem is false for $\mathfrak{g}$ and A.

Let X be an element of $\mathfrak{g}_\lambda$, and suppose that $X \neq 0$. Then $X = X_1 + (-1)^{\frac{1}{2}}X_2$ with X_1 and X_2 in $(\mathfrak{g}_\lambda + \mathfrak{g}_{\lambda^{-1}}) \cap \mathfrak{g}$. Hence $[X_1, X_2] = 0$.

(7) Set $\mathfrak{g}_1^X = \{Z \text{ in } \mathfrak{g} | [Z, X] = 0\}$. Then $\mathfrak{g}_1^X$ is maximal Abelian in $\mathfrak{g}$.

To prove (7) we note that $\mathfrak{g}_1^X$ is A-invariant. Thus $[\mathfrak{g}_1^X, \mathfrak{g}_1^X]$ is compact and A-invariant. The second inductive hypothesis implies that $[\mathfrak{g}_1^X, \mathfrak{g}_1^X] = (0)$. Now X_1 and X_2 are in $\mathfrak{g}_1^X$ and Z is in $\mathfrak{g}_1^X$ if and only if $[X_1, Z] = 0$ and $[X_2, Z] = 0$. Thus $\mathfrak{g}_1^X$ is indeed maximal Abelian in $\mathfrak{g}$.

(8) If X is in $\mathfrak{g}_\lambda$ and if $X \neq 0$ then X is a regular element of $\mathfrak{g}_C$. Let $\mathfrak{g}^X = \{Z | Z \text{ in } \mathfrak{g}_C, [X, Z] = 0\}$. If Z is in $\mathfrak{g}^X$ then $Z = Z_1 - (-1)^{\frac{1}{2}}Z_2$ with Z_1 and Z_2 in $\mathfrak{g}$, and $[X_1, Z_1] = [X_2, Z_2]$, $[X_1, Z_2] = -[X_2, Z_1]$. Thus $(\operatorname{ad} X_1)^2 Z_1 = -(\operatorname{ad} X_2)^2 Z_1$. But then

$$0 \leqslant B((\operatorname{ad} X_1)^2 Z_1, Z_1) = -B((\operatorname{ad} X_2)^2 Z_1, Z_1) \leqslant 0.$$

Hence since $(-1)^{\frac{1}{2}}Z$ is in $\mathfrak{g}^X$ we see that $\operatorname{ad} X_1 Z_2 = \operatorname{ad} X_2 Z_2 = 0$. Thus $\mathfrak{g}^X = \mathfrak{g}_1^X + (-1)^{\frac{1}{2}}\mathfrak{g}_1^X$, which proves (8).

(9) If $[\mathfrak{g}_\lambda, \mathfrak{g}_\mu] \neq (0)$ then $\mathfrak{g}_C = \sum \mathfrak{g}_{\lambda^s\mu^t}$. To see this we note that $\sum \mathfrak{g}_{\mu^s\lambda^t}$ is a subalgebra of $\mathfrak{g}_C$ which is invariant under A and the conjugation of $\mathfrak{g}_C$ relative to $\mathfrak{g}$. The intersection of its first derived algebra with $\mathfrak{g}$ is compact, nonzero, and A-invariant. The second inductive hypothesis implies that it must equal $\mathfrak{g}$. This proves (9).

Recall that we have assumed that the order of A is m.

(10) Let p be a prime dividing m. Then there is a γ so that $\mathfrak{g}_\gamma \neq (0)$ and γ is a pth root of unity. If $p = m$ there is nothing to prove. Otherwise A^p has lower order than A. Thus the first inductive hypothesis implies that A^p has a nonzero fixed vector. Clearly this proves (10).

(11) There exist λ and μ of prime order so that $[\mathfrak{g}_\lambda, \mathfrak{g}_\mu] \neq (0)$. Suppose that (11) is false. Let γ, δ be so that $[\mathfrak{g}_\lambda, \mathfrak{g}_\delta] \neq (0)$. Then for some a, b, γ^a and δ^b are of prime order. By (10) there are positive integers c, d so that $\mathfrak{g}_{\lambda^{ac}} \neq 0$ and $\mathfrak{g}_{\delta^{bd}} \neq 0$. By hypothesis $[\mathfrak{g}_{\lambda^{ac}}, \mathfrak{g}_{\delta^{bd}}] = 0$. Now $[\mathfrak{g}_{\lambda^{ac}}, \mathfrak{g}_\lambda] = [\mathfrak{g}_{\delta^{bd}}, \mathfrak{g}_\delta] = 0$, by (6). Since each of these spaces consists of regular elements we must conclude that $[\mathfrak{g}_\lambda, \mathfrak{g}_\delta] = 0$. This contradiction implies (11).

We now complete the proof of the theorem. Let λ, μ be of prime order p, q, respectively, so that $[\mathfrak{g}_\lambda, \mathfrak{g}_\mu] \neq (0)$. If $p \neq q$ then $\lambda\mu$ is a primative pq root of unity and (1) implies that $\mathfrak{g}_{\lambda\mu} \neq 0$. But then (6) and (9) imply that $\mathfrak{g}$ is Abelian. Since $\mathfrak{g}$ is not Abelian $p = q$. If $p = q$ then $m = p$. (6) again implies that $\mathfrak{g}$ is Abelian. We have thus (finally) run into the contradiction we were seeking.

3.9.2 *Corollary* Let $\mathfrak{g}$ be a compact Lie algebra. Let A be an automorphism of $\mathfrak{g}$. Then A leaves fixed a regular element of $\mathfrak{g}$.

PROOF Let G be the group of all automorphisms of $\mathfrak{g}$. Then G is a closed subgroup of the general linear group of $\mathfrak{g}$. Thus G is a Lie group. Let G_0 be the identity component of G. Then 3.4.6 implies that the Lie algebra of G_0 is $\operatorname{ad}(\mathfrak{g})$ which we identify with $\mathfrak{g}$. Thus 3.6.6 implies that G_0 is compact. Also A extends to an automorphism of G_0 by the formula $A(g) = AgA^{-1}$.

Let G_1 be the identity component of the fixed point set of A in G_0. Then the Lie algebra $\mathfrak{g}_1$ of G_1 is the fixed point set of A on $\mathfrak{g}$. Since G_0 is compact, G_1 is compact. Let T be a maximal torus in G_1. Let $\mathfrak{h}$ be the Lie algebra of T. Then $\mathfrak{h}$ is a maximal Abelian subalgebra of $\mathfrak{g}_1$. Let X be an element of $\mathfrak{h}$ so that the set $\{\exp tX | t \text{ in } R\}$ is dense in T. Let $\mathfrak{g}^X = \{Z | Z \text{ in } \mathfrak{g} \text{ and } [X, Z] = 0\}$. Then since $\mathfrak{g}$ is compact $\mathfrak{g}^X = \mathfrak{z} + [\mathfrak{g}^X, \mathfrak{g}^X]$ with $\mathfrak{z}$ the center of $\mathfrak{g}^X$ and $[\mathfrak{g}^X, \mathfrak{g}^X]$ compact. Clearly $\mathfrak{z} \supset \mathfrak{h}$. Now $[\mathfrak{g}^X, \mathfrak{g}^X] \cap \mathfrak{g}_1 \subset \mathfrak{g}^X \cap \mathfrak{g}_1 = \mathfrak{h} \subset \mathfrak{z}$. Thus $[\mathfrak{g}^X, \mathfrak{g}^X] \cap \mathfrak{g}_1 = (0)$. But $A[\mathfrak{g}^X, \mathfrak{g}^X] \subset [\mathfrak{g}^X, \mathfrak{g}^X]$. Hence 3.9.1 implies that $\mathfrak{g}^X$ is Abelian. Hence X is a regular element of $\mathfrak{g}$.

3.9.3 *Lemma* Let G be a compact connected Lie group. Let m be the dimension of a maximal torus of G. Let $\mathfrak{g}$ be the Lie algebra of G. If g is in G then the dimension of the fixed point set of $\operatorname{Ad}(g)$ is at least m.

PROOF Let for g in G, $\det(\operatorname{Ad}(g) - (t + 1)I) = \sum p_k(g)t^k$. We note that the lemma will be proved if we can show that $p_k(g) = 0$ for all g in G and all $k < m$. Let U_0 be a neighborhood of 0 in $\mathfrak{g}$ so that $U = \exp(U_0)$ is a neighborhood of e in G. If g is in U then g is contained in a maximal torus of G and since the maximal tori are conjugate in G we see that $p_k(g) = 0$ for all g in U and all $k < m$. Now p_k is a real analytic function on G. Hence for each g in G, $k < m$, $p_k(g) = 0$.

3.9.4 *Theorem* Let G be a compact connected Lie group and let g be an element of G. Then there is a maximal torus of G containing g.

PROOF We first show the following.

(1) If g is in G and if $\mathfrak{g}$ is the Lie algebra of G then there is a maximal Abelian subalgebra $\mathfrak{h}$ of $\mathfrak{g}$ so that $\operatorname{Ad}(g)|_{\mathfrak{h}} = I$.

$\mathfrak{g} = \mathfrak{z} + [\mathfrak{g}, \mathfrak{g}]$, with $\mathfrak{z}$, the center of $\mathfrak{g}$ and $[\mathfrak{g}, \mathfrak{g}] = \mathfrak{g}_0$ a compact ideal of $\mathfrak{g}$. If g is in G then $\operatorname{Ad}(g)|_{\mathfrak{z}}$ is the identity and $\operatorname{Ad}(g)$ leaves $\mathfrak{g}_0$ invariant. Thus to prove (1) we may assume that $\mathfrak{g} = \mathfrak{g}_0$. By 3.9.2 there is a regular element

X so that $\mathrm{Ad}(g)X = X$. Let $\mathfrak{h}$ be the maximal Abelian subalgebra of $\mathfrak{g}$ containing X. Clearly $\mathfrak{h}$ is $\mathrm{Ad}(g)$-invariant.

Let $\mathfrak{g}_C$ be the complexification of $\mathfrak{g}$ and let $\mathfrak{h}_C$ be the complexification of $\mathfrak{h}$ in $\mathfrak{g}_C$. Then $\mathfrak{h}_C$ is a Cartan subalgebra of $\mathfrak{g}_C$. Furthermore $(-1)^{\frac{1}{2}}X$ is contained in $\mathfrak{h}_R$. Let Δ be the root system of $\mathfrak{g}_C$ relative to $\mathfrak{h}_C$. Set $h_0 = (-1)^{\frac{1}{2}}X$. Let Δ^+ be the set of all the roots α so that $\alpha(h_0) > 0$. Since h_0 is regular $\Delta = \Delta^+ \cup (-\Delta^+)$. Furthermore if α and β are in Δ^+ and if $\alpha + \beta$ is in Δ then $\alpha + \beta$ is in Δ^+. Let for each α in Δ, $\sigma\alpha = \beta$ if $\mathrm{Ad}(g)\mathfrak{g}_\alpha = \mathfrak{g}_\beta$. σ is a permutation of Δ and $\sigma(\Delta^+) = \Delta^+$. Thus $\Delta^+ = \bigcup\{\beta\}$ the union taken over β in Δ^+, where $\{\beta\} = \{\beta, \sigma\beta, \ldots, \sigma^k\beta\}$ and $\sigma^{k+1}\beta = \beta$. Let for each β in Δ^+, $\mathfrak{g}^+_{\{\beta\}} = \sum \mathfrak{g}_{\sigma^s\beta}$ and $\mathfrak{g}^-_{\{\beta\}} = \sum \mathfrak{g}_{-\sigma^s\beta}$. Then $\mathrm{Ad}(g)\,(\mathfrak{g}^{\pm}_{\{\beta\}}) = \mathfrak{g}^{\pm}_{\{\beta\}}$. Let for each α in Δ, E_α be chosen so that E_α is in $\mathfrak{g}_\alpha$ and $B(E_\alpha, E_{-\alpha}) = 1$. Then $\mathrm{Ad}(g)E_{\sigma^s\beta} = c_sE_{\sigma^{s+1}\beta}$, $\mathrm{Ad}(g)E_{-\sigma^s\beta} = c_s^{-1}E_{-\sigma^{s+1}\beta}$ for $s \leqslant k$ and $\mathrm{Ad}(g)E_{\sigma^k\beta} = c_kE_\beta$, $\mathrm{Ad}(g)E_{-\sigma^k\beta} = c_k^{-1}E_{-\beta}$. Also $\mathrm{Ad}(\exp(tX))E_{\sigma^s\beta} = e^{(-1)^{\frac{1}{2}}t\beta(h_0)}E_{\sigma^s\beta}$. Thus

$$\det((\mathrm{Ad}(g\exp tX) - \lambda I)|_{\mathfrak{g}\{\beta\}}) = \lambda^{k+1} - c_0c_1 \cdots c_ke^{(-1)^{\frac{1}{2}}(k+1)t\beta(h_0)}$$

and

$$\det((\mathrm{Ad}(g\exp tX) - I)|_{\mathfrak{g}\{\beta\}}) = \lambda^{k+1} - c_0^{-1} \cdots c_k^{-1}e^{-(-1)^{\frac{1}{2}}(k+1)t\beta(h_0)}.$$

Thus t can be chosen so that $\mathrm{Ad}(g\exp tX)$ has no fixed vectors on any $\mathfrak{g}_{\{\beta\}}$. 3.9.3 now implies (1).

We return to the general case $\mathfrak{g} = \mathfrak{z} + \mathfrak{g}_0$. Let Z be the connected subgroup of G corresponding to $\mathfrak{z}$ and let G_0 be the connected subgroup of G corresponding to $\mathfrak{g}_0$. If g is in G then $g = zg_0$ with z in Z and g_0 in G_0. Since Z is contained in every maximal torus of G we need only show that g_0 is contained in a maximal torus of G_0. We may thus assume that $G = G_0$, $g = g_0$. By (1) there is a maximal torus T of G so that $gtg^{-1} = t$ for all t in T. Let $\mathfrak{h}$ be the Lie algebra of T and let $\mathfrak{g}_C$, $\mathfrak{h}_C$, Δ, Δ^+ be as above. We will show in 3.10.3 that there are elements $\alpha_1, \ldots, \alpha_m$ of Δ^+ so that they form a basis of $\mathfrak{h}_R^*$ and $E_{\alpha_1}, \ldots, E_{\alpha_m}, E_{-\alpha_1}, \ldots, E_{-\alpha_m}$ generate $\mathfrak{g}_C$. Now $\mathrm{Ad}(g)E_\alpha = c_\alpha E_\alpha$ for each α in Δ. Furthermore $|c_\alpha| = 1$ and $c_{\alpha+\beta} = c_\alpha c_\beta$ if α, β and $\alpha + \beta$ are in Δ. Let H be in H and such that $e^{\mathrm{ad}H}E_{\alpha_i} = c_{\alpha_i}E_{\alpha_i}$. Then $\mathrm{Ad}((\exp H)^{-1}g) = I$. Thus $(\exp H)^{-1}g$ is in the center of G. 3.8.3(2) implies that $(\exp H)^{-1}g$ is in T. Hence g is in T.

3.9.5 *Corollary* Let G be a compact, connected Lie group. Let T_1 be a torus contained in G and let g be an element of G so that $gtg^{-1} = t$ for all t in T_1. Then there exists a maximal torus in G containing T_1 and g.

PROOF Let H be the closure of the subgroup of G generated by T_1 and g. Then H is a closed Abelian subgroup of G. Let H_0 be the identity component of H. Since H is closed in G, H is compact. Hence H/H_0 is a finite Abelian group. If g is in H_0 then since $T_1 \subset H_0$ and H_0 is a torus, hence contained in a maximal torus, the result follows. We may thus assume that g is not in H_0. g^k is in H_0 for some positive integer k since H/H_0 is finite. Hence $g^k = \exp X$ for some X in the Lie algebra of H_0. Set $h = \exp((1/k)X)$, $h_1 = gh^{-1}$. Then $h_1^k = e$. Let t be an element of H_0 so that the group generated by t is dense in H_0 (see 3.6.9). Let Y be in the Lie algebra of H_0 so that $\exp Y = t$. Set $h_2 = h_1 \exp((1/k)Y)$. Let H_1 be the closure of the group generated by h_1. Then $H_1 \supset H$. On the other hand 3.9.4 implies that there is a maximal torus T of G containing h_2. T is clearly the desired maximal torus.

3.10 The Weyl Group

3.10.1 Let $\mathfrak{g}$ be a complex semisimple Lie algebra. Let $\mathfrak{h}$ be a Cartan subalgebra of $\mathfrak{g}$ and let Δ be the root system of $\mathfrak{g}$ relative to $\mathfrak{h}$. Let $\mathfrak{h}_R$ be as in 3.5.9(1) and let $\mathfrak{h}_R' = \{H|H \text{ in } \mathfrak{h}_R, \alpha(H) \neq 0 \text{ for all } \alpha \text{ in } \Delta\}$.

3.10.2 *Definition* A connected component P of $\mathfrak{h}_R$ is called a Weyl chamber of $\mathfrak{h}$. We note that P is convex and open and that if α is in Δ then α is either strictly positive or strictly negative on P. Let Δ_P^+ denote the set of all α in Δ that are strictly positive on P. An element α of Δ_P^+ is called simple if α cannot be written in the form $\gamma + \delta$ with γ and δ in Δ_P^+. Let π_P denote the set of all simple roots in Δ_P^+. π_P is called the simple system corresponding to P.

3.10.3 *Proposition* Let $\pi_P = \{\alpha_1, \ldots, \alpha_k\}$.

(1) If α is in Δ_P^+ then $\alpha = \sum n_i\alpha_i$ with n_i a nonnegative integer.

(2) Let for α in Δ, H_α be defined as in 3.5.7. Then $\alpha_i(H_{\alpha_j}) \leqslant 0$, for $i \neq j$.

(3) $\alpha_1, \ldots, \alpha_k$ is a basis for $\mathfrak{h}_R$.

PROOF If α is in Δ_P^+ and if α is in π_P then (1) is certainly true for α. If α is not in π_P then $\alpha = \gamma + \delta$ with γ, δ in Δ_P^+. If γ or δ is not in π_P then say γ splits into a sum of two elements of Δ_P^+. Continuing this process proves (1).

To prove (2) we note that if $i \neq j$ then $\alpha_i - \alpha_j$ is not in Δ. Indeed, if $\alpha_i - \alpha_j$ is in Δ then one of $\alpha_i - \alpha_j$ or $\alpha_j - \alpha_i$ is in Δ_P^+. If, say, $\alpha_i - \alpha_j$ is in

Δ^+ then $\alpha_i = (\alpha_i - \alpha_j) + \alpha_j$ contradicting the definition of simple root in Δ^+. 3.5.9(3) now implies (2).

By 3.5.9(1) we need only show that $\alpha_1, \ldots, \alpha_k$ are linearly independent to prove (3). Suppose that $H_{\alpha_k} = \sum a_j H_{\alpha_j}$, the sum taken from 1 to $k - 1$. By possibly relabeling the α_j we may assume that $a_1, \ldots, a_q$ are positive and that $a_{q+1}, \ldots, a_{k-1}$ are nonpositive. Set $H_1 = \sum a_j H_{\alpha_j}$, the sum taken from 1 to q. Set $H_2 = \sum a_j H_{\alpha_j}$, the sum taken from $q + 1$ to $k - 1$. If $H_1 = 0$ then α_k could not be in Δ^+. Hence $H_1 \neq 0$. Now $B(H_1, H_2) = \sum a_i a_j B(H_{\alpha_i}, H_{\alpha_j})$, the i sum from 1 to q and the j sum from $q + 1$ to $k - 1$. Thus (2) implies that $B(H_1, H_2) \geqslant 0$. Hence $B(H_{\alpha_k}, H_1) = B(H_1, H_1) + B(H_1, H_2)$ which by the above is positive. But $B(H_{\alpha_k}, H_1) = \sum a_j B(H_{\alpha_j}, H_{\alpha_k})$, the sum taken from 1 to q. Hence (2) implies that $B(H_{\alpha_k}, H_1)$ is negative. This contradiction proves (3).

3.10.4 If α is in Δ define a linear transformation s_α of $\mathfrak{h}$ to $\mathfrak{h}$ by $s_\alpha(H) = H - (2\alpha(H)/\alpha(H_\alpha))H_\alpha$. 3.5.9(3) implies that if γ is in Δ then $s_\alpha(H_\gamma) = H_\beta$ with β in Δ. Let $W(\Delta)$ be the group of linear transformations of $\mathfrak{h}$ generated by the s_α for α in Δ.

3.10.5 Let G be a compact and connected Lie group with Lie algebra $\mathfrak{g}_0$. Then $\mathfrak{g}_0 = \mathfrak{z} + [\mathfrak{g}_0, \mathfrak{g}_0]$ with $\mathfrak{g}_1 = [\mathfrak{g}_0, \mathfrak{g}_0]$ a compact Lie algebra. Let $\mathfrak{g}$ be the complexification of $\mathfrak{g}_1$. Then $\mathfrak{g}$ is a complex semisimple Lie algebra.

Let $\mathfrak{h}_0$ be a maximal Abelian subalgebra of $\mathfrak{g}_0$. Then $\mathfrak{z}$ is contained in $\mathfrak{h}_0$ and $\mathfrak{h}_0 = \mathfrak{z} + \mathfrak{h}_1$ with $\mathfrak{h}_1$ maximal Abelian in $\mathfrak{g}_1$. Let $\mathfrak{h}$ be the complexification of $\mathfrak{h}_1$ in $\mathfrak{g}$. Then $\mathfrak{h}$ is a Cartan subalgebra of $\mathfrak{g}$. Let Δ be the root system of $\mathfrak{g}$ relative to $\mathfrak{h}$.

Let T be the maximal torus of G corresponding to $\mathfrak{h}_0$. Then Δ is called the root system of G relative to T. Let $\mathfrak{g} = \mathfrak{h} + \sum \mathfrak{g}_\alpha$ be the root space decomposition of $\mathfrak{g}$ relative to $\mathfrak{h}$. If t is in T then $\mathrm{Ad}(t)\mathfrak{g}_\alpha = \mathfrak{g}_\alpha$ for each α in Δ. Since $\dim \mathfrak{g}_\alpha = 1$ for each α in Δ we see that each α in Δ defines a character of T, $t \mapsto t^\alpha$ defined by $\mathrm{Ad}(t)X = t^\alpha X$ for X in $\mathfrak{g}_\alpha$. Thus the root system of G relative to T may be looked upon as a collection of characters of T.

3.10.6 *Lemma* The center of G is exactly the set of all t in T so that $t^\alpha = 1$ for all α in Δ.

PROOF By 3.8.3(2) the center of G is contained in T. If t is in the center of G then $\mathrm{Ad}(t) = I$. Hence $t^\alpha = 1$ for all α in Δ. If $t^\alpha = 1$ for all α in Δ then $\mathrm{Ad}(t) = I$. Since G is connected, we see that t is central.

3.10.7 *Lemma* $\mathfrak{h}_1 = (-1)^{\frac{1}{2}}\mathfrak{h}_R$.

PROOF Since T is compact we see that $|t^\alpha| = 1$ for all t in T, α in Δ. Thus if H is in $\mathfrak{h}_1$, $\alpha(H)$ is pure imaginary for each α in Δ.

3.10.8 *Proposition* Let $N(T) = \{g \text{ in } G | gTg^{-1} \subset T\}$.

(1) If g is in G and if $gtg^{-1} = t$ for all t in T then g is in T.

(2) $N(T)/T = W(T)$ is a finite group called the Weyl group of G relative to T.

(3) If g is in $N(T)$ then $\mathrm{Ad}(g)\mathfrak{h} = \mathfrak{h}$, $\mathrm{Ad}(g)|_{\mathfrak{h}}$ depends only on gT and $\mathrm{Ad}(g)|_{\mathfrak{h}} = I$ if and only if g is in T. We may thus identify $W(T)$ with a group of linear transformations of $\mathfrak{h}$.

PROOF (1) follows from the argument of the last part of the proof of 3.9.4.

We now prove (2). Let (,) be an $\mathrm{Ad}(G)$-invariant inner product on $\mathfrak{g}_0$. Let $\mathfrak{n}$ be the Lie algebra of $N(T)$. Then $\mathfrak{n} = \mathfrak{h}_0 + \mathfrak{h}_0^\perp \cap \mathfrak{n}$ relative to (,). We note that if X is in $\mathfrak{n}$ and if Y is in $\mathfrak{h}_0$ then $[X, Y]$ is in $\mathfrak{h}_0$. Now if X is in $\mathfrak{h}_0^\perp \cap \mathfrak{n}$ and if Y is in $\mathfrak{h}_0$ then if Z is in $\mathfrak{h}_0$, $([X, Y], Z) = -(X, [Z, Y]) = 0$. Thus $[X, Y]$ is in $\mathfrak{h}_0^\perp \cap \mathfrak{n} \cap \mathfrak{h}_0 = (0)$. Hence $[X, Y] = 0$. But this implies that $[\mathfrak{n}, \mathfrak{h}_0] = (0)$. Since $\mathfrak{h}_0$ is maximal Abelian in $\mathfrak{g}_0$, this implies that $\mathfrak{n} = \mathfrak{h}_0$. Now $N(T)$ is compact, hence $N(T)/T$ is compact and discrete, hence finite.

The proof of (3) is an easy consequence of (1).

3.10.9 *Theorem* Let P be a fixed Weyl chamber of $\mathfrak{h}$.

(1) $W(\Delta)$ is generated by the s_α for α in π_P.

(2) $W(T)$ acts simply transitively on the Weyl chambers of $\mathfrak{h}$. (That is, $W(T)$ permutes the Weyl chambers and if s is in $W(T)$ and if $sQ = Q$ for some Weyl chamber Q, then $s = I$.)

(3) $W(T) = W(\Delta)$.

PROOF We first show that if α is in Δ then s_α is in $W(T)$. Let τ be the conjugation of $\mathfrak{g}$ relative to $\mathfrak{g}_1$. Let for each α in Δ, E_α be chosen in $\mathfrak{g}_\alpha$ so that $\tau E_\alpha = -E_{-\alpha}$ and $B(E_\alpha, E_{-\alpha}) = 1$. The proof of 3.7.5 says that this can be done. Let α in Δ be fixed. Let $U = 2^{-\frac{1}{2}}(E_\alpha - E_{-\alpha})$ and

$$V = -(-1)^{\frac{1}{2}}2^{-\frac{1}{2}}(E_\alpha + E_{-\alpha}).$$

If H is in $\mathfrak{h}_R$ then $(-1)^{\frac{1}{2}}H$ is in $\mathfrak{h}_1$, and

$$[(-1)^{\frac{1}{2}}H, U] = -\alpha(H)V, [(-1)^{\frac{1}{2}}H, V] = \alpha(H)U.$$

Also, $[U, V] = -(-1)^{\frac{1}{2}}H_\alpha$. Suppose that $\alpha(H) = 0$. We compute,

$$\mathrm{Ad}(\exp tU)H = e^{t\,\mathrm{ad}U}H = \sum t^n/n!(\mathrm{ad}\,U)^n H = H$$

since $\mathrm{ad}\,U(H) = 0$.

To proceed we need two observations.

(i) $$(\mathrm{ad}\,U)^{2n}H_\alpha = (-1)^n(\alpha(H_\alpha)^n)H_\alpha.$$

This observation can be proved directly by induction.

Applying $\mathrm{ad}(U)$ to both sides of (i) we see that

(ii) $$(\mathrm{ad}U)^{2n+1}H_\alpha = (-1)^{\frac{1}{2}}(-1)^{n+1}(\alpha(H_\alpha))^{n+1}V.$$

Combining these observations we see that

$$\begin{aligned}\mathrm{Ad}(\exp tU)H_\alpha &= \left(\sum (-1)^n \frac{\alpha(H_\alpha)^n}{(2n)!}t^{2n}\right)H_\alpha \\ &\quad -(-1)^{\frac{1}{2}}\left(\sum (-1)^n \frac{\alpha(H_\alpha)^{n+1}}{(2n+1)!}t^{2n+1}\right)V \\ &= \cos((\alpha(H_\alpha))^{\frac{1}{2}}t) - (-\alpha(H_\alpha))^{\frac{1}{2}}\sin((\alpha(H_\alpha))^{\frac{1}{2}}t)V.\end{aligned}$$

Hence $\mathrm{Ad}(\exp(\pi/(\alpha(H_\alpha))^{\frac{1}{2}})U)H_\alpha = -H_\alpha$. This implies that

$$\mathrm{Ad}(\exp(\pi/(\alpha(H_\alpha))^{\frac{1}{2}}U)|_{\mathfrak{h}} = s_\alpha.$$

Thus s_α is indeed in $W(T)$.

Suppose that s is in $W(T)$ and that $sQ = Q$ for some Weyl chamber Q. Let n be the order of s and let h be in Q. Set

$$H = (1/n)(h + sh + \cdots + s^{n-1}h).$$

Then $sH = H$. Furthermore since H is in Q, H is regular. Since s leaves fixed a regular element of $\mathfrak{h}_R$ the first part of the proof of 3.9.4 implies that $s = I$.

Let W' be the subgroup of $W(\Delta)$ generated by the s_α for α in π_P. To complete the proof of 3.10.9 we need only show that W' acts transitively on the Weyl chambers. Let Q be a Weyl chamber and let H, H_1 be, respectively, in P and Q. Suppose that the line segment $\{tH + (1-t)H_1 | 0 \leqslant t \leqslant 1\}$ intersects the hyperplane $\alpha(h) = 0$ for some α in π_P. Then

$$\|H - H_1\| > \|H - s_\alpha H_1\|$$

where $\|h\| = B(h, h)^{\frac{1}{2}}$ for h in $\mathfrak{h}_R$. Let s_0 in W' be so that $\|H - s_0H_1\|$ is minimal. Then the line segment joining H and s_0H_1 intersects no hyperplane $\alpha(h) = 0$ for α in π_P. But then $\alpha(s_0H_1) > 0$ for all α in π_P. 3.10.3(1) implies that $\alpha(s_0H_1) > 0$ for all α in H_P^+. Hence s_0H_1 is in P. Since the Weyl chambers are disjoint this implies that $s_0Q = P$. 3.10.9 is now completely proved.

3.10.10 *Lemma* Let α_i be in π_P. Then $s_{\alpha_i}\Delta_P^+ = \{\Delta_P^+ - \{\alpha_i\}\} \cup \{\alpha_i\}$.

PROOF Let α be in Δ_P^+, then $\alpha = \sum m_j\alpha_j$, the sum over the elements of π_P, and m_j is a nonnegative integer. We assert that if $\alpha \neq \alpha_i$ then $s_{\alpha_i}\alpha$ is in α_P^+. In fact, if $\alpha \neq \alpha_i$ then $m_j > 0$ for some $j \neq i$. Thus

$$s_{\alpha_i}\alpha = \sum_{i \neq j} m_j\alpha_j - m_i\alpha_i - \sum_{i \neq j} m_j \frac{2\alpha_j(H_{\alpha_i})}{\alpha_i(H_{\alpha_i})}\alpha_i.$$

Now, since $m_j > 0$ for some $j \neq i$ we see that $s_{\alpha_i}\alpha$ is in Δ_P^+. Since $s_{\alpha_i}\alpha_i = -\alpha_i$, the lemma is proved.

3.11 Exercises

3.11.1 Let A be an $n \times n$ complex matrix with nonzero determinant. Show that A can be written uniquely in the form $A = BC$ with B semisimple and $C - I$ nilpotent and $BC = CB$.

3.11.2 Show that the exponential mapping from $M_n(C)$ to $GL(n, C)$ is surjective.

3.11.3 Show that if $sl(2, R)$ is the Lie algebra of all 2×2 real matrices with trace zero and commutator $[X, Y] = XY - YX$ then $sl(2, R)$ is the Lie algebra of $SL(2, R)$, the group of all 2×2 matrices over R of determinant 1. Show that A is in $\exp(sl(2, R))$ if and only if $\det(A) = 1$ and $\mathrm{tr}A > -2$ or $A = -I$.

3.11.4 Let $\mathfrak{g}$ be a Lie subalgebra of $M_n(C)$. Let X be in $\mathfrak{g}$. If X is a nilpotent element of $M_n(C)$ then show that $\mathrm{ad}X$ is a nilpotent endomorphism of $\mathfrak{g}$.

3.11.5 If $\mathfrak{g}$ is a Lie algebra over a field K, define $D_1(\mathfrak{g}) = [\mathfrak{g}, \mathfrak{g}]$, and $D_{k+1}(\mathfrak{g}) = D_1(D_k(\mathfrak{g}))$. We say that $\mathfrak{g}$ is solvable if $D_k(\mathfrak{g}) = (0)$ for some k. Show that 3.3.2 is true for soluable Lie algebras over C.

3.11.6 Suppose that $K = R$ or C. Show that a Lie algebra is solvable if and only if $[\mathfrak{g}, \mathfrak{g}]$ is nilpotent.

3.11.7 Give an example of a Lie algebra over the complex numbers with (0) center but which is not semisimple.

3.11.8 Let $\mathfrak{g} = sl(n, C)$ be the Lie algebra of all $n \times n$ matrices over C of trace 0 and Lie bracket $[X, Y] = XY - YX$. Show that $\mathfrak{g}$ is semisimple. Show that the subspace of all diagonal elements of $\mathfrak{g}$ is a Cartan subalgebra of $\mathfrak{g}$. Find the roots of $\mathfrak{g}$ and the root space decomposition.

3.11.9 Let $su(n)$ be the space of all $n \times n$ skew Hermitian matrices of trace 0. Show that $su(n)$ is a compact real form of $sl(n, C)$. Let $A(X) = \bar{X}$ for X in $su(n)$. Extend A complex linear map of $sl(n, C)$ to $sl(n, C)$. Show that A so extended is an automorphism of $sl(n, C)$ and if $\mathfrak{h}$ is the subalgebra of all diagonal elements of $sl(n, C)$, show that $A|_{\mathfrak{h}} = -I$. If $\mathfrak{g}_u$ is the compact form of $sl(n, C)$ constructed as in 3.7.5 the $\mathfrak{g}_u = su(n)$.

3.11.10 What kind of Lie algebra over R has positive definite Killing form?

3.11.11 Suppose that $\mathfrak{g}$ is a Lie algebra over C and that $B_{\mathfrak{g}} = 0$. Show that $\mathfrak{g}$ is solvable (see 3.11.5).

3.11.12 Use the technique of the proof of 3.5.14 to show that if $\mathfrak{g}$ is a complex semisimple Lie algebra and if $\mathfrak{h}$ is a Cartan subalgebra of $\mathfrak{g}$ with root system Δ, if A is a linear isomorphism of $\mathfrak{h}$ to $\mathfrak{h}$ so that for each α in Δ there is a β so that $AH_\alpha = H_\beta$ and if A is an isometry relative to $B_{\mathfrak{g}}|_{\mathfrak{h}}$, then there is an automorphism of $\mathfrak{g}$ whose restriction to $\mathfrak{h}$ is A.

3.11.13 Show that if $\mathfrak{g}$ is a simple Lie algebra over C and if $\mathfrak{g}$ has a subalgebra of codimension 1 then $\mathfrak{g}$ is isomorphic with $sl(2, C)$.

3.11.14 Let G be a compact connected Lie group. Let T be a maximal torus of G. Show that $\dim H^2(G/T, R)$ is the number of simple roots of G relative to a Weyl chamber of T. (Hint: see the first part of the proof of 3.8.2.)

3.11.15 Use the technique of the proof of 3.8.2 to compute $H^p(S^n, R)$ where S^n is the n-sphere. (Hint: $S^n = SO(n + 1)/SO(n)$ with $SO(n) \subset SO(n + 1)$ by the map

$$k \mapsto \begin{bmatrix} k & 0 \\ 0 & 1 \end{bmatrix}.$$

The isotropy action is the standard action of $SO(n)$ on R^n. Show that there are no $SO(n)$ invariants in $\Lambda^k R^n$, $0 < k < n$.)

3.11.16 Let CP^n be as in 6.2.6 (see also 6.2.3 and 4 for pertinent notation). Compute $H^p(CP^n, R)$. (Hint: $CP^n = SU(n+1)/U(n)$ in the notation of 6.2. The isotropy representation of $U(n)$ is the usual action of $U(n)$ on C^n. Thus as a real representation it acts on R^{2n} under the map $U(n) \to SO(2n)$ given by

$$A + (-1)^{\frac{1}{2}} B \mapsto \begin{bmatrix} A & B \\ -B & A \end{bmatrix}.$$

Show that there are no $U(n)$ invariants in $\Lambda^{2p+1} R^{2n}$. Show that there is at least one nonzero $U(n)$ invariant in $\Lambda^{2p} R^{2n}$ if $0 \leqslant p \leqslant n$. Show that $\chi(CP^n) = n + 1$.)

3.11.17 Classify the Lie algebras over C of dimension $\leqslant 3$.

3.11.18 Let $\mathfrak{g}$ be a real or complex subalgebra of $sl(n, C)$. Suppose that if X is in $\mathfrak{g}$ then ${}^t\overline{X}$ is in $\mathfrak{g}$. Show that $\mathfrak{g} = \mathfrak{z} + [\mathfrak{g}, \mathfrak{g}]$ with $\mathfrak{z}$ the center of $\mathfrak{g}$ and $[\mathfrak{g}, \mathfrak{g}]$ semisimple. Use this result to show that the following Lie algebras are semisimple:

(1) $so(n, C) = \{X \text{ in } sl(n, C) | {}^tX = -X\}$.
(2) $sp(n, K) = \{X \text{ in } sl(2n, K) | XJ = -J{}^tX\}$ where $K = R$ or C and

$$J = \begin{bmatrix} 0 & I \\ -I & 0 \end{bmatrix}$$

with I the $n \times n$ identity matrix.

3.11.19 Let G be a finite subgroup of $GL(n, R)$. Let V be a convex subset of R^n such that $AV = V$ for all A in G. Show that there is a vector v in V so that $Av = v$ for all A in G.

3.11.20 Let G be either $SO(n)$ or $SU(n)$. Find a maximal torus of G and find the corresponding Weyl group. (Hint: the cases $SO(2n)$ and $SO(2n+1)$ will be different.)

3.11.21 Let $\mathfrak{g}$ be the Lie algebra of $SL(n, R)$. Let $\mathfrak{k} = \{X \text{ in } \mathfrak{g} | {}^tX = -X\}$, $\mathfrak{p} = \{X \text{ in } \mathfrak{g} | {}^tX = X\}$. Show that $\mathfrak{g} = \mathfrak{k} \oplus \mathfrak{p}$ is a Cartan decomposition of $\mathfrak{g}$.

3.11.22 Let J be as in 3.11.18. Let $Sp(n, R) = \{g \text{ in } SL(2n, R) | gJ^t g = J\}$. Show that $Sp(n, R) \cap O(2n)$ is Lie isomorphic with $U(n)$. (Hint: see the hint to 3.11.16.) Let $\mathfrak{g}$ be the Lie algebra of $Sp(n, R)$. Show that $\mathfrak{g} = sp(n, R)$ (see 3.11.18). Show that the Lie algebra of $Sp(n, R) \cap O(2n)$ is the "$\mathfrak{k}$" of a Cartan decomposition of $\mathfrak{g}$.

3.11.23 Let $\mathfrak{g}$ be a complex semisimple Lie algebra. Let $\mathfrak{g}_u$ be a compact real form of $\mathfrak{g}$. Let $\mathfrak{g}_R$ be the Lie algebra $\mathfrak{g}$ looked upon as a Lie algebra over R. Show that $\mathfrak{g}_R$ is semisimple and that $\mathfrak{g}_R = \mathfrak{g}_u \oplus (-1)^{\frac{1}{2}}\mathfrak{g}_u$ is a Cartan decomposition of $\mathfrak{g}_R$.

3.11.24 Find an example of a maximal Abelian subgroup of $SO(n)$ that is not a torus.

3.11.25 Let $\mathfrak{g}$ be a complex semisimple Lie algebra. Let $\mathfrak{h}$ be a nilpotent subalgebra of $\mathfrak{g}$ so that $\mathfrak{h}$ is its own normalizer in $\mathfrak{g}$. Show that $\mathfrak{h}$ is a Cartan subalgebra of $\mathfrak{g}$.

3.11.26 Let $\mathfrak{g}$ be a semisimple Lie algebra over R. Show that if $A: \mathfrak{g} \to \mathfrak{g}$ is an automorphism of $\mathfrak{g}$ so that $A^2 = I$ then there is a Cartan involution θ of $\mathfrak{g}$ such that $A\theta = \theta A$.

3.11.27 Let $\mathfrak{g}$ be a complex semisimple Lie algebra. Let X be an element of $\mathfrak{g}$ so that ad X is semisimple. Show that $\ker \operatorname{ad} X = \mathfrak{g}^X = \mathfrak{z} \oplus [\mathfrak{g}^X, \mathfrak{g}^X]$ with $\mathfrak{z}$ the center of $\mathfrak{g}^X$ and $[\mathfrak{g}^X, \mathfrak{g}^X]$ semisimple.

3.11.28 Use 3.11.27 and the technique of the proof of 3.9.1 to prove that if $\mathfrak{g}$ is a complex semisimple Lie algebra and if $A: \mathfrak{g} \to \mathfrak{g}$ is a semisimple automorphism of $\mathfrak{g}$ then A leaves fixed a nonzero element of $\mathfrak{g}$.

3.12 Notes

3.12.1 3.2.7 is due to Chevalley [1].

3.12.2 3.2.11 is due to Bourbaki [1].

3.12.3 The proofs of Section 3.5 are very strongly influenced by Helgason [1].

3.12.4 3.6.3 is due to Weil [1].

3.12.5 3.6.4 is taken from Hochschild [1] p. 37, Proposition 2.2. The proof of 3.6.6 using 3.6.4 follows Hochschild [1]. 3.6.6 is known as Weyl's theorem (see Weyl [1]).

3.12.6 The proof we use of 3.6.11(3) is due to Hunt [1]. 3.6.11 is due to Weyl [1].

3.12.7 The proof of 3.7.6 is essentially taken from Mostow [1]. The result is due to Cartan [1].

3.12.8 3.8.2 and 3.8.3 are due to Samelson [1].

3.12.9 3.9.1 is a theorem of Borel and Mostow [1] (see 3.11.28 for the full result).

3.12.10 The proof of 3.10.9 is essentially taken from Helgason [1].

CHAPTER 4

The Topology and Representation Theory of Compact Lie Groups

4.1 Introduction

In this chapter we classify and realize the irreducible unitary representations of compact Lie groups. We use this representation theory to show how the set of all compact Lie groups with the same Lie algebra can be found.

In Section 4.2 we develop the notion of the universal enveloping algebra of a Lie algebra. We use this formalism to prove the theorem of the highest weight. This theorem gives a complete classification of the finite dimensional irreducible representations of a complex semisimple Lie algebra. We then use the theorem of the highest weight to develop in Section 4.6 the Cartan–Stiefel theory of the topology of compact Lie groups.

In Section 4.7 we show how the compact theory can be used to find the holomorphic finite dimensional representations of a complex reductive Lie group.

In Sections 4.8 and 4.9 we prove the Weyl integral, character, and dimension formulas for compact Lie groups. In Section 4.10 we use the Weyl theory to give the basic structure of the ring of virtual representations of a compact Lie group.

4.2 The Universal Enveloping Algebra

4.2.1 Let $\mathfrak{g}$ be a Lie algebra over a field K. Let $T(\mathfrak{g})$ be the tensor algebra on $\mathfrak{g}$ (see A.3 for the pertinent definitions). Let $I(\mathfrak{g})$ be the two-sided ideal in $T(\mathfrak{g})$ generated by the elements of the form $X \otimes Y - Y \otimes X - [X, Y]$ for X, Y in $\mathfrak{g}$. We have identified $\mathfrak{g}$ with its image in $T(\mathfrak{g})$. Set $U(\mathfrak{g}) = T(\mathfrak{g})/I(\mathfrak{g})$. Then $U(\mathfrak{g})$ is an associative algebra over K with unit. Let ξ be the natural homomorphism of $T(\mathfrak{g})$ onto $U(\mathfrak{g})$. Let

$$j = \xi|_{\mathfrak{g}}, \text{ then } j([X, Y]) = j(X)j(Y) - j(Y)j(X).$$

The pair $(U(\mathfrak{g}), j)$ (or with j understood $U(\mathfrak{g})$) is called the universal enveloping algebra of $\mathfrak{g}$.

4.2.2 *Lemma* Let A be an associative algebra over K with unit. Let ρ be a Lie algebra homomorphism of $\mathfrak{g}$ into A (that is,

$$\rho([X, Y]) = \rho(X)\rho(Y) - \rho(Y)\rho(X)).$$

Then there is a unique unit preserving associative algebra homomorphism $\hat{\rho}$ of $U(\mathfrak{g})$ into A so that $\hat{\rho} \circ j = \rho$.

PROOF ρ is in particular a linear map of $\mathfrak{g}$ into A. Hence by the universal mapping property of $T(\mathfrak{g})$ (see A.3) ρ extends to an algebra homomorphism $\tilde{\rho}$ of $T(\mathfrak{g})$ into A preserving units. Now $\tilde{\rho}(X \otimes Y) = \tilde{\rho}(X)\tilde{\rho}(Y)$ for X, Y in $\mathfrak{g}$. Thus $\ker \tilde{\rho} \supset I(\mathfrak{g})$. This implies that $\tilde{\rho}$ induces an algebra homomorphism $\hat{\rho}$ of $U(\mathfrak{g})$ to A which is clearly unit preserving and $\hat{\rho} \circ j = \rho$. The uniqueness of $\hat{\rho}$ follows from the fact that $\mathfrak{g}$ generates $T(\mathfrak{g})$, hence $j(\mathfrak{g})$ generates $U(\mathfrak{g})$.

4.2.3 Let $X_1, \ldots, X_n$ be a basis for $\mathfrak{g}$. If $M = (m_1, \ldots, m_n)$ is an n-tuple of nonnegative integers, set $X^M = X_1^{m_1} \cdots X_n^{m_n}$ in $U(\mathfrak{g})$ ($X^{(0,\ldots,0)} = 1$). Also set $|M| = m_1 + \cdots + m_n$.

4.2.4 *Lemma* The set of all X^M spans $U(\mathfrak{g})$.

PROOF Let $\otimes^k\mathfrak{g}$ denote the k-fold tensor product of $\mathfrak{g}$ with itself. Set $T^i = \sum_{k \leq i} \otimes^k \mathfrak{g}$. Set $U^i = \xi(T^i)$. We prove that the X^M with $|M| \leqslant i$ span U^i.

Clearly $X^{(0,\ldots,0)}$ spans U^0. Suppose that we have proved the result for i. Let u be in $\otimes^{i+1}\mathfrak{g}$. Then u is a linear combination of elements of the form $X_{p_1} \otimes \cdots \otimes X_{p_{i+1}}$ with $1 \leqslant p_j \leqslant n$.

We note that

$$X_{p_1} \cdots X_{p_k} X_{p_{k+1}} \cdots X_{p_{i+1}} - X_{p_1} \cdots X_{p_{k+1}} X_{p_k} \cdots X_{p_{i+1}} = X_{p_1} \cdots [X_{p_k}, X_{p_{k+1}}] \cdots X_{p_{i+1}}$$

which is in U^i. Hence $X_{p_1} \cdots X_{p_{i+1}}$ is congruent mod U^i to one of the X^M with $|M| = i + 1$. This proves the lemma.

4.2.5 Let G be a real or complex Lie group (see A.5 for the pertinent information on complex Lie groups). Let $\mathfrak{g}$ be the Lie algebra of G. $\mathfrak{g}$ will be looked upon as a Lie algebra over R if G is a real Lie group; it will be looked upon as a Lie algebra over C if G is a complex Lie group. Let $C^\infty(G; K)$ denote the K-valued C^∞ functions on G with $K = R$ if G is a real Lie group, $K = C$ if G is a complex Lie group.

4.2.6 *Definition* A left invariant differential operator on G is a linear map $C^\infty(G; K)$ to $C^\infty(G; K)$, D, so that

(1) If g is in G then $D \circ L_g = L_g \circ D(L_g f(x) = f(g^{-1}x)$ for f in $C^\infty(G; K))$.

(2) If g is in G then there is a neighborhood U of G and a system of coordinates $\{x_1, \ldots, x_n\}$ (holomorphic if $K = C$) so that

$$Df|_U = \sum a_{i_1 \cdots i_n} \frac{\partial^{i_1 + \ldots + i_n}}{\partial x_1^{i_1} \cdots \partial x_n^{i_n}} f,$$

for f in $C^\infty(G; K)$, where $a_{i_1 \cdots i_n}$ is in $C^\infty(U; K)$ and depends only on D and U.

Let $D(G)$ denote the space of all left invariant differential operators on G. We note that $D(G)$ is an associative algebra with unit (the multiplication is composition of linear operators).

4.2.7 *Theorem* $U(\mathfrak{g})$ is naturally isomorphic with $D(G)$. Furthermore if $X_1, \ldots, X_n$ is a basis of $\mathfrak{g}$ over K then the monomials $X_1^{m_1} \cdots X_n^{m_n}$ form a basis of $U(\mathfrak{g})$ over K.

PROOF $D(G)$ is an associative algebra with unit. Clearly $\mathfrak{g}$ is contained in $D(G)$. Thus the injection of $\mathfrak{g}$ into $D(G)$ extends to an associative algebra homomorphism ρ of $U(\mathfrak{g})$ into $D(G)$. We show that ρ is bijective.

Let D be in $D(G)$. Define $D_e f = Df(e)$ for f in $C^\infty(G; K)$. 4.2.6(1) implies that

(1) D_e completely determines D.
Let $X_1, \ldots, X_n$ be a basis of $\mathfrak{g}$ over K and let

$$\psi(t_1, \ldots, t_n) = \exp t_1X_1 \cdots \exp t_nX_n.$$

Then $\psi_{*e}(X) = X_e$ for X in $\mathfrak{g}$. The inverse function theorem implies that there is a neighborhood U_0 of 0 in $\mathfrak{g}$ and a neighborhood U of e in G so that ψ is a diffeomorphism of U_0 onto U which is holomorphic if $K = C$.

(2) Let P be a polynomial in n indeterminates over K. Then there is a unique polynomial Q in n indeterminates over K so that $\deg Q = \deg P$ and $(P(\partial/\partial t_1, \ldots, \partial/\partial t_n)f)(e) = (Q(X_1, \ldots, X_n)f)(e)$ for all f in $C(G; K)$.

To see this we note that

$$\frac{\partial f}{\partial t_{j\psi(t_1 \cdots, t_n)}} = \frac{d}{dt} f(\exp t_1X_1 \cdots \exp t_jX_j \exp tX_j \exp t_{j+1}X_{j+1} \cdots \exp t_nX_n)\Big|_{t=0}$$

$$\frac{d}{dt} f(u_j(t_1, \ldots, t_n) \exp tX_j v_j(t_1, \ldots, t_n))\Big|_{t=0}$$

with

$$u_j(t_1, \ldots, t_n) = \exp t_1X_1 \cdots \exp t_jX_j \text{ and } v_j(t_1, \ldots, t_n) = \exp t_{j+1}X_{j+1} \cdots \exp t_nX_n.$$

Thus

$$\frac{\partial}{\partial t_{j_g}} f = \frac{d}{dt} f(g \exp t \operatorname{Ad}(v_j)^{-1}X_j)\Big|_{t=0} = (\operatorname{Ad}(v_j)^{-1}X_j)_g f.$$

Assertion (2) now follows by induction of the degree of P.

Clearly (2) implies that ρ is an isomorphism of $U(\mathfrak{g})$ onto $D(G)$. To prove the last statement, we note that

$$(X_1^{m_1} \cdots X_n^{m_n})f(e) = \frac{\partial^{m_1 + \cdots + m_n}}{\partial_{t_1}^{m_1} \cdots \partial_{t_n}^{m_n}} f(e) + \left(Q\left(\frac{\partial}{\partial t_1}, \ldots, \frac{\partial}{\partial t_n}\right) f\right)(e)$$

with Q a polynomial in n indeterminates over K of degree less than $m_1 + \cdots + m_n$.

4.2.8 In Hochschild [1] page 133, it is proved that if $\mathfrak{g}$ is a Lie algebra over R or C then there is a Lie group G with Lie algebra isomorphic with $\mathfrak{g}$. This implies the following theorem.

4.2.9 *Theorem* In the notation of 4.2.4 the X^M form a basis of $U(\mathfrak{g})$.

4.2.10 We will be applying 4.2.9 only to semisimple Lie algebras. If $\mathfrak{g}$ is semisimple then ad is a Lie algebra isomorphism of $\mathfrak{g}$ into $L(\mathfrak{g}, \mathfrak{g})$ (recall $\operatorname{ad}X(Y) = [X, Y]$). If K is either R or C and $GL(\mathfrak{g})$ is the group of all invertible elements of $L(\mathfrak{g}, \mathfrak{g})$, then the connected subgroup of $GL(\mathfrak{g})$ with Lie algebra $\operatorname{ad}(\mathfrak{g})$ has Lie algebra isomorphic with $\mathfrak{g}$. Thus we have completely proved 4.2.9 if $\mathfrak{g}$ is semisimple.

4.3 Representations of Lie Algebras

4.3.1 Let $\mathfrak{g}$ be a Lie algebra over a field K. Let V be a vector space over K (possibly infinite dimensional). A linear map ρ of $\mathfrak{g}$ into $L(V, V)$ so that $\rho[x, y] = \rho(x)\rho(y) - \rho(y)\rho(x)$ for x, y in $\mathfrak{g}$ is called a representation of $\mathfrak{g}$ on V. It is denoted (ρ, V). 4.2.2 implies that ρ extends to an algebra homomorphism of $U(\mathfrak{g})$ to $L(V, V)$ preserving units; we also denote this extension by ρ.

4.3.2 A representation (ρ, V) of $\mathfrak{g}$ is said to be irreducible if the only invariant subspaces of V are (0) and V. This says that V is irreducible if and only if for each v in V, $v \neq 0$, $\rho(U(\mathfrak{g}))v = V$.

4.3.3 Two representations (ρ, V) and (ρ, W) of $\mathfrak{g}$ are said to be equivalent if there is a linear bijection A of V to W so that $A\rho(X) = \rho(X)A$ for all X in $\mathfrak{g}$.

4.3.4 Let $\mathfrak{g}$ be a complex semisimple Lie algebra. Let $\mathfrak{h}$ be a Cartan subalgebra of $\mathfrak{g}$ and let Δ be the root system of $\mathfrak{g}$ relative to $\mathfrak{h}$. Let P be a Weyl chamber of $\mathfrak{h}$ and let Δ^+ and π be, respectively, the corresponding positive roots and simple roots. Then $\mathfrak{g} = \mathfrak{n}^- + \mathfrak{h} + \mathfrak{n}^+$ with $\mathfrak{n}^+ = \sum_{\alpha\in\Delta^+}\mathfrak{g}_\alpha$, $\mathfrak{n}^- = \sum_{\alpha\in\Delta^+}\mathfrak{g}_{-\alpha}$. From the definitions we see that $[\mathfrak{n}^+, \mathfrak{n}^+] \subset \mathfrak{n}^+$, $[\mathfrak{n}^-, \mathfrak{n}^-] \subset \mathfrak{n}^-$, and $[\mathfrak{h}, \mathfrak{n}^\pm] \subset \mathfrak{n}^\pm$.

4.3.5 Let (ρ, V) be a representation of $\mathfrak{g}$. If λ is in $\mathfrak{h}^*$ set $V_\lambda = \{v$ in V $\rho(H)v = \lambda(H)v$ for all H in $\mathfrak{h}\}$. λ is called a weight of (ρ, V) if $V_\lambda \neq (0)$.

4.3.6 *Lemma* $\sum V_\lambda$ is direct. Furthermore, if $W \subset V$ is an invariant subspace then $\sum W \cap V_\lambda = W \cap (\sum V_\lambda)$.

PROOF Let $\lambda_1, \ldots, \lambda_t$ be distinct weights of V and suppose that v_i is in V_{λ_i} and $v_1 + \cdots + v_t = 0$. We show by induction on t that $v_i = 0$ for $i = 1, \ldots, t$. If $t = 1$ the result is clear. Suppose true for $1 \leqslant t < m$. We check the case $t = m$. Let H be in $\mathfrak{h}$ so that $\lambda_1(H) \neq \lambda_2(H)$. Applying the inductive hypothesis to $0 = \rho(H) \sum v_i - \lambda_1(H) \sum v_i = \sum_{i=1} (\lambda_i(H) - \lambda_1(H)) v_i$ we find that $v_2 = 0$. Hence applying the inductive hypothesis again we find that $v_i = 0$ for $i = 1, \ldots, m$.

Now suppose that W is an invariant subspace of V. Let v_λ be in V_λ and suppose that $\sum v_\lambda \equiv 0 \bmod W$. Then since $v_\lambda + W$ is in $(V/W)_\lambda$ we see that $v_\lambda \equiv 0 \bmod W$ by the first part of the proof applied to V/W. Hence v_λ is in W. This completes the proof of the lemma.

4.3.7 *Lemma* If (ρ, V) is a nonzero finite dimensional representation of $\mathfrak{g}$ then there is a nonzero element v_0 in V and Λ in $\mathfrak{h}^*$ so that

(1) $\rho(H)v_0 = \Lambda(H)v_0$ for all H in $\mathfrak{h}^-$,
(2) $\rho(X)v_0 = 0$ for all X in $\mathfrak{n}^+$.

PROOF $\mathfrak{h}$ is Abelian, hence nilpotent. 3.3.2 implies that V has a weight λ. Partially order the weights of V so that $\mu > \gamma$ if $\mu - \gamma$ is a sum of positive roots. Let Λ be a maximal weight relative to this partial order. Let v_0 be in V_Λ and $v_0 \neq 0$. If X is in $\mathfrak{g}_\alpha$ with α in Δ^+ then

$$\rho(H)\rho(X)v_0 = \rho([H, X])v_0 + \rho(X)\rho(H)v_0 = (\alpha(H) + \Lambda(H))\rho(X)v_0.$$

This says that $\rho(X)V_\Lambda \subset V_{\Lambda+\alpha}$. But $\Lambda + \alpha > \Lambda$. Hence $\Lambda + \alpha$ is not a weight. This implies that v_0 satisfies (2).

4.3.8 *Definition* A three-dimensional simple Lie algebra (TDS) is a Lie algebra over C with basis h, e, f and commutation relations

$$[h, e] = 2e, \qquad [h, f] = -2f, \qquad [e, f] = h.$$

4.3.9 *Lemma* A TDS is a simple Lie algebra over C and is isomorphic with the Lie algebra of all 2×2 trace zero matrices over C.

PROOF Let $sl(2, C)$ denote the space of all 2×2 trace zero matrices over C. Let

$$h = \begin{bmatrix} 1 & 0 \\ 0 & -1 \end{bmatrix}, \qquad e = \begin{bmatrix} 0 & 1 \\ 0 & 0 \end{bmatrix}, \qquad f = \begin{bmatrix} 0 & 0 \\ 1 & 0 \end{bmatrix}.$$

A computation shows that h, e, f satisfy the commutation relations of a TDS.

Suppose that $\mathfrak{U}$ is an ideal of $sl(2,C)$. Since ad h is semisimple ad h diagonalizes on $\mathfrak{U}$. If $\mathfrak{U}$ is one-dimensional then $\mathfrak{U} = Ch$, Ce, or Cf. If $\mathfrak{U} = Ce$ then $[f, e] = h$ is in $\mathfrak{U}$, similarly if $\mathfrak{U} = Cf$ then h is in $\mathfrak{U}$. But if h is in $\mathfrak{U}$ then $[h, e] = 2e$ and $[h, f] = -2f$ are in $\mathfrak{U}$. Thus $\mathfrak{U}$ is at least two-dimensional. But then by the above argument h is in $\mathfrak{U}$. But then $\mathfrak{U} = sl(2, C)$.

4.3.10 *Proposition* Let $\mathfrak{g}$ be a TDS. Then up to equivalence every finite dimensional irreducible representation of $\mathfrak{g}$ is determined by its dimension. Furthermore, for each nonnegative integer n there is an $n + 1$ dimensional representation (π_n, V^n) with basis $v_0, v_1, \ldots, v_n$ so that

(1) $\pi_n(h)v_i = (n - 2i)v_i$,
(2) $\pi_n(e)v_i = -iv_{i-1}$ and $\pi_n(e)v_0 = 0$.
(3) $\pi_n(f)v_i = -(n - i)v_{i+1}$ and $\pi_n(f)v_n = 0$.

PROOF We first prove the existence of (π_n, V^n). Let us identify $\mathfrak{g}$ with $sl(2, C)$. Let $sl(2, C)$ act on C^2 by the natural matrix action. Let z, w be the usual coordinates on C^2. Let V^n be the space of all complex polynomials in z, w, homogeneous of degree n. If f is in V^n and if X is in $sl(2, C)$ define $(\pi_n(X)f)(v) = (d/dt)f(e^{-tX}v)|_{t=0}$. Then (π_n, V^n) defines a representation of $sl(2, C)$. If we take $v_j = z^j w^{n-j}$ then a straightforward computation shows that (1), (2), and (3) are satisfied. Suppose that $(0) \neq W \subset V^n$ and that W is invariant. Since $\pi_n(h)$ diagonalizes on V^n with distinct eigenvalues we see that v_j must be in W for some j. But then, repeated applications of $\pi_n(e)$ to v_j will according to (2) give a nonzero multiple of v_0. Now (3) implies that $W = V^n$. Hence (π_n, V^n) is indeed irreducible.

Let (ρ, V) be a finite dimensional irreducible representation of $\mathfrak{g}$. Let $\mathfrak{n}^+ = Ce$. Set $\mathfrak{h} = Ch$. Clearly $\mathfrak{h}$ is a Cartan subalgebra of $\mathfrak{g}$. Furthermore $\mathfrak{n}^+ = \mathfrak{g}_\alpha$ where α may be taken to be the positive root relative to a Weyl chamber. Let Λ, v_0 be as in 4.3.5. Let $\lambda = \Lambda(h)$. Now $\rho(\mathfrak{h} + \mathfrak{n}^+)v_0 \subset Cv_0$. Now, $\mathfrak{g}$ has basis e, f, h and thus 4.2.9 implies that $U(\mathfrak{g})$ has basis $f^k h^m e^p$. Hence $V = \rho(U(\mathfrak{g}))v_0 = \sum C\rho(f^k)v_0$.

Now

$$\begin{aligned}\rho(h)\rho(f^k)v_0 &= \rho(h)\rho(f)\rho(f^{k-1})v_0\\ &= -2\rho(f^k)v_0 + \rho(f)\rho(h)\rho(f^{k-1})v_0.\end{aligned}$$

An induction therefore shows that $\rho(h)\rho(f^k)v_0 = (\lambda - 2k)\rho(f^k)v_0$. Since $\rho(h)$ can have only a finite number of eigenvalues we see that $\varphi(f^{n+1})v_0 = 0$ and $\rho(f^n)v_0 \neq 0$ for some n. Now $\rho(h) = \rho(e)\rho(f) - \rho(f)\rho(e)$. Hence tr $\rho(h) = 0$. But then $(n + 1)\lambda - 2\sum_{k=0}^n k = 0$. This implies that $\lambda = n$.

Set $v_j = (-1)^j(n - j)!\rho(f^j)v_0$ for $j = 0, 1, \ldots, n$. Then $v_0, v_1, \ldots, v_n$ forms a basis of V, and relative to this basis the matrices of $\rho(h)$, $\rho(e)$, $\rho(f)$ are the same as the matrices of $\pi_n(h)$, $\pi_n(e)$, $\pi_n(f)$ as given in (1), (2), (3). Hence (ρ, V) is equivalent with (π_n, V^n).

4.4 *P*-Extreme Representations

4.4.1 In this section we take $\mathfrak{g}$ to be a complex semisimple Lie algebra, $\mathfrak{h}$ to be a Cartan subalgebra of $\mathfrak{g}$, and Δ to be the root system of $\mathfrak{g}$ relative to $\mathfrak{h}$. Let P be a Weyl chamber of $\mathfrak{h}$ and let Δ^+ and π be, respectively, the positive and simple roots corresponding to P. Let $\mathfrak{n}^+$ and $\mathfrak{n}^-$ be as in 4.3.4.

4.4.2 *Definition* A representation (ρ, V) of $\mathfrak{g}$ is said to be P-extreme with highest weight Λ in $\mathfrak{h}^*$ if

(1) there is v_0 in V with $v_0 \neq 0$ so that $\rho(X)v_0 = 0$ for all X in $\mathfrak{n}^+$,
(2) $\rho(H)v_0 = \Lambda(H)v_0$ for all H in $\mathfrak{h}$,
(3) $\rho(U(\mathfrak{g}))v_0 = V$.

4.4.3 *Lemma* Let (ρ, V) be a P-extreme representation with highest weight Λ. Then $V = \sum V_\mu$ the sum over the weights of V. $\dim V_\mu < \infty$ for each μ in $\mathfrak{h}^*$, $\dim V_\Lambda = 1$. If $V_\mu \neq (0)$ then $\mu = \Lambda - \sum n_i\alpha_i$ with n_i a nonnegative integer and $\pi = \{\alpha_1, \ldots, \alpha_l\}$.

PROOF Let $\Delta^+ = \{\beta_1, \ldots, \beta_r\}$. Let e_i be a nonzero element of $\mathfrak{g}_{\beta_i}$ and let f_i be a nonzero element of $\mathfrak{g}_{-\beta_i}$ for $i = 1, \ldots, r$. Let $h_1, \ldots, h_l$ be a basis of $\mathfrak{h}$. Then 4.2.9 implies that the elements

$$f_1^{m_1} \cdots f_r^{m_r} h_1^{k_1} \cdots h_l^{k_l} e_1^{n_1} \cdots e_r^{n_r}$$

span $U(\mathfrak{g})$. Now, (3) of 4.4.2 says that $V = \rho(U(\mathfrak{g}))v_0$. Hence (1) and (2) of 4.4.2 imply that V is spanned by elements of the form

$$\rho(f_1^{m_1} \cdots f_r^{m_r})v_0 = \rho(f^M)v_0$$

with $m_1, \ldots, m_r$ nonnegative integers. If H is in $\mathfrak{h}$ then

$$\rho(H)\rho(f^M) = \sum_i (f_1^{m_1} \cdots (Hf_i^{m_i} - f_r^{m_r}H) \cdots f_r^{m_r})v_0 + \Lambda(H)\rho(f^M)v_0.$$

The obvious induction shows that $Hf_i^m - f_i^m H = -m\beta_i(H)f_i^m$. Hence

$$\rho(H)\rho(f^M)v_0 = (\Lambda(H) - \sum m_i\beta_i(H))\rho(f^M)v_0.$$

This proves the first and the third assertions of the lemma.

If μ is in $\mathfrak{h}^*$ let $P(\mu)$ denote the number of ways that can be written as a sum of positive roots. By the above (with the understanding that $P(0) = 1$) we see that dim $V_\mu \leqslant P(\Lambda - \mu)$. The lemma is now completely proved.

4.4.4 Let $\mathfrak{b} = \mathfrak{h} + \mathfrak{n}^-$. Let $U(\mathfrak{b})$ be the universal enveloping algebra of $\mathfrak{b}$ injected in $U(\mathfrak{g})$ in the canonical fashion (the extension of the injection of $\mathfrak{b}$ into $\mathfrak{g}$). If Λ is in $\mathfrak{h}^*$ then set $\Lambda(H + X) = \Lambda(H)$ for H in $\mathfrak{h}$ and X in $\mathfrak{n}^-$. Then Λ is a homomorphism of $\mathfrak{b}$ into C and hence extends to a homomorphism Λ of $U(\mathfrak{b})$ into C. We define $T(\Lambda)$ to be the space of all f in $U(\mathfrak{g})^*$ so that $f(bg) = \Lambda(b)f(g)$ for b in $U(\mathfrak{b})$ and g in $U(\mathfrak{g})$. If x is in $\mathfrak{g}$, f is in $U(\mathfrak{g})^*$, and g is in $U(\mathfrak{g})$ then define $(\pi(x)f)(g) = f(gx)$. Then $(\pi, T(\Lambda))$ is a representation of $\mathfrak{g}$.

4.4.5 *Theorem* There is an element ξ_Λ of $T(\Lambda)$ so that $\xi_\Lambda(1) = 1$ and $\pi(\mathfrak{n}^+)\xi_\Lambda = 0$.

(1) If ξ is in $T(\Lambda)$ and if $\pi(\mathfrak{n}^+)\xi = 0$ then $\xi = \xi(1)\xi_\Lambda$.

(2) $\pi(U(\mathfrak{g}))\xi_\Lambda = V^\Lambda$ is an invariant, irreducible subspace of $T(\Lambda)$. Let π_Λ be the induced action of $\mathfrak{g}$ on V^Λ. Then (π_Λ, V^Λ) is P-extreme with highest weight Λ (aking ξ_Λ for v_0 in 4.4.2).

(3) If (ρ, V) is P-extreme with highest weight Λ then there is a linear map $A: V \to V^\Lambda$ so that $Av_0 = \xi_\Lambda$ and $A\rho(X) = \pi_\Lambda(X)A$ for all X in $\mathfrak{g}$. In particular (π_Λ, V^Λ) is the unique (up to equivalence) P-extreme irreducible representation of $\mathfrak{g}$ with highest weight Λ.

PROOF We retain the notation of the proof of 4.4.3.

Let ξ be in $T(\Lambda)$ and suppose that $\pi(\mathfrak{n}^+)\xi = 0$. Then by definition of $T(\Lambda)$, $\xi(f^M h^Q e^N) = 0$ if $|M| \neq 0$. By assumption on ξ, $\xi(H^Q X^N) = 0$ if $|N| \neq 0$. Furthermore $\xi(H^Q) = \Lambda(h_1)^{q_1} \cdots \Lambda(h_l)^{q_l}\xi(1)$.

Let ξ_Λ be defined by

$$\xi_\Lambda(f^M h^Q e^N) = \begin{cases} 0 \text{ if } |M| + |N| \neq 0 \\ \Lambda(h_1)^{q_1} \cdots \Lambda(h_l)^{q_l} \text{ if } |M| + |N| = 0. \end{cases}$$

Then ξ_Λ is in $T(\Lambda)$, $\pi(\mathfrak{n}^+)\xi_\Lambda = 0$, $\xi_\Lambda(1) = 1$, and clearly (1) is satisfied.

To prove (2), we first show that (π_Λ, V^Λ) satisfies conditions (1), (2), and (3) of 4.4.2. Taking $v_0 = \xi_\Lambda$, it is clear that (1) and (3) are satisfied. Let H be in $\mathfrak{h}$. If X is in $\mathfrak{n}^+$ then $\pi_\Lambda(X)\pi_\Lambda(H)\xi_\Lambda = \pi_\Lambda([X, H])\xi_\Lambda + \pi_\Lambda(H)\pi_\Lambda(X)\xi_\Lambda$. Now $\pi_\Lambda(X)\xi_\Lambda = 0$ and $[X, H]$ is in $\mathfrak{n}^+$, hence $\pi_\Lambda([X, H])\xi_\Lambda = 0$. Thus

$$\pi_\Lambda(X)(\pi_\Lambda(H)\xi_\Lambda) = 0$$

for all X in $\mathfrak{n}^+$. (1) now implies that

$$\pi_\Lambda(H)\xi_\Lambda = (\pi_\Lambda(H)\xi_\Lambda)(1)\xi_\Lambda.$$

But $(\pi_\Lambda(H)\xi_\Lambda)(1) = \xi_\Lambda(H) = \Lambda(H)$. Hence (π_Λ, V^Λ) is indeed P-extreme with highest weight Λ.

Suppose that $W \neq (0)$ and that W is an invariant subspace of V^Λ. Then 4.3.6 and 4.4.3 imply that $W = \sum W_\mu$, the sum taken over the weights of W. If $W_\mu \neq (0)$ then 4.4.3 implies that $\Lambda - \mu = \sum n_i\alpha_i$ with n_i a nonnegative integer. Let λ_0 be so that $W_{\lambda_0} \neq (0)$ and $\Lambda - \lambda_0 = \sum n_i\alpha_i$ with $\sum n_i$ minimal. Let ξ be in W_{λ_0} with $\xi \neq 0$. If α is in Δ^+ and if X is in $\mathfrak{g}$ then

$$\begin{aligned}\pi_\Lambda(H)\pi_\Lambda(X)\xi &= \pi_\Lambda([H, X])\xi + \pi_\Lambda(X)\pi_\Lambda(H)\xi\\ &= \alpha(H)\pi_\Lambda(X)\xi + \lambda_0(H)\pi_\Lambda(X)\xi.\end{aligned}$$

Hence $\pi_\Lambda(X)\xi$ is in $W_{\lambda_0+\alpha}$. But $\Lambda - \lambda_0 - \alpha = \sum m_i\alpha_i$ with $\sum m_i < \sum n_i$. Hence $W_{\lambda_0+\alpha} = (0)$. This implies that $\pi_\Lambda(X)\xi = 0$ for all X in $\mathfrak{n}^+$. (1) now tells us that $\xi = \xi(1)\xi_\Lambda$. Hence $\lambda_0 = \Lambda$. The definition of V^Λ implies that $V^\Lambda = W$. (2) is now completely proved.

Let (ρ, V) be a P-extreme representation of $\mathfrak{g}$ with highest weight Λ. Then 4.4.3 implies that $V = \sum V_\mu$, a direct sum over the weights of V. Let v_0 be in V so that $V_\Lambda = Cv_0$. Let δ be the element of V^* defined by $\delta|_{V_\mu} = 0$ if $\mu \neq \Lambda$ and $\delta(v_0) = 1$. If v is in V set $A(v)(g) = \delta(\rho(g)v)$. Then A is a linear map of V into $U(\mathfrak{g})^*$. We first note that if v is in V then $A(v)$ is in $T(\Lambda)$. Indeed, if h is in $\mathfrak{h}$ and if g is in $U(\mathfrak{g})$ then $A(v)(hg) = \delta(\rho(h)\rho(g)v)$. Now $\rho(g)v = \sum v_\mu$ with v_μ in V_μ, and $\rho(h)\rho(g)v = \sum \mu(h)v_\mu$. But then, $A(v)(hg) = \delta(\sum \mu(h)v_\mu) = \Lambda(h)\delta(v_\Lambda) = \Lambda(h)A(v)(g)$. If y is in $\mathfrak{n}^-$ then $\rho(y)V$ is contained in $\sum_{\mu\neq\Lambda} V_\mu$, thus $A(v)(yg) = 0$ for all y in $\mathfrak{n}^-$, g in $U(\mathfrak{g})$. We have thus shown that $A(V)$ is contained in $T(\Lambda)$.

Now $A(\rho(X)v)(g) = \delta(\rho(g)\rho(X)v) = \delta(\rho(gX)v) = A(v)(gX) = (\pi_\Lambda(X)A(v))(g)$ for g in $U(\mathfrak{g})$ and X in $\mathfrak{g}$. Hence $A\rho(X) = \pi_\Lambda(X)A$.

Finally, $A(\rho(X)v_0) = A(0) = 0$ for all X in $\mathfrak{n}^+$, and thus $\pi_\Lambda(X)A(v_0) = 0$ for all X in $\mathfrak{n}^+$. (1) implies that $A(v_0) = A(v_0)(1)\xi_\Lambda$. But $A(v_0)(1) = \delta(v_0) = 1$. Hence $Av_0 = \xi_\Lambda$. $A(V) = A(\rho(U(\mathfrak{g}))v_0) = \pi_\Lambda(U(\mathfrak{g}))Av_0 = \pi_\Lambda(U(\mathfrak{g}))\xi_\Lambda = V^\Lambda$. Thus A has all of the properties of (3).

4.5 The Theorem of the Highest Weight

4.5.1 We retain the notation of Section 4.4.

4.5.2 *Definition* Let Λ be in $\mathfrak{h}^*$. Let for each α in Δ, H_α be as in 3.5.7.

Λ is said to be dominant integral if $2\Lambda(H_\alpha)/\alpha(H_\alpha)$ is a nonnegative integer for each α in Δ^+.

4.5.3 *Theorem* (1) If Λ in $\mathfrak{h}^*$ is dominant integral then V^Λ is finite dimensional.

(2) If (π, V) is an irreducible finite dimensional representation of $\mathfrak{g}$ then (π, V) is equivalent with (π_Λ, V^Λ) with Λ dominant integral.

(3) If (π_Λ, V^Λ) is equivalent with (π_μ, V^μ) then $\Lambda = \mu$.

PROOF We prove (3) first. Suppose that A defines the equivalence between V^Λ and V^μ. Then A must map the set of all v in V^Λ so that, for all X in $\mathfrak{n}^+, \pi_\Lambda(X)v = 0$ to the set of all v in V^μ so that $\pi_\mu(X)v = 0$. But then we may assume that $A\xi_\Lambda = \xi_\mu$. Now $\Lambda(H)A\xi_\Lambda = A\pi_\Lambda(H)\xi_\Lambda = \pi_\mu(H)A\xi_\Lambda = \mu(H)\xi_\mu$, for all H in $\mathfrak{h}$. Hence $\mu = \Lambda$.

We next prove (2). 4.3.7 combined with the irreducibility of (π, V) implies that (π, V) is P-extreme. 4.4.5(3) implies that (π, V) is equivalent with (π_Λ, V^Λ) for some Λ in $\mathfrak{h}^*$. We must show that Λ is dominant integral.

Recall that $\pi = \{\alpha_1, \ldots, \alpha_l\}$. Let for each α in Δ, E_α be chosen in $\mathfrak{g}_\alpha$ so that $[E_\alpha, E_{-\alpha}] = H_\alpha$ for α in Δ^+. Let $E_i = E_{\alpha_i}$, $F_i = (2/\alpha_i(H_{\alpha_i}))E_{-\alpha_i}$ and let $H_i = (2/\alpha_i(H_{\alpha_i}))H_{\alpha_i}$. Let $\mathfrak{g}^i$ be the subalgebra of $\mathfrak{g}$ spanned by E_i, F_i, H_i, then $\mathfrak{g}^i$ is a TDS. Let $T^i = \pi_\Lambda(U(\mathfrak{g}^i))\xi_\Lambda$. We assert that T^i is irreducible. T^i is spanned by elements of the form $\pi_\Lambda(F_i^k)\xi_\Lambda$. Suppose that S is a nonzero invariant subspace of T^i. As in the proof of 4.3.10, S must contain $\pi_\Lambda(F^k)\xi_\Lambda$ for some k. Let k be the smallest integer such that $S \cap V_{\Lambda - k\alpha_i} \neq (0)$. Let v be an element of $S \cap V_{\Lambda - k\alpha_i}$. Then $\pi_\Lambda(E_i)v = 0$. If $i \neq j$ then $\pi_\Lambda(E_j)v$ is an element of $V_{\Lambda - k\alpha_i + \alpha_j} = (0)$ by 4.4.3. Now $E_1, \ldots, E_l$ generate $\mathfrak{n}^+$ as a Lie algebra. Thus $\pi_\Lambda(\mathfrak{n}^+)v = 0$. 4.4.5(1) implies that $k = 0$, hence $S = T^i$. Thus T^i is indeed irreducible.

Now T^i is irreducible and finite dimensional. Since $\mathfrak{g}^i$ is a TDS, 4.3.10 applies and we see that $\Lambda(H_i) = (\dim T^i - 1)$. Since $\dim T^i - 1$ is a nonnegative integer Λ is indeed dominant integral. This proves (2).

To prove (1) we let $\mathfrak{g}^i$ and T^i be as in the proof of (2). Then, as in the proof of (2), T^i is an irreducible representation of $\mathfrak{g}^i$. Set $n_i = 2\Lambda(H_{\alpha_i})/\alpha_i(H_{\alpha_i}) = \Lambda(H_i)$. Then n_i is a nonnegative integer. (2) combined with (3) and 4.3.10 implies that T^i is finite dimensional. Let W^i be the collection of all finite dimensional subspaces of V^Λ which are invariant under $\mathfrak{g}^i$. We have shown above that $W^i \neq \{(0)\}$. If M and N are in W^i then $M + N$ is in W^i. Thus the union of the elements of W^i, W_i, is a subspace of V^Λ. We note that if M is in W^i then $\pi_\Lambda(\mathfrak{g})M$ is in W^i. In fact, $\dim \pi_\Lambda(\mathfrak{g})M \leqslant \dim \mathfrak{g} \dim M < \infty$ and $\pi_\Lambda(\mathfrak{g}^i)\pi_\Lambda(\mathfrak{g})M \subset \pi_\Lambda([\mathfrak{g}^i, \mathfrak{g}])M + \pi_\Lambda(\mathfrak{g})\pi_\Lambda(\mathfrak{g}^i)M \subset \pi_\Lambda(\mathfrak{g})M$. This implies that W_i is a nonzero invariant subspace of V. The irreducibility of V^Λ implies that $W_i = V^\Lambda$.

Suppose that $V_\mu^\Lambda \neq (0)$: we show that $V_{s_{\alpha_i}\mu}^\Lambda \neq (0)$ for $i = 1, \ldots, l$. Here $s_{\alpha_i}\mu = \mu - \mu(H_i)\alpha_i = \mu \circ s_{\alpha_i}$ and s_{α_i} is the Weyl reflection defined in 3.10.4. Let M in W^i be so that $M \cap V_\mu^\Lambda \neq (0)$. Let v be a nonzero element of $M \cap V_\mu^\Lambda$. There is a positive integer k so that $\pi_\Lambda(E_i^k)v = (0)$ and $\pi_\Lambda(E_i^{k-1})v \neq (0)$(if not then dim $M = \infty$). $\pi_\Lambda(U(\mathfrak{g}^i))\pi_\Lambda(E_i^{k-1})v$ is extreme for $\mathfrak{g}^i$. 4.4.5(3) and 4.3.10 imply that $w = \pi_\Lambda(F^{\mu(H_i)+k-1})\pi_\Lambda(E_i^{k-1})v \neq 0$. But w is in $V_{s_{\alpha_i}\mu}$. This proves our assertion.

4.4.3 tells us that to complete the proof of (1), we need only show that there are only a finite number of μ in $\mathfrak{h}^*$ so that $V_\mu^\Lambda \neq (0)$. If μ is a weight of V_μ^Λ then there is s in $W(\Delta)$ so that $\mu \circ s$ is dominant integral. In fact, $\mu(H_i)$ is an integer for each $i = 1, \ldots, l$. Order the weights of V^Λ lexicographically (see the relative to the basis $H_1, \ldots, H_l$. Let $\tilde{\mu}$ be the largest weight of the form $\mu \circ s$. If $\tilde{\mu}(H_i) < 0$ then clearly $s_{\alpha_i}\tilde{\mu} > \tilde{\mu}$. Hence $\tilde{\mu}$ is indeed dominant integral. Since $W(\Delta)$ is a finite group, (1) now follows if we can show that there are only a finite number of dominant integral weights of V^Λ. (We have used 3.10.9(1) and the above to see that $\mu \circ s$ is a weight for each weight μ and each s in $W(\Delta)$.)

Let μ be a dominant integral weight of V^Λ. Then $\mu = \Lambda - \sum n_i\alpha_i$, with n_i a nonnegative integer. Set $\beta = \sum n_i\alpha_i$. Then $\mu = \Lambda - \beta$. Let (,) be the inner product on $\mathfrak{h}_R^*$ dual to the restriction of the Killing form to $\mathfrak{h}_R$. Then $(\Lambda, \Lambda) = (\mu + \beta, \mu + \beta) = (\mu, \mu) + 2(\mu, \beta) + (\beta, \beta) \geqslant (\mu, \mu)$ since $(\mu, \beta) \geqslant 0$. Thus μ is contained in the discrete set of all weights of V^Λ and the compact set of all μ so that $(\mu, \mu) \leqslant (\Lambda, \Lambda)$. Hence there can only be a finite number of dominant integral weights.

4.6 Representations and Topology of Compact Lie Groups

4.6.1 *Definition* Let T be a torus with Lie algebra $\mathfrak{h}$. Let $\Gamma_T = \{z \text{ in } \mathfrak{h} | \exp(z) = e\}$. Then Γ_T is called the unit lattice of T.

4.6.2 *Lemma* (1) If (π, V) is an irreducible representation of T then dim $V = 1$.

(2) Let λ be a complex valued linear form on $\mathfrak{h}$. There is a continuous homomorphism $t \mapsto t^\lambda$ of T to $C - \{0\}$ so that $\exp(H)^\lambda = e^{\lambda(H)}$ if and only if $\lambda(\Gamma_T) \subset (-1)^{\frac{1}{2}}2\pi Z$. ($Z$, as usual, denotes the ring of integers.)

PROOF (1) follows immediately from 2.3.4.

(2) Let χ be a continuous homomorphism of T into $C - \{0\}$. Then χ_* is a linear map of $\mathfrak{h}$ into C and $\chi(\exp(H)) = e^{\chi_*(H)}$. Furthermore if z is in Γ_T then $\chi(\exp z) = \chi(e) = 1$. Thus $e^{\chi_*(z)} = 1$. This implies that $\chi_*(z) = 2\pi(-1)^{\frac{1}{2}}k$ with k an integer. If λ is a linear form on $\mathfrak{h}$ so that $\lambda(\Gamma_T) \subset 2\pi(-1)^{\frac{1}{2}}Z$, then $e^{\lambda(z)} = 1$ for z in Γ_T. This implies that $\exp(H)^\lambda = e^{\lambda(H)}$ is well defined.

4.6.3 Let G be a connected compact Lie group. Let T be a maximal torus in G. Let $\mathfrak{g}$ and $\mathfrak{h}$ be, respectively, the Lie algebras of G and T. Let Δ be the root system of G relative to T, let P be a Weyl chamber of T, and let π be the corresponding system of simple roots. Set $\Gamma_T = \Gamma_G$. Then Γ is called the unit lattice of G.

4.6.4 *Proposition* Let G be a compact, connected, simply connected Lie group. Let $h_1, \ldots, h_l$ in $\mathfrak{h}_R$ be defined by $\alpha_i(h_n) = \delta_{ij}$, where $\pi = \{\alpha_1, \ldots, \alpha_l\}$ and $\mathfrak{h}_R$ is as in 3.5.9(2) in the complexification of $\mathfrak{g}$. Let $\Gamma^1 = \{z \text{ in } \mathfrak{h} | \exp z$ is in the center of $G\}$. Then $\Gamma^1 \supset \Gamma$.

(1) Γ^1 is the lattice generated by $2\pi(-1)^{\frac{1}{2}}h_1, \ldots, 2\pi(-1)^{\frac{1}{2}}h_l$.

(2) Let Z be the center of G, then Z is isomorphic with Γ^1/Γ.

(3) Let G_1 be a connected Lie group with Lie algebra $\mathfrak{g}$. Then

$$\Gamma_G \subset \Gamma_{G_1} \subset \Gamma^1 \quad \text{and} \quad \Gamma_{G_1}/\Gamma_G$$

is isomorphic with $\pi_1(G_1)$ (see A.2.5.3).

(4) If Γ_1 is a lattice in $\mathfrak{h}$ so that $\Gamma_G \subset \Gamma_1 \subset \Gamma^1$, then there is a connected Lie group with Lie algebra $\mathfrak{g}$ and unit lattice Γ_1.

PROOF The proof of (1) is an immediate consequence of 3.10.6.
The proof of (2) also follows from 3.10.6 and the definition of Γ_G.

(3) Since G and G_1 have the same Lie algebra and since G is simply connected we see that G is the universal covering group of G_1(see A.2.5). Let p be the covering mapping of G onto G_1. Then $\ker(p) = Z_1$ is a subgroup of Z. Clearly $\Gamma_{G_1} = \{z \text{ in } \mathfrak{h} | \exp z$ is in $Z_1\}$. Hence $\Gamma_G \subset \Gamma_{G_1} \subset \Gamma^1$. Now Z_1 is isomorphic to the fundamental group of G_1 (see A.2.5) and Z_1 is isomorphic with Γ_{G_1}/Γ_G. This proves (3).

(4) Let $Z_1 = \exp(\Gamma_1)$ in G. Then $G_1 = G/Z_1$ is the desired Lie group.

4.6.5 *Definition* Let G be a compact Lie group with Lie algebra $\mathfrak{g}$. Let T be a maximal torus of G with Lie algebra $\mathfrak{h}$. Then a complex valued linear form on $\mathfrak{h}$ that is the differential of a character of T is called a weight of G.

4.6.6 *Theorem* Let G be a compact, connected, simply connected Lie group and let $\mathfrak{g}$, T, $\mathfrak{h}$, P, π be as in 4.6.4.

(1) If Λ is a dominant integral form on $\mathfrak{h}_C$ then Λ is a weight of G.

(2) There is a bijective correspondence between the dominant integral forms on $\mathfrak{h}_C$ and the elements of $\hat{G}$ (recall that $\hat{G}$ is the set of all equivalence classes of irreducible finite dimensional representations of G).

PROOF We first prove (2). Let $\mathfrak{g}_C$ be the complexification of $\mathfrak{g}$. Then $\mathfrak{g}_C$ is a complex semisimple Lie algebra and $\mathfrak{h}_C$ is a Cartan subalgebra of $\mathfrak{g}_C$. 4.5.4 implies that there is a bijective correspondence between the equivalence classes of irreducible finite dimensional representations of $\mathfrak{g}_C$ and the dominant integral forms on $\mathfrak{h}_C$. Let (π, V) be an irreducible finite dimensional representation of G. Then (π_*, V) is an irreducible finite dimensional representation of $\mathfrak{g}_C$ (we denote by π_* the complex linear extension of the differential of π). Let (π, V) be an irreducible representation of $\mathfrak{g}_C$. Then π is a Lie algebra homomorphism of $\mathfrak{g}$ into $L(V, V)$. Thus there is a Lie homomorphism of G into $Gl(V)$ whose differential is π (see A.2.5.5). Since (π, V) and (π', V') are equivalent if and only if (π_*, V) and (π'_*, V') are equivalent, (2) follows.

To prove (1), let Λ be a dominant integral form on $\mathfrak{h}_C$. Let (π_Λ, V^Λ) be the corresponding finite dimensional irreducible representation of $\mathfrak{h}_C$. Then (π_Λ, V^Λ) extends to an irreducible representation of G, by the above. Now V^Λ_Λ is one dimensional, hence T acts on V^Λ_Λ by a character. Since the differential of the character must be Λ, (1) follows.

4.6.7 *Theorem* Let G be a compact connected, simply connected Lie group. We retain the notation of 4.6.4. Let $H_i = (2/\alpha_i(H_{\alpha_i}))H_{\alpha_i}$ for $i = 1, \ldots, l$. Then Γ_G is the lattice generated by $2\pi(-1)^{\frac{1}{2}}H_1, \ldots, 2\pi(-1)^{\frac{1}{2}}H_l$.

PROOF Let E_i, F_i be as in the proof of 4.5.4, and let $\mathfrak{g}^i$ be the Lie subalgebra of $\mathfrak{g}_C$ spanned by E_i, F_i, H_i. Then $\mathfrak{g}^i$ is a TDS. Hence $\mathfrak{g}^i$ is isomorphic with $sl(2, C)$ which is the complexification of the Lie algebra of $SU(2)$. Now $SU(2)$ is, as a topological space, homeomorphic with the three sphere which is simply connected. In the Lie algebra of $SU(2)$, $2\pi(-1)^{\frac{1}{2}}H_i$ corresponds to the matrix

$$\begin{bmatrix} 2\pi(-1)^{\frac{1}{2}} & 0 \\ 0 & -2\pi(-1)^{\frac{1}{2}} \end{bmatrix}$$

which clearly generates the unit lattice of $SU(2)$. This implies that for each i, $2\pi(-1)^{\frac{1}{2}}H_i$ is in Γ_G.

Now let Λ_i be defined by $\Lambda_i(H_j) = \delta_{ij}$. Then $\Lambda_1, \ldots, \Lambda_l$ are dominant integral. 4.6.6(1) implies that for each i, Λ_i is a weight of G. But then 4.6.2(2) implies that $\Lambda_i(\Gamma_G) \subset 2\pi(-1)^{\frac{1}{2}}Z$ for $i = 1, \ldots, l$. This implies that every element of Γ_G is an integral combination of $2\pi(-1)^{\frac{1}{2}}H_1, \ldots, 2\pi(-1)^{\frac{1}{2}}H_l$.

4.6.8 Let G be a compact connected Lie group. Let $\mathfrak{g}$ be the Lie algebra of G. Then $\mathfrak{g} = \mathfrak{z} + \mathfrak{g}_1$ with $\mathfrak{g}_1 = [\mathfrak{g}, \mathfrak{g}]$, a compact Lie algebra, and $\mathfrak{z}$ is the center of $\mathfrak{g}$ (see 3.6.1). Let Z_0 be the connected subgroup of G corresponding to $\mathfrak{z}$. Z_0 is the identity component of Z, the center of G, hence Z_0 is closed in G. Let G_1 be the connected subgroup of G corresponding to $\mathfrak{g}_1$. 3.6.6 implies that G_1 is closed in G.

4.6.9 *Lemma* (1) $G = Z_0 G_1$.
(2) G is Lie isomorphic with $(Z_0 \times G_1)/D$ where $D = \{(g^{-1}, g) | g \text{ in } Z_0 \cap G_1\}$. D is a finite subgroup of $Z_0 \times G_1$.

PROOF Let $\varphi\colon Z_0 \times G_1 \to G$ be defined by $\varphi(z, g) = zg$. Then φ is a Lie homomorphism. φ_* is a Lie algebra isomorphism, hence φ is surjective. This proves (1).

To prove (2), we note that if (z, g) is in $\ker \varphi$ then $zg = e$. This implies that $z = g^{-1}$ and g is in $G_1 \cap Z_0$. Now, G_1 is compact with finite center. $G_1 \cap Z_0$ is clearly central in G_1, hence it is finite. This implies that D is finite.

4.6.10 *Lemma* Let $\Gamma_G^1 = \{z \text{ in } \mathfrak{g} | \exp z \text{ is in } Z\}$. Then $\Gamma_G^1 = \mathfrak{z} \oplus \Gamma_{G_1}^1$.

PROOF Let $\mathfrak{h}_1 = \mathfrak{h} \cap \mathfrak{g}_1$. Let z be in Γ_G^1. Then $z = z_1 + z_2$ with z_1 in $\mathfrak{z}$ and z_2 in $\mathfrak{h}_1$. $\exp_G(z) = \exp_G(z_1)\exp_{G_1}(z_2)$. Hence $\exp_{G_1}(z_2)$ is in $Z \cap G_1$. This implies that z_2 is in $\Gamma_{G_1}^1$. This proves the lemma.

4.6.11 *Lemma* Let $\mathfrak{g} = \mathfrak{z} + \mathfrak{g}_1$, a direct sum of ideals, with $\mathfrak{z}$ the center of $\mathfrak{g}$ and $\mathfrak{g}_1$ a compact subalgebra. Let $\mathfrak{h} = \mathfrak{z} + \mathfrak{h}_1$ with $\mathfrak{h}_1$ maximal Abelian in $\mathfrak{g}_1$. Let Γ be the unit lattice of the compact connected, simply connected Lie group G_1 with Lie algebra $\mathfrak{g}_1$. Let Γ^1 be the lattice $\Gamma_{G_1}^1$. Let Γ_1 be any lattice in $\mathfrak{h}$ so that
(1) $\Gamma_1 \subset \mathfrak{z} \oplus \Gamma^1$,
(2) $\Gamma_1 \supset \Gamma$.
Then there exists a compact connected Lie group with Lie algebra $\mathfrak{g}$, unit lattice Γ_1, and fundamental group Γ_1/Γ.

PROOF Let $\mathfrak{z} \times G_1$ be the product Lie group, where $\mathfrak{z}$ is a Lie group under addition. Let $\tilde{\Gamma}_0 = \Gamma_1 \cap \mathfrak{z}$. Then $\mathfrak{z}/\tilde{\Gamma}_0$ is a torus. Let $\tilde{\Gamma}_1 = \Gamma_1 \cap \mathfrak{h}_1$. Then $\Gamma^1 \supset \tilde{\Gamma}_1 \supset \Gamma$. Thus, if $Z_1 = \exp_{G_1}(\tilde{\Gamma}_1)$ then Z_1 is a central subgroup of G_1. Let $Z = \exp(\Gamma_1)$. Then $Z/\tilde{\Gamma}_0 \times Z_1$ is a finite group and $G = (\mathfrak{z} \times G_1)/Z$ is covered by $\mathfrak{z}/\tilde{\Gamma}_0 \times G_1/Z_1$. Hence G is a compact connected Lie group with universal covering group $\mathfrak{z} \times G_1$. The result now follows.

4.6.12 *Theorem* Let G be a compact connected Lie group. Let T be a maximal torus of G and let Δ be the root system of G relative to T (see 3.10.5). Let P be a Weyl chamber of T and let Γ_G be the unit lattice of G. Let G_1 be as in 4.6.8 and let $T_1 = G_1 \cap T$. Let Γ_{G_1} be the unit lattice of G_1. Then $\hat{G}$ is in bijective correspondence with the linear forms Λ on $\mathfrak{h}$ such that

(1) $\Lambda(\Gamma_G) \subset 2\pi(-1)^{\frac{1}{2}}Z$,

(2) $\Lambda|_{\mathfrak{h}_1}$ is dominant integral.

PROOF Let (π, V) be an irreducible finite dimensional representation of G. Then as a representation of G_1, (π, V) is equivalent with $(\pi_{\Lambda_0}, V^{\Lambda_0})$ where Λ_0 is a dominant integral form on $\mathfrak{h}_1$. Furthermore $V^{\Lambda_0}_{\Lambda_0}$ is one-dimensional, hence Λ_0 extends to a character of T, Λ. Clearly Λ satisfies (1), (2). Furthermore, Λ uniquely determines the equivalence class of (π, V).

Suppose that Λ is a linear form on $\mathfrak{h}$ satisfying (1) and (2). Let $\Lambda_1 = \Lambda|_{\mathfrak{h}_1}$. Let $\tilde{G}_1$ be the universal covering group of G_1. Let $(\pi_{\Lambda_1}, V^{\Lambda_1})$ be the irreducible finite dimensional representation of $\tilde{G}_1$ with highest weight Λ_1. We assert that if p is the projection of $\tilde{G}_1$ onto G_1 then $\pi_{\Lambda_1}(\ker(p)) = (I)$. Indeed, $V^{\Lambda_1} = \sum V^{\Lambda_1}_\mu$, the sum over the weights of V^{Λ_1}. Now, μ is a weight of V^{Λ_1} only if $\mu = \Lambda_1 - \sum n_i\alpha_i$ with n_i nonnegative integers. If z is in $\ker(p)$ then z is in the center of $\tilde{G}_1$. Hence $z^{\alpha_i} = 1$ for $i = 1, \ldots, l$. Furthermore (1) implies that $z^{\Lambda_1} = 1$. Thus $\pi_{\Lambda_1}(z) = I$ as asserted. This implies that $(\pi_{\Lambda_1}, V^{\Lambda_1})$ is actually a representation of G_1.

Let Z_0 be the connected subgroup of G corresponding to $\mathfrak{z}$. Then 4.6.9(2) implies that $G = (Z_0 \times G)/D$. Let $\Lambda_0 = \Lambda|_{\mathfrak{z}}$. Then Λ_0 extends to a character of Z_0. Let $\pi_\Lambda(z, g) = z^{\Lambda_0}\pi_{\Lambda_1}(g)$. To complete the proof we need only show that $\pi_\Lambda|_D = I$. But, $D = \{(g^{-1}, g)|$ with g in $Z_0 \cap G_1\}$. If g is in $Z_0 \cap G_1$ then $\pi_{\Lambda_1}(g) = g^{\Lambda_1}I$ by Schur's lemma. Thus $\pi_\Lambda(g^{-1}, g) = g^{-\Lambda}g^\Lambda I = I$.

4.6.13 *Theorem* Let the notation be as in 4.6.9 and 4.6.12. Set $\rho_p = \frac{1}{2}\sum \alpha$, the sum taken over the elements of Δ_p^+. Then $\rho_p(H_i) = 1$ for $i = 1, \ldots, l$. Let $\pi_1(Z_0 \times G_1, G)$ be the group of deck transformations of the covering space $Z_0 \times G_1 \to G$. If $\pi_1(Z_0 \times G_1, G)$ has no 2-torsion (that is, there is no element whose square is the identity) then ρ_p is a weight of G.

PROOF 3.10.10 implies that $s_{\alpha_i}\rho = \frac{1}{2}\sum_{\alpha \neq \alpha_i, \alpha > 0} \alpha - \frac{1}{2}\alpha_i = \rho - \alpha_i$. Also $s\alpha_i = \rho - \rho(H_i)\alpha_i$. Thus $\rho(H_i) = 1$ as asserted.

$\mathfrak{h} = \mathfrak{z} + \mathfrak{h}_1$. Let Q be the corresponding projection of $\mathfrak{h}$ onto $\mathfrak{h}_1$. Then $\rho = \rho \circ Q$. Let Γ^1 and Γ be the lattices as in 4.6.4 for $\tilde{G}_1$. Set $\Gamma_0 = \mathfrak{z} \cap \Gamma_G$. Then $\Gamma_G \supset \Gamma_0 \oplus \Gamma$. It is not hard to show that $\Gamma_G/\Gamma_0 \oplus \Gamma$ is isomorphic to $\pi_1(Z_0 \times \tilde{G}_1, G)$.

To prove the last assertion of the theorem, we show that if ρ is not a weight of G then there is an element z of Γ_G so that z is not in $\Gamma_0 \oplus \Gamma$ but $2z$ is.

Let $u_1, \ldots, u_{r+l}$ be a Z-basis for Γ_G. Let $z_1, \ldots, z_r$ be a Z-basis for Γ_0. Set $z_{r+j} = 2\pi(-1)^{\frac{1}{2}}H_j$. Then $z_i = \sum b_{ji}u_j$ with b_{ij} integers. This implies that $u_j = \sum b^{ij}z_i$ with (b^{ij}) the inverse matrix to (b_{ij}). Cramer's rule implies that b^{ij} is a rational number.

If ρ is not a weight of G then there is γ in Γ_G so that $(2\pi(-1)^{\frac{1}{2}})^{-1}\rho(\gamma)$ is not an integer. $\gamma = (1/t)\sum s_i z_i$ with s_i, t integers, $t > 0$ having greatest common divisor 1. $\rho(u_i) = \frac{1}{2}\sum \alpha(u_i)$, the sum taken over the elements of Δ^+. But $\alpha(u_i)$ is in $2\pi(-1)^{\frac{1}{2}}Z$ (roots are weights!). Thus $\rho(\gamma) = \pi(-1)^{\frac{1}{2}}p$ with p an integer. The first part of this theorem implies that $\rho(\gamma) = (2\pi(-1)^{\frac{1}{2}}/t)\sum_{i=r+1} s_i$. Set $s = \sum_{i=r+1} s_i$. Then $2s = tp$. This implies that 2 divides either p or t. The assumption on γ implies that $t = 2t_1$. It is easy to see that if $z = t_1\gamma$ then z is in Γ_G but not in $\Gamma_0 \oplus \Gamma$. It is also clear that $2z$ is in $\Gamma_0 \oplus \Gamma$.

4.7 Holomorphic Representations

4.7.1 Let G be a connected, complex, reductive Lie group. That is, G is a complex Lie group (see A.5.3) with (complex) Lie algebra $\mathfrak{g}$ such that the adjoint representation of G on $\mathfrak{g}$ splits into a direct sum of irreducible representations.

4.7.2 *Lemma* $\mathfrak{g} = \mathfrak{z} \oplus \mathfrak{g}_1$ with $\mathfrak{z}$ the center of $\mathfrak{g}$ and $\mathfrak{g}_1 = [\mathfrak{g}, \mathfrak{g}]$ a complex semisimple ideal of $\mathfrak{g}$.

PROOF As a representation of G, $\mathfrak{g} = V_0 \oplus \sum_{j=1}^{r} V_j$ where V_0 is a multiple of the trivial representation of G and V_j is an irreducible, invariant subspace of $\mathfrak{g}$ on which G acts nontrivially for $j \geqslant 1$. Clearly, $V_0 = \mathfrak{z}$ the center of $\mathfrak{g}$ and $\mathfrak{g}_1 = \sum_{j=1}^{r} V_j$ is a semisimple ideal in $\mathfrak{g}$ so that $\mathfrak{g} = \mathfrak{z} \oplus \mathfrak{g}_1$. Since $\mathfrak{g}_1 = [\mathfrak{g}_1, \mathfrak{g}_1] = [\mathfrak{g}, \mathfrak{g}]$ the result follows.

4.7.3 Let $Z = \exp \mathfrak{z}$ and let G_1 be the connected subgroup of G corresponding to $\mathfrak{g}_1$. Then $G = ZG_1$.

4.7.4 Since $\mathfrak{g}_1$ is complex semisimple, $\mathfrak{g}_1$ has a compact form $\mathfrak{U}_1$. Let U_1 be the connected subgroup of G_1 corresponding to $\mathfrak{U}_1$. 3.6.6 implies that U_1 is compact.

4.7.5 *Theorem* Let (π, V) be a finite dimensional holomorphic representation of G (that is, (π, V) is a representation of G and $\pi: G \to GL(V)$ is holomorphic). Then there is a Hermitian inner product on V so that $\pi|_{U_1}$ is unitary and irreducible and there is a holomorphic mapping $\chi: Z \to C - \{0\}$ so that $\pi(z) = \chi(z)I$ for z in Z. Conversely, let (π, V) be a unitary representation of U_1. Suppose that there exists $\chi: Z \to C - \{0\}$ a holomorphic Lie homomorphism so that $\pi(z) = \chi(z)I$ for z in $U_1 \cap Z$. Then there exists a unique holomorphic representation (η, V) of G so that $\eta|_{U_1} = \pi$ and $\eta(z) = \chi(z)I$ for z in Z.

PROOF Let (π, V) be an irreducible holomorphic representation of G. Let $(\ , \ )$ be a Hermitian inner product on V. Define

$$\langle v, w \rangle = \int_{U_1} (\pi(k)v, \pi(k)w)\, dk.$$

$\langle \ , \ \rangle$ is an inner product on V relative to which $\pi|_{U_1}$ is unitary. Now (π, V) is irreducible, hence 2.3.4 implies that there is a holomorphic map $\chi: Z \to C - \{0\}$ so that $\pi(z) = \chi(z)I$ for z in Z. Thus (π, V) is irreducible if and only if $(\pi|_{G_1}, V)$ is irreducible. Suppose that W is a $\pi(U_1)$-invariant subspace of V. Then $\pi(\mathfrak{U}_1)W \subset W$. But π is holomorphic (hence complex linear on $\mathfrak{g}$), thus $\pi(\mathfrak{g}_1)W \subset W$. Hence $\pi(G_1)W \subset W$. This implies that $W = (0)$ or V. This proves the first part of the theorem.

Suppose that (π, V) and χ are as in the second assertion of the theorem. Let π also denote the corresponding representation of $\mathfrak{U}_1$ on V. Let $\tilde{\eta}$ denote the complex linear extension of π to $\mathfrak{g}_1$. Let $\tilde{G}_1$ be the universal covering group of G_1. Then $\tilde{\eta}$ induces a holomorphic representation of $\tilde{G}_1$ on V.

Let $\tilde{U}_1$ be the universal covering group of U_1. 3.6.6 implies that $\tilde{U}_1$ is compact. 2.8.4 implies that $\tilde{U}_1$ is Lie isomorphic with a Lie subgroup of $GL(n, C)$ for n sufficiently large. Let G_1 be the connected subgroup of $GL(n, C)$ corresponding to $\mathfrak{g}_1$ (identified with the complexification of the image of $\mathfrak{U}_1$ in $M_n(C)$). 7.2.5(2) implies that G_1 is simply connected. Hence $G_1 = \tilde{G}_1 \supset \tilde{U}_1$ and 7.2.5(1) implies that the center of G_1 is contained in U_1. This implies that $\tilde{\eta}$ actually defines a holomorphic representation of G_1. We have seen above that the center of G_1 is contained in U_1. Define $\mu: Z \times G_1 \to GL(V)$ by

$\mu(z, g) = \chi(z)\tilde{\eta}(g)$. Then μ is holomorphic. We assert that if $zg = z_1 g_1$, z, z_1 in Z and g, g_1 in G, then $\mu(z, g) = \mu(z_1, g_1)$. Indeed, $z^{-1}z_1 = gg_1^{-1}$. Thus $g_1 g^{-1}$ is in $G_1 \cap Z = U_1 \cap Z$. Hence $\tilde{\eta}(gg_1^{-1}) = \pi(gg_1^{-1}) = \chi(gg_1^{-1})I$. This says that $\tilde{\eta}(g) = \chi(gg_1^{-1})\tilde{\eta}(g_1)$. On the other hand $\chi(z^{-1}z_1) = \chi(gg_1^{-1})$. Thus $\chi(z) = \chi(g_1 g^{-1})\chi(z_1)$. Hence

$$\chi(z)\tilde{\eta}(g) = \chi(g_1 g^{-1})\chi(z_1)\chi(gg_1^{-1})\tilde{\eta}(g_1) = \chi(z_1)\tilde{\eta}(g_1).$$

Hence μ induces the desired holomorphic representation of G.

4.8 The Weyl Integral Formula

4.8.1 Let G be a connected, compact Lie group and let T be a maximal torus of G. Let $N(T)$ be the normalizer of T in G and let $N(T)/T$ be, as usual, the Weyl group of G relative to T. Let $w(G)$ be the order of $W(T)$. Let dg and dt be, respectively, normalized invariant measure on G and T (that is, $\int_G dg = \int_T dt = 1$). For unexplained notation see 3.10.

4.8.2 *Theorem* Let f be a continuous function on G. Then

$$w(G) \int_G f(g)dg = \int_T \det(I - \pi(t)) \int_G f(gtg^{-1})dg,$$

where π is the isotropy representation of T on G/T.

PROOF Let ψ: $G/T \times T \to G$ be defined by $\psi(gT, t) = gtg^{-1}$. Then ψ is a surjective C^∞ mapping. Let $T_r = \{t \text{ in } T | t^\alpha \neq 1 \text{ for all } \alpha \text{ in } \Delta\}$, where Δ is the root system of G relative to T. (Note that if $\Delta = \phi$ then $T_r = T$.) If g is in G then $\det(\mathrm{Ad}(g) - (\lambda + 1)I) = \sum p_k(g)\lambda^k$. Let $l = \dim T$. Then $p_k(g) = 0$ for $k < l$. We set $G_r = \{g \text{ in } G | p_l(g) \neq 0\}$. It is easy to see that $\psi(G/T \times T_r) = G_r$. $T - T_r$ is a finite union of tori of lower dimension than T, so that $\int_{G_r} f(g)dg = \int_G f(g)dg$.

Let $\mathfrak{g}$ be the Lie algebra of G and let $\mathfrak{h}$ be the Lie algebra of T. Let η and ω be, respectively, the left invariant q-form and n-form that define dt and dg. Let $h_1, \ldots, h_q$ be a basis of $\mathfrak{h}$ so that $\eta(h_1, \ldots, h_q) = 1$. Let (,) be an Ad($G$)-invariant inner product on $\mathfrak{g}$. Let $X_1, \ldots, X_m$ be a basis of the orthogonal complement to $\mathfrak{h}$ in $\mathfrak{g}$ so that $\omega(h_1, \ldots, h_q, X_1, \ldots, X_m) = 1$. Let $\eta_1, \ldots, \eta_q$, $\mu_1, \ldots, \mu_m$ be the dual basis to $h_1, \ldots, h_q, X_1, \ldots, X_m$. $\eta_1 \wedge \cdots \wedge \eta_q = \eta$. Let $\mu = \mu_1 \wedge \cdots \wedge \mu_m$. Then $\eta \wedge \mu = \omega$. We note that μ is T-invariant.

Thus if we identify $T(G/T)_{eT}$ with the orthogonal complement of $\mathfrak{h}$ in $\mathfrak{g}$, μ induces a G-invariant volume form on G/T, dx. We note that the definition of dx and Fubini's theorem A.4.2.8 imply that

$$\int_{G/T}\int_T f(gt)\,dt\,d(gT) = \int_G f(g)\,dg.$$

$\mu \wedge \eta$ defines a volume form on $G/T \times T$, which is invariant under the action of $G \times T$ in $G/T \times T$ given by $(g_0, t_0)(gT, t) = (g_0gT, t_0t)$. G is unimodular (see 2.5) so it is easy to see that $(\psi^*\omega)_{(gT,t)} = c(t)\,\eta \wedge \mu_{(gT,t)}$. We compute $\psi_{*(eT,t)}$. Let f be in $C^\infty(G)$.

(1) Let h be in $\mathfrak{h}$ (which we have identified with $T(T)_t$ for each t). Then

$$\psi_{*(eT,t)}(h)f = \frac{d}{ds}f(t\exp sh)|_{s=0} = h_t f.$$

(2) We identify the orthogonal complement of $\mathfrak{h}$ in $\mathfrak{g}$ with $T(G/T)_{eT}$. Let X be in $T(G/T)_{eT}$. Then

$$\psi_{*(eT,t)}f = \frac{d}{ds}f(\exp sXt\cdot\exp(-sX))\Big|_{s=0}$$

$$= \frac{d}{ds}f(t\exp s(\operatorname{Ad}(t^{-1})X)\exp(-sX))\Big|_{s=0} = (\operatorname{Ad}(t^{-1})X - X)_t f.$$

Using computations (1) and (2) we see that $c(t) = \det(\pi(t^{-1}) - I)$.

$W(T)$ acts naturally on the left on $G/T \times T$ via $s(gT, t) = (gn^{-1}T, ntn^{-1})$ for n in s. The action is free. In fact, if g is in G, t is in T and if n is in $N(T)$ and if $gnT = gT$ then clearly n is in T. Suppose that $\psi(g_0T, t_0) = \psi(gT, t)$, t, t_0 in T_r. Then $g_0t_0g^{-1} = gtg^{-1}$. Thus $g_0^{-1}gtg^{-1}g_0 = t_0$. Now, the centralizer in G of t_0 and t in G is T, hence $g_0^{-1}g = n$ in $N(T)$. Hence if g, g_0 are in G and if t, t_0 are in T_r, then $\psi(g_0T, t_0) = \psi(gT, t)$ if and only if $(gT, t) = s(g_0T, t_0)$ for some s in $W(T)$. Thus ψ is a $w(G)$-fold covering of G_r. This implies that if f is in $C^\infty(G)$, then

$$w(G)\int_G f(g)\,dg = \int_{G/T\times T_r}\psi^* f\omega$$

$$= \int_{T_r}\det(\pi(t^{-1}) - I)\int_{G/T} f(gtg^{-1})\,d(gT)\,dt$$

$$= \int_T \det(\pi(t^{-1}) - I)\int_G f(gtg^{-1})\,dg\,dt.$$

The result now follows from the observation that since T is connected $\det(\pi(t)) = 1$. Hence $\det(\pi(t^{-1}) - I) = \det(I - \pi(t))$.

4.8.3 *Definition* Let f be in $C(G)$. Then f is said to be central if for each g, x in G, $f(xgx^{-1}) = f(x)$. Let $C^{\#}(G)$ denote the space of all central functions on G.

4.8.4 *Corollary* Let G be a connected, compact Lie group. Let K be a connected, closed subgroup of G so that $\operatorname{rank}(K) = \operatorname{rank}(G)$. If f is in $C^{\#}(G)$ then

$$\frac{w(G)}{w(K)}\int_G f(g)\,dg = \int_K \det(I - \pi(k))f(k)\,dk,$$

where π is the isotropy representation of K on $T(G/K)_{eK}$.

PROOF Let T be a maximal torus of K. Let π_T be the isotropy representation of T on K/T. Then 4.8.2 inplies that

$$w(K)\int_K \det(I - \pi(k))f(k)\,dk$$
$$= \int_T \det(I - \pi_T(t))\det(I - \pi(t))f(t)\,dt.$$

Let π_G be the isotropy representation of T on G/T. Then $\pi_G = \pi_T \oplus \pi|_T$. Thus $\det(I - \pi_G(t)) = \det(I - \pi_T(t))\det(I - \pi(t))$. Hence,

$$w(K)\int_K \det(I - \pi(k))f(k)\,dk$$
$$= \int_T \det(I - \pi_G(t))f(t)\,dt = w(G)\int_G f(g)\,dg.$$

4.8.5 *Corollary* Let K be a compact connected subgroup of G such that $\operatorname{rank}(K) = \operatorname{rank}(G)$. Then $\chi(G/K) = w(G)/w(K)$.

PROOF This result is a direct consequence of 3.8.1(1) and 4.8.4.

4.9 The Weyl Character Formula

4.9.1 Let G be a compact connected Lie group and let T be a maximal torus of G. Let Δ be the root system of G relative to T. Let P be a Weyl chamber of T. We assume, for the sake of convenience, that $\rho = \frac{1}{2}\sum \alpha$, the sum taken over the elements of Δ_P^+, is a weight. If ρ were not a weight then there would be a compact covering group of G which had ρ for a weight (see 4.6.13) and hence our assumption is a mild one.

4.9.2 *Lemma* Set for t in T, $Q(t) = t^\rho \prod (1 - t^{-\alpha})$, the product taken over the elements of Δ_P^+. If π is the isotropy representation of T on G/T then $\det(I - \pi(t)) = Q(t)\overline{Q(t)}$. If s is in $W(T)$, then $Q(s \cdot t) = \det(s)Q(t)$, where $s \cdot t = ntn^{-1}$ for n in s, t in T and $\det(s)$ is the determinant of s as an endomorphism of the Lie algebra of T.

PROOF Let $\mathfrak{g}_C$ be the complexification of the Lie algebra of G. Then $\mathfrak{g}_C = \mathfrak{h}_C + \sum \mathfrak{g}_\alpha$, where $\mathfrak{h}_C$ is the complexification of the Lie algebra of T and the sum is taken over the root spaces of $\mathfrak{g}_C$ relative to $\mathfrak{h}_C$. Thus the complexified isotropy representation is equivalent to the sum of the root spaces. This says that $\det(I - \pi(t)) = \prod(1 - t^\alpha)$, the product taken over the elements of Δ. The first assertion is now clear. Let $\alpha_1, \ldots, \alpha_l$ be the simple roots of G relative to P. 4.6.13 implies that $s_{\alpha_i}\rho = \rho - \alpha_i$. 3.10.10 says that if α is in Δ_p^+ and $\alpha \neq \alpha_i$ then $s_{\alpha_i}\alpha$ is in Δ_p^+. Since $s_{\alpha_i}\alpha_i = -\alpha_i$, we see easily that $Q \circ s_{\alpha_i} = -Q$. Now, $\det(s_{\alpha_i}) = -1$ and the s_{α_i} generate $W(T)$ (see 3.10.9). The second assertion of the lemma now follows.

4.9.3 Let, for λ in $\hat{T}$, $A(\lambda)(t) = \sum \det(s)(s \cdot t)^\lambda = \sum \det(s)t^{s\lambda}$, the sum taken over the elements s of $W(T)$.

4.9.4 *Lemma* (1) If λ is in $\hat{T}$ and if $A(\lambda) \neq 0$ then $s\lambda \neq \lambda$ for all $W(T)$, $s \neq 1$.

(2) If λ, μ are in T and if $A(\lambda) \neq 0$ then

$$\int_T A(\lambda)\overline{A(\mu)}\,dt = \begin{cases} 0 \text{ if } \lambda \neq s\mu \text{ for any } s \text{ in } W(T) \\ \det(s)w(G) \text{ if } \lambda = s\mu. \end{cases}$$

PROOF The definition of $A(\lambda)$ implies that if s is in $W(T)$ then $A(s\lambda) = \det(s)A(\lambda)$. Let $\alpha_1, \ldots, \alpha_l$ be the simple roots relative to P. If $A(\lambda) \neq 0$ then $s_{\alpha_i}\lambda \neq \lambda$ for any i, since $\det(s_{\alpha_i}) = -1$. There is s' in $W(T)$ so that $s'\lambda$ is dominant integral (that is, $s'\lambda(H_{\alpha_i}) \geqslant 0$ for $i = 1, \ldots$). Again, since $A(s'\lambda) = \det(s')A(\lambda)$ we must have $(s'\lambda)(H_{\alpha_i}) > 0$ for each i. Suppose that $s\lambda = \lambda$ and that $s \neq 1$. Then $s's(s')^{-1}s'\lambda = s'\lambda$. Set $s'' = s's(s')^{-1}$. Since $s'' \neq 1$, there must be an i so that $(s'')^{-1}\alpha_i$ is negative on P (if not then s'' leaves invariant P, see 3.10.9(2)). But then, $0 < \lambda(H_{\alpha_i}) = (s''\lambda)(H_{\alpha_i}) = \lambda((s'')^{-1}H_{\alpha_i}) < 0$. This contradiction implies that if s is in $W(T)$ and if $s\lambda = \lambda$ then $s = 1$, which proves (1).

To prove (2) we note that $A(\lambda)(t)\overline{A(\mu)(t)} = \sum \det(s)\det(u)t^{s\lambda - u\mu}$, the sum taken over all s, u in $W(T)$. Now,

$$(*) \quad \int_T t^\gamma dt = 0 \text{ if } \gamma \neq 0.$$

The first equality of (2) follows from these observations. If $\lambda = s\mu$ then $A(\lambda) = \det(s)A(\mu)$. Using (*) again we find that the second equality of (2) is true.

4.9.5 *Lemma* $Q = A(\rho)$.

PROOF $Q(t) = \prod t^\rho(1 - t^{-\alpha})$ the product taken over the positive roots. If s is in $W(T)$ then $Q(s \cdot t) = \det(s)Q(t)$. Thus

$$(*) \quad Q = \sum c_\beta A(\rho - \beta)$$

where the sum is taken over certain β which are sums of distinct positive roots and the c_β are integers. If β is a sum of distinct positive roots and if s is in $W(T)$ then $s(\rho - \beta) = \rho - \beta'$ with β' a sum of distinct positive roots. Thus in the terms of the right-hand side of (*) we may assume that $\rho - \beta$ is dominant integral. But then

$$0 < 2(\rho - \beta)(H_{\alpha_i})/\alpha_i(H_{\alpha_i}) = 1 - 2\beta(H_{\alpha_i})/\alpha_i(H_{\alpha_i})$$

for each i. Now $2\beta(H_{\alpha_i})/\alpha_i(H_{\alpha_i})$ is an integer for each i. Thus if $A(\rho - \beta) \neq 0$ then $2\beta(H_{\alpha_i})/\alpha_i(H_{\alpha_i}) \leqslant 0$ for each i. Since β is a sum of positive roots $\beta = \sum k_i\alpha_i$ with $k_i \geqslant 0$ for each i. Thus $B(H_\beta, H_\beta) = \sum k_i\beta(H_{\alpha_i}) \leqslant 0$. Hence if $A(\rho - \beta) \neq 0$ then $\beta = 0$. We have thus shown that $Q = cA(\rho)$. But $\int_T Q(t)\overline{Q(t)}dt = w(G)$. Thus (2) of 4.9.4 implies that $Q = A(\rho)$.

4.9.6 *Theorem* Let (π, V) be an irreducible finite dimensional representation of G. Let Λ in $\hat{T}$ be the highest weight of (π, V) (see 4.6.12). Let χ_V be the character of (π, V). Then

$$A(\rho)\chi_V|_T = A(\rho + \Lambda).$$

PROOF As a T-representation $V = \sum V_\mu$ the sum taken over the weights of V. Thus $\chi_V|_T = \sum m_\mu t^\mu$ for t in T, where $m_\mu = \dim V_\mu$ and $m_\Lambda = 1$.

Partially order the weights of V by $\mu > \nu$ if $\mu - \nu$ is a sum of positive roots. If μ is a weight of V then $\mu = \Lambda - \beta$ with β a sum of positive roots. Thus if μ is a weight of V and if $s\mu \geqslant \Lambda$ then $s\mu = \Lambda$. This implies that $A(\rho)\chi_V|_T =$

$A(\rho + \Lambda) + \sum c_\beta A(\rho + \Lambda - \beta)$ with the sum taken over certain β which are sums of positive roots so that $A(\rho + \Lambda - \beta) \neq 0$, $\rho + \Lambda - \beta$ is dominant integral, and $s(\rho + \Lambda - \beta) \neq \Lambda + \rho$ for all s in $W(T)$. Thus $A(\rho)\chi_V|_T = A(\Lambda + \rho) + f$ with $\int_T A(\Lambda + \rho)(t)\overline{f(t)}dt = 0$.

Thus,

$$\int_T A(\rho)(t)\overline{A(\rho)}(t)\chi_V(t)\overline{\chi_V}(t)\,dt = w(G) + \int_T f(t)\overline{f(t)}\,dt.$$

Now 2.9.4, 4.8.2, and 4.9.2 combine to show that $\int_T A(\rho)(t)\overline{A(\rho)}(t)\chi_V(t)\overline{\chi_V}(t)dt = w(G)$. Hence $\int_T f(t)\overline{f(t)}dt = 0$. Thus $f = 0$ and the theorem is proved.

4.9.7 *Theorem* Let (π, V) and Λ be as in 4.9.6. Then

$$\dim V = \prod \frac{(\Lambda + \rho)(H_\alpha)}{\rho(H_\alpha)}$$

the product taken over the positive roots.

PROOF Clearly $\chi_V(e) = \dim V$. As usual, if λ is in $\mathfrak{h}_R^*$ let H_λ in $\mathfrak{h}_R$ be defined by $B(H_\lambda, H) = \lambda(H)$, for H in $\mathfrak{h}_R$ (B is the Killing form of the complexification of the Lie algebra of G). 4.9.6 implies that

$$\dim V = \lim_{t\to 0} \chi_V(\exp((-1)^{\frac{1}{2}}tH_\rho) = \lim_{t\to 0} \frac{A(\Lambda + \rho)(\exp((-1)^{\frac{1}{2}}tH_\rho)}{A(\rho)(\exp((-1)^{\frac{1}{2}}tH_\rho)}.$$

Now the definition of $A(\lambda)$ implies that

$$A(\gamma)(\exp((-1)^{\frac{1}{2}}tH_\mu) = A(\mu)((-1)^{\frac{1}{2}}tH_\gamma).$$

This implies that

$$\dim V = \lim_{t\to 0} \frac{A(\rho)(\exp((-1)^{\frac{1}{2}}tH_{\Lambda+\rho}))}{A(\rho)(\exp((-1)^{\frac{1}{2}}tH_\rho))}.$$

Now

$$\begin{aligned} A(\rho)(\exp((-1)^{\frac{1}{2}}tH) &= e^{(-1)^{\frac{1}{2}}t\rho(H)} \prod (1 - e^{-(-1)^{\frac{1}{2}}t\alpha(H)}) \\ &= (e^{\frac{1}{2}(-1)^{\frac{1}{2}}t\alpha(H)} - e^{-\frac{1}{2}(-1)^{\frac{1}{2}}t\alpha(H)}) \end{aligned}$$

all products taken over the positive roots. This implies that

$$\begin{aligned} \dim V &= \lim_{t\to 0} \prod \sin(\tfrac{1}{2}t\alpha(H_{\Lambda+\rho}))/\sin(\tfrac{1}{2}t\alpha(H_\rho)) \\ &= \lim_{t\to 0} \prod (t\alpha(H_{\Lambda+\rho}) + O(t^3))/(t\alpha(H_\rho) + O(t^3)) \\ &= \lim_{t\to 0} \prod (\alpha(H_{\Lambda+\rho}) + O(t^2))/(\alpha(H_\rho) + O(t^2)) \\ &= \prod (\alpha(H_{\Lambda+\rho}))/(\alpha(H_\rho)) \end{aligned}$$

all products taken over the positive roots. These computations prove the theorem.

4.10 The Ring of Virtual Representations

4.10.1 *Definition* Let G be a compact Lie group. Let $E(G)$ be the set of equivalence classes of finite dimensional irreducible representations of G. If (π, V) is a finite dimensional representation of G let $[\pi, V]$ denote its equivalence class. $E(G)$ is a commutative semiring under addition given by

$$[\pi_1, V_1] + [\pi_2, V_2] = [\pi_1 \oplus \pi_2, V_1 \oplus V_2]$$

and multiplication given by $[\pi_1, V_1] \cdot [\pi_2, V_2] = [\pi_1 \otimes \pi_2, V_1 \otimes V_2]$. Set $R(G) = K(E(G))$ (see 1.6). Then $R(G)$ is called the ring of virtual representations of G.

4.10.2 Let $C(G)$ be the ring of all complex valued continuous functions on G. Let $R^{\#}(G)$ be the subring of $C(G)$ generated by the characters of the finite dimensional representations of G.

4.10.3 *Lemma* The map $[\pi, V] \xrightarrow{\varphi} \chi_V$ (χ_V is the character of (π, V)) induces a ring isomorphism φ, between $R(G)$ and $R^{\#}(G)$.

PROOF The map φ is easily seen to be a semiring homomorphism (see 1.6). Hence φ induces a ring homomorphism $\bar{\varphi}$ of $R(G)$ into $R^{\#}(G)$. $\bar{\varphi}$ is clearly surjective. If γ is in $R(G)$ then $\gamma = [\pi_1, V_1] - [\pi_2, V_2]$ and if $\bar{\varphi}(\gamma) = 0$ then $\chi_{V_1} = \chi_{V_2}$. Thus $\gamma = 0$. Hence is injective.

4.10.4 If $[\pi, V]$ is in $E(G)$ define $\dim[\pi, V] = \dim V$. Then dim is a semiring homomorphism of $E(G)$ into Z. Hence dim induces a ring homomorphism of $R(G)$ into Z. (Note that $(\dim \circ \bar{\varphi}^{-1})(\chi) = \chi(e)$ for χ in $R^{\#}(G)$).

4.10.5 Let for $[\pi_1, V_1], [\pi_2, V_2]$ in $E(G)$, $\langle [\pi_1, V_1], [\pi_2, V_2] \rangle_G = \dim \mathrm{Hom}_G(V_1, V_2)$ (see 2.2.3 for notation). Then $\langle\ ,\ \rangle_G$ extends to a bilinear pairing of $R(G) \times R(G)$ with values in Z. We note that if χ_1, χ_2 are in $R^{\#}(G)$ then $\langle \bar{\varphi}^{-1}\chi_1, \bar{\varphi}^{-1}\chi_2 \rangle_G = \int_G \chi_1(g)\, \overline{\chi_2(g)}\, dg$. Thus if γ, δ are in $\hat{G}$ then $\langle \gamma, \delta \rangle_G = 0$ if $\gamma \neq \delta$, $\langle \gamma, \delta \rangle_G = 1$ if $\gamma = \delta$. This proves the following lemma.

4.10.6 *Lemma* $\hat{G}$ forms a free basis of $R(G)$ over Z.

4.10.7 Let G and H be compact Lie groups. Let φ be a continuous homomorphism of H into G. If $[\pi, V]$ is in $E(G)$ define $\varphi^*[\pi, V] = [\pi \circ \varphi, V]$ in $E(H)$. Then φ^* is a semiring homomorphism. Hence φ^* induces a ring homomorphism of $R(G)$ into $R(H)$ which we also denote φ^*.

4.10.8 If $[\pi, V]$ is in $E(G)$ define $\lambda^i([\pi, V]) = [\Lambda^i\pi, \Lambda^i V]$. Set $\lambda_{-1}[\pi, V] = \sum (-1)^i \lambda^i[\pi, V]$.

4.10.9 *Proposition* Let G be a compact, connected Lie group and let K be a connected, closed subgroup of maximal rank of G. Let (π, V) be the complexified isotropy representation of K on G/K. Set $\lambda_{-1}(G/K) = \lambda_{-1}([\pi, V])$. Let i be the canonical injection of K into G. If α, β are in $R(G)$ then

$$\frac{w(G)}{w(K)} \langle \alpha, \beta \rangle_G = \langle \lambda_{-1}(G/K) i^*\alpha, i^*\beta \rangle_K.$$

PROOF This proposition is just a restatement of 4.7.4 if one observes that $\chi_{\bar{\lambda}_{-1}(G/K)}(k) = \det(I - \pi(k))$ for k in K.

4.10.10 Let $\hat{R}(G)$ denote the set of all formal sums $\sum a_\gamma \gamma$, with a_γ in Z and the summation taken over the elements of $\hat{G}$ (there may possibly be an infinite number of a_γ nonzero). Then $\hat{R}(G)$ has a natural additive group structure.

4.10.11 Let G and H be compact Lie groups. Let φ be a continuous homomorphism of H into G. If μ is in $E(H)$ then set $\varphi_*(\mu) = \sum \langle \varphi^*\gamma, \mu \rangle_H \gamma$, the sum taken over the elements of $\hat{G}$. Then φ_* defines a semigroup homomorphism of $E(H)$ into $R(G)$. φ_* therefore extends to a homomorphism of $R(H)$ into $R(G)$.

4.10.12 *Proposition* Let G be a compact connected Lie group and let T be a maximal torus of G. Let i be the canonical injection of T into G. Then i^* is injective and Im i^* is the ring of invariants of $W(T)$ in $R(T)$.

PROOF The injectivity of i^* is a direct consequence of the conjugacy of maximal tori of G and the fact that every element of G is contained in some maximal torus of G. Clearly, $i^*(R(G))$ is contained in the $W(T)$ invariants of $R(T)$. Let P be a Weyl chamber of T. Let $H_i = (2/\alpha_i(H_{\alpha_i}))H_{\alpha_i}$, where $\alpha_1, \ldots, \alpha_l$ are the simple roots corresponding to P. Extend $H_1, \ldots, H_l$ to a basis of $(-1)^{\frac{1}{2}}\mathfrak{h}$, where $\mathfrak{h}$ is the Lie algebra of T. Order $\hat{T}$ lexicographically (see the proof of 3.5.14) relative to this basis. If γ is in $R(T)$ then it is easy to see that if γ is a $E(T)$ invariant then γ is an integral combination of elements of the form $S(\lambda)$ where $S(\lambda) = \sum[s \cdot \lambda]$, the summation taken over the elements of $W(T)$ and λ is in T. Now $S(\lambda) = S(s \cdot \lambda)$. Thus the Weyl group invariants in $R(T)$ are integral combinations of the elements $S(\lambda)$ with $\lambda(H_i) \geqslant 0$ and $\lambda(H_i)$ integers. Suppose that we have shown that $S(\mu)$ is in $i^*R(G)$ for all $\mu < \lambda$. Let (π_λ, V^λ) be the irreducible representation of G with highest weight λ (see 4.6.12). Then $i^*[V^\lambda] = S(\lambda) + \sum n_\mu S(\mu)$ the summation taken over certain μ with μ dominant integral and $\mu < \lambda$. $S(\lambda)$ is thus indeed in $i^*R(G)$.

4.10.13 *Theorem* Let G be a compact, connected Lie group and let H be a closed connected subgroup of G. Let i be the canonical injection of H into G. Extend $\langle\ ,\ \rangle_G$ to $\hat{R}(G) \times R(G)$ in the obvious fashion.

(1) If z is in $R(H)$, x is in $R(G)$ then $\langle i_* z, x\rangle_G = \langle i^* x, z\rangle_H$.

(2) $i_* R(H)$ is orthogonal to $\ker(i^*)$.

(3) If $\operatorname{rank}(H) < \operatorname{rank}(G)$ then the orthogonal complement to $\ker(i^*)$ in $R(G)$ is (0).

PROOF $i_* z = \sum \langle i^*\gamma, z\rangle_H \gamma$, the sum taken over the elements of G. If x is in $R(G)$ then $\langle i_* z, x\rangle_G = \sum \langle i^*\gamma, z\rangle_H \langle\gamma, x\rangle_G$, the sum taken over (a finite number of) the elements of G. Now $x = \sum \langle x, \gamma\rangle_G \gamma$, and hence $i^* x = \sum \langle x, \gamma\rangle_G i^*\gamma$. This implies that $\langle i^* x, z\rangle_H = \sum \langle x, \gamma\rangle_G \langle i^*\gamma, z\rangle_H = \langle i_* z, x\rangle_G$. We have thus proved (1).

(2) is an immediate consequence of (1).

We now prove (3). Let T_1 be a maximal torus of H and let T be a maximal torus of G containing T_1. Let i_T and i_{T_1} be, respectively, the canonical injections of T and T_1 into G and H. Let W and W_1 be, respectively, the Weyl groups of G and H relative to T and T_1. 4.9.12 says that $i_T^*(R(G))$ and $i_{T_1}^*(R(H))$ are, respectively, the rings of invariants, $I(G)$ and $I(H)$, in $R(T)$ and $R(T_1)$ relative to W and W_1. Let j be the canonical injection of T_1 into T. Clearly the following diagram is commutative:

$$\begin{array}{ccc} R(H) & \xleftarrow{\ i^*\ } & R(G) \\ {\scriptstyle i_T^*}\big\downarrow & & \big\downarrow{\scriptstyle i_T^*} \\ I(H) & \xleftarrow{\ j^*\ } & I(G)\,. \end{array}$$

Since $\dim T_1 < \dim T$ there is λ in $\hat{T}$ so that $\lambda \neq 1$ and $j^*\lambda = 1$. Set

$z = \prod (1 - s \cdot \lambda)$, the product taken over the elements of W. Then z is an element of $I(G)$, $z \neq 0$ and $j^*z = 0$.

If μ is in $\hat{T}$, let $\mu^* = \bar{\mu}$. Then $*$ extends to a Z linear map of $R(T)$ to $R(T)$. If u, v are in $R(T)$ then $\langle u, v\rangle_T = \langle 1, u^*v\rangle_T$.

zz^* is in $\ker(j^*)$ and $\langle 1, zz^*\rangle_T = \langle z^*, z^*\rangle_T$. Hence, $zz^* = a1 + \sum a_\mu \mu$, the sum taken over μ in $\hat{T}$, $\mu \neq 1$ and a is a positive integer.

Let $\mathfrak{h}$ be the Lie algebra of T. Let Γ be the unit lattice of $\mathfrak{h}$ ($\Gamma = (\exp_T)^{-1}(e)$). Then T may be identified with the set of all linear forms μ on $\mathfrak{h}$ so that $\mu(\Gamma) \subset (-1)^{\frac{1}{2}}Z$. Let $(\ ,\ )$ be an inner product on $\mathfrak{h}$ so that a free basis of Γ is an orthonormal basis of $\mathfrak{h}$. Then using $(\ ,\ )$ we may identify Γ with $\hat{T}$. Let for each c in R, $c > 0$, $U_c = \{x \text{ in } \mathfrak{h} | (x, x) \geqslant c\}$. Let $\psi_n(\mu)(t) = \mu(t^n)$ for μ in $\hat{T}$ and t in T. Then ψ_n extends to a linear map of $R(T)$ to $R(T)$. Clearly

$$\psi_n(I(G)) \subset I(G).$$

ψ_n is also defined on $R(T_1)$. It is also clear that $\psi_n \circ j^* = j^* \circ \psi_n$. Hence $\ker(i^*)$ is invariant under ψ_n.

We say that $U \subset \mathfrak{h}$ supports u in $R(T)$ if $u = \sum a_k\lambda_k$ with λ_k in $\hat{T}(= \Gamma)$ and λ_k is in U for each k.

Set $y = zz^* - a1$. Then y is supported by $U_{\frac{1}{2}}$. Hence $\psi_n y$ is supported by $U_{n/2}$. This implies that for each integer n, $\ker(i^*)$ contains an element of the form $a1 + y_n$, with y_n supported by U_n.

Suppose that u is in $R(G)$ and that $\langle u, \ker(i^*)\rangle_G = (0)$. Then $\langle u, (s1 + y_n)u\rangle_G = 0$ for all y_n as above. Hence $w(G)a\langle u, u\rangle_G + \langle \lambda_{-1}(G/T)i_T^*(u)i_T^*(u)^*, y_n\rangle_T = 0$. Now $\lambda_{-1}(G/T)i_T^*(u)i_T^*(u)^* = \sum a_k\lambda_k$ with λ_k in $\hat{T}$ and the summation taken over $k = 1, \ldots, m$. Let n be an integer greater than (λ_k, λ_k) for $k = 1, \ldots, m$. Then $\langle \lambda_{-1}(G/T)i_T^*(u)i_T^*(u)^*, y_n\rangle_T = 0$. Hence $\langle u, u\rangle_G = 0$. This implies that $u = 0$ and proves (3).

4.11 Exercises

4.11.1 Let $\mathfrak{g}$ be a Lie algebra over R or C. Let $U(\mathfrak{g})^*$ be the dual space of the universal enveloping algebra of $\mathfrak{g}$, $U(\mathfrak{g})$. Let $\mathfrak{g}$ act on $U(\mathfrak{g})^*$ by $xf(g) = f(gx)$ for f in $U(\mathfrak{g})^*$ and g in $U(\mathfrak{g})$. Let $U(\mathfrak{g})$ act on $U(\mathfrak{g})^* \otimes U(\mathfrak{g})^*$ by $x\cdot(f \otimes g) = (xf \otimes g + f \otimes xg)$. If f, g are in $U(\mathfrak{g})^*$ define $\varepsilon(f \otimes g) = f(1)g(1)$. Then ε defines a linear map of $U(\mathfrak{g})^* \otimes U(\mathfrak{g})^*$ into the base field.

(i) Show that relative to the multiplication $(fg)(z) = (z \cdot (f \otimes g))$, $U(\mathfrak{g})^*$ is a commutative associative algebra with unit.

(ii) Does $U(\mathfrak{g})^*$ have zero divisors?

4.11.2 Let G be a Lie group. Let for f in $C^\infty(G)$, $\eta(f)$ in $U(\mathfrak{g})^*$ be defined by $\eta(f)(z) = z \cdot f(e)$ for z in $U(\mathfrak{g}) = D(G)$ (see 4.2.7). Show that η is an algebra homomorphism of $C^\infty(G)$ into $U(\mathfrak{g})^*$.

4.11.3 Let $T(\Lambda)$ be as in 4.4.4. Let $\tilde{T}(\Lambda) = \sum_\mu T(\Lambda)_\mu$, the sum taken over the weights of $T(\Lambda)$.

(i) Show that $\tilde{T}(\Lambda)$ is an invariant subspace of $T(\Lambda)$.

(ii) Show that if $W \subset \tilde{T}(\Lambda)$ is an invariant subspace and if $W \neq (0)$ then $W \supset V^\Lambda$.

4.11.4 Using the maximal torus of $SO(n)(n \geqslant 3)$ in 3.11.20 compute the lattices Γ_G and Γ^1 (see 4.6). Show that $\pi_1(SO(n)) = Z_2$ if $n \geqslant 3$,

4.11.5 The universal covering group of $SO(n)$ is called Spin(n). Let G be a compact connected Lie group and let Ad be the adjoint representation of G on its Lie algebra, $\mathfrak{g}$. Let (,) be an invariant inner product on $\mathfrak{g}$. Using an orthonormal basis of $\mathfrak{g}$ relative (,), Ad induces a Lie homomorphism of G into $SO(n)$. Show that ρ (see 4.6.13) is a weight of G if and only if there is a Lie homomorphism π of G into Spin(n) so that π is the lift of Ad.

4.11.6 Let G be a compact connected Lie group and let T be a maximal torus of G. Let λ be in $\hat{T}$ and suppose that $\lambda(ntn^{-1}) = \lambda(t)$ for all t in T, n in $N(T)$. Show that λ extends to a character of G.

4.11.7 Let G be a compact connected Lie group. Let K be a closed subgroup of G. Show that if rank$(K) <$ rank(G) then $\chi(G/K) = 0$.

4.11.8 In this exercise we give an explicit construction of Spin(n). Let $V = R^n$. Let for v in V, $v^*(w) = \langle v, w\rangle$ for w in V. If z is in $\Lambda^k V$ (see A.3.1.15) then identifying $\Lambda^k V$ with the space of all alternating k-linear forms on V^* (see A.3.1.16), we define for v in V, $\varepsilon(v)z = v \wedge z$ and $i(v)z(v_1^*, \ldots, v_{k-1}^*) = z(v^*, v_1^*, \ldots, v_{k-1}^*)$. Then $\varepsilon(v)\colon \Lambda^k V \to \Lambda^{k+1} V$ and $i(v)\colon \Lambda^k V \to \Lambda^{k-1} V$. Argue as in 1.7.10 to show that $(\varepsilon(v) + i(v))^2 = -\langle v, v\rangle I$ on ΛV. Let $\mathfrak{g}$ be the linear span of all $(\varepsilon(v) + i(v))(\varepsilon(w) + i(w))$, $\langle v, w\rangle = 0$. Show that $\mathfrak{g}$ is a Lie subalgebra of $L(\Lambda V, \Lambda V)$. Show that $\mathfrak{g}$ is isomorphic as a Lie algebra with the Lie algebra $so(n)$ of $SO(n)$. (Hint: Let $e_1, \ldots, e_n$ be the standard basis of R^n. Let $E_{ij}e_k = \delta_{jk}e_i$. Let $\eta(E_{ij} - E_{ji}) = -\frac{1}{2}(\varepsilon(e_i) + i(e_i))(\varepsilon(e_j) + i(e_j))$ for $i \neq j$.) Let for v in V a unit vector $A_v w = w - 2\langle v, w\rangle v$. Show that every element of $SO(n)$ is a product of an even number of A_v. Let G be the connected subgroup of $GL(\Lambda V)$ corresponding to g. Show that every element of G is a product of an even number of elements of the form $(\varepsilon(v) + i(v))$ with $\langle v, v\rangle = I$. Show that the map $\varepsilon(v) + i(v) \mapsto A_v$ for $\langle v, v\rangle = 1$ induces a twofold covering map of G onto $SO(n)$. Use 4.11.4 to deduce that if $n \geqslant 3$ then G is Lie isomorphic with Spin(n).

4.11.9 Show that Spin(3) is Lie isomorphic with $SU(2)$ and that Spin(4) is Lie isomorphic with $SU(2) \times SU(2)$.

4.11.10 Show that a connected, compact, complex Lie group is a torus.

4.11.11 Let G be a connected complex semisimple Lie group (that is, G is a complex Lie group with semisimple Lie algebra). Show that every irreducible finite dimensional representation of G is a tensor product of an irreducible holomorphic representation with an irreducible antiholomorphic representation (that is, the complex conjugates of the matrix entries are holomorphic).

4.11.12 Let **K** be the algebra over R with basis $1, i, j, k$ and multiplication $ij = k$, $ik = -j$, $jk = i$. If $z = z_0 + z_1 i + z_2 j + z_3 k$ then set

$$\bar{z} = z_0 - z_1 i - z_2 j - z_3 k.$$

Show that $z\bar{z} = z_1^2 + z_1^2 + z_2^2 + z_3^2$. Use this to show that **K** is a division ring (**K** is known as the quaternions). Let $Sp(n)$ denote the group of all $n \times n$ matrices with entries in **K** such that $g^t\bar{g} = I$. Show that $Sp(n)$ is a compact Lie group. Show that $Sp(1)$ is isomorphic with $SU(2)$. Let

$$G_1 = \left\{ \left. \begin{bmatrix} a_1 & & 0 \\ & \ddots & \\ 0 & & a_n \end{bmatrix} \right| a_i \text{ in } Sp(1) \right\}.$$

Let $T \subset Sp(1) = SU(2)$ be a maximal torus. Show that

$$T = \left\{ \left. \begin{bmatrix} t_1 & & \\ & \ddots & \\ & & t_n \end{bmatrix} \right| t_i \text{ in } T \right\} \text{ is a maximal torus in } Sp(n).$$

Show that $Sp(n)$ is simply connected.

4.11.13 Noting that the Lie algebra of $SU(2)$, $su(2)$, is a compact form of $sl(2, C)$, use 4.3.10 to find the irreducible unitary representations of $SU(2)$.

Let T be the torus of diagonal elements of $SU(2)$. Use 4.9.6 to compute the restrictions of characters of $SU(2)$ to T. Find the multiplication table for $R(SU(2))$.

4.12 Notes

4.12.1 The representations (π_Λ, V^Λ) of 4.4.5 were first constructed in Harish-Chandra [1]. Our construction is based on Wallach [1].

4.12.2 4.5.3 is due to E. Cartan and H. Weyl. The proof we give is the simplification Cartier [1] gives of the proof in Harish-Chandra [1] (the first proof that did not use the classification of complex semisimple Lie algebras).

4.12.3 The basic results of 4.6 are due to Stiefel [1].

4.12.4 4.8 and 4.9 contain the germinal results of Weyl [1].

4.12.5 The idea of reformulating the Weyl theory of representations of compact Lie groups in terms of the ring of virtual representations is due to Bott [1]. 4.10.13 is also due to Bott [1].

CHAPTER 5

Harmonic Analysis on a Homogeneous Vector Bundle

5.1 Introduction

In this chapter we first develop the basic theory of homogeneous vector bundles and homogeneous differential operators. We then study the spectrum of the Laplacian on a homogeneous vector bundle over G/K where G is a compact Lie group. Using the growth properties of the spectrum of the Laplacian we develop the Sobolev theory of H^s spaces. In particular, we give sufficient conditions for convergence of Peter–Weyl series. We then use the Sobolev theory to study hypoellipticity for homogeneous differential operators. We then prove an index theorem (due to Bott) for globally hypoelliptic operators. We use the index theorem to give another proof of 3.8.2. As an appendix to this chapter, we give a proof of the Fourier integral theorem using the H^s theory.

5.2 Homogeneous Vector Bundles

5.2.1 *Definition* Let G be a Lie group and let K be a closed subgroup of G. Let $M = G/K$. A vector bundle E over M is called a homogeneous vector bundle if G acts on E on the left and the G action satisfies:

(1) $gE_x = E_{gx}$ for x in M, g in G.

(2) The mapping from E_x to E_{gx} induced by g is linear for g in G and x in M. (See 1.1 for unexplained notation.)

5.2.2 We give a basic construction that describes all homogeneous vector bundles over M. Let (τ, E_0) be a finite dimensional representation of K. Let K act on the right on $G \times E_0$ as follows: $(g, v)k = (gk, \tau(k)^{-1}v)$ for g in G, v in E_0, and k in K. We set $E = G \underset{\tau}{\times} E_0 = (G \times E_0)/K$. Set $[g, v] = (g, v)K$, and set $p([g, v]) = gK$. Then p is a well defined map from E to M. We give E the quotient topology and relative to this topology p is clearly continuous. If x is in M let U be an open neighborhood of x so that there is a C^∞ mapping $\psi\colon U \to G$ such that $\psi(x)K = x$ (see A.2.4.2). We define a map $\Psi\colon p^{-1}(U) \to U \times E_0$ by $\Psi(\psi(x),v) = (x, v)$. One sees easily that Ψ is well defined and surjective. It is also easy to see that Ψ is injective. By its definition Ψ is continuous and open. Hence E is indeed a vector bundle over M. Using U, a coordinate neighborhood of M and a basis of E_0, it is also not hard to see that the pairs (Ψ, U) as above define a C^∞ atlas for E so that E is a C^∞ vector bundle over M. We define a left action of G on $G \times E_0$ by $g_0 \cdot (g, v) = (g_0 g, v)$. This action of G on $G \times E_0$ induces an action of G on E making E into a homogeneous vector bundle.

5.2.3 *Lemma* Let E be a homogeneous vector bundle over M. Let $E_0 = E_{ek}$ and let $\tau(k)$ be the action of k on E_0 guaranteed by 5.2.1(2) for k in K. Then E is isomorphic as a homogeneous vector bundle with $G \underset{\tau}{\times} E_0$.

PROOF We define a map ξ from $G \times E_0$ to E by $\xi(g, v) = gv$. Suppose that $\xi(g, v) = \xi(g', v')$. Then $g \cdot v = g' \cdot v'$. Hence $g^{-1}g'v' = v$. In particular this says that $g^{-1}g' = k$ an element of K. Thus $g' = gk$ and $v' = \tau(k)^{-1}v$. Hence $\xi(g, v) = \xi(g', v')$ if and only if $[g, v] = [g', v']$. Thus ξ induces $\bar{\xi}$, a vector bundle isomorphism of $G \times E_0$ to E.

5.2.4 *Lemma* Let H be a closed subgroup of K. Let (τ, E_0) be a finite dimensional representation of K and let (τ_H, E_0) be the restriction of (τ, E_0) to H. If η is the natural map of G/H to $G/K(\eta(gH) = gK)$ then $\eta^*(G \underset{\tau}{\times} E_0) = G \times_{\tau_H} E_0$ [see 1.3.1 for the definition of $\eta^*(G \times E_0)$].

PROOF Set $E = G \underset{\tau}{\times} E_0$. Then $\eta^*E = \{(x, v) | v \text{ in } E_{\eta(x)}\}$. We define a G action on $\eta^*(E)$ by $g(x, v) = (gx, gv)$. This makes η^*E into a homogeneous vector bundle. It is also clear that the action of H on $(\eta^*E)_{eH}$ is just τ_H. 5.2.4 now follows from 5.2.3.

5.2.5 *Lemma* Let G be a Lie group and let K be a closed subgroup of G. Let (τ, E_0) and (σ, F_0) be finite dimensional representations of K. Let $E = G \underset{\sigma}{\times} E_0$ and let $F = G \underset{\tau}{\times} F_0$.

(1) $E \oplus F = G \underset{\tau \oplus \sigma}{\times} (E_0 \oplus F_0)$.

(2) $E \otimes F = G \underset{\tau \oplus \sigma}{\times} (E_0 \otimes F_0)$.

PROOF The lemma follows from 5.2.3 if we note that the action of K on $(E \oplus F)_{ek}$ is $\tau \oplus \sigma$ and that the action of K on $(E \otimes F)_{e_k}$ is $\tau \otimes \sigma$.

5.2.6 We note that a homogeneous vector bundle over M is the same thing as a G-vector bundle over M in the sense of 2.4.4. We will call a homogeneous vector bundle over M a unitary homogeneous vector bundle if it is a unitary G-vector bundle in the sense of 2.4.4.

5.2.7 *Lemma* Let (τ, E_0) be a finite dimensional representation of K. Let $\langle\ ,\ \rangle$ be a Hermitian inner product on E_0 so that relative to $\langle\ ,\ \rangle$, (τ, E_0) is a unitary representation of K. Then there is a unique unitary structure on $G \underset{\tau}{\times} E_0 = E$ that is homogeneous and is such that if we identify E_{ek} with E_0 in the canonical manner, then the inner product on E_0 so induced agrees with $\langle\ ,\ \rangle$.

PROOF Set $([g, v], [g, w])_{gk} = \langle v, w\rangle$ for v, w in E_0, g in G. We leave it to the reader to show that that $(\ ,\ )_{gk}$ is well defined and hence is a unitary structure on E satisfying the conditions of the lemma. We also leave it to the reader to check that if $(\ ,\ )$ is a homogeneous unitary structure on E, then $(\ ,\ )$ is given as above if we take for $\langle\ ,\ \rangle$ the inner product on E_{ek}.

5.3 Frobenius Reciprocity

5.3.1 Let G be a compact Lie group and let K be a closed subgroup of G. We assume that G/K is orientable. Let ω be a G-invariant volume element on M. ω may be chosen so that if f is in $C(G/K)$ then

$$\int_G f(gK)\,dg = \int_{G/K} f\omega.$$

5.3.2 Let E be a unitary homogeneous vector bundle over $M = G/K$.

Let G act on ΓE by $(\pi(g)f)(x) = gf(g^{-1}x)$ for g in G, f in ΓE, and x in G/K. 2.4.6 implies that $(\pi, \Gamma E)$ extends to a unitary representation $(\pi, L^2(E))$ of G. In the next paragraph we give an alternate realization of this unitary representation of G.

5.3.3 According to 5.2.7, $E = G \times_\tau E_0$ with (τ, E_0) a unitary finite dimensional representation of K. Let $C(G; \tau)$ be the space of all continuous mappings of G into E_0, f, so that $f(gk) = \tau(k)^{-1}f(g)$ for g in G and k in K. If f is in $C(G; \tau)$ define $(\tilde{\pi}(g_0)f)(g) = f(g_0^{-1}g)$. Let for f_1, f_2 in $C(G; \tau)$, $(f_1, f_2) = \int_G \langle f_1(g), f_2(g)\rangle dg$. Applying 2.4.6 to $G \times E_0$ we see that $\tilde{\pi}$ extends to a unitary representation $(\tilde{\pi}, L^2(G; \tau))$ of G. If f is in ΓE define $\tilde{f}(g) = g^{-1}f(gK)$ for g in G. Using the canonical identification of E_0 with E_{ek} we see that $\tilde{f}$ is in $C(G; \tau)$.

5.3.4 *Lemma* The map $f \mapsto \tilde{f}$ of ΓE to $C(G; \tau)$ extends to a unitary equivalence of $(\pi, L^2(E))$ with $(\tilde{\pi}, L^2(G; \tau))$.

PROOF Let f_1, f_2 be in ΓE. Then $(f_1, f_2) = \int_{G/K} \langle f_1, f_2\rangle\omega$.

$$\begin{aligned}(\tilde{f}_1, \tilde{f}_2) &= \int_G \langle \tilde{f}_1(g), \tilde{f}_2(g)\rangle dg \\ &= \int_G \langle g^{-1}f_1(gK), g^{-1}f_2(gK)\rangle dg = \int_G \langle f_1(gK), f_2(gK)\rangle\, dg \\ &= \int_M \langle f_1, f_2\rangle\omega = (f_1, f_2).\end{aligned}$$

The map $f \mapsto \tilde{f}$ therefore extends to a unitary mapping of $L^2(E)$ to $L^2(G; \tau)$, which we call A. We note that A is surjective, since if h is in $C(G; \tau)$ set $f(gK) = [g, h(g)]$. Then f is well defined and hence defines a cross section of E. Clearly $\tilde{f} = h$. To complete the proof we need to show that A intertwines π and $\tilde{\pi}$.

$$\begin{aligned}(\pi(g_0)f)^{\sim}(g) &= g^{-1}(\pi(g_0)f)(gK) \\ &= g^{-1}g_0f(g_0^{-1}gK) = \tilde{f}(g_0^{-1}g) = (\tilde{\pi}(g_0)\tilde{f})(g).\end{aligned}$$

We have therefore proved the lemma.

5.3.5 Let $\hat{G}$ denote the set of all equivalence classes of finite dimensional irreducible representations of G. Let for each γ in $\hat{G}$, (π_γ, V_γ) be a representative of γ.

Let (τ, E_0) be a representation of K. To each γ in G we assign a map A_γ from $V_\gamma \otimes \mathrm{Hom}_K(V_\gamma, E_0)$ to $C(G; \tau)$ by the rule $A_\gamma(v \otimes L)(g) = L(g^{-1}v)$. We note that $(\tilde{\pi}(g)A_\gamma(v \otimes L)) = A_\gamma(\pi_\gamma(g)v \otimes L)$.

5.3.6 *Theorem* Let (τ, E_0) be a finite dimensional unitary representation of K. In the notation of 5.3.5, the unitary representation $(\pi, L^2(G \times E_0))$ is the unitary direct sum

$$\sum_{\gamma} V_\gamma \otimes \mathrm{Hom}_K(V_\gamma, E_0),$$

the sum taken over the elements of G.

Furthermore the algebraic direct sum

$$\sum A^{-1}(A_\gamma(V_\gamma \otimes \mathrm{Hom}_K(V_\gamma, E_0)))$$

is uniformly dense in $\Gamma(G \underset{\tau}{\times} E_0)$ relative to the uniform topology (here A is the mapping of 5.3.4).

PROOF Let $C(G; E_0)$ be the space of all continuous functions from G to E_0. Let $G \times G$ act on $C(G; E_0)$ by $(\eta(g, h)f)(x) = f(g^{-1}xh)$. Then $(\eta, C(G; E_0))$ extends to a unitary representation $(\eta, L^2(G; E_0))$ relative to the L^2 inner product. 2.8.2 implies that as a unitary representation of $G \times G$, $L^2(G; E_0) = \sum V_\gamma \otimes V_\gamma^* \otimes E_0$. Here

$$\eta(g, h)(v \otimes \lambda \otimes w) = \pi_\gamma(g)v \otimes \pi_\gamma^*(g)\lambda \otimes w$$

for g, h in G, v in V_γ, λ in V_γ^*, and w in E_0.

We identify $V_\gamma^* \otimes E_0$ with $L(V_\gamma, E_0)$ in the canonical fashion. Then $\eta(g, h)(v \otimes L) = \pi_\gamma(g)v \otimes L\pi_\gamma(h)^{-1}$.

$L^2(G; \tau)$ is a closed invariant subspace of $L^2(G; E_0)$ relative to the action $\pi(g) = \eta(g, e)$ on $L^2(G; E_0)$. Hence $L^2(G; \tau) = \sum (V_\gamma \otimes L(V_\gamma, E_0)) \cap L^2(G; \tau)$ as a representation of G. Suppose that $f = v \otimes L$ is in $L^2(G; \tau)$ for v in V_γ, $v \neq 0$, and L in $L(V_\gamma, E_0)$. Then $f(g) = L(\pi_\gamma(g^{-1})v)$. Now $f(gk) = \tau(k)^{-1}f(g)$ for k in K and g in G. Hence $L(\pi_\gamma(k)^{-1}\pi_\gamma(g)v) = \tau(k)^{-1}L(\pi_\gamma(g)^{-1}v)$. The irreducibility of V implies that $L(\pi_\gamma(k)w) = \tau(k)L(w)$ for all w in V and k in K. Tracing this line of reasoning backwards shows that

$$V \otimes L(V_\gamma, E_0) \cap L^2(G; \tau) \supset V_\gamma \otimes \mathrm{Hom}_K(V_\gamma, E_0).$$

The multiplicity of V_γ in $L^2(G; \tau)$ is $\dim \mathrm{Hom}_G(V_\gamma, C(G; \tau))$. If B is in $\mathrm{Hom}_G(V_\gamma, C(G; \tau))$ then define $B^\#(v) = B(v)(e)$. Then $B^\#$ maps V to E_0. Furthermore,

$$\begin{aligned} B^\#(\pi_\gamma(k)v) &= B(\pi_\gamma(k)v)(e) \\ &= B(v)(k^{-1}) = \tau(k)B(v)(e) = \tau(k)B^\#(v). \end{aligned}$$

Thus $B^\#$ is in $\mathrm{Hom}_K(V_\gamma E_0)$. If $B^\# = 0$ then $B(v)(e) = 0$ for all v in V. Hence, if v is in V_γ and g is in G, then $0 = B(\pi_\gamma(g^{-1})v)(e) = (\pi(g^{-1})B(v))(e) = B(v)(g)$. This implies that if $B^\# = 0$ then $B = 0$. The map $B \mapsto B^\#$ of

$$\mathrm{Hom}_G(V_\gamma, C(G; \tau))$$

into $\mathrm{Hom}_K(V_\gamma, E_0)$ is thus injective. This clearly implies that

$$(V_\gamma \otimes L(V_\gamma, E_0)) \cap L^2(G; \tau) = V_\gamma \otimes \mathrm{Hom}_K(V_\gamma E_0).$$

This proves the first assertion of the theorem.

The second assertion of the theorem follows from the fact that $C(G; \tau)$ is closed in $C(G; E_0)$ relative to the uniform topology and 2.8.5.

5.3.7 *Proposition* Let (τ, E_0) be a finite dimensional unitary representation of K. Set $E = G \times E_0$. If γ is in $\hat{G}$ set

$$\Gamma_\gamma E = A^{-1}(A_\gamma(V_\gamma \otimes \mathrm{Hom}_K(V_\gamma, E_0))).$$

Then $L^2(E)$ is the unitary direct sum $\sum \Gamma_\gamma E$. Furthermore if P_γ is the projection of $L^2(E)$ onto $\Gamma_\gamma E$ then

$$P_\gamma(f) = d(\gamma)\pi(\bar{\chi}_\gamma)f$$

for f in $L^2(E)$. Here, $d(\gamma) = \dim V_\gamma$, χ_γ is the character of V_γ, and $\pi(\bar{\chi}_\gamma)$ is defined as in 2.7.1.

PROOF It is enough to show that if γ and μ are in $\hat{G}$ and if v is in V_γ, L is in $\mathrm{Hom}_K(V_\gamma, E_0)$, then

$$(*)\quad d(\gamma) \int_G \bar{\chi}_\gamma(g) L(\pi_\gamma(g^{-1}g_0)^{-1}v)\, dg = \delta_{\gamma,\mu} L(\pi_\mu(g_0)^{-1}v).$$

But this follows from 2.9.5 since $\bar{\chi}_\mu$ is the character of V_μ^* and if λ is in E_0^* then the map $g \mapsto \lambda(L\pi_\gamma(g^{-1})v)$ is a matrix entry of the representation V_γ^*.

5.3.8 *Definition* Let f be in $L^2(E)$ (the notation is as in 5.3.7). Define $\hat{f}(\gamma) = P_\gamma f$ in $\Gamma_\gamma E$ for each γ in $\hat{G}$. We call $\hat{f}$ the Fourier transform of f. We will sometimes identify $\hat{f}(\gamma)$ with the corresponding element of

$$V_\gamma \otimes \mathrm{Hom}_K(V_\gamma, E_0).$$

5.4 Homogeneous Differential Operators

5.4.1 *Definition* Let G be a Lie group and let K be a closed subgroup of G. Let $M = G/K$. Let E and F be homogeneous vector bundles over M. A differential operator D from E to F is called a homogeneous differential operator if $gDf = Dgf$ for g in G and f in E. Here $gf(x) = gf(g^{-1}x)$ for g in G and f in $\Gamma^{\infty}E$.

5.4.2 Our purpose in this section is to give an explicit description of the homogeneous differential operators from E to F.

5.4.3 Let $U(\mathfrak{g})$ be the universal enveloping algebra of $\mathfrak{g}$, the Lie algebra of G. Let $U^i(\mathfrak{g})$ be the subspace U^i of $U(\mathfrak{g})$ defined in the proof of 4.2.4. Let G act on $\mathfrak{g}$ by the adjoint representation. Let ξ be the natural map of $T(\mathfrak{g})$ onto $U(\mathfrak{g})$. We let G act on $T(\mathfrak{g})$ by extending $\mathrm{Ad}(g)$ to an algebra homomorphism for each g in G. Clearly, $\mathrm{Ad}(g)$ preserves $\ker \xi$ and hence Ad induces an action of G on $U(\mathfrak{g})$ so that $(\mathrm{Ad}, U^i(\mathfrak{g}))$ is a representation of G for each i. (Note that $(\mathrm{Ad}, U(\mathfrak{g}))$ is not a representation in our sense since we have not given $U(\mathfrak{g})$ a Hilbert space structure.)

5.4.4 Let (τ, E_0) and (σ, F_0) be finite dimensional representations of $\hat{G}$. Let $E = G \underset{\tau}{\times} E_0$ and $F = G \underset{\sigma}{\times} F_0$. Let $C^{\infty}(G; \tau)$ (resp. $C^{\infty}(G; \sigma)$) be the space of all C^{∞} elements of $C(G; \tau)$ (resp. $C(G; \sigma)$). If f is in $\Gamma^{\infty}E$ then $\tilde{f}$ is in $C^{\infty}(G; \tau)$ (see 5.3.3 for the definition of $\tilde{f}$). Let $f \mapsto f^0$ be the inverse map to $f \mapsto \tilde{f}$.

Let D be a homogeneous differential operator from E to F. Define $\tilde{D}$ by $\tilde{D}f = (Df^0)^{\sim}$ for f in $C^{\infty}(G; \tau)$. Then $\tilde{D}$ is a linear map of $C^{\infty}(G; \tau)$ to $C^{\infty}(G; \sigma)$.

5.4.5 Let $C^{\infty}(G; E_0)$ (resp. $C^{\infty}(G; F_0)$) denote the space of all C^{∞} elements of $C(G; E_0)$ (resp. $C(G; F_0)$). $\mathfrak{g}$ acts on $C^{\infty}(G; E_0)$ by

$$Xf(g) = (d/dt)f(g \exp tX)|_{t=0}.$$

This action induces a representation of $U(\mathfrak{g})$ on $C^{\infty}(G; E_0)$. Let $L(E_0, F_0)$ denote (as usual) the space of all linear maps of E_0 to F_0. If L is in $L(E_0, F_0)$ and if X is in $U(\mathfrak{g})$ then set $(L \otimes X)f = L \cdot Xf$. This rule assigns to each element of $L(E_0, F_0) \otimes U(\mathfrak{g})$ a differential operator from $G \times E_0$ to $G \times F_0$.

5.4.6 Let K act on $L(E_0, F_0) \otimes U(\mathfrak{g})$ as follows: $\mu(k)(L \otimes X) = \sigma(k)L\tau(k)^{-1} \otimes \mathrm{Ad}(k)X$ for L in $L(E_0, F_0)$, X in $U(\mathfrak{g})$. Then

$$(\mu, L(E_0, F_0) \otimes U^j(\mathfrak{g}))$$

is a representation of K for each j. Let

$$\begin{aligned}(L(E_0, F_0) &\otimes U(\mathfrak{g}))^K \\ &= \{D \text{ in } L(E_0, F_0) \otimes U(\mathfrak{g}) \mid \mu(k)D = D \text{ for all } k \text{ in } K\}.\end{aligned}$$

5.4.7 *Lemma* If D is in $(L(E_0, F_0) \otimes U(\mathfrak{g}))^K$ then

$$DC^\infty(G; \tau) \subset C^\infty(G; \sigma).$$

If D is in $L(E_0, F_0) \otimes U(\mathfrak{g})$ and if $DC^\infty(G; \tau) \subset C^\infty(G; \sigma)$ then $\mu(k)D|_{C^\infty(G,\tau)} = D|_{C^\infty(G,\tau)}$ for all k in K. Furthermore D induces a differential operator D^0 from E to F by the rule $D^0 f = (D\tilde{f})^0$.

PROOF If f is in $C^\infty(G; E_0)$, X is in $U(\mathfrak{g})$ and if k is in K then a direct computation shows that $R_k \cdot Xf = \mathrm{Ad}(k)X \cdot R_k \cdot f$. Thus if f is in $C^\infty(G; \tau)$ and if D is in $L(E_0, F_0) \otimes U(\mathfrak{g})$ then $Df(gk) = \sigma(k)^{-1}(\mu(k)D)f(g)$ for all g in G, k in K. This proves the first two assertions of the lemma. The last assertion is a direct consequence of the definitions of the terms involved.

5.4.8 *Definition* Let $\mathfrak{k}$ be the Lie algebra of K. If there is an $\mathrm{Ad}(K)$-invariant complement to $\mathfrak{k}$ in $\mathfrak{g}$ then G/K is called a reductive coset space.

5.4.9 Suppose that G/K is a reductive coset space and that $\mathfrak{p}$ is an $\mathrm{Ad}(K)$-invariant complement to $\mathfrak{k}$ in $\mathfrak{g}$. Let $Y_1, \ldots, Y_m$ be a basis of $\mathfrak{k}$ and let $X_1, \ldots, X_n$ be a basis of $\mathfrak{p}$. Let $\psi(t_1, \ldots, t_n) = (\exp t_1X_1 \ldots \exp t_nX_n)K$. Then ψ is a C^∞ map from R^n to G/K. By the definition of $\mathfrak{p}$, ψ_{*0} is a linear isomorphism of R^n onto $T(G/K)_{eK}$. The inverse function theorem implies that there is a neighborhood U of eK in G/K so that $t_1, \ldots, t_n$ defines a system of coordinates on U. Also Let

$$\phi(t_1, \ldots, t_n, s_1, \ldots, s_m) = \exp t_1X_1 \ldots \exp t_nX_n \exp s_1Y_1 \ldots \exp s_mY_m.$$

Then as above $t_1, \ldots, t_n, s_1, \ldots, s_m$ define a system of local coordinates in a neighborhood of e in G. Proof (2) of 4.2.7 immediately implies the following lemma.

5.4.10 *Lemma* Let P be a polynomial in n indeterminates. Then there is a polynomial Q in $n + m$ indeterminates so that $\deg P = \deg Q$ and

$$P\left(\frac{\partial}{\partial t_1}, \ldots, \frac{\partial}{\partial t_n}\right)f(e) = Q(X_1, \ldots, X_n, Y_1, \ldots, Y_m)f(e).$$

5.4.11 *Proposition* Let $M = G/K$ be a reductive coset space. Let $E = G \times_\tau E^0$ and let $F = G \times_\gamma F^0$ be homogeneous vector bundles over M. Let D be a homogeneous differential operator from E to F of order j. Let $\tilde{D}$ be the corresponding map from $C^\infty(G; \tau)$ to $C^\infty(G; \sigma)$. Then there is an element $\bar{D}$ in $L(E_0, F_0) \otimes U^j(\mathfrak{g})$ so that $\bar{D}|_{C_\infty(G,\tau)} = \tilde{D}$. If K is compact then $\bar{D}$ may be taken to be in $(L(E_0, F_0) \otimes U^j(\mathfrak{g}))^K$.

PROOF We retain the notation of 5.4.9. If $\beta = (\beta_1, \ldots, \beta_m)$ and $\alpha = (\alpha_1, \ldots, \alpha_n)$ are, respectively, m-tuples and n-tuples of nonnegative integers then set $X^\alpha Y^\beta = X_1^{\alpha_1} \cdots X_n^{\alpha_n} Y_1^{\beta_1} \cdots Y_m^{\beta_m}$. 5.4.10 immediately implies that $(\tilde{D}f)(e) = \sum B_{\alpha,\beta}(X^\alpha Y^\beta f)(e)$ with $B_{\alpha,\beta}$ in $L(E_0, F_0)$, for f in $C^\infty(G; \tau)$. Now, $\tilde{D}$ is G-invariant, thus $(\tilde{D}f)(g) = \sum B_{\alpha,\beta}(X^\alpha Y^\beta f)(g)$. The first assertion of the proposition follows by taking $\bar{D} = \sum B_{\alpha,\beta} X^\alpha Y^\beta$.

5.4.10 implies that $\mu(k)\bar{D}|_{C_\infty(G,\tau)} = \bar{D}|_{C_\infty(G,\tau)}$. Let $D_1 = \int_K \mu(k)\bar{D}dk$. Then by the above $D_1|_{C_\infty(G,\tau)} = \tilde{D}$.

5.5 The Symbol and Formal Adjoint of a Homogeneous Differential Operator

5.5.1 *Lemma* Let G be a Lie group and let K be a closed subgroup of G. Let E and F be homogeneous vector bundles over G/K. Let D be a homogeneous differential operator from E to F. If p is in G/K and if λ is in $T(M)_p^*$ then $\sigma(D, g^*\lambda) = g^{-1}\sigma(D, \lambda)g$ (see 1.7.6).

PROOF Let φ be in $C^\infty(M)$ so that $d\varphi_p = \lambda$ and let f be in $\Gamma^\infty E$. Then

$$\sigma(D, d\varphi)f = \frac{1}{m!}\frac{d^m}{dt^m} e^{-t\varphi} D e^{t\varphi} f|_{p=0}.$$

$$e^{-tg\varphi} D e^{tg\varphi} gf = e^{-tg\varphi} Dg(e^{t\varphi}f) = e^{-tg\varphi} g(De^{t\varphi}f) = g(e^{-t\varphi}De^{t\varphi}f).$$

Differentiating we find that

$$(g \cdot \sigma(D, d\varphi)f)(x) = g \cdot \sigma(D, d\varphi_{g^{-1}x})f(g^{-1}x) = g\sigma(D, d\varphi_{g^{-1}x})g^{-1}(g \cdot f)(x).$$

This implies that $\sigma(D, d(g\varphi)) = g\sigma(D, d\varphi_{g^{-1}x})g^{-1}$ as was to be proved.

5.5.2 *Lemma* Let D be a homogeneous differential operator from E to F. Let D^* be its formal adjoint. Then D^* is a homogeneous differential operator from F to E. Let $\bar{D}$ be the element of $(L(E_0, F_0) \times U(\mathfrak{g}))^K$ associated with D as in 5.4.11. Let $\bar{D}^*$ be the formal adjoint of $\bar{D}$ as a differential operator from $G \times E_0$ to $G \times F_0$. Then $\bar{D}^*$ is in $(L(F_0, E_0) \times U(\mathfrak{g}))^K$ and $(\bar{D})^* = (D^*)^-$.

PROOF This lemma follows directly from the definitions of the terms involved in its statement.

5.5.3 Let $\mathfrak{g}_C$ be the complexification of $\mathfrak{g}$, the Lie algebra of G. We look upon $U(\mathfrak{g}_C)$ as the space of all left invariant (complex) differential operators on G. Let for $X = X_1 + (-1)^{\frac{1}{2}}X_2$, X_1, X_2 in $\mathfrak{g}$, $\delta(X) = -X_1 + (-1)^{\frac{1}{2}}X_2$. Extend δ to be an antiautomorphism of $U(\mathfrak{g}_C)$. That is, $\delta(xy) = \delta(y)\delta(x)$ and δ is R linear. If L is in $L(E_0, F_0)$ set $\eta(L) = L^*$.

5.5.4 *Lemma* If D is in $L(E_0, F_0) \otimes U(\mathfrak{g}_C)$ then the formal adjoint of D as a differential operator from $G \times E_0$ to $G \times F_0$ is $D^* = (\eta \otimes \delta)D$.

PROOF If X is in $\mathfrak{g}$ and if f is in $C_0^\infty(G)$ then Stokes' theorem implies that $\int_G Xf(g)dg = 0$.

Let $D = L \otimes X$ with X in $\mathfrak{g}_C$, L in $L(E_0, F_0)$. $X = X_1 + (-1)^{\frac{1}{2}}X_2$ with X_1, X_2 in $\mathfrak{g}$. Let f be in $C_0^\infty(G; E_0)$ and let h be in $C_0^\infty(G; F_0)$. Then

$$\begin{aligned}\int_G \langle Df, h\rangle dg &= \int_G \langle (X_1 + (-1)^{\frac{1}{2}}X_2)f, L^*h\rangle dg \\ &= \int_G \langle X_1 f, L^*h\, dg\rangle + (-1)^{\frac{1}{2}} \int_G \langle X_2 f, L^*h\rangle\, dg \\ &= -\int_G \langle f, X_1 L^*h\rangle\, dg - (-1)^{\frac{1}{2}} \int_G \langle f, X_2 L^*h\rangle\, dg \\ &= \int_G \langle f, L^*(-X_1 + (-1)^{\frac{1}{2}}X_2)h\rangle\, dg.\end{aligned}$$

The lemma now follows.

5.6 The Laplacian

5.6.1 Let G be a compact connected Lie group. Let $\mathfrak{g}$ be the Lie algebra of G. Then $\mathfrak{g} = \mathfrak{z} \oplus \mathfrak{g}_1$ with $\mathfrak{z}$ the center of $\mathfrak{g}$ and $\mathfrak{g}_1 = [\mathfrak{g}, \mathfrak{g}]$ a compact Lie algebra. Let $\langle\ ,\ \rangle$ be an inner product on $\mathfrak{g}$ satisfying

(1) $\langle \mathfrak{g}_1, \mathfrak{z}\rangle = 0$,

(2) $\langle\ ,\ \rangle|_{\mathfrak{g}_1 \times \mathfrak{g}_1} = -B_{\mathfrak{g}_1}$ (see 3.4.3 and 3.6.2),

(3) Let $\Gamma_0 = \{z \text{ in } \mathfrak{z} |\ \exp z = e\}$. Then there is a free basis $z_1, \ldots, z_r$ of Γ_0 so that $\langle z_i, z_j\rangle = \delta_{ij}$.

Let $U(\mathfrak{g})$ be the universal enveloping algebra of $\mathfrak{g}$. Let $x_1, \ldots, x_n$ be an orthonormal basis of $\mathfrak{g}$. Set $\Omega = \sum x_i^2$ in $U(\mathfrak{g})$. Ω is called a Laplacian for G.

5.6.2 *Lemma* Ω is independent of the choice of orthonormal basis of $\mathfrak{g}$. If g is in G then $\mathrm{Ad}(g)\Omega = \Omega$.

PROOF If $y_1, \ldots, y_j$ is an orthonormal basis of $\mathfrak{g}$ then $y_i = \sum a_{ji}x_j$ with $(a_{ij}) = A$, a matrix satisfying ${}^tA = A^{-1}$. Now

$$\sum y_i^2 = \sum_{i,j,k} a_{ji}a_{ki}x_jx_k = \sum \delta_{jk}x_jx_k = \Omega.$$

If g is in G then set $y_i = \mathrm{Ad}(g)x_i$. The definition of $\langle\ ,\ \rangle$ implies that $y_1, \ldots, y_n$ is an orthonormal basis of $\mathfrak{g}$. Thus $\mathrm{Ad}(g)\Omega = \sum y_i^2 = \Omega$.

5.6.3 Let K be a closed subgroup of G. Let (τ, E_0) be a finite dimensional unitary representation of K. We note that $I \otimes \Omega$ is in $(L(E_0, E_0) \otimes U(\mathfrak{g}))^K$. Let $E = G \underset{\tau}{\times} E_0$. 5.4.7 implies that $I \otimes \Omega$ induces a homogeneous differential operator from E to E which we denote Ω_E. Ω_E is called a Laplacian for E. We will usually employ the symbol Ω for Ω_E.

5.6.4 *Lemma* Let T be a maximal torus of G. Let G_1 be the connected subgroup of G corresponding to $\mathfrak{g}_1$ and let Z be the connected subgroup of G corresponding to $\mathfrak{z}$. Let $\mathfrak{h}$ be the Lie algebra of T. Extend the inner product $\langle\ ,\ \rangle$ on $\mathfrak{h}$ to a Hermitian inner product on the space of all complex valued linear forms on $\mathfrak{h}$. Let γ be in $\hat{G}$. Then $\Omega|_{\Gamma_\gamma E}$ (see 5.3.7 for notation) $= -c(\gamma)I$ with $c(\gamma) = \langle\Lambda_\gamma + \rho, \Lambda_\gamma + \rho\rangle - \langle\rho, \rho\rangle$ where Λ_γ, ρ are respectively the highest weight of γ (see 4.6.12) and half the sum of the positive roots of G relative to a Weyl chamber of T.

PROOF We identify (as in 5.3.7) ΓE with $C(G; \tau)$. Then $\Gamma_\gamma E$ corresponds to the space spanned by the functions $(v \otimes L)(g) = L(\pi_\gamma(g)^{-1}v)$ for L in $\mathrm{Hom}_K(V_\gamma, E_0)$ and v in V_γ.

If X is in $\mathfrak{g}$ then

$$(X\cdot(L \otimes v))(g) = \frac{d}{dt}(v \otimes L)(g \exp tX)\bigg|_{t=0}$$

$$= \frac{d}{dt}(L(\pi_\gamma(\exp -tX)\pi_\gamma(g)^{-1}v)\bigg|_{t=0} = -L(\pi_\gamma(X)\pi_\gamma(g)^{-1}v).$$

Hence

$$\Omega \cdot (L \otimes v)(g) = L(\pi_\gamma(\Omega)\pi_\gamma(g)^{-1}v) = L(\pi_\gamma(g)^{-1}\pi_\gamma(\Omega)v)$$

(here we have used 5.6.2). Thus to prove the lemma we need only show that $\pi_\gamma(\Omega) = -c(\gamma)I$.

Now, $\pi_\gamma(g)\pi_\gamma(\Omega)\pi_\gamma(g)^{-1} = \pi_\gamma(\mathrm{Ad}(g)\Omega) = \pi_\gamma(\Omega)$. Hence 2.3.4 implies that $\pi_\gamma(\Omega) = -cI$ for some complex number c. It remains to compute c.

Let $x_1, \ldots, x_r$ be an orthonormal basis of $\mathfrak{z}$ and let $y_1, \ldots, y_m$ be an orthonormal basis of $\mathfrak{g}_1$. Set $\Omega' = \sum x_i^2$ and set $\Omega'' = \sum y_i^2$. Then $\Omega = \Omega' + \Omega''$. Clearly $\mathrm{Ad}(g)\Omega' = \Omega'$ for all g in G. Hence $\mathrm{Ad}(g)\Omega'' = \Omega''$ for all g in G. 2.3.4 implies that there is a linear form λ on $\mathfrak{z}$ so that if z is in $\mathfrak{z}$ then $\pi_\gamma(z)v = \lambda(z)v$ for all v in V. Thus $\pi_\gamma(\Omega') = \sum \lambda(x_i)^2 I$.

Let Δ be the root system of G relative to T. Let $\mathfrak{h}_1 = \mathfrak{h} \cap \mathfrak{g}_1$. Let P be a Weyl chamber of T. $\mathfrak{g}_{1_C} = \mathfrak{h}_{1_C} + \sum \mathfrak{g}_\alpha$. Let E_α in $\mathfrak{g}_\alpha$ be chosen so that $B(E_\alpha, E_{-\alpha}) = 1$ (B is the Killing form of $\mathfrak{g}_{1_C}$, the complex bilinear extension of $B_{\mathfrak{g}_1}$). Let $h_1, \ldots, h_l$ be an orthonormal basis of $(-1)^{\frac{1}{2}}\mathfrak{h}_1$. Then $-\Omega'' = \sum h_i^2 + \sum (E_\alpha E_{-\alpha} + E_{-\alpha}E_\alpha)$, the sum taken over the elements of Δ_p^+. Now $E_\alpha E_{-\alpha} = [E_\alpha, E_{-\alpha}] + E_{-\alpha}E_\alpha = H_\alpha + E_{-\alpha}E_\alpha$. Thus $-\Omega'' = \sum h_i^2 + \sum H_\alpha + 2\sum E_{-\alpha}E_\alpha$, the sums taken over the elements of Δ_p^+. Let $\mathfrak{n}^+ = \sum \mathfrak{g}_\alpha$, the sum taken over the elements of Δ_p^+. Let v in V_γ be so that $v \neq 0$ and $\pi_\gamma(\mathfrak{n}^+)v = 0$. If H is in $\mathfrak{h}_{1_C}$ then $\pi_\gamma(H)v = \Lambda(H)v$ where Λ is a linear form on $\mathfrak{h}_{1_C}$ (see 4.5.3(1)). Furthermore $\Lambda_\gamma = \lambda + \Lambda$.

Now, $\pi_\gamma(\Omega'') = -c''I$. Thus to compute c'' we need only compute $\pi_\gamma(\Omega'')v$. But $\pi_\gamma(E_\alpha)v = 0$ for all α in Δ_p^+. Hence

$$\begin{aligned}\pi_\gamma(\Omega'')v &= -\sum \pi_\gamma(h_i)^2 v - \sum \pi_\gamma(H_\alpha)v \\ &= \{-\langle\Lambda, \Lambda\rangle - 2\langle\Lambda, \rho\rangle\}v.\end{aligned}$$

This implies the asserted formula for $c(\gamma)$.

5.6.5 Let $x_1, \ldots, x_r$ be a free basis of Γ_0 which is orthonormal relative to $\langle\ ,\ \rangle$. Let $\lambda_1, \ldots, \lambda_r$ be complex valued linear forms on $\mathfrak{z}$ defined by $\lambda_i(x_j) = 2\pi(-1)^{\frac{1}{2}}\delta_{ij}$. Let P be a Weyl chamber of T. Let $\Lambda_1, \ldots, \Lambda_l$ be defined by $2\Lambda_i(H_{\alpha j})/\alpha_j(H_{\alpha_j}) = \delta_{ij}$ where $\alpha_1, \ldots, \alpha_l$ are the simple roots relative to P. Let γ be in $\hat{G}$ and let Λ_γ correspond to γ as in 5.6.4, then $\Lambda_\gamma = \sum n_i\lambda_i + \sum m_j\Lambda_j$ with the n_i integers and the m_j nonnegative integers. Set $\|\gamma\|^2 = \sum n_i^2 + \sum m_j^2$.

5.6.6 *Lemma* There are positive constants c_1, c_2 so that

$$c_1\|\gamma\|^2 \leqslant c(\gamma) \leqslant c_2\|\gamma\|^2$$

for all γ in $\hat{G}$.

PROOF 5.6.4 says that $c(\gamma) = \langle\Lambda_\gamma, \Lambda_\gamma\rangle + 2\langle\rho, \Lambda_\gamma\rangle$. $\langle\Lambda_\gamma, \rho\rangle \geqslant 0$ by definition of ρ. Thus

$c(\gamma) \geqslant \langle\Lambda_\gamma, \Lambda_\gamma\rangle \geqslant c_1\|\gamma\|^2$. Also $2\langle\Lambda_\gamma, \rho\rangle \leqslant \langle\Lambda_\gamma, \Lambda_\gamma\rangle + \langle\rho, \rho\rangle$.

Hence $c(\gamma) \leqslant 2\langle\Lambda_\gamma, \Lambda_\gamma\rangle + \langle\rho, \rho\rangle$. There are only a finite number of γ in G so that $\langle\Lambda_\gamma, \Lambda_\gamma\rangle < \langle\rho, \rho\rangle$ and since $\Lambda_\gamma = 0$ if and only if γ is the trivial representation we see that $\langle\rho, \rho\rangle \leqslant c_3\langle\Lambda_\gamma, \Lambda_\gamma\rangle$ for γ nontrivial. Hence $c(\gamma) \leqslant (2 + c_3)\langle\Lambda_\gamma, \Lambda_\gamma\rangle \leqslant c_2\|\gamma\|^2$.

5.6.7 *Lemma* Let for γ in $\hat{G}$, $d(\gamma) = \dim V_\gamma$. Then the series $\sum d(\gamma)^2(1 + \|\gamma\|^2)^{-s}$, the sum taken over the elements of $\hat{G}$, converges if $s > n/2$ where $n = \dim G$.

PROOF 4.9.7 says that $d(\gamma) = \prod[(\Lambda_\gamma + \rho)(H_\alpha)/\rho(H_\alpha)]$, the product taken over the elements of Δ^+. Now

$$\begin{aligned}(\Lambda_\gamma + \rho)(H_\alpha)/\rho(H_\alpha) &= \Lambda_\gamma(H_\alpha)/\rho(H_\alpha) + 1\\ &= \sum m_i\Lambda_i(H_\alpha)/\sum\Lambda_i(H_\alpha) + 1 \leqslant \sum m_i + 1\end{aligned}$$

since $\sum\Lambda_i(H_\alpha) \geqslant \Lambda_i(H_\alpha)$ for each i. There are $(n - l - r)/2$ elements of Δ_p^+. Hence $d(\gamma) \leqslant (\sum m_i + 1)^{\frac{1}{2}(n-l-r)}$. This implies that there is a constant c_3 so that $d(\gamma) \leqslant c_3\|\gamma\|^{\frac{1}{2}(n-l-r)}$.

Let $\hat{G}_j = \{\gamma \text{ in } \hat{G} | \max(|n_i|, m_k) = j\}$. If γ is in $\hat{G}_j$ then $(1 + \|\gamma\|^2)^{-s}d(\gamma)^2 \leqslant c_4 j^{n-l-r}(1 + j^2)^{-s} \leqslant c_5 j^{-2s+n-l-r}$ where c_4 and c_5 are positive constants depending only on s. The number of elements of G_j is at most

$$(2r + l)(2j + 1)^{r+l-1}.$$

Thus $\sum_{\gamma\in G_j} d(\gamma)^2(1 + \|\gamma\|^2)^{-s} \leqslant c_6 j^{-2s+n-1}$ with c_6 depending only on s. Now,

$$\begin{aligned}&\sum d(\gamma)^2(1 + \|\gamma\|^2)^{-s}\\ &\quad = 1 + \sum_{j=1}^{\infty}\sum_{\gamma\in G_j} d(\gamma)^2(1 + \|\gamma\|^2)^{-s}\\ &\quad \leqslant 1 + c_6\sum_{j=1}^{\infty} j^{-2s+n-1}.\end{aligned}$$

We therefore see by the comparison theorem that the sum converges if $-2s + n - 1 < -1$. That is, $s > n/2$.

5.7 The Sobolev Spaces

5.7.1 We retain the notation of 5.6. Let for γ in $\hat{G}$, $V_\gamma \otimes V_\gamma^*$ be identified with a space of continuous functions on G by the rule $v \otimes w^*(g) = w^*(\pi_\gamma(g)^{-1}v)$

for g in G, v in V_γ and w^* in V_i^*. We put the inner product $\langle\ ,\ \rangle$ on $V_\gamma \otimes V_\gamma^*$ given by

$$\langle f, h\rangle = \int_G f(g)\overline{h(g)}\, dg.$$

2.9.3 implies that if $(\ ,\ )$ is a G-invariant inner product on V_γ and if we put the dual inner product on V_γ^* then $\langle\ ,\ \rangle = d(\gamma)^{-1}$ (the tensor product inner product on $V_\gamma \otimes V_\gamma^*$).

5.7.2 Let (τ, E_0) be a unitary representation of K. Let $E = G \underset{\tau}{\times} E_0$ be the corresponding unitary homogeneous vector bundle over $M = G/K$. Put the tensor product inner product on $V_\gamma \otimes V_\gamma^* \otimes E_0$ (here we use the inner product defined in 5.7.1 on $V_\gamma \otimes V_\gamma^*$; this inner product is clearly just the L^2 inner product on $V_\gamma \otimes V_\gamma^* \otimes E_0$ looked upon as a space of continuous functions from G to E_0).

We identify in the canonical manner $V_\gamma \times \mathrm{Hom}_K(V_\gamma, E_0)$ with a subspace of $V_\gamma \otimes V_\gamma^* \times E_0$. This clearly puts an inner product on $V_\gamma \otimes \mathrm{Hom}_K(V_\gamma, E_0)$. Identifying $V_\gamma \times \mathrm{Hom}_K(V_\gamma, E_0)$ with $\Gamma_\gamma E$ we leave it to the reader to check that $\langle f, h\rangle = \int_{G/K} (f, h)\omega$ where $(\ ,\ )$ is the unitary structure on E induced by the K-invariant inner product on E_0 and ω is as in 5.3.1.

5.7.3 Let $P(E)$ be the collection of all sequences $\{u(\gamma)\}$ with $u(\gamma)$ in $\Gamma_\gamma E = V_\gamma \times \mathrm{Hom}_K(V_\gamma, E_0)$ for γ in $\hat{G}$. Let $P(E_0)$ be the collection of all sequences $\{u(\gamma)\}$ with $u(\gamma)$ in $V_\gamma \otimes V_\gamma^* \times E_0$. Clearly $P(E_0) \supset P(E)$.

5.7.4 If u is in $P(E_0)$ and if r is a real number define

$$\|u\|_r^2 = \sum (1 + \|\gamma\|^2)^r \|u(\gamma)\|^2,$$

the sum taken over the elements of G. Let $H^r(E_0) = \{u \text{ in } P(E_0) \mid \|u\|_r < \infty\}$. Set $H^r(E) = H^r(E_0) \cap P(E)$.

5.7.5 *Lemma* Let u be in $H^s(E_0)$. If $s > n/2$ then $\sum_{\gamma\in\hat{G}} u(\gamma)$ converges absolutely and uniformly to an element of $C(G; E_0)$.

PROOF Let f be in $V_\gamma \otimes V_\gamma^* \otimes E_0$. Let $v_1, \ldots, v_{d(\gamma)}$ be an orthonormal basis of V_γ and let $v_1^*, \ldots, v_{d(\gamma)}^*$ be the dual basis. Let $w_1, \ldots, w_q$ be an orthonormal basis of E_0. Then $f = \sum a_{ijk} v_i \otimes v_j^* \otimes w_k$. Let A_k be the $d(\gamma) \times d(\gamma)$ matrix (a_{ijk}). Then

$$f(g) = \sum_{ijk} a_{ijk} v_j^*(\pi_\gamma(g)^{-1} v_i) w_k$$
$$= \sum_k \operatorname{tr}({}^t A_k \pi_\gamma(g)^{-1}) w_k.$$

Thus

$$\|f(g)\|^2 = \sum_k |\operatorname{tr} {}^t A_k \pi_\gamma(g)^{-1}|^2 \leqslant d(\gamma) \sum_k \|A_k\|^2$$

by the Schwarz inequality (here we have taken the inner product on the $d(\gamma) \times d(\gamma)$ matrices to be $(A, B) = \operatorname{tr} A^t \bar{B}$). On the other hand we have observed in 5.7.1 that $\|f\|^2 = d(\gamma)^{-1} \sum |a_{ijk}|^2 = d(\gamma)^{-1} \sum \|A_k\|^2$. We therefore see that $\|f(g)\|^2 \leqslant d(\gamma) \sum \|A_k\|^2 = d(\gamma)^2 \|f\|^2$. Hence $\|f(g)\| \leqslant d(\gamma)\|f\|$ for all g in G

To prove the lemma we need only show that the series $\sum d(\gamma)\|u(\gamma)\|$ (the sum taken over the elements of $\hat{G}$) converges.

Now,

$$\sum_{\|\gamma\|^2 \leqslant N^2} d(\gamma)\|u(\gamma)\|$$
$$= \sum_{\|\gamma\|^2 \leqslant N^2} d(\gamma)(1 + \|\gamma\|^2)^{-s/2}(1 + \|\gamma\|^2)^{s/2}\|u(\gamma)\|$$
$$\leqslant \Big(\sum_{\|\gamma\|^2 \leqslant N^2} d(\gamma)^2(1 + \|\gamma\|^2)^{-s}\Big)^{\frac{1}{2}}\Big(\sum_{\|\gamma\|^2 \leqslant N^2} (1 + \|\gamma\|^2)^s\|u(\gamma)\|^2\Big)^{\frac{1}{2}}$$
$$\leqslant \|u\|_s\Big(\sum_{\|\gamma\|^2 \leqslant N^2} d(\gamma)^2(1 + \|\gamma\|^2)^{-s}\Big)^{\frac{1}{2}}.$$

(by Schwarz inequality) 5.6.7 now implies that $\sum d(\gamma)\|u(\gamma)\|$ converges if $n/2 > s$.

5.7.6 Let f be in $C(G; E_0)$. We set $\hat{f}(\gamma) = d(\gamma) \int_G \bar{\chi}_\gamma(g) f(g)\, dg$ for each γ in $\hat{G}$. $\hat{f}(\gamma)$ is in $V_\gamma \otimes V_\gamma^* \otimes E_0$ for each γ in $\hat{G}$.

5.7.7 *Lemma* Let $C^k(G; E_0)$ denote the space of all elements of $C(G; E_0)$ of class C^k. If f is in $C^k(G; E_0)$ and if x is in $U^k(\mathfrak{g})$ then $(xf)\hat{}\,(\gamma) = x \cdot \hat{f}(\gamma)$.

PROOF It is enough to prove this result for x in $\mathfrak{g}$ and $k = 1$.

$$(xf)\hat{}\,(\gamma)(g_0) = d(\gamma) \int_G \bar{\chi}_\gamma(g)(xf)(g^{-1}g_0)\, dg$$
$$= -d(\gamma) \int_G (x \cdot \bar{\chi}_\gamma)(g) f(g^{-1}g_0)\, dg.$$

Now $x \cdot \chi_\gamma$ is in $V_\gamma \otimes V_\gamma^*$, hence

$$\int_G (x \cdot \bar{\chi}_\gamma)(g) f(g^{-1} g_0)\, dg$$

$$= \int_G (x\bar{\chi}_\gamma)(g) \hat{f}(\gamma)(g^{-1} g_0)\, dg$$

$$= -\int_G \bar{\chi}_\gamma(g) x\hat{f}(\gamma)(g^{-1} g_0)\, dg.$$

The lemma now follows.

5.7.8 *Proposition* If f is in $C^r(G;\ E_0)$ then f is in $H^{r-n/2-\varepsilon}(E_0)$ for every positive ε.

PROOF Suppose that f is in $C^r(G;\ E_0)$. Let $x_1, \ldots, x_n$ be an orthonormal basis of $\mathfrak{g}$. Then $\Omega = \sum x_i^2$. Hence $\Omega^r = \sum x_{i_1}^2 x_{i_2}^2 \cdots x_{i_r}^2$, the sum taken over all $1 \leqslant i_1, \ldots, i_r \leqslant n$. If $r = 2p$ then set $z_{i_1 \cdots i_r} = x_{i_1}^2 \cdots x_{i_p}^2$ and set $w_{i_1 \cdots i_r} = x_{i_{p+1}}^2 \cdots x_{i_r}^2$. If $r = 2p + 1$ then set $z_{i_1 \cdots i_r} = x_{i_1}^2 \ldots x_{i_p}^2 x_{i_{p+1}}$ and set $w_{i_1 \cdots i_r} = x_{i_{p+1}} x_{i_{p+2}}^2 \cdots x_{i_r}^2$. Then $\Omega^r = \sum z_{i_1 \cdots i_r} w_{i_1 \cdots i_r}$.

Now

$$|\sum \langle (w_{i_1 \cdots i_r} f)^\wedge(\gamma), (z_{i_1 \cdots i_r} f)^\wedge(\gamma)\rangle|$$

$$= \left| \sum d(\gamma)^2 \iiint \bar{\chi}_\gamma(g)\chi_\gamma(y)\langle (w_{i_1 \cdots i_r} f)(g^{-1}x), (z_{i_1 \cdots i_r} f)(y^{-1}x)\rangle\, dg\, dy\, dx \right|$$

$$\leqslant d(\gamma)^2 \sum \iiint |\bar{\chi}_\gamma(g)||\chi_\gamma(y)||\langle \pi(g)(w_{i_1 \cdots i_r} f)(x), \pi(y)(z_{i_1 \cdots i_r} f)(x)\rangle| dg\, dy\, dx$$

$$\leqslant d(\gamma)^2 \sum \|w_{i_1 \cdots i_r} f\|\, \|z_{i_1 \cdots i_r} f\| = C_r d(\gamma)^2.$$

On the other hand

$$|\sum \langle (w_{i_1 \cdots i_r} f)^\wedge(\gamma), (z_{i_1 \cdots i_r} f)^\wedge(\gamma)\rangle|$$

$$= |\sum \langle (z_{i_1 \cdots i_r} w_{i_1 \cdots i_r} \hat{f}(\gamma), \hat{f}(\gamma)\rangle$$

$$= |\langle \Omega^r \hat{f}(\gamma), \hat{f}(\gamma)\rangle|$$

$$= c(\gamma)^r \langle \hat{f}(\gamma), \hat{f}(\gamma)\rangle.$$

We therefore see that

$$c(\gamma)^r \|\hat{f}(\gamma)\|^2 \leqslant C_r d(\gamma)^2 \tag{1}$$

with C_r a constant depending only on r and f.

Equation (1) combined with 5.6.6 implies that

(2) $$\|\hat{f}(\gamma)\|^2 \leqslant (1 + \|\gamma\|^2)^{-r} C_r' d(\gamma)^2$$

with C_r' depending only on f and r. If $r = n/2 + k + \varepsilon$ then

$$(1 + \|\gamma\|^2)^k \|\hat{f}(\gamma)\|^2 \leqslant C_r' d(\gamma)^2 (1 + \|\gamma\|^2)^{-n/2-\varepsilon}.$$

The proposition now follows from 5.6.7.

5.7.9 *Corollary* If f is in $C^\infty(G; E_0)$ then $\sum d(\gamma)^2 \|f(\gamma)\| < \infty$.

PROOF Since f is in $C^s(G; E_0)$ for each s, we may assign to each s a positive real number C_s so that

$$\|\hat{f}(\gamma)\|^2 \leqslant C_s(1 + \|\gamma\|^2)^{-s} d(\gamma)^2.$$

The proof of 5.6.7 implies that there is a constant C' so that

$$d(\gamma)^2 \leqslant C'(1 + \|\gamma\|^2)^{(n-q)/2}$$

where $n = \dim G$ and $q = \dim T$.

The result now follows from 5.6.7.

5.7.10 *Corollary* If f is in $C^\infty(G; E_0)$ then $\hat{f}$ is in $\bigcap_{s=-\infty}^{\infty} H^s(E_0)$. In particular, $\sum \hat{f}(\gamma)$ converges absolutely and uniformly to f. Furthermore, if X is in $U(\mathfrak{g})$ then $\sum X\hat{f}(\gamma)$ converges absolutely and uniformly to Xf.

PROOF The first statement is an immediate consequence of 5.7.8. 5.7.5 implies that $\sum \hat{f}(\gamma)$ converges absolutely and uniformly to h, a continuous function on G. $h - f$ is orthogonal to each of the $V_\gamma \otimes V_\gamma^* \otimes E_0$. Hence, $h - f = 0$. The third statement follows from the second and 5.7.7.

5.7.11 5.7.10 says that the map $f \mapsto \hat{f}$ is an injection of $C^\infty(G; E_0)$ into $H^s(E_0)$ for each s. We may, therefore, identify $C^\infty(G; E_0)$ with its image in $H^s(E_0)$ for each s.

5.7.12 *Lemma* $C^\infty(G; E_0)$ is dense in $H^s(E_0)$ for each s.

PROOF If u is in $P(E_0)$ and if $u(\gamma)$ is 0 for all but a finite number of γ in $\hat{G}$ then u is in $C^\infty(G; E_0)$. Clearly the set of all such u is dense in $H^s(E_0)$.

5.7.13 *Lemma* If u is in $H^s(E_0)$ and if x is in $U^j(\mathfrak{g})$, set $(xu)(\gamma) = xu(\gamma)$. Then xu is in $H^{s-j}(E_0)$.

PROOF It is enough to prove that if X is in $\mathfrak{g}$ then Xu is in $H^{s-1}(E_0)$. Now,

$$\begin{aligned}\|Xu(\gamma)\|^2 &= -\langle X^2u(\gamma), u(\gamma)\rangle \\ &\leqslant -\langle X, X\rangle \langle \Omega u(\gamma), u(\gamma)\rangle \\ &= c(\gamma)\langle X, X\rangle \|u(\gamma)\|^2.\end{aligned}$$

Thus

$$(1 + \|\gamma\|^2)^{s-1}\|Xu(\gamma)\|^2 \leqslant C(1 + \|\gamma\|^2)^s\|u(\gamma)\|^2$$

with C depending only on X and s.

5.7.14 *Corollary* Let (σ, F_0) be a finite dimensional unitary representation of K. Let D be in $(L(E_0, F_0) \times U^j(\mathfrak{g}))^K$. Then the corresponding differential operator from E to F extends to a bounded linear operator from $H^s(E)$ to $H^{s-j}(F)$ for each s.

PROOF It is enough to show that if A is in $L(E_0, F_0)$ and if X is in $\mathfrak{g}$ then $A \otimes X$ induces a bounded linear map of $H^s(E_0)$ to $H^{s-1}(F_0)$. The proof of 5.7.13 implies that if u is in $H^s(E_0)$ then $\|Xu\|_{s-1} \leqslant C\|u\|_s$. Also if v is in $H^{s-1}(E_0)$ then $\|Av\|_{s-1} \leqslant \|A\|\,\|v\|_{s-1}$.

5.7.15 *Proposition* If k is a nonnegative integer and if u is in $H^{n/2+\varepsilon+k}(E_0)$ with ε a positive real number, then $\sum u(\gamma)$ converges absolutely and uniformly to an element of $C^k(G; E_0)$.

PROOF It is enough to show that if u is in $H^{\frac{1}{2}n+\varepsilon+1}$ with $\varepsilon > 0$ then $\sum u(\gamma)$ converges to an element of $C^1(G; E_0)$.

5.7.5 says that $\sum u(\gamma)$ converges uniformly and absolutely to an element f of $C(G; E_0)$. 5.7.5 combined with 5.7.13 implies that $\sum Xu(\gamma)$ converges absolutely and uniformly to an element h^X of $C(G; E_0)$ for each X in $\mathfrak{g}$. The proposition will be proved if we can show that

$$\lim_{t\to 0}\frac{f(\exp(-tX)g) - f(g)}{t} = h^X(g) \tag{*}$$

for all X in $\mathfrak{g}$ and g in G.

Set $v_\gamma(t) = u(\gamma)(\exp(-tX)g)$ with g and X fixed. Then $\sum v_\gamma(t)$ converges uniformly to $f(\exp(-tX)g)$. Also $\sum v_\gamma'(t)$ converges uniformly and absolutely to $h^X(\exp(-tX)g)$. Now $v_\gamma(t) = v_\gamma(0) + \int_0^t v_\gamma'(s)\,ds$. Thus

$$\sum v_\gamma(t) = \sum v_\gamma(0) + \sum \int_0^t v'_\gamma(s)\, ds.$$

Uniform and absolute convergence allow us to interchange summation and integration. We therefore see that

$$f(\exp(-tX)g) = f(g) + \int_0^t h^X(\exp(-sX)g)\, ds.$$

This clearly proves (*) and hence the proposition.

5.7.16 *Corollary* $C^\infty(G; E_0) = \bigcap_{r=-\infty}^{\infty} H^r(E_0)$. $\Gamma^\infty E = \bigcap_{r=-\infty}^{\infty} H^r(E)$.

PROOF The corollary follows directly from 5.7.14, 5.7.8, and 5.7.10.

5.7.17 If f is in $C^\infty(G; E_0)$ and if u is in $H^r(E_0)$ for some r, then 5.7.16 implies that $\sum \langle \hat{f}(\gamma), u(\gamma)\rangle$ converges. Thus, for each r the elements of $H^r(E_0)$ define linear forms on $C^\infty(G; E_0)$. Set $\mathscr{D}'(E_0) = \bigcup_{r=-\infty}^{\infty} H^r(E_0)$ and set $\mathscr{D}'(E) = \bigcup_{r=-\infty}^{\infty} H^r(E)$. We note that since $H^r(E_0) \subset H^s(E_0)$ for $r > s$, $\mathscr{D}'(E_0)$ and $\mathscr{D}'(E)$ are subspaces of $P(E_0)$. By the above we may look upon $\mathscr{D}'(E)$ as a space of linear forms on $\Gamma^\infty E$. An element of $\mathscr{D}'(E)$ is called a generalized cross section of E. 5.7.14 implies that if E and F are homogeneous vector bundles over G/K and if D is a homogeneous differential operator from E to F, then D induces a linear map of $\mathscr{D}'(E)$ to $\mathscr{D}'(F)$.

5.7.18 *Lemma* Let $\{u_j\}$ be a sequence of elements of $H^t(E)$ with $\|u_i\|_t \leqslant 1$ for all j. If $s < t$ then there is a subsequence $\{u_{j_k}\}$ which converges in $H^s(E)$.

PROOF The hypothesis on the u_j implies that for each fixed γ, $\|u_j(\gamma)\| \leqslant (1 + \|\gamma\|^2)^{-t/2}$. Thus if we take a subsequence of the u_j using the diagonal procedure we may assume that $u_j(\gamma)$ converges for each γ. If $N > 0$ is an integer set $u_j^N(\gamma) = u_j(\gamma)$ if $\|\gamma\|^2 \leqslant N^2$, $u_j^N(\gamma) = 0$ if $\|\gamma\|^2 > N^2$. Set $v_j^N = u_j - v_j^N$. If j, k are arbitrarily given then it is easily seen that

$$\|v_j^N - v_k^N\|_s^2 > 4(1 + N^2)^{s-t}$$

(we are assuming that $s < t$). Choose N so large that $4(1 + N^2)^{s-t} < \varepsilon^2/4$ where $\varepsilon > 0$ is chosen in advance. Then $\|u_j - u_k\|_s < \varepsilon/2 + \|u^K - u^K\|_s$. By the above it is clear that there exists C a positive number such that if $j, k > C$ then $\|u_j^N - u_k^N\|_s < \varepsilon/2$. This shows that $\{u_j\}$ is a Cauchy sequence in $H^s(E)$.

5.8 Globally Hypoelliptic Differential Operators

5.8.1 Let G be a compact Lie group and let K be a closed subgroup of G so that G/K is orientable. We retain the notation of the previous section.

5.8.2 *Definition* Let E and F be homogeneous vector bundles over G/K. Let D be a homogeneous differential operator from E to F. Then D is said to be globally hypoelliptic if whenever Du is in $\Gamma^\infty F$ for u in $\mathscr{D}'(E)$ then u is in $\Gamma^\infty E$.

5.8.3 *Theorem* Let E and F be homogeneous vector bundles over G/K. Set $\hat{G}(E) = \{\gamma \text{ in } \hat{G} | \Gamma_\gamma E \neq (0)\}$. If D is a homogeneous differential operator from E to F then set for γ in $\hat{G}(E)$

$$m_\gamma(D) = \inf\{\|Df\|_0 | f \text{ in } \Gamma_\gamma E, \|f\|_0 = 1\}.$$

Then D is globally hypoelliptic if and only if there are constants $C > 0$ and k so that

$$m_\gamma(D) \geqslant C(1 + \|\gamma\|^2)^{k/2} \tag{1}$$

for all but a finite number of γ in $\hat{G}(E)$.

PROOF Suppose that D satisfies (1). Let u be in $\mathscr{D}'(E)$ and suppose that Du is in $\Gamma^\infty F$. Let V be the set of all γ in $\hat{G}(E)$ for which (1) is satisfied and let V' be the complement to V in $G(E)$. Then V' is a finite set. Now

$$\begin{aligned}\|Du\|_s^2 &= \sum (1 + \|\gamma\|^2)^s \|Du(\gamma)\|^2 \\ &\geqslant \sum_{\gamma \in V'} (1 + \|\gamma\|^2)^s \|Du(\gamma)\|^2 \\ &\quad + C^2 \sum (1 + \|\gamma\|^2)^{s+k} \|u(\gamma)\|^2.\end{aligned}$$

Thus $\sum (1 + \|\gamma\|^2)^{s+k}\|u(\gamma)\|^2 < \infty$ for all s. 5.7.16 implies that u is in $\Gamma^\infty E$.

Suppose that D does not satisfy (1). We find u in $H^{-\frac{1}{2}n-1}(E)$ so that u is not in $H^0(E)$ and Du is in $\Gamma^\infty F$. In fact, to each C, k with $C > 0$ there are an infinite number of γ in $\hat{G}(E)$ so that (1) is not satisfied. Thus to each positive integer j we may assign a γ_j in $\hat{G}(E)$ and u_j in $\Gamma_{\gamma_j}(E)$ so that $\|u_j\|_0 = 1$ and $\|Du_j\|_0 < (1/2^j)(1 + \|\gamma\|^2)^{-j}$. We may also assume that $\gamma_j \neq \gamma_k$ if $j \neq k$. Let u be defined by $u(\gamma) = 0$ if $\gamma \neq \gamma_j$ for any j and let $u(\gamma) = u_j$ if $\gamma = \gamma_j$. Then u is the desired element of $\mathscr{D}'(E)$.

5.8.4 *Corollary* If D is a homogeneous differential operator from E to F then D is globally hypoelliptic if and only if

(1) $\{u \text{ in } \mathscr{D}'(E) | Du = 0\}$ is a finite dimensional subspace of $\Gamma^\infty E$.

(2) There is a constant k so that if s is given there is a constant C_s such that if u is in $\mathscr{D}'(E)$ there is u' in $\mathscr{D}'(E)$ so that $Du = Du'$ and

$$\|u'\|_s \leqslant C_s \|Du\|_{s-k}.$$

PROOF This is just a reformulation of 5.8.3: the constant k may be taken to be the same as that of 5.8.3(1). The varying C_s is just to compensate for the $H^s(E)$ norm on the (finite dimensional) space ker D.

5.8.5 *Corollary* Let D be a homogeneous differential operator from E to F. Then D is globally hypoelliptic if and only if

(1) $\{u \text{ in } \mathscr{D}'(E) | Du = 0\}$ is contained in $\Gamma^\infty E$,

(2) $D\Gamma^\infty E$ is closed in $\Gamma^\infty F$ relative to $\|\cdots\|_s$ for each s.

PROOF We first show that if D is globally hypoelliptic then D satisfies (1) and (2). (1) is clear. Suppose that $\{u_j\}$ is a sequence in $\Gamma^\infty E$, that f is in $\Gamma^\infty F$, and that $\lim_{j\to\infty} Du_j = f$ in $H^s(F)$. Applying 5.8.4(2) we find a sequence $\{u_j'\}$ in $\mathscr{D}'(E)$ and a constant k so that $Du_j' = Du_j$ and $\|u_j'\|_{s+k} \leqslant C\|Du_j\|_s$. Thus $\{u_j'\}$ is bounded in $H^{s+k}(E)$. Using 5.7.18 we may assume that there is a $t \leqslant s$ and an element u in $H^t(E)$ so that $\lim_{j\to\infty} u_j' = u$ in $H^t(E)$. But then $f = \lim_{j\to\infty} Du_j = Du$ in $H^{t-p}(F)$ where p is the degree of D. But D is globally hypoelliptic, hence u is in $\Gamma^\infty E$. This proves (2).

Suppose that D satisfies (1) and (2). Let u be in $\mathscr{D}'(E)$ and suppose that Du is in $\Gamma^\infty F$. u is in $H^s(E)$ for some s. Let $\{u_j\}$ be a sequence in $\Gamma^\infty E$ so that $\lim_{j\to\infty} u_j = u$ ($\Gamma^\infty E$ is dense in $H^s(E)$). Then $\lim_{j\to\infty} Du_j = Du$ in $H^{s-p}(F)$ where p is the order of D. But then Du is in the closure of $D\Gamma^\infty E$ in $\Gamma^\infty F$ relative to $\| \quad \|_{s-p}$. Condition (2) implies that there is f in $\Gamma^\infty E$ so that $Df = Du$. Thus $D(u - f) = 0$. Condition (1) now implies that $u - f$ is in $\Gamma^\infty E$. Hence u is in $\Gamma^\infty E$.

5.8.6 *Definition* Let D be a homogeneous hypoelliptic differential operator from E to F. Let k be such that D satisfies 5.8.3(1) for some given $C > 0$ and all but a finite number of γ in $\hat{G}(E)$. Then D is said to have degree of hypoellipticity k.

5.8.7 Let E be a homogeneous vector bundle over G/K, then Ω defines a differential operator from E to F. 5.6.4 combined with 5.6.6 implies that Ω has degree of hypoellipticity 2.

5.8.8 *Proposition* If D is a homogeneous globally hypoelliptic differential operator from E to F with degree of hypoellipticity k and if D_1 is a homogeneous differential operator from E to F of order $< k$, then $D + D_1$ is globally hypoelliptic.

PROOF Let for γ in $\hat{G}(E)$, $M_\gamma(D_1) = \sup\{\|D_1 f\|_0 | f \text{ in } \Gamma_\gamma E, \|f\|_0 = 1$. Then $m_\gamma(D + D_1) \geqslant m_\gamma(D) - M_\gamma(D_1)$. Now $M_\gamma(D_1) \leqslant C_1(1 + \|\gamma\|^2)^{p/2}$ where p is the order of D_1. The result now follows from 5.8.3.

5.8.9 *Proposition* Let D be a homogeneous globally hypoelliptic differential operator from E to E. Then D^* is globally hypoelliptic. Furthermore

$$\Gamma^\infty E = D\Gamma^\infty E \oplus \ker(D^*),$$

orthogonal direct sum relative to the L^2 inner product. dim ker(D) and dim ker(D^*) are finite dimensional.

PROOF The global hypoellipticity of D^* and the finite dimensionality of ker(D) and ker(D^*) are immediate consequences of 5.8.3. Let V be the orthogonal complement to $D\Gamma^\infty E$ in $H^0(E)$. If f is in V then by definition of D^*, $D^*f = 0$. Thus $V \subset \Gamma^\infty E$ due to the global hypoellipticity of D^*. Thus $H^0(E) = \overline{D\Gamma^\infty E} \oplus \ker D^*$. If f is in $\Gamma^\infty E$ then $f = f_1 + f_2$ with f_1 in $\overline{D\Gamma^\infty E}$ and f_2 in ker (D^*). By the above f_2 is in $\Gamma^\infty E$. Thus f_1 is in $\Gamma^\infty E$ and in the closure of $\overline{D\Gamma^\infty E}$ relative to $\| \quad \|_0$. 5.8.5 now implies that f_1 is in $D\Gamma^\infty E$.

5.8.10 *Corollary* Let $\{E_i\}$ be a sequence (finite or infinite) of homogeneous vector bundles over G/K. Let D_i be a homogeneous differential operator from E_i to E_{i+1}. Suppose that

(1) $D_{i+1} \circ D_i = 0$,

(2) $\Delta_i = D_{i-1} \circ D_{i-1}^* + D_i^* \circ D_i$ is globally hypoelliptic. Then $\Gamma^\infty E_i = D_{i-1}\Gamma^\infty E_{i-1} \oplus D_i^*\Gamma^\infty E_{i+1} \oplus \ker(\Delta_i)$, an orthogonal direct sum relative to the L^2 inner product, $\ker(D_i) = D_{i-1}\Gamma^\infty E_{i-1} \oplus \ker(\Delta_i)$, an orthogonal direct sum. Furthermore dim $\ker(\Delta_i) < \infty$.

PROOF 5.8.9 implies that $\Gamma^\infty E_i = \Delta_i\Gamma^\infty E_i + \ker(\Delta_i)$, an orthogonal direct sum. Using the formula for Δ_i it is clear that

$$\Gamma^\infty E_i = D_{i-1}\Gamma^\infty E_{i-1} + D_i^*\Gamma^\infty E_{i+1} + \ker(\Delta_i).$$

If h is in $\ker(\Delta_i)$ then $\langle \Delta_i h, h\rangle = 0$. Hence $\langle D_{i-1}^* h, D_{i-1}^* h\rangle + \langle D_i h, D_i h\rangle = 0$. This implies that $D_i h = 0$ and $D_{i-1}^* h = 0$. If f is in $\Gamma^\infty E_{i-1}$ then $\langle h, D_{i-1} f\rangle = \langle D_{i-1}^* h, f\rangle = 0$. If g is in $\Gamma^\infty E_{i+1}$ then $\langle h, D_i^* g\rangle = \langle D_i h, g\rangle = 0$. Finally $\langle D_{i-1} f, D_i^* g\rangle = \langle D_i D_{i-1} f, g\rangle = 0$. This implies that the decomposition $\Gamma^\infty E_i = D_{i-1}\Gamma^\infty E_{i-1} + D_{i-1}^*\Gamma^\infty E_{i+1} + \ker(\Delta_i)$ is indeed an orthogonal direct sum. Suppose that u is in $\ker(D_i)$, then $u = D_{i-1} f + D_i^* g + h$ as above. Since $D_i u = 0$ we see that $D_i D_i^* g = 0$. Hence

$$0 = \langle D_i D_i^* g, g\rangle = \langle D_i^* g, D_i^* g\rangle.$$

Thus $D_i^* g = 0$. The corollary now follows.

5.8.11 *Proposition* Let E and F be homogeneous vector bundles over G/K. Let D be a homogeneous differential operator from E to F. If D and D^* are globally hypoelliptic then $\Gamma^\infty F = D\Gamma^\infty E \oplus \ker(D^*)$, an orthogonal direct sum relative to the L^2 inner product. Furthermore, $\dim \ker(D) < \infty$ and $\dim \ker(D^*) < \infty$. $\ker(D)$ and $\operatorname{coker}(D) = \Gamma^\infty F/D\Gamma^\infty E$ define finite dimensional unitary representations of G. We set

$$\operatorname{ind}_h(D) = [\ker(D)] - [\operatorname{coker}(D)] \text{ in } R(G) \text{ and } \operatorname{ind}(D) = \dim \operatorname{ind}_h(D).$$

PROOF The proof of this proposition is exactly the same as that of 5.8.9.

5.8.12 *Example* Let $\mathfrak{g}$ be the Lie algebra of G and let $\langle\ ,\ \rangle$ be an inner product on $\mathfrak{g}$ as in 5.6 (it was used to define Ω). Let $\mathfrak{k}$ be the Lie algebra of K and let $\mathfrak{p}$ be the orthogonal complement to $\mathfrak{k}$ in $\mathfrak{g}$ relative to $\langle\ ,\ \rangle$. Let p be the natural map of G onto G/K. Let $\langle\ ,\ \rangle$ denote the Riemannian structure on G/K that makes the map p_{*e} from $\mathfrak{p}$ to $T(G/K)_{eK}$ an isometry (here we have taken the inner product $\langle\ ,\ \rangle$ restricted to $\mathfrak{p}$). d defines a homogeneous differential operator from $\Lambda^i T(G/K)^* \otimes C$ to $\Lambda^{i+1} T(G/K)^* \otimes C$. Giving the bundles unitary structures as in 1.7.10 we find that the operator defined in 5.8.10 is the same as the Hodge Laplacian defined in 1.7.10. Furthermore the computation of the symbol of Δ shows that $\sigma(\Delta) \equiv \sigma(\Omega)$. Hence 5.8.8 applies to show that d satisfies 5.8.10. Let $\Lambda_e(M) = \sum \oplus \Lambda^{2k} T(M)^* \otimes C$ and let $\Lambda_0(M) = \sum \oplus \Lambda^{2k+1} T(M)^* \otimes C$. Let D be the differential operator $d + d^*$ from $\Lambda_e(G/K)$ to $\Lambda_0(G/K)$. Then by the above D and D^* are globally hypoelliptic. Combining the de Rham theorem with 5.8.10 (the Hodge theorem) we find that $\operatorname{ind}(D)$ is the Euler characteristic of G/K.

5.9 Bott's Index Theorem

5.9.1 *Theorem* Let G be a compact Lie group and let K be a closed subgroup of G so that G/K is orientable. Let (σ, E_0) and (μ, F_0) be finite dimensional unitary representations of K. Let $E = G \times E_0$ and $F = G \times F_0$ be the corresponding homogeneous vector bundles over G/K (see 5.2.2 for notation). Let D be a homogeneous differential operator from E to F so that D and D^* are globally hypoelliptic. Let i be the canonical injection of K in G. Then

(1) $i_*([E_0] - [F_0])$ is in $R(G)$ (see 4.10.11 for notation).
(2) $\operatorname{ind}_h(D) = i_*([E_0] - [F_0])$.

PROOF Applying 5.8.11 we see that $[\operatorname{coker}(D)] = [\ker(D^*)]$. Also

$$\ker(D) = \sum_\gamma \Gamma_\gamma E \cap \ker(D), \ \ker(D^*) = \sum_\gamma \Gamma_\gamma F \cap \ker(D^*).$$

Now, $[\Gamma_\gamma E] = [\ker(D|_{\Gamma_\gamma E})] + [D\Gamma_\gamma E]$ and $[\Gamma_\gamma F] = [D\Gamma_\gamma E] + [\ker(D^*|_{\Gamma_\gamma F})]$. Thus $[\ker D|_{\Gamma_\gamma E}] - [\ker D^*|_{\Gamma_\gamma F}] = [\Gamma_\gamma E] - [\Gamma_\gamma F]$ and thus $[\Gamma_\gamma E] - [\Gamma_\gamma F]$ is zero for all but a finite number of γ in $\hat{G}$. Thus

$$\operatorname{ind}_h(D) = \sum_\gamma [\ker(D|_{\Gamma_\gamma E})] - [\ker(D^*|_{\Gamma_\gamma F})] = \sum_\gamma ([\Gamma_\gamma E] - [\Gamma_\gamma F]).$$

5.3.6 says that $[\Gamma_\gamma E] = \langle i^*\gamma, [E_0]\rangle_K \gamma$ and $[\Gamma_\gamma F] = \langle i^*\gamma, [F_0]\rangle_K \gamma$. Thus

$$\begin{aligned}\operatorname{ind}_h(D) &= \sum_\gamma (\langle i^*\gamma, [E_0]\rangle_K \gamma - \langle i^*\gamma, [F_0]\rangle_K \gamma) \\ &= \sum_\gamma (\langle i^*\gamma, [E_0] - [F_0]\rangle_K \gamma) = i_*([E_0] - [F_0]).\end{aligned}$$

5.9.2 *Corollary* Let G, K, E, F, and D be as in 5.9.1. If $\operatorname{rank}(K) < \operatorname{rank}(G)$ then $\operatorname{ind}_h(D) = 0$.

PROOF 5.9.1 says that $\operatorname{ind}_h(D) = i_*([E_0] - [F_0])$ in $R(G)$. The corollary now follows from 4.10.13(2) and (3).

5.9.3 We continue with example 5.8.12. Let $\Lambda_e(G/K) = G \underset{\sigma}{\times} E_0$ and let $\Lambda_0(G/K) = G \underset{\mu}{\times} F_0$. Then $[E_0] - [F_0] = \lambda_{-1}(G/K)$ (see 4.10.8). Let D be as in 5.8.12. Then we have seen that $\operatorname{ind}(D) = \chi(G/K)$, the Euler characteristic of G/K. Applying 5.9.2 we see that if rank$(K) <$ rank(G) then ind$(D) = 0$. Thus $\chi(G/K) = 0$. If rank$(K) =$ rank(G) then

$$\begin{aligned}\operatorname{ind}_h(D) &= i_*\left(\lambda_{-1}\frac{G}{K}\right)\\ &= \sum\left\langle i^*\gamma, \lambda_{-1}\frac{G}{K}\right\rangle_K \gamma\\ &= \sum\frac{w(G)}{w(K)}\langle\gamma, 1\rangle_G\gamma\end{aligned}$$

by 4.10.9 (here $w(G)$ and $w(K)$ are, respectively, the orders of the Weyl groups of G and K). Thus ind$_h(D) = (w(G)/w(K))1$. This gives alternate proofs of 3.8.1 and 4.8.5.

5.A Appendix: The Fourier Integral Theorem

5.A.1 *Definition* Let f be in $C_0^\infty(R^n)$. Let t be in R^n, then the Fourier transform of f at t is defined to be

$$\hat{f}(t) = (2\pi)^{-n/2}\int_{R^n} f(x)e^{-(-1)^{\frac{1}{2}}(x,t)}\,dx$$

where $dx = dx_1 \wedge \cdots \wedge dx_n$ is the usual volume element on R^n and $(x, t) = x_1t_1 + \cdots + x_nt_n$.

5.A.2 *Lemma* If f is in $C_0^\infty(R^n)$, then $\hat{f}$ is continuous and

$$\lim |t_i|^k|\hat{f}(t)| = 0$$

for all nonnegative integers k and $i = 1, \ldots, n$.

PROOF Let for $a > 0$, $I_a = \{x \text{ in } R^n |\ |x_i| \leqslant a\}$. We suppose that supp$(f) \subset I_a$. Let $C = \sup\{|f(x)|\ |\ x \text{ in } R^n\}$. A direct computation shows that $|\hat{f}(t)| \leqslant (2a)^n(2\pi)^{n/2}C$ for all t in R^n.

Let h and t be in R^n then

$$
\begin{aligned}
&|\hat{f}(t+h) - \hat{f}(t)| \\
&= \left|\int_{R^n} f(x) e^{-(-1)^{\frac{1}{2}}(x,t)}\left(e^{-(-1)^{\frac{1}{2}}(x,h)} - 1\right) dx\right| \\
&\leqslant \int_{R^n} |f(x)|\, |e^{-(-1)^{\frac{1}{2}}(x,h)/2} - e^{(-1)^{\frac{1}{2}}(x,h)/2}|\, dx \\
&\leqslant C \int_{I_a} |\sin((x,h)/2)|\, dx.
\end{aligned}
$$

This clearly proves the continuity.

Integrating by parts one sees that

$$\left(\frac{\partial^k f}{\partial x_i^k}\right)^{\wedge}(t) = (-(-1)^{\frac{1}{2}} t_i)^k \hat{f}(t).$$

This implies that

$$|t^k|\, |\hat{f}(t)| = |t_i|^{-1} \left| \left(\frac{\partial^{k+1}}{\partial x_i^{k+1}}\right)^{\wedge} f(t) \right|.$$

By the above $|(\partial^{k+1} f/\partial x_i^{k+1})^{\wedge}(t)| \leqslant M$, with M a positive constant. The lemma now follows.

5.A.3 Let $A(R)$ be the space of all continuous functions f on R so that $\lim_{|t|\to\infty} |t|^k |f(t)| = 0$ for all nonnegative integers k.

5.A.4 *Lemma* If f is in $A(R)$ then for all positive real numbers w $\sum_{n=-\infty}^{\infty} f(n/w)$ converges. Furthermore,

$$\lim_{w\to\infty} \frac{1}{w} \sum_{n=-Nw}^{Nw} f\left(\frac{n}{w}\right) = \int_{-\infty}^{\infty} f(t)\, dt.$$

PROOF The first assertion is clear. Let $\varepsilon > 0$ be given. Let m be a positive integer so that if $N \geqslant m$, then

$$\left| \int_{-N}^{N} f(t)\, dt - \int_{-\infty}^{\infty} f(t)\, dt \right| < \frac{\varepsilon}{3}.$$

Here we note that if f is in $A(R)$, then $\int_{-\infty}^{\infty} |f(t)|\, dt < \infty$.

Using Riemann sums we see that

$$\lim_{w\to\infty} \frac{1}{w} \sum_{n=-Nw}^{Nw} f\left(\frac{n}{w}\right) = \int_{-N}^{N} f(t)\, dt.$$

This implies that for each N, $N \geqslant m$, there is w_N so that if $w \geqslant w_N$ then

$$\left|\frac{1}{w}\sum_{n=-Nw}^{Nw} f\left(\frac{n}{w}\right) - \int_{-N}^{N} f(t)\,dt\right| < \frac{\varepsilon}{3}.$$

Now,

$$\left|\frac{1}{w}\sum_{n=Nw+1}^{\infty} f\left(\frac{n}{w}\right)\right| \leqslant \frac{1}{w}\sum_{n=Nw}^{\infty} |f(nw)|.$$

But $|x|^2|f(x)| < C$, where C is a fixed positive constant. Hence

$$\frac{1}{w}\sum_{n=Nw}^{\infty}\left|f\left(\frac{n}{w}\right)\right| < \frac{C}{w}\sum_{n=Nw}^{\infty}\frac{w^2}{n^2} < \frac{C}{N}$$

by the integral test.

Let N be chosen so that $N \geqslant m$ and $C/N < \varepsilon/6$. Then if $w \geqslant w_N$,

$$\begin{aligned}
&\left|\frac{1}{w}\sum_{n=-\infty}^{\infty} f\left(\frac{n}{w}\right) - \int_{-\infty}^{\infty} f(t)\,dt\right| \\
&\qquad \leqslant \left|\frac{1}{w}\sum_{n=-Nw}^{Nw} f\left(\frac{n}{w}\right) - \int_{-N}^{N} f(t)\,dt\right| \\
&\qquad\quad + \left|\int_{-N}^{N} f(t)\,dt - \int_{-\infty}^{\infty} f(t)\,dt\right| \\
&\qquad\quad + \left|\frac{1}{w}\sum_{n=Nw+1}^{\infty} f\left(\frac{n}{w}\right)\right| \\
&\qquad\quad + \left|\frac{1}{w}\sum_{n=Nw+1}^{\infty} f\left(-\frac{n}{w}\right)\right| \\
&\qquad < \frac{\varepsilon}{3} + \frac{\varepsilon}{3} + \frac{\varepsilon}{6} + \frac{\varepsilon}{6} = \varepsilon.
\end{aligned}$$

5.A.5 *Theorem* Let f be in $C_0(R^n)$. Then

$$f(x) = (2\pi)^{-n/2}\int_{R^n}\hat{f}(t)e^{(-1)^{\frac{1}{2}}(x,t)}\,dt.$$

PROOF Suppose that $\operatorname{supp} f \subset I_w$ (see the proof of 5.A.2). Then we can extend f to be a C^∞ function on R^n so that $f(x + 2wk) = f(x)$ where $k = (k_1, \ldots, k_n)$ denotes an n-tuple of integers. Thus f induces a C^∞ function on the n-torus T^n with normalized Haar measure $(2w)^{-n}dx_1 \ldots dx_n$. With these coordinates the characters of T^n are given by $x \mapsto e^{\pi(-1)^{\frac{1}{2}}(x,n)/w}$. Applying 5.7.5 and 5.7.8 to T^n we find that

(1) $$f(x) = (2w)^{-n} \sum_k \int_{-w}^{w} \cdots \int_{-w}^{w} f(t) e^{-(-1)^{\frac{1}{2}}\pi(k,t)/w} \, dt_1 \ldots dt_n e^{\pi(-1)^{\frac{1}{2}}(k,x)/w}.$$

The convergence in (1) is absolute and uniform for x in I_w.

Applying 5.A.4 we find, by taking the limit as w goes to in the right-hand side of (1) that if x is in supp f then

$$f(x) = (2)^{-n} \int_{-\infty}^{\infty} \cdots \int_{-\infty}^{\infty} \hat{f}(t) e^{(-1)^{\frac{1}{2}}\pi(s,x-t)} \, dt_1 \ldots dt_n ds_1 \ldots ds_n$$

$$= (2\pi)^{-n} \int_{-\infty}^{\infty} \cdots \int_{-\infty}^{\infty} f(t) e^{(-1)^{\frac{1}{2}}(s,x-t)} \, dt_1 \ldots dt_n ds_1 \ldots ds_n$$

(here we have used the transformation $s \mapsto (1/\pi)s$)

$$= (2\pi)^{n/2} \int_{-\infty}^{\infty} \cdots \int_{-\infty}^{\infty} \hat{f}(s) e^{(-1)^{\frac{1}{2}}(s,x)} \, ds_1 \ldots ds_n$$

by Fubini's theorem.

If x is not in supp f then the left-hand side of (1) is 0, hence the limit of the right-hand side of (1) is 0. Hence

$$f(x) = (2\pi)^{n/2} \int_{-\infty}^{\infty} \cdots \int_{-\infty}^{\infty} \hat{f}(s) e^{(-1)^{\frac{1}{2}}(s,x)} \, ds_1 \ldots ds_n$$

for all x in R^n. Using 5.A.2 we see that f is absolutely integrable. Hence Fubini's theorem now implies 5.A.5.

5.10 Exercises

5.10.1 Let G be a compact Lie group. Let K be a closed subgroup of G. Let $[\pi, V]$ be an equivalence class of a finite dimensional representation of K. Let $\mu[\pi, V] = [G \times_\pi V]$ (see 5.2.2) in $K(G/K)$ (see 1.6.3). Show that μ induces a ring homomorphism of $R(K)$ (4.10.1) into $K(G/K)$.

5.10.2 Use 5.5.1 to show that if E and F are homogeneous vector bundles over G/K with G a compact Lie group and K a closed connected subgroup of G so that $\text{rank}(K) < \text{rank}(G)$ and if D is a homogeneous elliptic operator from $\Gamma^\infty E$ to $\Gamma^\infty F$, then E and F are isomorphic as homogeneous vector bundles.

5.10.3 Let T^2 be the 2-torus and let X_1, X_2 be a basis of the Lie algebra of T^2 so that ker exp $= \{n_1X_1 + n_2X_2 \mid n_1, n_2 \text{ integers}\}$. Show that $X = X_1 + (2)^{\frac{1}{2}}X_2$ is globally hypoelliptic.

5.10.4 Show that if G is a compact connected Lie group such that there is a left invariant globally hypoelliptic vector field on G then G is a torus.

5.10.5 Let $G = SU(2)$. Let $\mathfrak{g}$ be the Lie algebra of G and let X, Y, Z be a basis of $\mathfrak{g}$ so that $[X, Y] = Z$, $[X, Z] = -Y$, $[Y, Z] = X$. Show that $X^2 + Y^2$ defines a globally hypoelliptic differential operator on G with degree of hypoellipticity 1. (Hint: $X^2 + Y^2 + Z^2 = \Omega$ and we may assume that Z spans the Lie algebra of a maximal torus of G. $\mathfrak{g}_C$ is a TDS. Use 4.3.10 with $Z = (-1)^{\frac{1}{2}}h$.)

5.10.6 Let G be a compact Lie group and let K be a closed subgroup of G. Let φ be in $C^\infty(G/K)$ and let E, F be homogeneous vector bundles over G/K. Let $D_\varphi\colon \Gamma^\infty E \to \Gamma^\infty F$ be defined by $D_\varphi f = \varphi f$. Show that $\|D_\varphi f\|_s \leqslant C_s\|f\|_s$ with C_s a constant depending only on φ and s. Show that this implies that if D is a differential operator from E to F of order m then D induces a continuous operator from $H^s(E)$ to $H^{s-m}(F)$. Show that if D is a differential operator from E to E and if $\sigma(D, \cdot\,) = \sigma(\Omega, \cdot\,)$, then D is hypoelliptic.

5.10.7 Let μ be a measure on G, a compact Lie group (see A.7.1.3). Show that if f is in $V_\gamma \times V_\gamma^*$, then $\mu(f) = \langle f, u(\gamma)\rangle$ for some $u(\gamma)$ in $V_\gamma \times V_\gamma^*$. Show that $\{u(\gamma)\}_{\gamma\in\hat{K}}$ is in $H^{-n+(l+r)/2+\varepsilon}$ (n, l, r as in 5.6.5) for each $\varepsilon > 0$. In particular deduce that measures are in $\mathscr{D}'(G)$.

5.10.8 Let G be a compact Lie group. Let $\|f\|^2 = \int_G |f(g)|^2\, dg$ for f in $C(G)$. Let $X_1, \ldots, X_n$ be a basis of the Lie algebra of G. Let

$$X^M = X_1^{m_1} \cdots X_n^{m_n},\ M = (m_1, \ldots, m_n)$$

and $|M| = \sum m_i$. Show that if $(\|f\|_s')^2 = \sum_{|M|\leqslant s} \|X^M f\|^2$, then there are constants c_1, c_2 so that if f is in $C^\infty(G)$ then

$$c_1\|f\|_s' \leqslant \|f\|_s \leqslant c_2\|f\|_s'$$

c_1, c_2 depending only on s. (Hint: See the proofs of 5.7.8, 5.7.13, and 5.7.14.)

5.10.9 Use 5.10.8 to show that if $D\colon \Gamma^\infty E \to \Gamma^\infty F$ is a second-order elliptic, invariant differential operator then D has degree of hypoellipticity 2.

5.10.10 Let G be a compact connected Lie group. Let Ω be a Laplacian for G. Let for γ in $\hat{K}$, $-c(\gamma)I = \pi_\gamma(\Omega)$. Let

$$g(t, x) = \sum_{\gamma \in K} e^{-c(\gamma)t} d(\gamma)\chi_\gamma(x) \qquad (d(\gamma) = \dim V_\gamma, \chi_\gamma \text{ the character of } \gamma).$$

Use the results of 5.6 and 5.7 to show that $g(t, x)$ converges absolutely and uniformly for $t > 0$. Use the results of 5.7 to show that $g\colon (0, \infty) \times G \to C$ is C^∞. Show that $(\partial g/\partial t)(t, x) = \Omega g(t, x)$. Furthermore, show that if f is a continuous function on G and if

$$\hat{f}(t, g) = \int_G g(t, x) f(x^{-1}g)\, dx$$

then $(\partial \hat{f}/\partial t)(t, g) = \Omega \hat{f}(t, g)$ and $\lim_{t \to 0} \hat{f}(t, g) = f(g)$.

5.10.11 Let G be a conneced compact Lie group. Let K be a closed subgroup of G. Let $M = G/K$. Let S^n be the unit sphere in R^{n+1}. Show that there is an equivariant imbedding of G/K into S^n for n sufficiently large. (An equivariant imbedding of M into S^n is an injective C^∞ map f of M into S^n so that f_{*p} is injective for each p in M and such that there is a Lie homomorphism ψ of G into $O(n + 1)$ such that $f(gx) = \psi(g)f(x)$.) (Hint: Use 5.3.6 with τ the trivial representation of K and argue as in 2.8.4.)

5.11 Notes

5.11.1 The material of 5.2 is taken from Bott [1].

5.11.2 The material on the Sobolev spaces in 5.7 follows the spirit of the development of the Sobolev spaces on an n-torus in F. Warner [1].

5.11.3 The material of 5.8 is an outgrowth of joint work with S. Greenfield. (See Greenfield and Wallach [1].) Similar results were obtained in Cerezo and Rouviere [1].

5.11.4 The results of 5.9 were taken from Bott [1].

CHAPTER 6

Holomorphic Vector Bundles over Flag Manifolds

6.1 Introduction

The main purpose of this chapter is to show how to use complex analysis to give realizations of the irreducible representations of compact, semisimple Lie groups (see 6.3.7, 6.3.8). To make the theory more palatable we give the basic structure theory for generalized flag manifolds in Section 6.2. In particular, we show that every generalized flag manifold has a holomorphic imbedding into a complex projective space of sufficiently high dimension (combined with Chow's theorem this implies that generalized flag manifolds are projective varieties). The projective imbeddings given are generalizations of Segre's projective imbeddings of products of projective spaces. In the last section we show how to use the index theorem (5.9.1) to prove a result of Borel, and Hirzbruch giving the Euler class of a homogeneous holomorphic line bundle over a flag manifold.

6.2 Generalized Flag Manifolds

6.2.1 *Definition* Let G be a compact connected Lie group and let T_1 be a torus contained in G. Let $C(T_1)$ be the centralizer of T_1 in G. That is,

$C(T_1) = \{g \text{ in } G | gtg^{-1} = t \text{ for all } t \text{ in } T_1\}$. The manifold $G/C(T_1)$ is called a generalized flag manifold.

6.2.2 *Lemma* Let G be a compact Lie group and let T_1 be a torus in G. Then $C(T_1)$ is connected.

PROOF Let g be in $C(T_1)$. Then there is a maximal torus T of G so that T contains g and T_1 (see 3.9.5). Clearly $T \subset C(T_1)$.

6.2.3 *Example* Let $G = SU(n)$, the group of all $n \times n$ complex matrices Z so that ${}^t\bar{Z}Z = I$ and $\det(Z) = 1$. Let T be the maximal torus of G consisting of diagonal matrices in G. 3.10.8(1) says that $C(T) = T$. Thus G/T is a generalized flag manifold.

A flag in C^n is an increasing collection of complex subspaces V_i of C^n, $\mathbf{f} = \{V_1 \subset V_2 \subset \cdots \subset V_{n-1}\}$ where $\dim V_i = i$. Let F^n be the set of all flags in C^n. We let $SU(n)$ act on F^n by setting $g\mathbf{f} = \{gV_1 \subset \cdots \subset gV_{n-1}\}$. Let $\mathbf{f}^0$ be the flag gotten from the standard basis of C^n by setting $V_i^0 = Ce_1 + \cdots + Ce_i$. If $\boldsymbol{f} = \{V_1 \subset \cdots \subset V_{n-1}\}$ is an arbitrary flag then let v_1 be a unit vector in V_1; if $v_1, \ldots, v_k$ have been defined let v_{k+1} be a unit vector in V_{k+1} orthogonal to V_k. This procedure gives a set $v_1, \ldots, v_{n-1}$ of orthonormal unit vectors. Let v_n be a unit vector orthogonal to V_{n-1} and so that if $v_i = \sum a_{ji}e_j$ then $g = (a_{ij})$ is in $SU(n)$. Then clearly, $g\boldsymbol{f}^0 = \boldsymbol{f}$. If g is in $SU(n)$ and if $g\boldsymbol{f}^0 = \boldsymbol{f}^0$ then we leave it to the reader to check that g is in T. We may thus identify G/T with F^n. F^n is called the manifold of flags in C^n.

6.2.4 *Example* Let $k_1, \ldots, k_r$ be a set of positive integers so that $k_1 + \cdots + k_r = n$. Let $F(k_1, \ldots, k_r)$ be the set of all

$$\boldsymbol{f} = \{V_1 \subset V_2 \subset \cdots \subset V_r\}$$

with $\dim V_j = k_1 + k_2 + \cdots + k_j$. We let $SU(n)$ act on $F(k_1, \ldots, k_r)$ by $g\boldsymbol{f} = \{gV_1 \subset \cdots \subset gV_r\}$. Set $t_j = k_1 + \cdots + k_j$ and set $V_j^0 = Ce_1 + \cdots + Ce_{t_j}$. Let $\boldsymbol{f}^0 = \{V_1^0 \subset \cdots \subset V_r^0\}$. Then proceeding in the same manner as in 6.2.3 we find that if $\boldsymbol{f}$ is in $F(k_1, \ldots, k_r)$ then there is g in $SU(n)$ so that $g\boldsymbol{f}^0 = \boldsymbol{f}$.

If g is in $SU(n)$ and if $g\boldsymbol{f}^0 = \boldsymbol{f}^0$ then

$$g = \begin{bmatrix} A_1 & & 0 \\ & \ddots & \\ 0 & & A_r \end{bmatrix}$$

with A_i a $k_i \times k_i$ matrix so that ${}^t\bar{A}_i A_i = 1$, that is, A_j is in $U(k_j)$ and

$$\det(A_1)\cdots\det(A_r) = 1.$$

Let us denote by $S(U(k_1) \times \cdots \times U(k_r))$ the group of all $(A_1, \ldots, A_r)$ with A_i in $U(k_i)$ and $\det(A_1)\ldots\det(A_r) = 1$. Thus

$$SU(n)/S(U(k_1) \times \cdots \times U(k_r)) = F(k_1, \ldots, k_r).$$

Let $T_1 = T(k_1, \ldots, k_r)$ be the subgroup of $SU(n)$ consisting of matrices of the form

$$t = \begin{bmatrix} c_1 I_1 & & 0 \\ & \ddots & \\ 0 & & c_r I_r \end{bmatrix}$$

with I_j the $k_j \times k_j$ identity matrix, c_j in C with $|c_j| = 1$, and $c_1^{k_1}\ldots c_r^{k_r} = 1$. Then T_1 is a torus and $C(T_1) = S(U(k_1) \times \cdots \times U(k_r))$. Thus $F(k_1, \ldots, k_r)$ is a generalized flag manifold. Clearly $F^n = F(1, \ldots, 1)$.

6.2.5 Using exercise 6.5.1 we see that the $F(k_1, \ldots, k_r)$ exhaust the generalized flag manifolds corresponding to $SU(n)$.

6.2.6 *Example* A particularly important special case of 6.2.4 is $F(1, n-1)$ which we denote by CP^{n-1}. Notice that if $\boldsymbol{f}$ is in $F(1, n-1)$ then $\boldsymbol{f} = \{V_1 \subset V_2 = C^n\}$. Thus $\boldsymbol{f}$ is completely determined by V_1. This says that $F(1, n-1)$ is the collection of all complex lines through 0 in C^n. Furthermore if we identify $U(n-1)$ with the subgroup $S(U(n-1) \times U(1))$ in $SU(n)$ then $CP^{n-1} = F(1, n-1) = SU(n)/U(n-1)$.

Let $C_*^n = C^n - \{0\}$. If v is in C_*^n then Cv is a one-dimensional subspace of C^n. Thus we may set $[v] = Cv$ in CP^{n-1}. This defines a map of C_*^n onto CP^{n-1}. If $(z_1, \ldots, z_n)$ is in C_*^n we set $[(z_1, \ldots, z_n)] = [z_1, \ldots, z_n]$ and call $z_1, \ldots, z_n$ the homogeneous coordinates of the point $[z_1, \ldots, z_n]$. Let $U_j = \{[z_1, \ldots, z_n] | z_j \neq 0\}$. Then we may define a bijective map $\Psi_j\colon U_j \to C^{n-1}$ by

$$\Psi_j([z_1, \ldots, z_n]) = (z_1 z_j^{-1}, \ldots, z_{j-1} z_j^{-1}, z_{j+1} z_j^{-1}, \ldots, z_n z_j^{-1}).$$

If $(u_1, \ldots, u_{n-1})$ is in $\Psi_j(U_i \cap U_j)$ and if $i < j$ then

$$\Psi_i \Psi_j^{-1}(u_1, \ldots, u_{n-1})$$
$$= (u_1 u_i^{-1}, \ldots, u_{i-1} u_i^{-1}, u_{i+1} u_i^{-1}, \ldots, u_{j-1} u_i^{-1}, u_i^{-1}, u_j u_i^{-1}, \ldots, u_{n-1} u_i^{-1}).$$

Thus $\Psi_i \Psi_j^{-1}; \Psi_j(U_i \cap U_j) \to \Psi_i(U_i \cap U_j)$ is a holomorphic diffeomorphism. This gives CP^{n-1} the structure of a complex manifold. Clearly $SU(n)$ acts on CP^{n-1} by holomorphic diffeomorphisms. We will show that every generalized flag manifold $G/C(T_1)$ has a complex structure so that G acts by a group of holomorphic diffeomorphisms. We will also show that relative to each G-invariant complex structure on $G/C(T_1) = M$ there is a holomorphic imbedding of M into CP^N for N sufficiently large.

6.2.7 *Example* Let $G = SO(n)$ the group of all $n \times n$ real matrices g so that ${}^t gg = I$ and $\det(g) = 1$. Let T_1 be the torus in G consisting of all

$$t = \begin{bmatrix} A & 0 \\ 0 & I \end{bmatrix}$$

where

$$A = \begin{bmatrix} \cos\theta & \sin\theta \\ -\sin\theta & \cos\theta \end{bmatrix}$$

for θ in R and I is the $n-2 \times n-2$ identity matrix.

If g is in $C(T_1)$ then

$$g = \begin{bmatrix} g_1 & 0 \\ 0 & g_2 \end{bmatrix}$$

with g_1 in $SO(2)$ and g_2 in $SO(n-2)$.

An oriented 2-plane of R^n is a pair (V, ω) of a real two-dimensional subspace V of R^n and an orientation ω of V (that is, ω is an equivalence class of nonzero elements of $\Lambda^2 V^*$ where $\eta \sim \mu$ if and only if $\eta = c\mu$ with $c > 0$). Let $G_2(n)$ be the space of all oriented 2-planes in R^n. We assume that $n \geqslant 4$. If (V, ω) is in $G_2(n)$ and g is in $SO(n)$ then set $g(V, \omega) = (gV, (g^{-1})^*\omega)$.

Let $e_1, \ldots, e_n$ be the standard basis of R^n. Let $V_0 = Re_1 + Re_2$ and let ω_0 be the orientation of V_0 so that if μ is in ω_0 then $\mu(e_1, e_2) > 0$. Arguing as above we find that $SO(n)$ acts transitively on $G_2(n)$. $g(V_0, \omega_0) = (V_0, \omega_0)$ if and only if g is in $SO(2) \times SO(n-2) = C(T_1)$. We may thus identify $G/C(T_1)$ with $G_2(n)$.

We give an alternate description of $G_2(n)$. Let (V, ω) be in $G_2(n)$. Let u_1, u_2 be a positive orthonormal basis of V relative to ω (that is, $\mu(u_1, u_2) > 0$ for all μ in ω). We assert that the element $[u_1 + (-1)^{\frac{1}{2}}u_2]$ in CP^{n-1} depends only on (V, ω). Indeed if u_1', u_2' is another positive orthonormal basis of (V, ω) then $u_1' + (-1)^{\frac{1}{2}}u_2' = e^{(-1)^{\frac{1}{2}}\theta}(u_1 + (-1)^{\frac{1}{2}}u_2)$ with θ the angle between u_1' and u_1. This defines a map of $G_2(n)$ into CP^{n-1}.

Let for $z = (z_1, \ldots, z_n)$, $w = (w_1, \ldots, w_n)$ in C^n, $z \cdot w = z_1 w_1 + \cdots + z_n w_n$. We note that if u_1, u_2 is a positive orthonormal basis of (V, ω) and if $v = u_1 + (-1)^{\frac{1}{2}}u_2$ then $v \cdot v = 0$. Set $Q_{n-2} = \{[v] | v \text{ in } C^n_*, v \cdot v = 0\}$. We have defined a mapping of $G_2(n)$ into Q_{n-2} which is easily seen to be injective. Suppose that $[v]$ is in Q_{n-2}. Then $v = u_1 + (-1)^{\frac{1}{2}}u_2$ with u_1 and u_2 in R^n and $0 = v \cdot v = u_1 \cdot u_1 - u_2 \cdot u_2 + 2(-1)^{\frac{1}{2}}u_1 \cdot u_2$. Thus $u_1 \cdot u_2 = 0$ and

$u_1 \cdot u_1 = u_2 \cdot u_2$. This clearly implies that the map of $G_2(n)$ into Q_{n-2} is surjective. We therefore have the identifications $G/C(T_1) = G_2(n) = Q_{n-2}$. Considering $SO(n)$ as a subgroup of $SU(n)$, then $Q_{n-2} = SO(n)[e_1 + (-1)^{\frac{1}{2}}e_2]$ in CP^{n-1}.

6.2.8 Let G be a compact connected Lie group and let T_1 be a torus in G. Let $C(T_1)$ be the centralizer of T_1 in G. Let T be a maximal torus of G containing T_1. Clearly $T \subset C(T_1)$. Let Δ be the root system of G relative to T and let Δ_1 be the root system of $C(T_1)$ relative to T. Let P be a Weyl chamber of T relative to G. Then P is said to be T_1-admissible if there is a Weyl chamber P_1 of T relative to $C(T_1)$ so that if Δ^+ and Δ_1^+ are, respectively, the positive roots of Δ and Δ_1 relative to P and P_1 then

(1) $\Delta^+ \cap \Delta_1 = \Delta_1^+$,

(2) if α is in $\Delta^+ - \Delta_1^+$, β is in Δ_1, and if $\alpha + \beta$ is in Δ then $\alpha + \beta$ is in $\Delta^+ - \Delta_1^+$.

6.2.9 *Lemma* There always exist T_1-admissible Weyl chambers. If P is a T_1-admissible Weyl chamber, if π is the set of simple roots of Δ relative to P, and if π_1 is the set of simple roots of Δ_1 relative to P_1, then $\pi_1 \subset \pi$.

PROOF Let $\mathfrak{g}$ be the Lie algebra of G and let $\mathfrak{h}$ be the Lie algebra of T. Let $\mathfrak{h}_1$ be the Lie algebra of T_1. Then $\mathfrak{h} \supset \mathfrak{h}_1$. We have seen that the roots of G relative to T take real values on $(-1)^{\frac{1}{2}}\mathfrak{h}$. Let $h_1, \ldots, h_r$ be a basis of $(-1)^{\frac{1}{2}}\mathfrak{h}_1$ and extend this basis of $(-1)^{\frac{1}{2}}\mathfrak{h}_1$ to a basis of $(-1)^{\frac{1}{2}}\mathfrak{h}$, $h_1, \ldots, h_l$, so that $B(h_i, h_{r+j}) = 0$ where B is the Killing form of $\mathfrak{g}$. Let us say that α in Δ is positive if $\alpha(h_1) > 0$ or, if $\alpha(h_1) = 0$, then $\alpha(h_i) = 0$ for $i \leqslant j$ and $\alpha(h_j) > 0$. Let Δ^+ be the set of positive roots. Let P be the corresponding Weyl chamber. We show that P is T_1-admissible.

We first make the observation that $\Delta_1 = \{\alpha \text{ in } \Delta | t^\alpha = 1 \text{ for all } t \text{ in } T_1\}$. In fact, if α is in Δ then $t^\alpha = 1$ for all t in T_1 if and only if every element of $\exp((\mathfrak{g}_\alpha + \mathfrak{g}_{-\alpha}) \cap \mathfrak{g})$ commutes with every element of T_1.

Thus if α is in Δ_1 then $\alpha(h_i) = 0$ for $i = 1, \ldots, r$. Hence if α is in Δ_1 and $\alpha > 0$ then $\alpha(h_{r+j}) > 0$ for some j and $\alpha(h_i) = 0$ for all $i \leqslant r$. Let P_1 be the corresponding Weyl chamber of T relative to $C(T_1)$. Then clearly $\Delta_1^+ = \Delta_1 \cap \Delta^+$.

Suppose that α is in $\Delta^+ - \Delta_1^+$ and that β is in Δ_1. If α is in $\Delta^+ - \Delta_1^+$ then $\alpha(h_i) \neq 0$ for some $i \leqslant r$. Let j be the smallest such i. Then $\alpha(h_j) > 0$. If β is in Δ_1 then $\beta(h_i) = 0$ for $i = 1, \ldots, r$. Thus the first i so that $(\alpha + \beta)(h_i) \neq 0$ is j and $(\alpha + \beta)(h_j) = \alpha(h_j) > 0$. Thus P satisfies (1) and (2) of 6.2.8.

Suppose that P is a T_1-admissible Weyl chamber of T. Let π_1 be the simple system of Δ_1^+. If we can show that if α is in π_1 then α is simple in Δ^+ then the lemma will be proved. Suppose that α is in π_1 and that $\alpha = \gamma + \delta$ with γ, δ in Δ^+. Then say γ is in $\Delta^+ - \Delta_1^+$. Using 6.2.8(2) we see that $\delta = \alpha - \gamma$ cannot be in Δ^+. Thus α is indeed simple in Δ^+.

6.2.10 We now give another construction of generalized flag manifolds motivated by 6.2.9. Let G be a compact, connected Lie group. Let T be a maximal torus of G and let Δ be the root system of G relative to T. Let P be a Weyl chamber of T, let Δ^+ be the corresponding set of positive roots and let π be the corresponding system of simple roots. Let π_1 be a subset of π. Let $\mathfrak{g}_C$ be the complexification of $\mathfrak{g}$, the Lie algebra of G. Then

$$\mathfrak{g}_C = \mathfrak{h}_C + \sum \mathfrak{g}_\alpha$$

where $\mathfrak{h}_C$ is the complexification of the Lie algebra $\mathfrak{h}$ of T and $\mathfrak{g}_\alpha$ is the root space corresponding to α in Δ.

Let $\Delta(\pi_1) = \{\alpha \text{ in } \Delta | \alpha = \sum k_\gamma \gamma,\ k_\gamma$ an integer and the summation is over the γ in $\pi_1\}$. Set $\mathfrak{k}(\pi_1)_C = \mathfrak{h}_C + \sum \mathfrak{g}_\alpha$, the sum over α in $\Delta(\pi_1)$. Let $\mathfrak{k}(\pi_1) = \mathfrak{k}(\pi_1)_C \cap \mathfrak{g}$. Set $\mathfrak{h}_1 = \{h \text{ in } \mathfrak{h} | \gamma(h) = 0 \text{ for all } \gamma \text{ in } \pi_1\}$. Let T_1 be the connected subgroup of G corresponding to $\mathfrak{h}_1$. Then $T_1 = \{t \text{ in } T | t^\gamma = 1 \text{ for all } \gamma \text{ in } \pi_1\}$. We assert that $\mathfrak{k}(\pi_1)$ is the Lie algebra of $C(T_1)$. Let $\mathfrak{k}$ be the Lie algebra of $C(T_1)$. Then clearly $\mathfrak{k}(\pi_1) \subset \mathfrak{k}$. Since $\mathfrak{h} \subset \mathfrak{k}$ we see that if $\mathfrak{k}_C$ is the complexification of $\mathfrak{k}$ then $\mathfrak{k}_C = \mathfrak{h}_C + \sum \mathfrak{g}_\alpha$, the summation taken over a subset Δ_1 of Δ. Clearly $\Delta_1 \supset \Delta(\pi_1)$. To prove that $\mathfrak{k}(\pi_1) = \mathfrak{k}$ we need only show that $\Delta_1 = \Delta(\pi_1)$.

Let $\mathfrak{h}_1^\perp$ be the orthogonal complement to $\mathfrak{h}_1$ in $\mathfrak{h}$ relative to the Killing form of $\mathfrak{g}$. Since the center of $\mathfrak{g}$ is contained in $\mathfrak{h}_1$ we see that $\mathfrak{h}_1 \cap \mathfrak{h}_1^\perp = (0)$. Let for α in Δ, H_α be the element of $[\mathfrak{g}_\alpha, \mathfrak{g}_{-\alpha}] \cap (-1)^{\frac{1}{2}}\mathfrak{h}$ such that $B(H, H_\alpha) = \alpha(H)$ for all H in $\mathfrak{h}_C$; here B is the Killing form of $\mathfrak{g}_C$. Let $\pi_1 = \{\alpha_1, \ldots, \alpha_r\}$ and let $\pi = \{\alpha_1, \ldots, \alpha_l\}$. If $H_{\alpha_1}, \ldots, H_{\alpha_r}$ do not form a basis for $(-1)^{\frac{1}{2}}\mathfrak{h}_1^\perp$ then there is H in $(-1)^{\frac{1}{2}}\mathfrak{h}_1^\perp$ orthogonal to $H_{\alpha_1}, \ldots, H_{\alpha_r}$. This implies that $\alpha_i(H) = 0$ for all $i = 1, \ldots, r$. But then H is in $(-1)^{\frac{1}{2}}\mathfrak{h}_1 \cap (-1)^{\frac{1}{2}}\mathfrak{h}_1^\perp = (0)$.

We note that, by definition of $\mathfrak{k}$, if α is in Δ_1 then H_α is in $(-1)^{\frac{1}{2}}\mathfrak{h}_1^\perp$. Thus if α is in Δ_1, $\alpha = \sum_{i=1}^r k_i\alpha_i$ with k_i in R. But then k_i is in Z for each i. This implies that $\Delta_1 = \Delta(\pi_1)$.

6.2.11 *Theorem* Let $M = G/C(T_1)$ be a generalized flag manifold. Then to each T_1-admissible Weyl chamber of T, a maximal torus containing T_1, there exists a complex structure on M so that G acts as a group of holomorphic diffeomorphisms of M.

PROOF Let Z be the center of G. Then, clearly, $Z \subset C(T_1)$. Thus $M = (G/Z)/(C(T_1)/Z)$. Let $T \supset T_1$ be a maximal torus of G. Let be the root system of G relative to T and let Δ_1 be the root system of $C(T_1)$ relative to T. Let P be a T_1-admissible Weyl chamber of T. Let π be the corresponding simple system of roots of G relative to T and let π_1 be the corresponding simple system of roots of $C(T_1)$ relative to T. Let Δ^+ and Δ_1^+ be the corresponding positive roots. Set $\Phi = \Delta^+ - \Delta_1^+$. Let $\mathfrak{k}(\pi_1)_C$ be as in 6.2.10. Then

$$\mathfrak{g}_C = \sum_{\alpha\in\Phi} \mathfrak{g}_\alpha \oplus \mathfrak{k}(\pi_1)_C + \sum_{\alpha\in\Phi} \mathfrak{g}_{-\alpha}.$$

Set

$$\mathfrak{n}^+(\pi_1) = \sum_{\alpha\in\Phi} \mathfrak{g}_\alpha, \; \mathfrak{n}^-(\pi_1) = \sum_{\alpha\in\Phi} \mathfrak{g}_{-\alpha}.$$

Using 6.2.8(2) we see that $[\mathfrak{k}(\pi_1)_C, \mathfrak{n}^+(\pi_1)] \subset \mathfrak{n}^+(\pi_1)$. Also 6.2.9 implies that $\mathfrak{n}^+(\pi_1)$ and $\mathfrak{n}^-(\pi_1)$ are closed under the Lie bracket.

Now $\mathfrak{g} = \mathfrak{z} \oplus \mathfrak{g}_1$ with $\mathfrak{z}$ the Lie algebra of Z and $\mathfrak{g}_1 = [\mathfrak{g}, \mathfrak{g}]$, a compact Lie algebra. Thus $\mathfrak{g}_C = \mathfrak{z}_C \oplus \mathfrak{g}_{1_C}$. Furthermore, $\mathfrak{z}_C \subset \mathfrak{k}(\pi_1)_C$, thus $\mathfrak{k}(\pi_1)_C = \mathfrak{z}_C \oplus \mathfrak{k}'(\pi_1)_C$ with $\mathfrak{k}'(\pi_1)_C = \mathfrak{g}_{1_C} \cap \mathfrak{k}(\pi_1)_C$. This implies that $\mathfrak{g}_{1_C} = \mathfrak{n}^+(\pi_1) + \mathfrak{k}'(\pi_1)_C + \mathfrak{n}^-(\pi_1)$.

Ad maps G into $GL(\mathfrak{g}_1)$ and has kernel Z. Thus $\mathrm{Ad}(G)$ is Lie isomorphic with G/Z. We may thus identify G/Z with a subgroup G' of $GL(n, R) \subset GL(n, C)$ with $n = \dim \mathfrak{g}_1$. $\mathfrak{g}_1$ may be identified with the Lie algebra of G'. Let G_C be the connected subgroup of $GL(n, C)$ corresponding to $\mathfrak{g}_{1_C}$. The connected subgroup K' of G_C corresponding to $\mathfrak{k}'(\pi_1) = \mathfrak{k}'(\pi_1)_C \cap \mathfrak{g}_1$ is Lie isomorphic with $C(T_1)/Z$. Thus $G'/K' = M$. Let K'_C be the connected subgroup of G'_C corresponding to $\mathfrak{k}'(\pi_1)_C$. Let N^+ and N^- be the connected subgroups of G'_C corresponding to $\mathfrak{n}^+(\pi_1)$ and $\mathfrak{n}^-(\pi_1)$, respectively.

(1) $K'_C N^- = B$ is a closed subgroup of G_C.

To prove (1) we set $\mathfrak{b} = \mathfrak{k}'(\pi_1)_C + \mathfrak{n}^-(\pi_1)$ and we assert that B is the identity component of $\tilde{B} = \{g \text{ in } G'_C | \mathrm{Ad}(g)\mathfrak{b} \subset \mathfrak{b}\}$. To show this we need only show that if X is in $\mathfrak{g}_{1_C}$ and if $[X, \mathfrak{b}] \subset \mathfrak{b}$ then X is in $\mathfrak{b}$. If X is in $\mathfrak{g}_{1_C}$ then $X = Y + W$ with Y in $\mathfrak{n}^+(\pi_1)$ and W in $\mathfrak{b}$. Thus, if $[X, \mathfrak{b}] \subset \mathfrak{b}$ we see that $[Y, \mathfrak{b}] \subset \mathfrak{b}$. Let $\mathfrak{h}_C$ be the complexified Lie algebra of T/Z. If H is in $\mathfrak{h}'_C$ then H is in $\mathfrak{k}'(\pi_1)_C$. Thus $[Y, H]$ is in $\mathfrak{b}$. But $[H, Y]$ is in $\mathfrak{n}^+(\pi_1)$. Hence $[H, Y]$ is in $\mathfrak{b} \cap \mathfrak{n}^+(\pi_1) = (0)$. Now $Y = \sum_\alpha Y_\alpha$ with Y_α in $\mathfrak{g}_\alpha$ and α in Φ. By the above $\sum \alpha(H) Y_\alpha = 0$ for all H in $\mathfrak{h}'_C$. But if $Y_\alpha \neq 0$ then $\alpha(H) = 0$ for all H in $\mathfrak{h}'_C$. This clearly implies that $Y_\alpha = 0$ for all α in Ω. Hence $Y = 0$. We have thus proved (1).

Order Δ lexicographically (see the beginning of the proof of 6.2.9) relative to the ordered basis $\pi = \{\alpha_1, \ldots, \alpha_l\}$ of $((-1)^{\frac{1}{2}}\mathfrak{h}')^*$. Let $\Delta^+ = \{\beta_1 > \cdots > \beta_k\}$. We take a basis $z_1, \ldots, z_n$ of $\mathfrak{g}_{1_C}$ so that z_i is in $\mathfrak{g}_{-\alpha_i}$ for $i = 1, \ldots, k$, z_{k+i} is in $\mathfrak{h}'$ for $i = 1, \ldots, l$, and $z_{2k+l+1-i}$ is in $\mathfrak{g}_{\beta_i}$ for $i = 1, \ldots, k$.

If X is in $\mathfrak{n}^-(\pi_1)$ then relative to this basis $\mathrm{ad}(X)$ is upper triangular with zeros along the diagonal. If X is in $\mathfrak{n}^+(\pi_1)$ then $\mathrm{ad}(X)$ is lower triangular with zeros along the diagonal. Let, for X in $\mathfrak{g}_{1_C}$, $\bar{X}$ be its conjugate relative to the real form $\mathfrak{g}_1$. If X is in $\mathfrak{g}_\alpha$ then $\bar{X}$ is in $\mathfrak{g}_{-\alpha}$. We may therefore take $z_{k+l+j} = \bar{z}_{k-j+1}$ for $j = 1, \ldots, k$. If g is in $GL(n, C)$ define $\bar{g}z = \overline{(g, \bar{z})}$. If g is in G'_C then $\bar{g}$ is in G'_C and $G' = \{g \text{ in } G'_C | \bar{g} = g\}$, $K' = \{g \text{ in } K'_C | \bar{g} = g\}$.

(2) $K'_C N^+ \cap K'_C N^- = K'_C$.

To prove (2) we first note that if g is in $K'_C N^+$ then $\mathrm{Ad}(g)\mathfrak{n}^+(\pi_1) \subset \mathfrak{n}^+(\pi_1)$, if g is in $K'_C N^-$ then $\mathrm{Ad}(g)\mathfrak{n}^-(\pi_1) \subset \mathfrak{n}^-(\pi_1)$. Suppose that n is in N^+ and $\mathrm{Ad}(n)\mathfrak{n}^-(\pi_1) \subset \mathfrak{n}^-(\pi_1)$. Using the fact that $\mathrm{ad}(\mathfrak{n}^+(\pi_1))$ consists of lower triangular matrices with zeros along the diagonal we see easily that exp: $\mathfrak{n}^+(\pi_1) \to N^+$ is onto. Thus $\mathrm{Ad}(n) = e^{\mathrm{ad}X}$ for some X in $\mathfrak{n}^+(\pi_1)$. But then $e^{\mathrm{ad}X}\mathfrak{n}^-(\pi_1) \subset \mathfrak{n}^-(\pi_1)$. Observing that $e^{\mathrm{ad}X} = I + U$ with U lower triangular with zeros on the diagonal we see that the power series for $\log(I + U)$ is actually a finite sum and is equal to $\mathrm{ad}\, X$. Hence $\mathrm{ad}\, X\mathfrak{n}^-(\pi_1) \subset \mathfrak{n}^-(\pi_1)$. This implies that $X = 0$. If g is in $K'_C N^+ \cap K'_C N^-$ then $g = g_1 n$ with g_1 in K'_C, n in N^+. This implies that $\mathrm{Ad}(n)\mathfrak{n}^-(\pi_1) \subset \mathfrak{n}^-(\pi_1)$ and hence $n = 1$. This proves (2).

If g is in $K'_C N^-$ and if $\bar{g} = g$ then g is in $K'_C N^- \cap K'_C N^+$. Hence (2) implies that g is in K'_C and $\bar{g} = g$. This proves that

(3) $G' \cap K'_C N^- = K'$.

Equation (3) says that the orbit of eB relative to G' in G'_C/B is diffeomorphic with $G'/K' = M$. By comparing dimensions of M and G'_C/B we see that $G'eB$ is open in G'_C/B. Since G' is compact we see that $G'eB$ is closed in G'_C/B. We have therefore shown

(4) G'_C/B is diffeomorphic with M.

Now G'_C/B is the quotient of a complex Lie group by a closed complex subgroup. A.5.3.5 applies and hence G'_C/B is a complex manifold with G'_C acting as a group of holomorphic diffeomorphisms. This gives the desired complex structure on M with G acting as a group of holomorphic diffeomorphisms.

6.2.12 Let $G/C(T_1) = M$ be a generalized flag manifold. Let T, π, π_1, $\mathfrak{k}(\pi_1)_C$, $\mathfrak{n}^+ = \mathfrak{n}^+(\pi_1)$, $\mathfrak{n}^- = \mathfrak{n}^-(\pi_1)$, G_C, N^+, N^-, and B be as in the proof of 6.2.11. We study the complex structure on M corresponding to π_1.

Let $\psi: N^+ \to M$ be defined by $\psi(n) = nB$. Then ψ is a holomorphic diffeomorphism of N^+ onto an open subset of M (this follows from 6.2.11(2) and the inverse function theorem). Thus ψ_{*e} is a real linear isomorphism of $\mathfrak{n}^+(\pi_1)$ onto $T(M)_e$ where $\bar{e} = eB$ so that if J is the almost complex structure on M then $\psi_{*e}((-1)^{\frac{1}{2}}X) = J\psi_{*e}(X)$.

Let $\mathfrak{k}$ be the Lie algebra of $C(T_1)$ in $\mathfrak{g}$, the Lie algebra of G, and let $\mathfrak{p}$ be the orthogonal complement to $\mathfrak{k}$ in $\mathfrak{g}$ relative to the Killing form of $\mathfrak{g}$. We may identify $\mathfrak{p}$ with $T(M)_{\bar{e}}$ relative to the linear isomorphism $p_{*e}: \mathfrak{p} \to T(M)_{\bar{e}}$ where $p: G \to M$ is the natural map. We may thus identify $\mathfrak{p}_C$ with $T^C(M)_{\bar{e}}$ as representations of $C(T_1)$ (see A.5.2.2).

As a representation of $C(T_1)$, $\mathscr{T}(M)_{\bar{e}}$ is by the above equivalent with $(\mathrm{Ad}_{C(T_1)}, \mathfrak{n}^+)$. Thus under the identification of $T^C(M)_e$ with $\mathfrak{p}_C$ we see that $\mathscr{T}(M)_e$ is identified with $\mathfrak{n}^+$. Since $\overline{\mathfrak{n}^+} = \mathfrak{n}^-$ we see that $\bar{\mathscr{T}}(M)_e$ is identified with $\mathfrak{n}^-$. We have proved the following lemma.

6.2.13 *Lemma* In the notation of 6.2.11, A.5.2.2, and 5.2.2,

(1) $\mathscr{T}(G/C(T_1)) = G \underset{\mathrm{Ad}|_{C(T_1)}}{\times} \mathfrak{n}^+(\pi_1)$.

(2) $\bar{\mathscr{T}}(G/C(T_1)) = G \underset{\mathrm{Ad}|_{C(T_1)}}{\times} \mathfrak{n}^-(\pi_1)$.

Furthermore there is a system of holomorphic coordinates $z_1, \ldots, z_n$ on $\psi(N^+)$ (ψ is defined in 6.2.12) so that the *complex* vector fields on M corresponding to the elements of $\mathfrak{n}^+(\pi_1)$ are linear combinations of the $\partial/\partial z_j$ and the complex vector fields corresponding to the elements of $\mathfrak{n}^-(\pi_1)$ are linear combinations of the $\partial/\partial \bar{z}_j$.

6.2.14 *Theorem* Let $M = G/C(T_1)$ be a generalized flag manifold. Let T be a maximal torus of G and let P be a T_1-admissible Weyl chamber of T and let M have the corresponding complex structure. Then there is a Lie homomorphism ρ of G into $SU(n)$, for n sufficiently large, and a holomorphic imbedding ψ of M into CP^{n-1} so that $\psi(gx) = \rho(g)\psi(x)$ for g in G and x in M.

PROOF Since the center Z of G is contained in $C(T_1)$, we have seen that $M = (G/Z)/(C(T_1)/Z)$. We may thus assume that G is the universal covering group of G/Z. 2.8.4 says that we may assume that G is a closed subgroup of $GL(n, \mathbf{C})$ for n sufficiently large. Let $\mathfrak{g}_\mathbf{C}$ be the complexification of the lie algebra of G, $\mathfrak{g}$, in $M_n(\mathbf{C})$. Let $G_\mathbf{C}$ be the connected subgroup of $GL(n, \mathbf{C})$ corresponding to $\mathfrak{g}_\mathbf{C}$.

Let Δ be the root system of G relative to T and let Δ_1 be the root system of $C(T_1)$ relative to T. Let π be the simple set of roots relative to P and let $\pi_1 = \pi \cap \Delta_1$. Let $\pi = \{\alpha_1, \ldots, \alpha_l\}$ and let us assume that $\pi - \pi_1 = \{\alpha_1, \ldots, \alpha_s\}$. Let $\lambda_1, \ldots, \lambda_l$ in $(\mathfrak{h}_C)^*$ be defined by $2\alpha_i(H_{\alpha_j})/\alpha_j(H_{\alpha_j}) = \delta_{ij}$ for $i, j = 1, \ldots, l$. Let $(\pi_{\lambda_j}, V^{\lambda_j})$ be the irreducible holomorphic representation of G_C with lowest weight λ_j. Set $V = V^{\lambda_1} + \cdots + V^{\lambda_s}$. Let for g in G_C, $\mu(g) = \pi_{\lambda_1}(g) \oplus \cdots \oplus \pi_{\lambda_s}(g)$. Let v_i in V^{λ_i} be a nonzero element of the λ_i weight space in V^{λ_i}. Set $v = v_1 + \cdots + v_s$.

Let $q_i \colon V^{\lambda_i} - \{0\} \to CP^{n_i}$, where $n_i = \dim V^{\lambda_i} - 1$, be the canonical map. Then

$$q = q_1 \times \cdots \times q_s \colon (V^{\lambda_1} - 0) \times \cdots \times (V^{\lambda_s} - 0) \to CP^{n_1} \times \cdots \times CP^{n_s}.$$

We now define a holomorphic imbedding ζ of $CP^{n_1} \times \cdots \times CP^{n_s}$ into

CP^N for $N = (n_1 + 1) \dots (n_s + 1) - 1$ by the formula $\zeta([w_1], \dots, [w_s]) = [w_1 \otimes \cdots \otimes w_s]$. We leave it to the reader to check that ζ is a holomorphic imbedding.

Let for g in G_C, $\Phi(g) = \mu(g)v$. Then clearly Φ is a holomorphic mapping of G_C into $(V^{\lambda_1} - \{0\}) \times \cdots \times (V^{\lambda_s} - \{0\})$. Suppose that g and h are in G_C and that $q(\Phi(gh)) = q(\Phi(g))$. Then $\mu(gh)v = \sum z_i \mu(g)v_i$ with z_i in $C - \{0\}$. Set $\tilde{B} = \{g \text{ in } G_C | \mu(g)v = \sum z_i v_i \text{ with } z_i \text{ in } C\}$. Then $q \circ \Phi$ induces a C^∞ imbedding of $G_C/\tilde{B}$ into $CP^{n_1} \times \cdots \times CP^{n_s}$. If we can show that $\tilde{B} = B$ then using we see that $q \circ \Phi$ is a holomorphic imbedding of M. Then $\zeta \circ q \circ \Phi$ would be the desired holomorphic imbedding of M into CP^N. Thus, to prove the theorem we need only show that $\tilde{B} = B$.

Let α be a positive root. Let $\mathfrak{g}^\alpha = \mathfrak{g}_\alpha + CH_\alpha + \mathfrak{g}_{-\alpha}$. Then $\mathfrak{g}^\alpha$ is a TDS (see 4.3.9). Let $k_i = 2\lambda_i(H_\alpha)/\alpha(H_\alpha)$. If $k_i < 0$ then $\pi_{\lambda_i}(X)v_i \neq 0$ for all $X \neq 0$ in $\mathfrak{g}_\alpha$. If $k_i = 0$ then $\pi_{\lambda_i}(X)v_i = 0$ for all X in $\mathfrak{g}^\alpha$.

Let $\tilde{\mathfrak{b}}$ be the Lie algebra of $\tilde{B}$. Then it is clear that $\mathfrak{h}_C \subset \mathfrak{b}$. Let $\tilde{\Delta} = \{\alpha \text{ in } \Delta | \mathfrak{g}_\alpha \subset \mathfrak{b}\}$. Then $\tilde{\mathfrak{b}} = \mathfrak{h}_C + \sum_{\alpha \in \tilde{\Delta}} \mathfrak{g}_\alpha$. If α is in Δ_1 then $k_i = 0$ for $i = 1, \dots, s$. Thus we see that if $\mathfrak{b}$ is the Lie algebra of B then $\mathfrak{b} \subset \tilde{\mathfrak{b}}$.

Let α be in $\tilde{\Delta}$ and suppose that $\alpha > 0$. If $k_i < 0$ then $\pi_{\lambda_i}(X)v_i \neq 0$ for all X in $\mathfrak{g}_\alpha$. But the definition of $\tilde{\mathfrak{b}}$ implies that if $i \leqslant s$ then $\pi_{\lambda_i}(X)v_i = z_i v_i$. Since $\pi_{\lambda_i}(X)v_i$ is in the $\lambda_i + \alpha$ weight space of V^{λ_i} we see that this implies that $k_i = 0$ for α in $\tilde{\Delta}$, $\alpha > 0$. This clearly implies that $\mathfrak{b} = \tilde{\mathfrak{b}}$.

If we can show that B is connected then we have clearly shown that $B = \tilde{B}$. Now G acts transitively on G_C/B, hence on $G_C/\tilde{B}$. Thus if we can show that $\tilde{B} \cap G = C(T_1)$ we will have completed the proof of the theorem. Let b be in $\tilde{B} \cap G$. Since $C(T_1)$ is connected there is k_1 in $C(T_1)$ so that if $u_1 = k_1 b$ then $\mathrm{Ad}(u_1)\mathfrak{h} = \mathfrak{h}$. Thus $\mathrm{Ad}(u_1)$ permutes the Weyl chambers of T relative to $C(T_1)$. There is thus an element k_2 of $C(T_1)$ so that $\mathrm{Ad}(k_1)\mathfrak{h} = \mathfrak{h}$ and so that if $u_2 = k_2 u_1$, then $\mathrm{Ad}(u_2)$ preserves the Weyl chamber of T relative to $C(T_1)$ corresponding to π_1. Now u_2 is in $B \cap C(T_1)$. Thus $\mathrm{Ad}(u_2)\mathfrak{n}^- \subset \mathfrak{n}^-(\mathfrak{n}^- = \mathfrak{n}^-(\pi_1))$. This implies that $\mathrm{Ad}(u_2)$ preserves the Weyl chamber P. But then u_2 is in T since we have assumed that G is connected. This proves that u_2 is in $C(T_1)$ and hence b is in $C(T_1)$.

6.3 Holomorphic Vector Bundles over Generalized Flag Manifolds

6.3.1 Let G be a compact connected Lie group and let T_1 be a torus in G. Let $C(T_1)$ be the centralizer of T_1 in G and let $M = G/C(T_1)$ be the corresponding generalized flag manifold. Let T be a maximal torus of G containing T_1. Let P be a T_1-admissible Weyl chamber for T. We give M the complex structure corresponding to P (see 6.2.11). We may (and do) assume that T_1 is the connected component of the identity of the center of $C(T_1)$.

6.3.2 Let Z be the center of G. Then Z is contained in $C(T_1)$. Hence $M = (G/Z)/(C(T_1)/Z)$. Let $\tilde{G}$ be the universal covering group of G/Z. 3.6.6 implies that $\tilde{G}$ is compact. Let $\tilde{T}_1$ be the torus in $\tilde{G}$ corresponding to T_1/Z. Then $C(\tilde{T}_1)$ is the connected subgroup of $\tilde{G}$ corresponding to $C(T_1)/Z$. We therefore see that $\tilde{G}/C(\tilde{T}_1) = M$. We may thus assume in addition to the assumptions of 6.3.1 that G is simply connected.

6.3.3 *Proposition* Let (σ, E_0) be a finite dimensional unitary representation of $C(T_1)$. Then the homogeneous vector bundle $E = G \underset{\sigma}{\times} E_0$ has the structure of a holomorphic vector bundle over M with G acting as a group of holomorphic diffeomorphisms of E to E.

PROOF Let $\mathfrak{g}$ be the Lie algebra of G. Let $\mathfrak{g}_C$ be the complexification of $\mathfrak{g}$ and let G_C be the simply connected Lie group with Lie algebra $\mathfrak{g}_C$. Then we may identify G with a connected subgroup of G_C. Let $\mathfrak{k}$ be the Lie algebra of $C(T_1)$ and let $\mathfrak{k}_C$ be its complexification. Let K_C be the connected subgroup of G_C corresponding to $\mathfrak{k}_C$. Then (σ, E_0) extends to a holomorphic representation of K_C (see 4.7.5), which we denote by $(\tilde{\sigma}, E_0)$. Let $\mathfrak{n}^- = \mathfrak{n}^-(\pi_1)$ be as in 6.2.11. Let $B = K_C N^-$ where N^- is the connected subgroup of G_C corresponding to $\mathfrak{n}^-$. Extend $\tilde{\sigma}$ to B by setting $\tilde{\sigma}(kn) = \tilde{\sigma}(k)$ for k in K_C and n in N^-. Then using the results of the proof of 6.2.11 we see that $\tilde{\sigma}$ is well defined and defines a holomorphic representation of B. As G-homogeneous vector bundles over M we have $G_C \underset{\tilde{\sigma}}{\times} E_0 = G \underset{\sigma}{\times} E_0$. We leave it to the reader to check that a modification of the technique of 6.2.11 that gives G_C/B a complex structure shows that $G_C \underset{\tilde{\sigma}}{\times} E_0$ is a holomorphic vector bundle with G_C acting on it by holomorphic mappings.

6.3.4 Let $\langle\ ,\ \rangle$ be the inner product $-B_{\mathfrak{g}}$ on $\mathfrak{g}$, the Lie algebra of G. Let $\mathfrak{k}$ be the Lie algebra of $C(T_1)$ and let $\mathfrak{p}$ be the orthogonal complement to $\mathfrak{k}$ in $\mathfrak{g}$. Let $\langle\ ,\ \rangle$ denote the corresponding Riemannian structure on M (see 5.8.12). Then $\langle\ ,\ \rangle$ induces a Hermitian structure on $\mathcal{T}(M)$ and hence on $\Lambda^k(\mathcal{T}(M))^*$ for each k (see 1.8). Let $E = G \underset{\sigma}{\times} E_0$ be a holomorphic vector bundle over M. Let $V_k = E \otimes \Lambda^k(\mathcal{T}(M)^*)$. Let $D_k = \bar{\partial}|_{\Gamma^\infty(V_k)}$. Here, $\bar{\partial}$ is defined as in 1.8. Then the D_k, V_k satisfy the hypotheses of 5.8.10. In fact, $\Box_k = D_{k-1}D_{k-1}^* + D_k^*D_k = \Box$ as defined in 1.8. In 1.8 the symbol of $\Box$ was computed. In this case we find that $\sigma(\Box) = -\sigma(\Omega)$. Thus 5.8.8 implies that $\Box$ is globally hypoelliptic. We therefore have the following theorem.

6.3.5 *Theorem* Let E be a holomorphic vector bundle $E = G \underset{\sigma}{\times} E_0$ over M as in 6.3.3. Set

$$\mathscr{H}^k(E) = \ker(\bar{\partial}|_{\Gamma^\infty(E \otimes \Lambda^k(\bar{\mathscr{T}}(M)^*))})/\bar{\partial}(\Gamma^\infty(E \otimes \Lambda^{k-1}(\bar{\mathscr{T}}(M)^*)).$$

Set

$$\mathscr{H}^k_\Box(E) = \{f \text{ in } \Gamma^\infty(E \otimes \Lambda^k(\bar{\mathscr{T}}(M)^*)| \Box f = 0\}.$$

Then $\dim \mathscr{H}^k(E) = \dim \mathscr{H}^k_\Box(E) < \infty$ and $[\mathscr{H}^k(E)] = [\mathscr{H}^k_\Box(E)]$ in $R(G)$.

6.3.6 In 1.8.5 it was shown that $\mathscr{H}^0(E)$ is the space of all holomorphic cross sections of E. Thus, in particular, 6.3.5 says that the space of all holomorphic cross sections of E is finite dimensional.

6.3.7 *Theorem* Let (σ, E_0) be an irreducible unitary representation of $C(T_1)$. If there exists no irreducible holomorphic representation (τ, V) of G_C so that the representation $(\tau|_{C(T_1)}, V^{N^+})$ is equivalent with (σ, E_0) then $\mathscr{H}^0(E) = (0)$ (here N^+ is the subgroup of G_C corresponding to $\mathfrak{n}^+(\pi_1)$ in the proof of 6.2.11 and $V^{N^+} = \{v \text{ in } V | \tau(n)v = v \text{ for all } n \text{ in } N^+\}$). If there exists such a (τ, V) then $[\mathscr{H}^0(E)] = [\tau|_G, V]$ as elements of $R(G)$ (see 4.10).

PROOF As in 5.3 we identify $\Gamma^\infty(E)$ with $C^\infty(G; \sigma)$ as representations of G. Thus we may look upon $\mathscr{H}^0(E)$ as a space of functions f from G to E_0 so that $f(gk) = \sigma(k)^{-1}f(g)$ for g in G and k in $C(T_1)$.

We also note that if $\tilde{\sigma}$ is as in 6.3.4 then $G_C \underset{\sigma}{\times} E_0 = E$. We may thus identify $\Gamma^\infty(E)$ with $C^\infty(G_C; \tilde{\sigma})$. Now $\mathscr{H}^0(E)$ is a finite dimensional representation of G; it therefore has (by 4.7.5) a unique extension to a holomorphic representation of G_C. We may thus identify $\mathscr{H}^0(E)$ with a subspace of $C^\infty(G_C; \tilde{\sigma})$ with G_C acting by $g \cdot f(x) = f(g^{-1}x)$.

Suppose that f is in $(\mathscr{H}^0(E))^{N^+}$. Let $B = K_C N^-$ as in 6.3.3. If n is in N^+, b is in B, then $f(nb) = \tilde{\sigma}(b)^{-1}f(e)$. Thus f is completely determined on N^+B by its value at e. Now f is holomorphic and N^+B is open in G_C and hence the map $(\mathscr{H}^0(E))^{N^+} \to E_0$, $\varepsilon(f) = f(e)$ is injective. Furthermore, if k is in $C(T_1)$ then $kf(e) = f(k^{-1}) = \sigma(k)f(e)$. Thus ε is in $\mathrm{Hom}_{C(T_1)}((\mathscr{H}^0(E))^{N^+}, E_0)$. Since (σ, E_0) is irreducible we find that $\varepsilon = 0$ or ε is bijective. We assert that this implies that as a representation of G_C, $\mathscr{H}^0(E)$ is irreducible. In fact, if $\mathscr{H}^0(E)$ were reducible then $\mathscr{H}^0(E) = V_1 \oplus V_2$ with V_i invariant and nonzero. The theorem of the highest weight implies that $V_i^{N^+} \neq (0)$ for $i = 1, 2$. Hence

$(\mathscr{H}^0(E))^{N^+} = V_1^{N^+} \oplus V_2^{N^+}$. But $(\mathscr{H}^0(E))^{N^+}$ is irreducible as a $C(T_1)$ representation. This contradiction implies that $\mathscr{H}^0(E)$ is indeed irreducible. This clearly shows that if there is no (τ, V) satisfying the conditions of the theorem then $\mathscr{H}^0(E) = (0)$.

Suppose that such a (τ, V) exists. Let (τ^*, V^*) be the contragradient holomorphic representation of G_C. Let $(V^*)^{N^-} = \{v^* \text{ in } V^* | \tau^*(n)v = v \text{ for all } n \text{ in } N^-\}$. If v is in V we define a map $\alpha(v)$ of G_C into $((V^*)^{N^-})^*$ by $\alpha(v)(g)(v^*) = v^*(\tau(g)^{-1}v)$. Then

$$\begin{aligned}\alpha(\tau(g_0)v)(g)(v^*) &= v^*(\tau(g)^{-1}\tau(g_0)v)\\ &= v^*(\tau(g_0^{-1}g)^{-1}v) = \alpha(v)(g_0^{-1}g)(v^*).\end{aligned}$$

Let $\mathscr{A}(G_C; (V^{*N^-})^*)$ be the space of all holomorphic mappings of G_C into $(V^{*N^-})^*$ with G_C acting by $gf(x) = f(g^{-1}x)$. We have shown that α defines a G_C intertwining operator from V into $\mathscr{A}(G_C; (V^{*N^-})^*)$. Clearly $\alpha \neq 0$. Thus, since V is irreducible α is injective.

Let n^+ be in N^+, k be in K_C, and let n^- be in N^-. If v is in V^{N^+} then

$$\alpha(v)(n^+kn^-)(v^*) = v^*(\tau(k)^{-1}v) = \tau^*(k)v^*(v) = (\delta(k)^{-1}(\alpha(v)(e)))(v^*)$$

where δ is the contragradient representation to the representation $(\tau^*|_{K_C}, V^{*N^-})$. Hence the map V^{N^+} into $(V^{*N^-})^*$ given by $v \to \alpha(v)(e) = \iota(v)$ is injective. It is also clear that ι intertwines $(\tau|_{K_C}, V^{N^+})$ and $(\delta, (V^{*N^-})^*)$. Interchanging the roles of V and V^* and N^+ and N^- we find that, as representations of B, $\tilde{\delta}$ and $\tilde{\sigma}$ are equivalent. We may thus identify $\tilde{\delta}$ and $\tilde{\sigma}$. It is easy to see that if v is in V then $\alpha(v)$ is in $C^\infty(G_C, \tilde{\delta}) = C^\infty(G_C, \tilde{\sigma})$ and is holomorphic. Thus $\alpha(v)$ induces an element of $\mathscr{H}^0(E)$. We have thus proved that as elements of $R(G)$, $[\mathscr{H}^0(E)] = [V]$.

6.3.8 *Corollary to the proof of* **6.3.5** Let (τ, V) be an irreducible finite dimensional holomorphic representation of G_C. Then the representations $(\tau|_{K_C}, V^{N^+})$ and $((\tau^*|_{K_C})^*, (V^{*N^-})^*)$ are equivalent irreducible representations of K_C.

6.3.9 *Corollary* Let G be a compact simply connected Lie group. Let T be a maximal torus of G. Let P be a Weyl chamber of T. Let G/T have the corresponding complex structure. Let λ be in $\hat{T}$ and let $L_\lambda = G \underset{\lambda}{\times} C$ be the corresponding holomorphic line bundle over G/T. Then

$$[\mathscr{H}^0(L_\lambda)] = \begin{cases}[0] \text{ if } \lambda \text{ is not dominant integral}\\ [V^\lambda] \text{ if } \lambda \text{ is dominant integral.}\end{cases}$$

Here (π_λ, V^λ) is the irreducible finite dimensional representation of G with highest weight λ.

6.4 An Alternating Sum Formula

6.4.1 Let $M = G/C(T_1)$ be a generalized flag manifold. Give M a complex structure as in 6.2.11. Let E be a homogeneous, holomorphic vector bundle over M. We set $\chi_h(E) = \sum(-1)^k[\mathscr{H}^k(E)]$ in $R(G)$ where $\mathscr{H}^k(E)$ is defined as in 6.3.5.

6.4.2 Let $\bar{\partial}$ and $\bar{\partial}^*$ be as in 6.3.4. Let $F^+ = \sum \Lambda^{2r}\mathscr{T}(M)^*$ and let $F^- = \sum \Lambda^{2r+1}\mathscr{T}(M)^*$. Then $D = \bar{\partial} + \bar{\partial}^*$ defines a homogeneous differential operator from $E \times F^+$ to $E \times F^-$. Since D^*D and DD^* are given by $\Box$, 5.8.11 applies and ker D and coker D are finite dimensional.

6.4.3 *Lemma* $\operatorname{ind}_h(D) = \chi_h(E)$.

PROOF $\operatorname{ind}_h(D) = [\ker D] - [\ker D^*]$. If $Du = 0$ then $D^*Du = 0$. Hence $\Box u = 0$. If $\Box u = 0$ then $\bar{\partial}u$ and $\bar{\partial}^*u = 0$. Hence $Du = 0$. Thus $[\ker D] = [\ker \Box|_{\Gamma^\infty(E\otimes F^+)}] = \sum[\mathscr{H}^{2k}(E)]$. Similarly $[\ker D^*] = \sum[\mathscr{H}^{2k+1}(E)]$.

6.4.4 *Theorem* Let G be a connected, simply connected, compact Lie group. Let T be a maximal torus of G. Let P be a Weyl chamber of T and let G/T have the corresponding complex structure. Let $\rho = \frac{1}{2}\sum \alpha$, the sum over the positive roots relative to P. Let λ be in $\hat{T}$ and let $L_\lambda = G \underset{\lambda}{\times} C$ be the corresponding homogeneous, holomorphic line bundle.

(i) If there is no s in W (the Weyl group of T) so that $s(\lambda + \rho) - \rho$ is dominant integral then $\chi_h(L_\lambda) = 0$.

(ii) If there is an s in W so that $s(\lambda + \rho) - \rho$ is dominant integral then $\chi_h(L_\lambda) = \det(s)[V^{s(\lambda+\rho)-\rho}]$ where $(\pi_{s(\lambda+\rho)-\rho}, V^{s(\lambda+\rho)-\rho})$ is the irreducible representation of G with highest weight $s(\lambda + \rho) - \rho$.

PROOF In 6.2.13 we saw that $\mathscr{T}(G/T) = G \underset{\mathrm{Ad}|_T}{\times} \mathfrak{n}^-$. Now as a representation of T, $(\mathrm{Ad}|_T, \mathfrak{n}^+)$ is equivalent with $((\mathrm{Ad}|_T)^*, (\mathfrak{n}^-)^*)$. Hence if $L_\lambda \otimes F^+ = G \underset{\tau}{\times} W$ and if $L_\lambda \otimes F^- = G \underset{\sigma}{\times} V$ then $[W] - [V] = [\lambda] \cdot \lambda_{-1}(\mathfrak{n}^+)$ in the notation of 4.10.8. Combining 6.4.3 and 5.9.1 we find that

$$\chi_h(L_\lambda) = i_*([\lambda]\cdot\lambda_{-1}(\mathfrak{n}^+))$$

where i is the canonical injection of T into G. Using the definition of i_* we therefore have

$$i_*([\lambda]\lambda_{-1}(\mathfrak{n}^+)) = \sum_{\gamma\in\hat{G}} \langle[\lambda]\lambda_{-1}(\mathfrak{n}^+), i^*\gamma\rangle_T\gamma.$$

Now $\lambda_{-1}(\mathfrak{n}^+) = \prod_{\alpha\in\Delta^+}(1 - [\alpha])$. Thus

$$\langle[\lambda]\lambda_{-1}(\mathfrak{n}^+), i^*\gamma\rangle_T = \int_T \prod_{\alpha\in\Delta^+}(1 - t^\alpha)t^\lambda\bar{\chi}_\gamma(t)\,dt$$

where χ_γ is the character of γ.

4.8.6 says that $A(\rho)\chi_{\gamma|T} = A(\rho + \Lambda)$ where Λ is the highest weight of γ and $A(\mu)$ is as in 4.8.3. 4.8.5 says that $\prod_{\alpha>0}(1 - t^\alpha) = t^\rho\overline{A(\rho)}(t)$. Hence

$$(*) \qquad \int_T \prod_{\alpha>0}(1 - t^\alpha)t^\lambda\chi_\gamma(t)\,dt = \sum_{\alpha>0}\det(s)\int_T t^\rho t^\lambda t^{-s(\Lambda+\rho)}\,dt.$$

Thus (*) is nonzero only if $\lambda + \rho = s(\Lambda + \rho)$. This proves (i).

Suppose that Λ and Λ' are dominant integral and that $\lambda + \rho = s(\Lambda + \rho) = s'(\Lambda' + \rho)$. We show that this implies that $\Lambda = \Lambda'; s = s'$. In fact, let us assume that $\Lambda \geqslant \Lambda'$. Set $s'' = s^{-1}s'$. Then $\Lambda + \rho = s''\Lambda' + s''\rho$. Hence $\Lambda - s''\Lambda' = s''\rho - \rho$. But, $\Lambda - s''\Lambda' \geqslant 0$ and $s''\rho - \rho \leqslant 0$. Hence $s''\rho = \rho$. This implies that $s'' = 1$. Hence (*) also implies (ii).

6.5 Exercises

6.5.1 Let $G = SU(n)$ (see 6.2.4). Let $T_1 \subset G$ be a torus. Show that there exists g in G and $k_1, \ldots, k_r$ so that $k_i > 0$, k_i an integer, $k_1 + \cdots + k_r = n$ such that $gT_1g^{-1} = T(k_1, \ldots, k_r)$ (see 6.2.4).

6.5.2 Let $M = CP^n$ with the complex structure of 6.2.6. Let $\mathcal{T}(M)^*$ be the holomorphic dual bundle to the holomorphic tangent bundle of M. Let V be the symmetric square of $\mathcal{T}(M)^*$. A holomorphic cross section of V is called a quadratic differential on M. Show that M has no nonzero quadratic differentials.

6.5.3 Let $G = SU(n)$. Let P^k be the space of all polynomials in $z_1, \ldots, z_n$ that are homogeneous of degree k. If p is in P^k and if g is in G let $\pi_k(g)p(z) = p(g^{-1}z)$. Let L be the line bundle over CP^{n-1} given by: L_x is the complex line in C^n corresponding to x. Let for $k \geqslant 0$, L^k be the k-fold tensor product of

L with itself. Let L^{-k} be the k-fold tensor product of L^* with itself. Compute the class in $R(G)$ corresponding to $\mathscr{H}^0(L^k)$ in terms of the (π_l, P^l).

6.5.4 Let $G_{n,k}$ and $E_{n,k}$ for $K = C$ be as in 1.9.4. Show that $G_{n,k}$ is a generalized flag manifold corresponding to $SU(n)$ and that $E_{n,k}$ is a homogeneous vector bundle over $G_{n,k}$. Find $\mathscr{H}^0(E_{n,k})$.

6.5.5 Let L^k be as in 6.5.3. Compute $\chi(L^k)$.

6.5.6 Give a direct proof of 6.3.7.

6.6 Notes

6.6.1 It was proved by Wang [1] and Borel]2] that the only simply connected homogeneous Kahler manifolds (see Helgason [1], or Kobayashi and Nomizu [1] for the pertinent definitions) are the generalized flag manifolds with the complex structures given in 6.2.

6.6.2 6.3.7 is usually called the Borel–Weil theorem (see Serre [1]). The proof given here is taken from Wallach [2].

6.6.3 Relative to 6.2.14 see also Borel [1] III.10.3.

6.6.4 6.4.4 is due to Borel and Hirzbruch [1]. In Borel and Hirzbruch [1] 6.4.4 is proved using Hirzbruch's Riemann–Roch theorem (see Hirzbruch [1]).

CHAPTER 7

Analysis on Semisimple Lie Groups

7.1 Introduction

In this chapter the basic structure theory of semisimple Lie groups is developed. We prove the Cartan, Iwasawa, and Bruhat decompositions of a semisimple Lie group. Using the Bruhat decomposition we derive the Gelfand–Naimark decomposition. We derive the integral formulas for these decompositions. We then study the adjoint action of a semisimple Lie group on itself and its Lie algebra. We derive the corresponding integral formulas. In Section 10, we study differential operators on a reductive Lie algebra and derive Harish-Chandra's formula for the restriction of a constant coefficient invariant differential operator to a Cartan subalgebra. In Sections 11 and 12 the results of the previous sections are used to derive some basic formulas for semisimple Lie groups with one conjugacy class of Cartan subalgebra. From Section 5 to Section 12 almost every result or proof is either due to Harish-Chandra or uses techniques originating with Harish-Chandra.

We should warn the reader that there are many notions of "F_f" (see 7.7.11) in the literature. The "F_f" defined in 7.7.11 is a generalization of that used by Harish-Chandra in [8] (see also Helgason [1], Chapter 10).

7.2 The Cartan Decomposition of a Semisimple Lie Group

7.2.1 *Definition* Let G be a Lie group with Lie algebra $\mathfrak{g}$. Then G is said to be semisimple if $\mathfrak{g}$ is semisimple.

7.2.2 Let $\mathfrak{g}$ be a semisimple Lie algebra over R. 3.7.9 says that $\mathfrak{g}$ has a Cartan involution. That is, there is an involutive automorphism, θ, from $\mathfrak{g}$ to $\mathfrak{g}$ so that if $\mathfrak{k} = \{X \text{ in } \mathfrak{g} | \theta X = X\}$ and if $\mathfrak{p} = \{X \text{ in } \mathfrak{g} | \theta X = -X\}$ and if $B = B_{\mathfrak{g}}$ then $B|_{\mathfrak{k}\times\mathfrak{k}}$ is negative definite and $B|_{\mathfrak{p}\times\mathfrak{p}}$ is positive definite.

Let $\mathfrak{g}_C$ be the complexification of $\mathfrak{g}$. Let $\mathfrak{g}_u = \mathfrak{k} \oplus (-1)^{\frac{1}{2}}\mathfrak{p}$ in $\mathfrak{g}_C$. Then $\mathfrak{g}_u$ is a compact form of $\mathfrak{g}_C$. $B_{\mathfrak{g}_C}$ is the complex bilinear extension of $B_{\mathfrak{g}}$ (or $B_{\mathfrak{g}_u}$). On $\mathfrak{g}_C$ define the Hermitian inner product $(X, Y) = -B_{\mathfrak{g}_C}(X, \tau Y)$ where τ is conjugation in $\mathfrak{g}_C$ relative to $\mathfrak{g}_u$ (note that $\tau|_{\mathfrak{g}} = \theta$).

7.2.3 *Lemma* Relative to the inner product (,) on $\mathfrak{g}_C$, if X is in $\mathfrak{k}$ then ad X is skew Hermitian as an endomorphism of $\mathfrak{g}_C$. If X is in $\mathfrak{p}$ then ad X is Hermitian.

PROOF If X is in $\mathfrak{k}$ and if Y, Z are in $\mathfrak{g}_C$ then

$$\begin{aligned}(\mathrm{ad}(X)Y, Z) &= -B(\mathrm{ad}(X)Y, \tau Z) = B(Y, \mathrm{ad}(X)\tau Z)\\ &= B(Y, [X, \tau Z]) = (B(Y, \tau[\tau X, Z)\\ &= B(Y, \tau\mathrm{ad}(X)Z) = -(Y, \mathrm{ad}(X)Z).\end{aligned}$$

If X is in $\mathfrak{p}$ then since $\tau X = -X$, the same argument shows that $(\mathrm{ad}(X)Y, Z) = (Y, \mathrm{ad}(X)Z)$.

7.2.4 *Lemma* Let G be a connected semisimple Lie group with Lie algebra $\mathfrak{g}$. Let $\mathfrak{g} = \mathfrak{k} \oplus \mathfrak{p}$ be a Cartan decomposition of $\mathfrak{g}$. Let K be the connected subgroup of G corresponding to $\mathfrak{k}$. Then $\mathrm{Ad}(K)$ is a compact subgroup of $\mathrm{Ad}(G) \subset GL(\mathfrak{g})$.

PROOF Let Aut($\mathfrak{g}$) be the group of automorphisms of $\mathfrak{g}$. Then Aut($\mathfrak{g}$) is a closed (hence Lie) subgroup of $GL(\mathfrak{g})$. Since every derivation of $\mathfrak{g}$ is of the form ad X with X in $\mathfrak{g}$ (3.4.6), we see that $\mathrm{Ad}(G)$ is the identity component of Aut($\mathfrak{g}$). If X is in $\mathfrak{k}$ then ad X is a skew Hermitian endomorphism of $\mathfrak{g}_C$ relative to (,). Since $\mathrm{Ad}(K)$ is the connected subgroup of $\mathrm{Ad}(G)$ corresponding to ad($\mathfrak{k}$) we see that $\mathrm{Ad}(K)$ is a subgroup of the unitary group of (,). If we can show that $\mathrm{Ad}(K)$ is closed we will have proved the lemma.

Let $\tilde{K} = \{g \text{ in } \mathrm{Ad}(G) | g\,\mathfrak{k} \subset \mathfrak{k}\}$. Let $\tilde{\mathfrak{k}}$ be the Lie algebra of $\tilde{K}$. If V is in $\tilde{\mathfrak{k}}$ then V is a derivation of $\mathfrak{g}$. Hence $V = \mathrm{ad}\, X$ for some X in $\mathfrak{g}$. Now, $X = X_1 + X_2$ with X_1 in $\mathfrak{k}$ and X_2 in $\mathfrak{p}$. Thus $\mathrm{ad}(X_1 + X_2)\mathfrak{k} \subset \mathfrak{k}$. But then $\mathrm{ad}(X_2)\mathfrak{k} \subset \mathfrak{k}$. But $[\mathfrak{k}, \mathfrak{k}] \subset \mathfrak{k}$, $[\mathfrak{k}, \mathfrak{p}] \subset \mathfrak{p}$, and $[\mathfrak{p}, \mathfrak{p}] \subset \mathfrak{k}$. Hence $\mathrm{ad}(X_2)\mathfrak{k} = (0)$. $\mathfrak{p}$ is the orthogonal complement to $\mathfrak{k}$ in $\mathfrak{g}$ relative to $B_{\mathfrak{g}}$. Hence ad $X_2\mathfrak{p} \subset \mathfrak{p}$. This implies that ad $X_2\mathfrak{p} = (0)$. Hence ad $X_2 = 0$. $\mathfrak{g}$ is semisimple, and thus $X_2 = 0$. We have thus shown that $\tilde{\mathfrak{k}} = \operatorname{ad}\mathfrak{k}$. Thus Ad($K$) is the identity component of $\tilde{K}$ and is thus closed in $GL(\mathfrak{g}_C)$.

7.2.5 *Theorem* Let G be a connected semisimple Lie group with Lie algebra $\mathfrak{g}$. Let $\mathfrak{g} = \mathfrak{k} \oplus \mathfrak{p}$ be a Cartan decomposition of $\mathfrak{g}$. Let K be the connected subgroup of G corresponding to $\mathfrak{k}$. Let Z be the center of G. Then

(1) $Z \subset K$,

(2) The map $K \times \mathfrak{p} \to G$ given by $(k, X) \mapsto k \exp(X)$ is a diffeomorphism of $K \times \mathfrak{p}$ onto G.

PROOF Let (,) be the Hermitian inner product on $\mathfrak{g}_C$ corresponding to the given Cartan decomposition of $\mathfrak{g}$ (see 7.2.2). Let $Z_1, \ldots, Z_n$ be an orthonormal basis of $\mathfrak{g}_u$ relative to (,). Then Ad defines a Lie homomorphism of G into $GL(n, C)$ relative to the above basis. We note that the differential of Ad is ad and since $\mathfrak{g}$ is semisimple ker(ad) is (0). This implies that ker(Ad) $= Z$ is discrete.

If Y is in $\mathfrak{g}$ and if U, V are in $\mathfrak{g}_C$ then

$$(\operatorname{ad} Y\, U, V) = -B([Y, U], \tau V) = B(U, [Y, \tau V])$$
$$= B(U, \tau[\tau Y, V]) = -(U, \operatorname{ad}(\tau Y)V).$$

Thus ${}^t\overline{\operatorname{ad} Y} = (\operatorname{ad} Y)^* = -\operatorname{ad}(\tau Y)$. Thus if g is in Ad(G) then ${}^t\bar{g}$ is in Ad(G).

Let g be in Ad(G). Then $g = Ue^X$ with U in $U(n)$ and X Hermitian (see 3.2.10). Thus ${}^t\bar{g}g = e^{2X}$ is in Ad(G) by the above. The statement "$W\mathfrak{g} \subset \mathfrak{g}$ is a homomorphism of $\mathfrak{g}$" for W in $M_n(C)$ can be written as a set of polynomial equations $F_i(W) = 0$. Now, $F_i(e^{2kX}) = 0$ for all integral k. Hence 3.2.7 implies that $F_i(e^{2tX}) = 0$ for all t in R. This implies that $X\mathfrak{g} \subset \mathfrak{g}$ and X is a derivation of $\mathfrak{g}$. Hence $X = \operatorname{ad} Y$ for some Y in $\mathfrak{g}$. But $\operatorname{ad}(\theta Y) = -(\operatorname{ad} Y)^* = -\operatorname{ad} Y$. Thus $\theta Y = -Y$. This implies that Y is in $\mathfrak{p}$.

We now know that U is in Ad(G). Also $U^* = U^{-1}$ and $U^* = \tau U^{-1}\tau$ since U is in Ad(G). This implies that $U\mathfrak{g} \subset \mathfrak{g}$, $U\mathfrak{g}_u \subset \mathfrak{g}_u$. Hence $U\mathfrak{k} \subset \mathfrak{k}$. We therefore see that U is in $\tilde{K}$ (see the proof of 7.2.4). Let $t \mapsto g(t)$ be a curve in Ad(G) so that $g(0) = e$ and $g(1) = g$. Then $g(t) = U(t)e^{\operatorname{ad}Y(t)}$ with $t \mapsto Y(t)$ a curve in $\mathfrak{p}$ so that $Y(0) = 0$ and $Y(1) = Y$. Thus $U(t) = g(t)e^{-\operatorname{ad}Y(t)}$ is a curve in $\tilde{K}$ joining e and U. This implies that U is in the identity component of $\tilde{K}$ which is Ad(K) (see the proof of 7.2.4). We have therefore proved that the map Ad(K) $\times\, \mathfrak{p} \to$ Ad(G) given by $(k, X) \mapsto ke^{\operatorname{ad}X}$ is a surjective diffeomorphism (see 3.2.10). This implies that the map $X \mapsto e^{\operatorname{ad}X} = \operatorname{Ad}(\exp X)$ of $\mathfrak{p}$ into Ad(G) is a diffeomorphism into. Thus the map $X \to \exp X$ of $\mathfrak{p}$ into G is an into diffeomorphism.

We now prove (1). Let $H = \operatorname{Ad}^{-1}(\operatorname{Ad}(K))$. Then $H = KZ$. Now G/H is diffeomorphic with $\mathrm{Ad}(G)/\mathrm{Ad}(K)$ which (by the above) is diffeomorphic with $\mathfrak{p}$. Hence G/H is simply connected. This implies that H is connected and hence $H = K$ (Z is discrete). We have thus shown that $Z \subset K$.

Now let g be in G. Then $\mathrm{Ad}(g) = \mathrm{Ad}(k)e^{\mathrm{ad}X}$ with k in K and X in $\mathfrak{p}$. Hence $v = zk\exp(X)$ with z in Z. (1) now implies that the map $K \times \mathfrak{p} \to G$ is surjective. Suppose that $k_1 \exp(X_1) = k_2\exp(X_2)$ with k_1, k_2 in K and X_1, X_2 in $\mathfrak{p}$. Then $\mathrm{Ad}(k_1)e^{\mathrm{ad}X_1} = \mathrm{Ad}(k_2)e^{\mathrm{ad}X_2}$. Applying (2) for $\mathrm{Ad}(G)$ again we see that $X_1 = X_2$ and hence $k_1 = k_2$. To complete the proof we must show that the differential of $\phi: K \times \mathfrak{p} \to G$ given by $\phi(k, X) = k\exp(X)$ is nonsingular. If $\pi: G \to G/Z$ is the natural progection then we have shown above that $\pi \circ \phi$ is everywhere regular. Hence ϕ is everywhere regular.

7.3 The Iwasawa Decomposition of a Semisimple Lie Algebra

7.3.1 Let $\mathfrak{g}$ be a semisimple Lie algebra over R. Let $\mathfrak{g} = \mathfrak{k} \oplus \mathfrak{p}$ be a Cartan decomposition of $\mathfrak{g}$. Let $\mathfrak{a}$ be a subspace of $\mathfrak{p}$ which is maximal subject to the condition that $[X, Y] = 0$ for X, Y in $\mathfrak{a}$.

7.3.2 If X is in $\mathfrak{a}$ then 7.2.3 implies that ad X is a diagonalizable endomorphism of $\mathfrak{g}$ with real eigenvalues. Thus the elements of $\mathrm{ad}(\mathfrak{a})$ are simultaneously diagonalizable with real eigenvalues. If λ is in $\mathfrak{a}^* - \{0\}$ we set $\mathfrak{g}_\lambda = \{X \text{ in } \mathfrak{g} | \mathrm{ad}\, HX = \lambda(H)X \text{ for all } H \text{ in } \mathfrak{a}\}$. We set $\mathfrak{m} = \{X \text{ in } \mathfrak{k} | [X, \mathfrak{a}] = 0\}$. Let $\Lambda = \{\lambda \text{ in } \mathfrak{a}^* - \{0\} | \mathfrak{g}_\lambda \neq 0\}$.

7.3.3 *Definition* Λ is called the restricted root system of $\mathfrak{g}$ relative to $\mathfrak{a}$.

7.3.4 *Lemma* $\mathfrak{g} = \mathfrak{a} \oplus \mathfrak{m} \oplus \sum_{\lambda \in \Lambda} \mathfrak{g}_\lambda$, a direct sum of subspaces.

PROOF Set $\mathfrak{g}_0 = \{X \text{ in } \mathfrak{g} | [X, \mathfrak{a}] = (0)\}$. Then clearly $\mathfrak{g} = \mathfrak{g}_0 \oplus \sum_{\lambda \in \Lambda} \mathfrak{g}_\lambda$, a direct sum of subspaces. If H is in $\mathfrak{a}$ then $\theta(H) = -H$ (here θ is the Cartan involution of $\mathfrak{g}$ associated with the given Cartan decomposition of $\mathfrak{g}$). Thus $\theta\mathfrak{g}_0 = \mathfrak{g}_0$. This implies that $\mathfrak{g}_0 = \mathfrak{g}_0 \cap \mathfrak{k} \oplus \mathfrak{g}_0 \cap \mathfrak{p}$. The defintion of $\mathfrak{a}$ implies that $\mathfrak{g}_0 \cap \mathfrak{p} = \mathfrak{a}$. Clearly $\mathfrak{g}_0 \cap \mathfrak{k} = \mathfrak{m}$.

7.3.5   Let $H_1, \ldots, H_l$ be a basis of $\mathfrak{a}$. Order $\mathfrak{a}^*$ lexicographically (see the proof of 3.5.14) relative to the ordered basis $(H_1, \ldots, H_l)$ of $\mathfrak{a}$. Let $\Lambda^+ = \{\lambda \text{ in } \Lambda | \lambda > 0\}$ and let $\Lambda^- = \{\lambda \text{ in } \Lambda | \lambda < 0\}$. Then $\Lambda = \Lambda^+ \cup \Lambda^-$ and $\Lambda^+ \cap \Lambda^- = \phi$. Also if λ, ν are in Λ^+ and if $\lambda + \nu$ is in Λ then $\lambda + \nu$ is in Λ^+. We also note that since $\theta(H) = -H$ for H in $\mathfrak{a}$, $\Lambda^- = \{-\lambda | \lambda \text{ in } \Lambda^+\}$. Set $\mathfrak{n} = \sum_{\lambda \in \Lambda^+} \mathfrak{g}_\lambda$. Then $\mathfrak{n}$ is a subalgebra of $\mathfrak{g}$. Furthermore since there are only a finite number of elements of Λ^+ and if λ, ν are in Λ^+ then $\lambda + \nu$ is larger than λ or ν and we see that $\mathfrak{n}$ is nilpotent. Set $\bar{\mathfrak{n}} = \sum_{\lambda \in \Lambda^-} \mathfrak{g}_\lambda$. Clearly $\bar{\mathfrak{n}} = \theta(\mathfrak{n})$. It is also clear that $\mathfrak{g} = \bar{\mathfrak{n}} \oplus \mathfrak{m} \oplus \mathfrak{a} \oplus \mathfrak{n}$.

7.3.6   *Proposition* $\mathfrak{g} = \mathfrak{k} \oplus \mathfrak{a} \oplus \mathfrak{n}$, a direct sum of subalgebras.

PROOF   Suppose that X is in $\mathfrak{k}$, H is in $\mathfrak{a}$, and Y is in $\mathfrak{n}$ and that $X + H + Y = 0$.   Then $0 = \theta(X + H + Y) = X - H + \theta Y$.   Thus $2H + Y - \theta Y = 0$. But Y is in $\mathfrak{n}$, θY is in $\bar{\mathfrak{n}}$, and $\mathfrak{a} \cap (\mathfrak{n} + \bar{\mathfrak{n}}) = (0)$. Thus $H = 0$ and $Y - \theta Y = 0$. But $\bar{\mathfrak{n}} \cap \mathfrak{n} = (0)$, hence $Y = 0$. This clearly implies that $X = 0$. We have thus shown that the sum $\mathfrak{k} \oplus \mathfrak{a} \oplus \mathfrak{n}$ is direct.

We assert that the map of $\mathfrak{m} + \bar{\mathfrak{n}}$ to $\mathfrak{k}$ given by $X + Y \mapsto X + Y + \theta Y$ for X in $\mathfrak{m}$, Y in $\bar{\mathfrak{n}}$, is bijective. The map of $\mathfrak{g}$ to $\mathfrak{k}$ given by $Z \mapsto Z + \theta Z$ is surjective. Thus if Y is in $\mathfrak{k}$, $Y = Z + \theta Z$ for some Z in $\mathfrak{g}$. Now $Z = U + V + H + W$ with U in $\mathfrak{n}$, V in $\mathfrak{m}$, H in $\mathfrak{a}$, and W in $\bar{\mathfrak{n}}$. $U = \theta U'$ for some U' in $\bar{\mathfrak{n}}$. Thus $Y = V + (U' + W) + \theta(U' + W)$ as asserted. The directness of the sum $\bar{\mathfrak{n}} + \mathfrak{m} + \mathfrak{a} + \mathfrak{n}$ implies that the map $\mathfrak{m} + \bar{\mathfrak{n}}$ to $\mathfrak{k}$ is injective. Hence $\dim \mathfrak{k} + \dim \mathfrak{a} + \dim \mathfrak{n} = \dim \mathfrak{m} + \dim \bar{\mathfrak{n}} + \dim \mathfrak{a} + \dim \mathfrak{n} = \dim \mathfrak{g}$. This proves that $\mathfrak{k} \oplus \mathfrak{a} \oplus \mathfrak{n} = \mathfrak{g}$.

7.4   The Iwasawa Decomposition of a Semisimple Lie Group

7.4.1   *Lemma*   Let $\mathfrak{n}_p$ be the Lie algebra of all $p \times p$ real matrices that are upper triangular with zeros on the diagonal. Let N_p be the Lie group of all upper triangular $p \times p$ matrices with 1's on the diagonal. Then exp is a diffeomorphism of $\mathfrak{n}_p$ onto N_p.

PROOF   If X is in $\mathfrak{n}_p$ set

$$\log(I + X) = \sum_{n=1}^{p} \frac{(-1)^{n-1}}{n} X^n = \sum_{n=1}^{\infty} \frac{(-1)^{n-1}}{n} X^n.$$

We note that if X is in $\mathfrak{n}_p$ then $X(I + X + \cdots + X^p) = (I + X + \cdots + X^p) - I$ since $X^{p+1} = 0$. This says that $(I - X)^{-1} = (I + X + \cdots + X^p)$. Now

$$\frac{d}{dt}\log(I + tX) = \sum_{n=1}^{p} \frac{(-1)^{n-1}}{n} nt^{n-1}X^n = X\sum_{n=1}^{p} (-tX)^{n-1} = X(I + tX)^{-1}.$$

We therefore find that

$$\frac{d}{dt}\log(e^{tX}) = \frac{d}{dt}\log(I + Z(t))\left(\text{with } Z(t) = tX + \frac{t^2X^2}{2!} + \cdots + \frac{t^pX^p}{p!}\right)$$

$$= Z'(t)(I + Z(t))^{-1} = \left(\frac{d}{dt}e^{tX}\right)e^{-tX} = X.$$

This implies that $\log(e^{tX}) = tX + C$ with C in $\mathfrak{n}_p$. But $\log(e^0) = \log(I) = 0$. Thus $C = 0$. We therefore see that $\log(e^X) = X$ for all X in $\mathfrak{n}_p$.

Since $\exp_{*0} = I$ we see that there is a neighborhood V_0 of 0 in $\mathfrak{n}_p$ and a neighborhood V of I in N so that exp defines a diffeomorphism of V_0 onto V. We therefore see that for n in V, $e^{\log(n)} = n$. Now, the map $n \mapsto e^{\log(n)}$ is a mapping that is polynomial in the matrix entries of n. Thus $e^{\log(n)} = n$ for all n in N_p. This implies that exp defines a homeomorphism of $\mathfrak{n}_p$ onto N_p. Also $\exp_{*X}(Z) = (((I - e^{-\operatorname{ad}X})/\operatorname{ad}X)Z)_{\exp X}$ (see A.2.2.8). Since ad X is nilpotent we see that exp is regular at each point of $\mathfrak{n}_p$, hence exp is a diffeomorphism.

7.4.2 *Lemma* Let $O(p)$ be the group of all $p \times p$ orthogonal matrices. Let $D(p)$ be the group of all $p \times p$ real diagonal matrices with positive entries on the diagonal. Then the map $O(p) \times D(p) \times N_p \to GL(p, R)$ given by $(k, a, n) \to kan$ for k in $O(p)$, a in $D(p)$, and n in N_p is a surjective homeomorphism.

PROOF Suppose that $g = kan = k'a'n'$ with k, k' in $O(p)$, a, a' in $D(p)$, and n, n' in N_p. Then $(k')^{-1}k = a'n'n^{-1}a^{-1}$ which is upper triangular with positive diagonal entries. But $(k')^{-1}k$ is in $O(p)$. Hence $k' = k$. Hence $an = a'n'$. But then $(a')^{-1}a = n'n^{-1}$. Since $n'n^{-1}$ is upper triangular with 1's on the diagonal we see that $a' = a$. Hence $n = n'$.

To complete the proof we define a continuous map of $GL(p, R)$ to $O(p) \times D(p) \times N_p$ which defines the inverse to our given mapping. Let g be in $GL(p, R)$. Let $e_1, \ldots, e_p$ be the standard basis of R^p. Set $v_i = g^{-1}e_i$. We apply the Gram-Schmidt process to $v_1, \ldots, v_p$. That is, set $u_1 = \|v_1\|^{-1}v_1$, if u_k has been defined for $j \leqslant q$ set

$$u_{q+1} = \left\|v_{q+1} - \sum_{j=1}^{q} \langle v_{q+1}, u_j\rangle u_j\right\|^{-1}\left(v_{q+1} - \sum_{j=1}^{q} \langle v_{q+1}, u_j\rangle u_j\right).$$

Then $u_i = \sum_{j \leqslant i} a_{ji} v_j$ with $a_{ii} > 0$ for each i. Set

$$a(g) = \begin{bmatrix} a_{11} & & 0 \\ & \ddots & \\ 0 & & a_{pp} \end{bmatrix},$$

$$n(g) = a(g)^{-1} \begin{bmatrix} a_{11} & \cdots & a_{1p} \\ & \ddots & \vdots \\ 0 & & a_{pp} \end{bmatrix}.$$

Then $a(g)n(g)v_i = u_i$. Let $k(g)$ in $O(p)$ be the unique matrix defined by $k(g)u_i = e_i$. Then $k(g)a(g)n(g)g^{-1}e_i = e_i$ for each i. Thus $g = k(g)a(g)n(g)$ and since $g \mapsto k(g)$, $g \mapsto a(g)$, and $g \mapsto n(g)$ are clearly continuous, the lemma follows.

7.4.3 *Theorem* Let G be a connected semisimple Lie group. Let $\mathfrak{g}$ be its Lie algebra and let $\mathfrak{g} = \mathfrak{k} \oplus \mathfrak{a} \oplus \mathfrak{n}$ be an Iwasawa decomposition of $\mathfrak{g}$. Let K, A, N be, respectively, the connected subgroups of G corresponding to $\mathfrak{k}$, $\mathfrak{a}$, $\mathfrak{n}$. Then

(1) exp: $\mathfrak{a} \to A$ is a Lie isomorphism,

(2) exp: $\mathfrak{n} \to$ is a surejective diffeomorphism,

(3) the map $K \times A \times N \to G$ given by $(k, a, n) \mapsto kan$ (k in K, a in A, n in N) is a surjective diffeomorphism.

The decomposition $G = KAN$ is called an *Iwasawa decomposition of G*.

PROOF Let $\mathfrak{g} = \mathfrak{k} \oplus \mathfrak{p}$ be a Cartan decomposition of $\mathfrak{g}$ so that $\mathfrak{a} \subset \mathfrak{p}$. Let θ be the corresponding Cartan involution of $\mathfrak{g}$. Set $B = B_{\mathfrak{g}}$, the Killing form of $\mathfrak{g}$. Let $\langle X, Y \rangle = -B(X, \theta Y)$. Then $\langle\ ,\ \rangle$ is a positive definite inner product on $\mathfrak{g}$. Let Λ be the set of restricted roots of $\mathfrak{g}$ relative to $\mathfrak{a}$. We pick a linear order on $\mathfrak{a}^*$ as in 7.3.5. Then $\mathfrak{g} = \mathfrak{a} \oplus \mathfrak{m} \oplus \bar{\mathfrak{n}} \oplus \mathfrak{n}$ in the notation of 7.3.5. Let $\Lambda^+ = \{\lambda_1 < \cdots < \lambda_r\}$. Let $X_1, \ldots, X_{p_1}$ be an orthonormal basis of $\mathfrak{g}_{\lambda_r}, \ldots$; let $X_{p_1 + \cdots + p_{r-1}+1}, \ldots, X_{p_1 + \cdots + p_r}$ be an orthonormal basis of $\mathfrak{g}_{\lambda_1}$. Set $q = p_1 + \cdots + p_r$. Let $X_{q+1}, \ldots, X_{q+m}$ be an orthonormal basis of $\mathfrak{m} \oplus \mathfrak{a}$. Let $X_{q+m+j} = \theta(X_{q-j+1})$, $j = 1, \ldots, q$. Then $X_1, \ldots, X_n$ is an orthonormal basis of $\mathfrak{g}$ ($n = 2q + m$). Relative to this basis ad$X$ is, respectively, skew-symmetric, diagonal, or upper triangular with zeros on the diagonal if X is in $\mathfrak{k}$, $\mathfrak{a}$ or in $\mathfrak{n}$.

Now ad: $\mathfrak{g} \to M_n(R)$ is injective and $\mathrm{Ad}(\exp X) = e^{\mathrm{ad}X}$. (1) now follows since the map $X \to e^{\mathrm{ad}X}$ of $\mathfrak{a}$ into $D(n)$ (see 7.4.2) is an injective diffeomorphism. Now set $G_* = \mathrm{Ad}(G)$. Then G_* is a closed subgroup of $GL(n, R)$ (see the proof of 7.2.4). ad($\mathfrak{n}$) is a closed subset of $\mathfrak{n}_r$ (see 7.4.2), thus 7.4.1 implies that $e^{\mathrm{ad}(\mathfrak{n}_r)}$ is closed in $GL(n, R)$ and hence in G_*. Now (2) follows from 7.4.1.

Let Z be the center of G. Then $Z \subset K$ and $G/Z = G_*$ (see 7.2.5). Now $\mathrm{Ad}(A) \subset D(n)$ and $\mathrm{Ad}(N) \subset N_n$. We assert that this implies that $\mathrm{Ad}(A)\mathrm{Ad}(N)$ is closed in G_*. In fact, this follows from 7.4.2 and the fact that $\mathrm{Ad}(A)$ and $\mathrm{Ad}(N)$ are closed in G_*. 7.2.4 implies that $\mathrm{Ad}(K)$ is compact. Thus

$$\mathrm{Ad}(K)\,\mathrm{Ad}(A)\,\mathrm{Ad}(N)$$

is a closed subset of G_*. Using 7.4.2 again we find that the map $\psi(k, a, n) = kan$ of $\mathrm{Ad}(K) \times \mathrm{Ad}(A) \times \mathrm{Ad}(N)$ into G_* is injective. We compute ψ_*. Let X be in $\mathrm{ad}(\mathfrak{k})$, Y in $\mathrm{ad}(\mathfrak{a})$, Z in $\mathrm{ad}(\mathfrak{n})$ and let f be in $C^\infty(G_*)$. Then

(i)
$$\begin{aligned}\psi_{*(k,a,n)}(X, 0, 0)f &= \frac{d}{dt}f(k \exp tXan)\Big|_{t=0} \\ &= \frac{d}{dt}f(kan \exp(\mathrm{Ad}(an)^{-1}tX)\Big|_{t=0} \\ &= (\mathrm{Ad}(an^{-1}X)_{kan}f.\end{aligned}$$

(ii)
$$\begin{aligned}\psi_{*(k,a,n)}(0, Y, 0)f &= \frac{d}{dt}f(ka \exp tYn)\Big|_{t=0} \\ &= \frac{d}{dt}f(kan \exp t\, \mathrm{Ad}(n)^{-1}Y)\Big|_{t=0} \\ &= (\mathrm{Ad}(n)^{-1}Y)_{kan}f.\end{aligned}$$

(iii)
$$\psi_{*(k,a,n)}(0, 0, Z) = Z_{kan}.$$

We therefore have the formula

(iv)
$$\psi_{*(k,a,n)}(X, Y, Z) = (\mathrm{Ad}(an)^{-1}X + \mathrm{Ad}(n)^{-1}Y + Z)_{kan}.$$

If $\psi_{*(k,a,n)}(X, Y, Z) = 0$ then $\mathrm{Ad}(an)^{-1}X + \mathrm{Ad}(n)^{-1}Y + Z = 0$. But then $X = \mathrm{Ad}(a)Y + \mathrm{Ad}(na)Z$. Now $\mathrm{Ad}(a)Y + \mathrm{Ad}(na)Z$ is in $\mathfrak{a} + \mathfrak{n}$ and X is in $\mathfrak{k}$. 7.3.6 now implies that $X = 0$ and $Y + \mathrm{Ad}(n)Z = 0$. But Y is in $\mathfrak{a}$ and $\mathrm{Ad}(n)Z$ is in $\mathfrak{n}$. Applying 7.3.6 again we find that $Y = 0$ and $Z = 0$. Thus ψ is everywhere regular. Hence using 7.3.6 we see that $\mathrm{Ad}(K)\mathrm{Ad}(A)\mathrm{Ad}(N)$ is open in G_*. Thus ψ is a surjective diffeomorphism.

Let g be in G. Then $\mathrm{Ad}(g) = \mathrm{Ad}(k)\mathrm{Ad}(a)\mathrm{Ad}(n)$ with k in K, a in A, and n in N. Thus $g = zkan$ with z in Z. But 7.2.5 (1) says that $Z \subset K$. Hence the map of $K \times A \times N$ to G given by $\Phi(k, a, n) = kan$ is surjective. Suppose that k, k', a, a', n, n' are, respectively, in K, A, or N and that $kan = k'a'n'$. Then $\mathrm{Ad}(k)\mathrm{Ad}(a)\mathrm{Ad}(n) = \mathrm{Ad}(k')\mathrm{Ad}(a')\mathrm{Ad}(n')$. Thus $\mathrm{Ad}(a) = \mathrm{Ad}(a')$ and $\mathrm{Ad}(n) = \mathrm{Ad}(n')$. Hence $a = a'$ and $n = n'$ (this follows from (1) and (2) of this theorem). We have thus shown that Φ is a bijection. Replacing ψ by Φ in the computations (i)–(iv) we find that Φ is everywhere regular. This implies (3) and proves the theorem.

7.5 The Fine Structure of Semisimple Lie Groups

7.5.1 Let G be a semisimple Lie group with Lie algebra $\mathfrak{g}$. Let $\mathfrak{g} = \mathfrak{k} \oplus \mathfrak{p}$ be a Cartan decomposition of $\mathfrak{g}$. Let $\mathfrak{a} \subset \mathfrak{p}$ be as in 7.3.2. Let Λ be the restricted root system of $\mathfrak{g}$ relative to $\mathfrak{a}$ (see 7.3.2 and 7.3.3).

7.5.2 Let K be the connected subgroup of G corresponding to $\mathfrak{k}$. Let M be the centralizer of $\mathfrak{a}$ in K. in K. That is, $M = \{k \text{ in } K | \text{Ad}(k)|_{\mathfrak{a}} = I\}$. Let M^* be the normalizer of $\mathfrak{a}$ in K. That is, $M^* = \{k \text{ in } K | \text{Ad}(k)\mathfrak{a} \subset \mathfrak{a}\}$. Set $W = W(\mathfrak{a}) = W(A) = M^*/M$ where A is the connected subgroup of G corresponding to $\mathfrak{a}$. W is called the Weyl group of (G, A).

7.5.3 Let $\mathfrak{a}' = \{H \text{ in } \mathfrak{a} | \lambda(H) \neq 0 \text{ for all } \lambda \text{ in } \Lambda\}$. Then $\mathfrak{a}'$ is a union of open convex subsets of $\mathfrak{a}$.

7.5.4 *Lemma* If H is in $\mathfrak{a}'$ then $\mathfrak{m} + \mathfrak{a} = \{X \text{ in } \mathfrak{g} | [X, H] = 0\}$. Here $\mathfrak{m}$ is the Lie algebra of M.

PROOF Let X be in $\mathfrak{g}$ and suppose that $[H, X] = 0$. $X = H_1 + Y + \sum X_\lambda$ with X_λ in $\mathfrak{g}_\lambda$, H_1 in $\mathfrak{a}$, Y in $\mathfrak{m}$, and $\mathfrak{g}_\lambda$ is as in 7.3.2. Now $0 = [H, X] = \sum \lambda(H)X_\lambda$. Since $\lambda(H) \neq 0$ for all λ in Λ we see that $X_\lambda = 0$ for all λ in Λ. Thus $X = H_1 + Y$.

7.5.5 *Proposition* Let $\mathfrak{a}_1$ and $\mathfrak{a}_2$ be two maximal Abelian subalgebras of $\mathfrak{p}$. Then there is k in K so that $\text{Ad}(k)\, \mathfrak{a}_1 = \mathfrak{a}_2$.

PROOF 7.2.4 implies that $\text{Ad}(K)$ is a compact connected Lie group. Let H_i be in $\mathfrak{a}_i$ for $i = 1, 2$. Let B be the Killing form of $\mathfrak{g}$. Recall that $B|_{\mathfrak{k}\times\mathfrak{k}}$ is negative definite and $B|_{\mathfrak{p}\times\mathfrak{p}}$ is positive definite. Let for u in $\text{Ad}(K)$, $f(u) = B(u\cdot H_1, H_2)$. Then f is a C^∞ function from $\text{Ad}(K)$ to R. f takes a minimum, say, $f(u_0)$. If X is in $\mathfrak{k}$ then

$$0 = \frac{d}{dt} B(u_0 e^{t\,\text{ad}X} H_1, H_2)\Big|_{t=0} = B(u_0[X, H_1], H_2)$$

$$= B([u_0 X, u_0 H_1], H_2) = B(u_0 X, [u_0 H_1, H_2]).$$

But $u_0 X$ is an arbitrary element of $\mathfrak{k}$ and $[H_2, u_0 H_1]$ is in $\mathfrak{k}$. Hence

$[u_0 H_1, H_2] = 0$. 7.5.4 now implies that $u_0 H_1$ is in $\mathfrak{a}_2$. Thus $u_0 \mathfrak{a}_1 \subset \mathfrak{a}_2$. Since $\mathfrak{a}_1$ is maximal $u_0 \mathfrak{a}_1 = \mathfrak{a}_2$.

7.5.6 *Lemma* $W(A)$ is a finite group. Furthermore the map $W(A) \to GL(\mathfrak{a})$ given by $sH = \mathrm{Ad}(m^*)H$ for H in $\mathfrak{a}$, m^* in s, is injective.

PROOF Suppose that $sH = H$ for all H in $\mathfrak{a}$. If m^* is in s then $\mathrm{Ad}(m^*)H = H$ for all H in $\mathfrak{a}$. But then m^* is in M. Hence $s = I$.

We now show that $\mathfrak{m}$ is the Lie algebra of M^*. Let $\tilde{\mathfrak{m}}$ denote the Lie algebra of M^*. Clearly, $\tilde{\mathfrak{m}} \supset \mathfrak{m}$. If X is in $\tilde{\mathfrak{m}}$ then $X = \sum X_\lambda + Y + H_1$ with X_λ in $\mathfrak{g}_\lambda$, Y in $\mathfrak{m}$, H_1 in $\mathfrak{a}$. If H is in $\mathfrak{a}$ then $[H, X] = \sum \lambda(H) X_\lambda$. But $[H, X]$ is in $\mathfrak{a}$. Thus $[H, X] = 0$ for all H in $\mathfrak{a}$. Now $\tilde{\mathfrak{m}} \subset \mathfrak{k}$, thus $\tilde{\mathfrak{m}} \subset \mathfrak{m}$.

We therefore see that M^*/M is a discrete group. On the other hand the map $M^*/M \to GL(\mathfrak{a})$ is injective. Thus $M^*/M = \mathrm{Ad}(M^*)/\mathrm{Ad}(M)$, but $\mathrm{Ad}(M^*)$ is compact. Hence $W(A)$ is indeed finite.

7.5.7 Let $\mathfrak{h}^-$ be a maximal Abelian subalgebra of $\mathfrak{m}$. Set $\mathfrak{h}_0 = \mathfrak{h}^- \oplus \mathfrak{a}$.

7.5.8 *Lemma* Let $\mathfrak{g}_C$ be the complexification of $\mathfrak{g}$. Let $\mathfrak{h}$ be the complexification of $\mathfrak{h}_0$ in $\mathfrak{g}_C$. Then $\mathfrak{h}$ is a Cartan subalgebra of $\mathfrak{g}_C$.

PROOF We must show that $\mathfrak{h}$ is an Abelian subalgebra of $\mathfrak{g}_C$ maximal subject to the condition that if X is in $\mathfrak{g}_C$, ad$X$ is diagonalizable. We show first of all that $\mathfrak{h}$ is maximal Abelian. Let X be in $\mathfrak{g}_C$ and suppose that $[X, \mathfrak{h}] = 0$. Now $X = X_1 + (-1)^{\frac{1}{2}} X_2$, X_i in $\mathfrak{g}$, $i = 1, 2$. Thus $[X_i, \mathfrak{h}_0] = 0$, $i = 1, 2$. But $X_i = Y_i + Z_i$ with Y_i in $\mathfrak{k}$, Z_i in $\mathfrak{p}$, $i = 1, 2$. If H is in $\mathfrak{a}$ then $0 = [X_i, H] = [Y_i, H] + [Z_i, H]$, $[Y_i, H]$ is in $\mathfrak{p}$, $[Z_i, H]$ is in $\mathfrak{k}$. Thus $[Y_i, H] = 0$, $[Z_i, H] = 0$. Hence Z_i is in $\mathfrak{a}$ and Y_i is in $\mathfrak{m}$. But then $[Y_i, \mathfrak{h}^-] = 0$. Hence Y_i is in $\mathfrak{h}^-$. This shows that X is in $\mathfrak{h}$.

If X is in $\mathfrak{h}^-$ then X is in $\mathfrak{k}$ and thus ad X is a diagonalizable endomorphism of $\mathfrak{g}_C$. If X is in $\mathfrak{a}$ then we have seen that ad X is diagonalizable. Now if X is in $\mathfrak{h}^-$ and H is in $\mathfrak{a}$ then $[X, H] = 0$, thus ad X and ad H are simultaneously diagonalizable. Thus ad X + ad H is diagonalizable. We see that if Y is in $\mathfrak{h}_0$ then ad Y is diagonalizable. This implies the result.

7.5.9 Let Δ be the root system of $\mathfrak{g}_C$ relative to $\mathfrak{h}$. We note that $\mathfrak{h}_R = \{H \text{ in } \mathfrak{h} | \alpha(H) \text{ in } R \text{ for all } \alpha \text{ in } \Delta\} = (-1)^{\frac{1}{2}}\mathfrak{h}^- + \mathfrak{a}$. Let $H_1, \ldots, H_l$ be a basis of $\mathfrak{a}$ and let Λ^+ be the corresponding system of positive restricted roots (see 7.3.5). Let $H_{l+1}, \ldots, H_q$ be a basis of $(-1)^2\mathfrak{h}^-$. Let $>$ be the lexicographic order on $\mathfrak{h}_R^*$ corresponding to $H_1, \ldots, H_q$. Let Δ^+ be the corresponding system of positive roots in Δ.

7.5.10 Let σ be the conjugation of $\mathfrak{g}_C$ relative to $\mathfrak{g}$. That is, $\sigma(X + (-1)^{\frac{1}{2}}Y) = X - (-1)^{\frac{1}{2}}Y$ for X, Y in $\mathfrak{g}$. Then $\sigma\mathfrak{h} = \mathfrak{h}$ and $\sigma[X, Y] = [\sigma X, \sigma Y]$ for X, Y in $\mathfrak{g}$. We therefore see that if α is in Δ, $\sigma^*\alpha = \alpha\circ\sigma$ is in Δ (as a linear functional on $\mathfrak{h}_R$).

7.5.11 *Lemma* (1) If α is in Δ^+ and if $\sigma^*\alpha \neq -\alpha$ then $\sigma^*\alpha$ is in Δ^+.
(2) Let $\Delta_1 = \{\alpha \text{ in } \Delta | \sigma^*\alpha = -\alpha\}$. Then $\mathfrak{m}_C = \mathfrak{h}_C + \sum_{\alpha\in\Delta_1}(\mathfrak{g}_C)_\alpha$.
(3) Set $\Sigma = \Delta^+ - \Delta_1$. Then $\mathfrak{n}_C = \sum_{\alpha\in\Sigma}(\mathfrak{g}_C)_\alpha$.

PROOF (1) If H is in $(-1)^{\frac{1}{2}}\mathfrak{h}^-$ then $\sigma H = -H$ and if H is in $\mathfrak{a}$ then $\sigma H = H$. Thus if $\sigma^*\sigma \neq -\alpha$ then $\alpha|_\mathfrak{a} \neq 0$. Hence if $\alpha > 0$ and $\sigma^*\alpha \neq -\alpha$ then $\sigma^*\alpha > 0$.

(2) Let X be in $\mathfrak{g}_C$. Then $X = H + \sum X_\alpha$ with X_α in $(\mathfrak{g}_C)_\alpha$ and H in $\mathfrak{h}$. Suppose that $[X, \mathfrak{a}] = 0$. Then $\sum\alpha(\mathfrak{a})X_\alpha = (0)$. Hence if $X_\alpha \neq 0$, $\alpha|_\mathfrak{a} = 0$. Thus $\sigma^*\alpha = -\alpha$ if $X_\alpha \neq 0$. This proves (2).

(3) If α is in Σ then $\sigma^*\alpha$ is in Σ and $\sigma(\mathfrak{g}_C)_\alpha = (\mathfrak{g}_C)_{\sigma^*\alpha}$. Set $\mathfrak{n}_1 = \sum_{\alpha\in\Sigma}(\mathfrak{g}_C)_\alpha$. Then $\sigma\mathfrak{n}_1 = \mathfrak{n}_1$. Set $\mathfrak{n}_2 = \sum_{\alpha\in\Sigma}(\mathfrak{g}_C)_{-\alpha}$. (2) implies that $\mathfrak{g} = (\mathfrak{n}_2 \cap \mathfrak{g}) \oplus \mathfrak{m} \oplus \mathfrak{a} \oplus (\mathfrak{n}_1 \cap \mathfrak{g})$. Thus $\dim \mathfrak{n}_1 \cap \mathfrak{g} = \dim \mathfrak{n}$. Since it is clear that $\mathfrak{n}_1 \cap \mathfrak{g} \subset \mathfrak{n}$ we see that $\mathfrak{n} = \mathfrak{n}_1 \cap \mathfrak{g}$.

7.5.12 7.5.11(3) implies that $\Lambda = \{\alpha|_\mathfrak{a} | \alpha \text{ in } \Delta\}$. This is the reason why the elements of Λ are called restricted roots.

7.5.13 *Lemma* Set $\mathfrak{b} = \mathfrak{m} \oplus \mathfrak{a} \oplus \mathfrak{n}$. If X is in $\mathfrak{b}$ and if ad X has only real characteristic values, then X is in $\mathfrak{a} + \mathfrak{n}$.

PROOF $X = Y + H + Z$ with Y in $\mathfrak{m}$, H in $\mathfrak{a}$, and Z in $\mathfrak{n}$. Let $\mathfrak{h}^-$ be chosen so that Y is in $\mathfrak{h}^-$ (this is possible since Ad(M) is compact). But then relative to a basis of $\mathfrak{g}_C$, ad X is upper triangular with diagonal entries equal to 0 or $\alpha(Y + H)$ for α in Δ. If $\alpha(Y + H)$ is in R for all α in Δ then $Y + H$ is in $(-1)^{\frac{1}{2}}\mathfrak{h}^- + \mathfrak{a}$. Thus $Y = 0$.

7.5.14 Set $\mathfrak{m}_1 = \mathfrak{m} \oplus \mathfrak{a}$. If X is in $\mathfrak{m}_1$ then ad $X \cdot \mathfrak{n} \subset \mathfrak{n}$. Let $\mathfrak{m}_1' = \{X \text{ in } \mathfrak{m}_1 | \det(\text{ad } X|_\mathfrak{n}) \neq 0\}$.

7.5.15 *Lemma* If H is in $\mathfrak{m}_1'$ then the map $n \to \mathrm{Ad}(n)H$ is a bijection of N onto $H + \mathfrak{n} = \{H + Z | Z \text{ in } \mathfrak{n}\}$.

PROOF $H = H_- + H_+$ with H_- in $\mathfrak{m}$, H_+ in $\mathfrak{a}$. We choose $\mathfrak{h}^-$ so that H_- is in $\mathfrak{h}^-$. Then the condition that H be in $\mathfrak{m}_1'$ is just that $\alpha(H) \neq 0$ for α in Σ. Let $>$ be the order of Δ as in 7.5.9.

If n in N then $n = \exp X$ with X in $\mathfrak{n}$ by 7.4.3(2). Now $\mathrm{Ad}(n) = e^{\mathrm{ad} X} H = H + \sum_{k=1} (k!)^{-1} (\mathrm{ad}\, X)^k H$. But $\mathrm{ad}\, X \cdot H$ is in $\mathfrak{n}$. Thus $\mathrm{Ad}(n)H$ is indeed an element of $H + \mathfrak{n}$.

(1) If n is in N and $\mathrm{Ad}(n)H = H$ then $n = e$.

To prove (1) we note that if X is in $\mathfrak{n}$ and $X \neq 0$ then $X = X_\alpha + \sum_{\beta > \alpha} X_\beta$ with X_β in $(\mathfrak{g}_C)_\beta$, $X_\alpha \neq 0$, α in Σ, and the summation being over the elements of Σ. If $n \neq e$ then $X \neq 0$. Hence $\mathrm{Ad}(n)H - H \equiv -\alpha(H) X_\alpha \bmod \sum_{\alpha < \beta} (\mathfrak{g}_C)_\beta$. Since H is in $\mathfrak{m}_1'$ we see that $\alpha(H) \neq 0$. Thus $\mathrm{Ad}(n)H \neq H$. This proves (1).

If $\mathrm{Ad}(n_1)H = \mathrm{Ad}(n_2)H$ then $\mathrm{Ad}(n_1^{-1} n_2)H = H$. Thus (1) implies that $n_1^{-1} n_2 = e$. Thus $n_1 = n_2$. Thus to complete the proof of the lemma we need only show that the map $N \to H + \mathfrak{n}$ is surjective.

Let V be the set of all Z in $\mathfrak{n}$ not of the form $\mathrm{Ad}(n)H - H$ for any n in N. Let α in Σ be the largest element so that there is Z in V, $Z = X_\alpha + \sum_{\beta > \alpha} X_\beta$ with X_γ in $(\mathfrak{g}_C)_\gamma$, $X_\alpha \neq 0$. Since $\mathrm{ad}\, H(\mathfrak{n}) = \mathfrak{n}$ there is W in $\mathfrak{n}$ so that $[H, W] = Z$. Set $n_1 = \exp W$. Then $\mathrm{Ad}(n_1)(H + Z) - H \equiv [W, H] + Z \bmod \sum_{\beta > \alpha} (\mathfrak{g}_C)_\beta$. Thus $\mathrm{Ad}(n_1)(H + Z) - H$ is in $\sum_{\beta > \alpha} (\mathfrak{g}_C)_\beta$. Set $\mathrm{Ad}(n_1)(H + Z) = H + Z'$. Then by the definition of α, there is n_2 in N so that $\mathrm{Ad}(n_2)H = H + Z'$. Thus $\mathrm{Ad}(n_1^{-1} n_2)H = H + Z$. This contradiction completes the proof of the lemma.

7.5.16 Let $B = MAN$. M is clearly a closed subgroup of K. Since G is diffeomorphic with $K \times A \times N$ we see that MAN is closed in G. Thus B is a closed subgroup of G. B is called a Borel or minimal parabolic subgroup of G.

7.5.17 Let for s in $W(A)$, m_s be chosen in M^* so that $m_s M = s$. We note that the subset Bm_sB of G depends only on s. We write $Bm_sB = BsB$.

7.5.18 *Theorem* $G = \bigcup_{s \in W(A)} BsB$ and the union is disjoint.

PROOF We first show that if k is in K then $\mathrm{Ad}(k)\mathfrak{b} \cap \mathfrak{b} + \mathfrak{n} = \mathfrak{b}$. Obviously, we need only show that $\dim\{(\mathrm{Ad}(k)\mathfrak{b} \cap \mathfrak{b} + \mathfrak{n}\} = \dim \mathfrak{b}$. Let $\langle X, Y \rangle = -B(X, \theta Y)$ for X, Y in $\mathfrak{g}$ and θ is the Cartan involution of $\mathfrak{g}$ corresponding to $\mathfrak{k}$. $\langle\ ,\ \rangle$ is a positive definite $\mathrm{Ad}(K)$-invariant inner product on $\mathfrak{g}$.

$\bar{\mathfrak{n}}$ is the orthogonal complement of $\mathfrak{b}$ in $\mathfrak{g}$ relative to $\langle\ ,\ \rangle$. If V is a subspace of $\mathfrak{g}$ let $V^\perp$ be the orthogonal complement of V in $\mathfrak{g}$ relative to $\langle\ ,\ \rangle$.

$$(\mathfrak{b} + \mathrm{Ad}(k)\mathfrak{b})^\perp = \theta(\mathfrak{n} \cap \mathrm{Ad}(k)\mathfrak{n}) = \theta(\mathfrak{n} \cap \mathrm{Ad}(k)\mathfrak{b} \cap \mathfrak{b}).$$

Clearly

$$\dim \mathfrak{g} = \dim(\mathfrak{b} + \mathrm{Ad}(k)\mathfrak{b}) + \dim(\mathfrak{b} + \mathrm{Ad}(k)\mathfrak{b})^\perp.$$

Also

$$\dim(\mathfrak{b} \cap \mathrm{Ad}(k)\mathfrak{b}) + \mathfrak{n}) = \dim(\mathfrak{b} \cap \mathrm{Ad}(k)\mathfrak{b}) + \dim \mathfrak{n} - \dim(\mathfrak{b} \cap \mathrm{Ad}(k)\mathfrak{b} \cap \mathfrak{n}).$$

But

$$\begin{aligned}\dim \mathfrak{b} \cap \mathrm{Ad}(k)\mathfrak{b} \cap \mathfrak{n} &= \dim \mathfrak{g} - (\dim \mathfrak{b} + \mathrm{ad}(k)\mathfrak{b}) \\ &= \dim \mathfrak{g} - 2 \dim \mathfrak{b} + \dim(\mathfrak{b} \cap \mathrm{Ad}(k)\mathfrak{b}).\end{aligned}$$

Hence $\dim(\mathfrak{b} \cap \mathrm{Ad}(k)\mathfrak{b} + \mathfrak{n}) = \dim \mathfrak{n} - \dim \mathfrak{g} + 2 \dim \mathfrak{b} = \dim \mathfrak{b}$ since $\dim \mathfrak{g} = \dim \mathfrak{n} + \dim \mathfrak{b}$.

We now show that if x is in G then there is s in $W(A)$ so that x is in BsB. Now $x = kan$ with k in K, a in A, n in N. Thus if we can show that k is in BsB for some s we will know that x is in BsB. Let H be in $\mathfrak{a}'$ (see 7.5.3). Clearly $\mathfrak{a}' = \mathfrak{m}_1' \cap \mathfrak{a}$. Let X in $\mathfrak{n}$ be so that $H + X$ is in $\mathrm{Ad}(k)\mathfrak{b} \cap \mathfrak{b}$ (this is possible since $\mathrm{Ad}(k)\mathfrak{b} \cap \mathfrak{b} + \mathfrak{n} = \mathfrak{b}$). 7.5.15 implies that there is a unique element n in N so that $H + X = \mathrm{Ad}(n)H$. Thus $\mathrm{Ad}(k^{-1}n)H$ is in $\mathfrak{b}$. But ad H has only real eigenvalues. Thus $\mathrm{Ad}(k^{-1}n)H$ has only real eigenvalues. We therefore see that 7.5.13 implies $\mathrm{Ad}(k^{-1}n)H = H_1 + X_1$ with H_1 in $\mathfrak{a}'$ and X_1 in $\mathfrak{n}$. Applying 7.5.15 again we see that there is n_1 in N so that $\mathrm{Ad}(n_1)(H_1 + X_1) = H_1$. Thus $\mathrm{Ad}(n_1k^{-1}n)H = H_1$. But H, H_1 are in $\mathfrak{a}'$. 7.5.4 implies that $\mathrm{Ad}(n_1k^{-1}n)\mathfrak{m}_1 \subset \mathfrak{m}_1$. But 7.5.13 now implies that $\mathrm{Ad}(n_1k^{-1}n)\mathfrak{a} \subset \mathfrak{a}$.

Set $y = n^{-1}kn_1^{-1}$. Then $\mathrm{Ad}(y)\mathfrak{a} \subset \mathfrak{a}$. $y = k^1a^1n^1$ with k^1 in K, a^1 in A, and n^1 in N. Thus $\mathrm{Ad}(k^1)^{-1}\mathfrak{a} = \mathrm{Ad}(a^1n^1)\mathfrak{a}$. But $\mathrm{Ad}(a^1n^1) \subset \mathfrak{a} + \mathfrak{n}$. Thus $\mathrm{Ad}(k^1)^{-1}\mathfrak{a} \subset \mathfrak{a} + \mathfrak{n}$. But $\mathrm{Ad}(k^1)^{-1}\mathfrak{a} \subset \mathfrak{p}$ and $\mathfrak{p} \cap (\mathfrak{a} + \mathfrak{n}) = \mathfrak{a}$. Thus $\mathrm{Ad}(k^{-1})^{-1}\mathfrak{a} \subset \mathfrak{a}$. Hence k^1 is in M^*. Thus $n^{-1}kn_1^{-1}(a^1n^1)^{-1}$ is in M^*. Let s be the corresponding element of $W(A)$. Then k is in BsB.

Suppose that $BsB = Bs'B$ for s, s' in $W(A)$. Then if m_1^* is in s, m_2^* is in s', there are elements b_1, b_2 in B so that $b_1m_1^* = m_1^*b_2$. Since $B = MAN$ we may assume that b_1, b_2 are in AN. If H is in $\mathfrak{a}$ then $\mathrm{Ad}(b_1m_1^*)H = \mathrm{Ad}(m_1^*)\, H + X_1$ with X_1 in $\mathfrak{n}$ and $\mathrm{Ad}(m_2^*b_2)H = \mathrm{Ad}(m_2^*)(H + X_2)$ with X_2 in $\mathfrak{n}$. But $\mathrm{Ad}(m_2^*)X_2 = X_3 + X_4$ with X_3 in $\mathfrak{n}$ and X_4 in $\bar{\mathfrak{n}}$. Since $\mathfrak{n} \cap \bar{\mathfrak{n}} = (0)$ we see that $\mathrm{Ad}(m_2^*)H = \mathrm{Ad}(m_1^*)H$ for all H in $\mathfrak{a}$. Thus $s = s'$.

7.5.19 7.5.18 is the celebrated Bruhat lemma. The first proof of 7.5.18 in full generality was given by Harish–Chandra. We have reproduced Harish Chandra's proof.

7.5.20 *Corollary* Let $\bar{N}$ be the connected subgroup of G corresponding to $\bar{\mathfrak{n}}$. Then $\bar{N}MAN$ is open and its compliment is a finite union of submanifolds of strictly lower dimension. Furthermore the map $\bar{N} \times M \times A \times N \to G$ given by $(\bar{n}, m, a, n) \to nman$ is a diffeomorphism.

PROOF Suppose that $\bar{n}'m'a'n' = \bar{n}man$ with $\bar{n},\bar{n}'$ in $\bar{N}$, m,m' in M, a,a' in A, n,n' in N. Then $\bar{n}^{-1}\bar{n}'$ is in B. On the other hand we may replace N by $\bar{N}$ in 7.5.15 if we replace $\mathfrak{n}$ by $\bar{\mathfrak{n}}$. Thus if $\bar{n}_1 = \bar{n}^{-1}\bar{n}$ and H is in $\mathfrak{m}_1$ then $\mathrm{Ad}(\bar{n}_1)H = H + Z$, Z in $\bar{\mathfrak{n}}$. But $\bar{n}_1$ is in B, thus $\mathrm{Ad}(\bar{n}_1)H$ is in $\mathfrak{b}$. But the sum $\mathfrak{g} = \bar{n} \oplus \mathfrak{b}$ is direct. Hence $\mathrm{Ad}(\bar{n}_1)H = H$. Hence $\bar{n}_1 = e$. This implies that $\bar{n} = \bar{n}'$. Now the Iwasawa decomposition (8.4.3) implies that $m' = m$, $a = a'$, $n = n'$.

Let s be in $W(A)$ and let m^* be in s. Let

$$\Sigma_s^+ = \{\alpha \text{ in } \Sigma | s\alpha \text{ in } \Sigma\},\ \Sigma_s^- = \{-\alpha | \alpha \text{ in } \Sigma,\ s\alpha > 0\}.$$

Then $s^{-1}\Sigma = \Sigma^+ \cup \Sigma^-$. Thus $\mathrm{Ad}(m^*)^{-1}\mathfrak{n} = \mathfrak{n}_s + \bar{\mathfrak{n}}_s$ where $\mathfrak{n}_s = (\sum_{\alpha\in\Sigma_s^+}\mathfrak{g}_\alpha) \cap \mathfrak{g}$, $\bar{\mathfrak{n}}_s = (\sum_{\alpha\in\Sigma_s^-}\mathfrak{g}_\alpha) \cap \mathfrak{g}$. Both $\mathfrak{n}_s$ and $\bar{\mathfrak{n}}_s$ are subalgebras of $\mathfrak{g}$. $\mathfrak{n}_s$ is a subalgebra of $\mathfrak{n}$, $\bar{\mathfrak{n}}_s$ a subalgebra of $\bar{\mathfrak{n}}$.

Let $N_s = m^{*-1}Nm^*$. Then N_s is a Lie subgroup of G with Lie algebra $\mathfrak{n}_s + \bar{\mathfrak{n}}_s$. Let $\psi\colon \mathfrak{n}_s + \bar{\mathfrak{n}}_s \to N_s$ be defined by $\psi(X + Y) = \exp X \exp Y$, X in $\mathfrak{n}_s$, Y in $\bar{\mathfrak{n}}_s$. Then $\psi_{*e} = I$. There are therefore neighborhoods U_0^- and U_0^+ of 0 in $\mathfrak{n}_s$ and $\bar{\mathfrak{n}}_s$, respectively, so that if $U_e^- = \exp(U_0^-)$, $U_e^+ = \exp(U_0^+)$. $U_e^- U_e^+$ is a neighborhood of e in N_s. Let $\phi\colon G \to G/B$ be the natural mapping. Then $\phi(BsB) = m^*\phi(N_s)$. Now $\phi(U_e^- U_e^+) = \phi(U_e^-)$. Thus using the open covering $nU_e^- U_e^+$, n in N_s we see that $\dim \phi(BsB) = \dim \bar{\mathfrak{n}}_s$. Thus $\dim \phi(BsB) = \dim \bar{\mathfrak{n}}_s$. This implies that if $\bar{\mathfrak{n}}_s \neq \bar{\mathfrak{n}}$ then $\dim \phi(BsB) < \dim G/B$. Since $\bigcup_{s\in W(A)} \phi(BsB) = G/B$ and $W(A)$ is finite we conclude that there is s in $W(A)$ so that $\bar{\mathfrak{n}}_s = \bar{\mathfrak{n}}$. We assert that s is unique. Suppose $\bar{\mathfrak{n}}_{s'} = \bar{\mathfrak{n}}$. Then $(s')^{-1}s\Sigma = \Sigma$. Let $C \subset \mathfrak{a}'$ be the set of all H in $\mathfrak{a}'$ so that $\alpha(H) > 0$ for all α in Λ^+. Then C is convex and open in $\mathfrak{a}'$. Clearly if $s_1 = (s')^{-1}s$, $s_1C = C$. Let m_1^* be in s_1. Then $\mathrm{Ad}(m_1^*)$ as an automorphism of $\mathfrak{m}$ leaves fixed a regular element H_- of $\mathfrak{m}$. Without changing s_1 we may assume that H_- is in $\mathfrak{h}^-$ (this is due to the fact that $\mathrm{Ad}(M)$ is compact and hence there is m in M so that $\mathrm{Ad}(m)H_-$ is in $\mathfrak{h}^-$ and $mm^*m^{-1}M = s$). Now $s_1^k = I$ for some k, and $(1/k)(H + s_1H + \cdots + s_1^kH)$ is in C for each H in C. Set $H_+ = (1/k)(H + s_1H + \cdots + s_1^kH)$ for some fixed H in C. Then $\mathrm{Ad}(m_1^*)(H_- + H_+) = H_- + H_+$. But $H_- + H_+$ is regular in $\mathfrak{h}_C$. Applying the argument of the proof of 3.9.4(1) we see that $\mathrm{Ad}(m_1^*)|_{\mathfrak{a}} = I$. But then $s_1 = I$.

We have shown that

(1) There is a unique element s_0 in $W(A)$ so that $s_0\Lambda^+ = \Lambda^-$.

(1) also shows that dim $BsB <$ dim G if $s \neq s_0$. Thus Bs_0B is open and dense in G. But then if m^* is in s_0 then $m^{*-1}Bs_0B$ is open and dense in G. But $m^{*-1}Bs_0B = \bar{N}B$.

Let $\xi: \bar{N} \times M \times A \times N \to G$ be defined by $\xi(\bar{n}, m, a, n) = \bar{n}man$. Then

(2) $\xi_{*(\bar{n},m,a,n)}(X, Y, H, Z) = (\mathrm{Ad}(man)^{-1}X + \mathrm{Ad}(an)^{-1}Y + \mathrm{Ad}(n)^{-1} H + Z)_{\bar{n}man}$,

for X in $\bar{\mathfrak{n}}$, Y in $\mathfrak{Q}$, H in $\mathfrak{a}$, and Z in $\mathfrak{n}$. Thus ξ is everywhere regular. The corollary now follows.

7.5.21 Let $M_1 = MA$. If m is in M_1 then $\mathrm{Ad}(m)\mathfrak{n} \subset \mathfrak{n}$. Let $M_1' = \{m \in M_1 | \mathrm{Ad}(m)Z \neq Z \text{ for any } Z \text{ in } \mathfrak{n}\}$. We note that $A' = \exp \mathfrak{a}' \subset M_1'$.

7.5.22 *Lemma* Let h be in M_1'. If n is in N then $h^{-1}nhn^{-1}$ is in N and the map $n \to h^{-1}nhn^{-1}$ is a diffeomorphism of N onto N.

PROOF Let H be in $\mathfrak{a}'$ so that if λ is in Λ^+ then $\lambda(H) > 0$. Let $a_t = \exp(-tH)$.

(1) If n is in N then $\lim_{t\to\infty} ana_t^{-1} = e$.

To prove (1) we note that 7.4.3(2) implies that if n is in N then $n = \exp(X)$ for a unique X in $\mathfrak{n}$. Now $X = \sum_{\lambda\in\Lambda^+} X_\lambda$ with $[H, X_\lambda] = \lambda(H)X_\lambda$. Thus $\mathrm{Ad}(a_t)X = \sum e^{-t\lambda(H)}X_\lambda$. But $\lambda(H) > 0$ for all λ in Λ^+, hence

$$\lim_{t\to\infty} \sum e^{-t\lambda(H)}X_\lambda = 0.$$

On the other hand

$$\lim_{t\to\infty} a_tna_t^{-1} = \lim_{t\to\infty} a_t \exp Xa_t^{-1} = \lim_{t\to\infty} \exp(\mathrm{Ad}(a_t)X)$$
$$= \exp(\lim_{t\to\infty} \mathrm{Ad}(a_t)X) = e.$$

This proves (1).

Let $\psi(n) = h^{-1}nhn^{-1}$. Then $\psi(n)$ is in M_1N. Thus $\psi(n) = mn_1$ with m in M_1 and n_1 in N. Hence (1) implies that $\lim_{t\to\infty} a_t\psi(n)a_t^{-1} = m$. On the other hand $a_th^{-1}nhn^{-1}a_t^{-1} = h^{-1}a_tna_t^{-1}h(a_tna^{-1})^{-1}$. Thus (1) implies $\lim_{t\to\infty} a_t\psi(n)a_t^{-1} = e$. Hence $\psi(n)$ is indeed in N.

We now compute the differential of ψ. Let X be in $\mathfrak{n}$. Let f be in $C^\infty(N)$. Then

$$\psi_{*n}(X)f = \frac{d}{dt} f(h^{-1}n \exp tXh \exp(-tX)n^{-1})\Big|_{t=0}$$

$$= \frac{d}{dt} f(h^{-1}nhn^{-1} \exp(t\,\mathrm{Ad}(nh^{-1})X) \exp(-t\,\mathrm{Ad}(n)X)\Big|_{t=0}$$

$$= (\mathrm{Ad}(n)\{\mathrm{Ad}(h^{-1})X - X\}_{\psi(n)}f.$$

We therefore see that

(2) $\psi_{*n}(X) = (\mathrm{Ad}(n)\{\mathrm{Ad}(h^{-1})X - X\})_{\psi(n)}$

for n in N, X in $\mathfrak{n}$.

The definition of M_1' (7.5.21) combined with (2) now implies that ψ is everywhere regular. In particular there is a neighborhood U of e in N so that $\psi(U) = V$ is a neighborhood of e in N and ψ: $U \to V$ is a diffeomorphism. Now $a_t\psi(n)a_t^{-1} = \psi(a_t n a_t^{-1})$. (1) implies that $\bigcup_{t>0} a_t^{-1}Va_t = N$. But then $\psi(N) = N$. Hence ψ is surjective. If $\psi(n_1) = \psi(n_2)$ then let $t > 0$ be so large that $a_t n_1 a_t^{-1}$ and $a_t n_2 a_t^{-1}$ are in U. Then $a_t\psi(n_1)a_t^{-1} = a_t\psi(n_2)a_t^{-1}$. Hence $\psi(a_t n_1 a_t^{-1}) = \psi(a_t n_2 a_t^{-1})$. But ψ is injective on U. Thus $a_t n_1 a_t^{-1} = a_t n_2 a_t^{-1}$. Hence $n_1 = n_2$. Thus ψ: $N \to N$ is a surjective diffeomorphism as asserted.

7.5.23 *Lemma* There is a neighborhood U of 0 in $\mathfrak{m}_1$ so that $\exp$: $U + \mathfrak{n} \to B = M_1N$ is injective and $\exp_{*X}$: $\mathfrak{g} \to T(G)_{\exp(X)}$ is nonsingular for all X in $U + \mathfrak{n}$.

PROOF There is a neighborhood V_0 of 0 in $\mathfrak{g}$ so that $\exp$: $V_0 \to V_e = \exp V_0$ is a diffeomorphism onto V_e, an open neighborhood of e in G. Let $U = V_0 \cap \mathfrak{m}_1$.

Let a_t be as in the proof of 7.5.22. If X is in $\mathfrak{n}$, Y is in U, then there is $t > 0$ so that $Y + \mathrm{Ad}(a_t)X$ is in V_0. Thus exp is regular at $Y + \mathrm{Ad}(a_t)X = \mathrm{Ad}(a_t)(Y + X)$. But then exp is regular at $Y + X$.

Suppose that $\exp(H_1 + Z_1) = \exp(H_2 + Z_2)$ with H_1, H_2 in U, Z_1, Z_2 in $\mathfrak{n}$. Then

$$\lim_{t\to\infty} a_t \exp(H_1 + Z_1)a_t^{-1} = \lim_{t\to\infty} a_t \exp(H_2 + Z_2)a_t^{-1}.$$

Thus $\exp H_1 = \exp H_2$. But exp is injective on U. Thus $H_1 = H_2$. Set $H = H_1$. Let $t > 0$ be so large that $H + \mathrm{Ad}(a_t)Z_i$ is in V_0 for $i = 1, 2$. Then

$$a_t \exp(H + Z_i)a_t^{-1} = \exp(H + \mathrm{Ad}(a_t)Z_i), \qquad i = 1, 2.$$

Hence $\exp(H + \mathrm{Ad}(a_t)Z_1) = \exp(H + \mathrm{Ad}(a_t)Z_2)$. But then $H + \mathrm{Ad}(a_t)Z_1 = H + \mathrm{Ad}(a_t)Z_2$. Hence $Z_1 = Z_2$. This proves the lemma.

7.5.24 *Lemma* Let $U + \mathfrak{n}$ be as in 7.5.23. If $V \subset U$ is compact then $\exp(V + \mathfrak{n})$ is closed in G.

PROOF We assert that $\exp(V + \mathfrak{n}) = \exp V \cdot N$. In fact, let h be in V, n in N. Let a_t be as in the proof of 7.5.22. If t is sufficiently large then $a_t(\exp h)na_t^{-1}$ is in V_e (here V_0 and V_e are as in the proof of 7.5.23). Thus $a_t(\exp h)na_t^{-1}$ is in $\exp(V + \mathfrak{n})$. Thus $(\exp h)n$ is in $a_t^{-1} \exp(V + \mathfrak{n})a_t = \exp(V + \mathfrak{n})$. Thus $\exp(V + \mathfrak{n}) = \exp V \cdot N$ as asserted. But $\exp V$ is compact and N is closed in G. Thus $\exp V \cdot N$ is closed in G.

7.6 The Integral Formula for the Iwasawa Decomposition

7.6.1 Let G be a connected semisimple Lie group and let $G = KAN$ be an Iwasawa decomposition of G (see 7.4.3).

7.6.2 *Lemma* Let $S = AN$. Let da and dn be left invariant measures on A and N, respectively. Then the left invariant measure on S, ds, can be normalized so that if f is in $C_0(S)$ then

$$\int_S f(s)ds = \int_{A \times N} f(an)\, da\, dn.$$

PROOF The map $A \times N \to S$ given by $(a, n) \mapsto an$ is a surjective diffeomorphism by 7.4.3. There is therefore a continuous function $h: A \times N \to R$ so that if f is in $C_0(S)$ then

$$\int_S f(s)\, ds = \int_{A \times N} f(an)h(a, n)\, da\, dn.$$

Let $\psi(a, n) = an$. Then $\psi^*ds = hdadn$. Let for a_0 in A, $\tau(a_0)(a, n) = (a_0a, n)$ for a in A, n in N. Then $\psi \circ \tau(a_0) = L_{a_0} \circ \psi$. Thus $(\psi \circ \tau(a_0))^*ds = (L_{a_0} \circ \psi)^*ds = \psi^*L_{a_0}ds = \psi^*ds$. Thus $h(a_0a, n) = h(a, n)$ for all a_0, a in A, n in N. This implies that h is a function of N alone.

Now, if n is in N, $R_n^*ds = \det(\mathrm{Ad}(n)|_{\mathfrak{a}+\mathfrak{n}})ds$. But $(\mathrm{Ad}(n) - I)^k = 0$ for some k. Thus $\det(\mathrm{Ad}(n)|_{\mathfrak{a}+\mathfrak{n}}) = 1$. Hence $R_n^*ds = ds$. Let for n_0 in N, $\mu(n_0)(a, n) = (a, nn_0)$ for a in A, n in N. Then $\psi \circ \mu(n_0) = R_{n_0} \circ \psi$. Arguing as above we now find $h(nn_0) = h(n)$ for all n, n_0 in N. Thus h is constant.

7.6.3 *Lemma* Let $d_R s$ be right invariant measure on S. Let $\rho(H) = \frac{1}{2}\text{tr}(\text{ad}(H)|_{\mathfrak{n}})$ for H in $\mathfrak{a}$ (here $\mathfrak{a}$, $\mathfrak{n}$ have the same meaning as in 7.4.3). Let $\log: A \to \mathfrak{a}$ be the inverse to $\exp: \mathfrak{a} \to A$ (see 7.4.3(1)). If f is in $C_0(S)$ then

$$\int_S f(s)d_R s = \int_{A\times N} f(an)e^{2\rho(\log a)}\, da\, dn.$$

PROOF $d_R s = \delta(s)ds$ where δ is the modular function of S (see 2.5). Now $\delta(s) = \det(\text{Ad}(s)|_{\mathfrak{a}+\mathfrak{n}})$ for s in S. If s is in S then $s = an$, a in A and n in N. We have seen that $\det(\text{Ad}(n)|_{\mathfrak{a}+\mathfrak{n}}) = 1$ in the proof of 7.5.2. Hence $\delta(an) = \det(\text{Ad}(a)|_{\mathfrak{a}+\mathfrak{n}}) = \det(\text{Ad}(a)|_{\mathfrak{n}})$ since $\text{Ad}(a)|_{\mathfrak{a}} = I$. Now $\det(\text{Ad}(a)|_{\mathfrak{a}}) = \exp(\text{tr}\,\text{ad}(\log a)|_{\mathfrak{n}} = e^{2\rho(\log a)}$. The result now follows from 7.6.2.

7.6.4 *Proposition* The invariant measure dg on G can be normalized so that if dk is normalized invariant measure on K, da is invariant measure on A and dn is invariant measure on N and if f is in $C_0(G)$ then

$$\int_G f(g)dg = \int_{K\times A\times N} f(kan)e^{2\rho(\log a)}\, dk\, da\, dn$$

(see 7.6.3 for notation).

PROOF Let $d_R s$ be right invariant measure on S. Let $\psi: K \times S \to G$ be defined by $\psi(k, s) = ks$. Then $\psi^* dg = h(k, s)dk d_R s$. Now G is unimodular, hence dg is both left and right invariant. We can therefore argue just as in the proof of 7.6.2 (using the fact that $\psi^* dg$ is invariant under the right action of S on $K \times S$ and the left action of K on $K \times S$) to see that h is constant. The result now follows from 7.6.3.

7.6.5 Let Z be the center of G. Then $Z \subset K$ and K/Z is compact (see 7.2.4 and 7.2.5). Let M be the centralizer of A in K. Then clearly $Z \subset M$. Thus S/M is compact. Let ω be the K invariant volume element on K/M so that $\int_{K/M} \omega = 1$. The Iwasawa decomposition says that $G/MAN = K/M$. Thus G acts on K/M as a group of diffeomorphisms.

7.6.6 *Lemma* If g is in G, f is in $C(K/M)$ then

$$\int_{K/M} f(gkM)e^{-2\rho(H(gk))}\, d(kM) = \int_{K/M} f(x)dx$$

where $H: G \to \mathfrak{a}$ is defined by $H(kan) = \log a$ for k in K, a in A, n in N.

PROOF Let f be in $C(K/M)$. Then since $K/M = G/MAN$ there is f_1 in $C_0(G)$ (see 3.6.3) so that

$$f(gMAN) = \int_{MAN} f_1(gman)\, dm\, da\, dn.$$

Here if $B = MAN$ then a slight modification of the proof of 7.6.3 shows that $db = dmdadn$ if db is left invariant measure on B, dm left invariant measure on M, since the elements of M and A commute. Set $f_2(k) = \int_{A\times N} f_1(kan)dadn$. Then $f(kM) = \int_M f_2(km)dm$. This implies that we may assume that $\int_{K/M} f(x)dx = \int_K f_2(k)dk$. Hence $\int_{K/M} f(x)dx = \int_{K\times A\times N} f_1(kan)dk\, da\, dn$.

Now let g be in G. Then $(g^*f)(x) = f(gx)$. We see that

$$\int_{K/M} (g^*f)(x)dx = \int_{K\times A\times N} f_1(gkan)\, dk\, da\, dn.$$

Now $gkan = k(gk)\exp(H(gk))aa^{-1}n(gk)an$ where

$$gk = k(gk)\exp(H(gk))n(gk)$$

with $k(gk)$ in K, $n(gk)$ in N. Thus

$$\int_{K\times A\times N} f_1(gkan) = \int_{K\times A\times N} f_1(k(gk)\exp(H(gk))an)\, dk\, da\, dn.$$

But

$$\int_{K\times A\times N} f_1(k(gk)\exp(H(gk))an)\, e^{-2\rho(H(gk))}e^{2\rho(\log a)}\, dk\, da\, dn$$

$$= \int_{K\times A\times N} f_1(kan)e^{2\rho(\log a)}\, dk\, da\, dn.$$

Thus

$$\int_{K\times A\times N} f_1(k(gk)\exp(H(gk))an)e^{-2\rho(H(gk))}\, dk\, da\, dn$$

$$= \int_{K\times A\times N} f_1(kan)\, dk\, da\, dn = \int_{K/M} f(x)\, dx.$$

Hence

$$\int_{K/M} (g^*f)(kM)e^{-2\rho(H(gk))}d(kM) = \int_{K/M} f(x)dx.$$

This implies that

$$\int_{K/M} (g^*f)(x)dx = \int_{K/M} f(kM)e^{-2\rho(H(gk))}\, d(kM),$$

as was to be proved.

7.6.7 *Proposition* Let $\bar{N}$, M, A, N be as in 7.5.20. Let dg, $d\bar{n}$, dm, da, and dn denote invariant measures on G, $\bar{N}$, M, A, and N, respectively. Then dg can be normalized so that if f is in $C_0(G)$ then

$$\int_G f(g)dg = \int_{\bar{N}\times M\times A\times N} f(\bar{n}man)e^{2\rho(\log a)}\, d\bar{n}\, dm\, da\, dn.$$

PROOF Let $\psi: \bar{N} \times M \times A \times N \to G$ be given by $\psi(\bar{n}, m, a, n) = \bar{n}man$. Then 7.5.20 implies that $\psi^*dg = h(\bar{n}, m, a, n)d\bar{n}dmdadn$. Let for $\bar{n}_0$ in N, $\tau(\bar{n}_0)(\bar{n}, m, a, n) = (\bar{n}_0\bar{n}, m, a, n)$. Let for s_0 in $S = AN$, $\mu(s_0)(\bar{n}, m, s) = (\bar{n}, m, ss_0)$. Then $\psi \circ \tau(\bar{n}_0) \circ = L_{n_0} \circ \psi$, $\psi \circ \mu(s_0) = R_{s_0} \circ \psi$. Arguing as in 7.6.4 we find that $h(\bar{n}, m, a, n) = e^{2\rho(\log a)}h(m)$. Let $\alpha(m_0)(\bar{n}, m, a, n) = (\bar{n}, m_0m, a, n)$, then $\psi(\alpha(m_0)(\bar{n}, m, a, n) = \bar{n}m_0man = m_0(m^{-1}\bar{n}m_0)man$. But $d\bar{n}$ is invariant under $\bar{n} \to m_0^{-1}\bar{n}m_0$ since $\mathrm{Ad}(M_0)$ is compact. Thus h is constant. The result now follows from 7.5.20.

7.6.8 *Lemma* Let $\psi: \bar{N} \to K/M$ be defined by $\psi(\bar{n}) = \bar{n}MAN$. Then $\psi(\bar{N})$ is open and its compliment is a finite union of submanifolds of lower dimension. If f is in $C(K/M)$ then $\int_{K/M} f(x)dx = \int_{\bar{N}} f(\psi(\bar{n}))e^{-2\rho(H(\bar{n}))}d\bar{n}$ where $d\bar{n}$ can be (and is) normalized so that $\int_{\bar{N}} e^{-2\rho(H(\bar{n}))}d\bar{n} = 1$.

PROOF We normalize dg so that $dg = e^{2\rho(\log a)}dmd\bar{n}dadn$. Now

$$m\bar{n}an = mk(\bar{n}) \exp(H(\bar{n}))n(\bar{n}))an = mk(\bar{n}) \exp(H(\bar{n}) + \log a)a^{-1}n(\bar{n})an.$$

Hence $dg = e^{2\rho(H(\bar{n}))}e^{2\rho(\log a)}dmdk(\bar{n})dad\bar{n}$. Thus $dk = e^{-2\rho(H(\bar{n}))}dk(\bar{n})dm$. This implies that $\int_{K/M} f(x)dx = \int_{\bar{N}} f(\bar{n})e^{-2\rho(H(\bar{n}))}d\bar{n}$. This formal argument is justified using the techniques of the proof of 7.6.6.

7.7 Integral Formulas for the Adjoint Action

7.7.1 We retain the notation of 7.5. Let $C = \{g \text{ in } G | \mathrm{Ad}(g)|_{\mathfrak{h}_0} = I\}$. Let

$$C^* = \{g \in G | \mathrm{Ad}(g)\mathfrak{h}_0 \subset \mathfrak{h}_0\}.$$

7.7.2 Let A_- be the centralizer of $\mathfrak{h}_0$ in K and A_-^* be the normalizer of $\mathfrak{h}_0$ in K. That is, $A_- = K \cap C$, $A_-^* = K \cap C^*$.

7.7.3 *Lemma* $C = A_-A$ and $C^* = A_-^*A$.

PROOF Let Δ be the root system of $\mathfrak{g}_C$ relative to $\mathfrak{h}$. Let $\mathfrak{h}_0' = \{H \text{ in } \mathfrak{h}_0 | \alpha(H) \neq 0 \text{ for all } \alpha \text{ in } \Delta\}$. If H is in $\mathfrak{h}_0'$ then $\mathfrak{h}_0 = \{X \text{ in } \mathfrak{g} | [H, X] = 0\}$. Suppose that g is in G and $\mathrm{Ad}(g)\mathfrak{h}_0 \subset \mathfrak{h}_0$. Then $g = kan$, k in K, a in A, and n in N. Now $\mathrm{Ad}(k^{-1})\mathfrak{h}_0 = \mathrm{Ad}(an)\mathfrak{h}_0 \subset \mathfrak{h}_0 + \mathfrak{n}$. On the other hand $\mathrm{Ad}(k^{-1})\mathfrak{a} \subset \mathfrak{p}$ and $\mathfrak{p} \cap (\mathfrak{h}_0 + \mathfrak{n}) = \mathfrak{a}$. Also $\mathrm{Ad}(k^{-1})\mathfrak{h}^- \subset \mathfrak{k}$ and $\mathfrak{k} \cap (\mathfrak{h}_0 + \mathfrak{n}) = \mathfrak{h}^-$. Thus $\mathrm{Ad}(k^{-1})\mathfrak{h}_0 \subset \mathfrak{h}_0$. This implies that k is in A_-^*. Also $\mathrm{Ad}(an)\mathfrak{h}_0 \subset \mathfrak{h}_0$. Thus $\mathrm{Ad}(n)\mathfrak{h}_0 \subset \mathfrak{h}_0$. Let H be in $\mathfrak{h}_0'$. Then $\mathrm{Ad}(n)H = H + Z$ with Z in $\mathfrak{n}$ by 7.5.15. But then $Z = 0$ ($\mathrm{Ad}(n)\mathfrak{h}_0 \subset \mathfrak{h}_0$). Thus 7.5.15 implies $n = e$. Hence $C^* = A_-^* A$.

Suppose now g is in C. Then $g = ma$ with m in A_-^*, a in A. But then $\mathrm{Ad}(m)|_{\mathfrak{h}_0} = \mathrm{Ad}(g)|_{\mathfrak{h}_0}$. Hence $\mathrm{Ad}(m)|_{\mathfrak{h}_0} = I$. Thus m is in A_-.

7.7.4 *Lemma* C^*/C is a finite group.

PROOF 7.7.3 implies that $C^*/C = A_-^*/A_- = \mathrm{Ad}(A_-^*)/\mathrm{Ad}(A_-)$ since Z, the center of G, is contained in A_-. Thus C^*/C is compact. We leave it to the reader to argue as in 7.5.6 to see that C and C^* have the same Lie algebra. But then C^*/C is discrete and compact, hence finite.

7.7.5 Set $W = W(\mathfrak{h}_0) = C^*/C$. Also if x is in C let us call x regular if $\mathfrak{h}_0 = \{x \text{ in } \mathfrak{g} | \mathrm{Ad}(x)X = X\}$. Let C' be the set of regular elements of C.

7.7.6 *Lemma* There exists a G-invariant volume element on G/C.

PROOF We leave it to the reader to check that the Lie algebra of C is $\mathfrak{h}_0$. We may thus identify $T(G/C)_{eC} \otimes C$ with $\sum_{\alpha\in\Delta} (\mathfrak{g}_C)_\alpha$. If x is in C then $\mathrm{Ad}(x)(\mathfrak{g}_C)_\alpha \subset (\mathfrak{g}_C)_\alpha$. Thus $\mathrm{Ad}(x)|_{(\mathfrak{g}_C)_\alpha} = c_\alpha I$. Since $\mathrm{Ad}(x)$ leaves invariant the Killing form of $\mathfrak{g}_C$ we see that $c_{-\alpha} = c_\alpha^{-1}$. Let μ be the isotopy representation of C on $T(G/C)_{eC}$. Then $\det \mu(x) = \prod_{\alpha\in\Delta} c_\alpha = 1$. This clearly implies that there is an invariant volume.

7.7.7 If x is in G set $\bar{x} = xC$ in G/C. Let $d\bar{x}$ denote the G-invariant volume element on G/C.

7.7.8 *Lemma* $d\bar{x}$ can be normalized so that if f is in $C_0(G/C)$ then

$$\int_{G/C} f(\bar{x})d\bar{x} = \int_{K_* \times N} f(knC)\, dk_* dn$$

where $K_* = K/Z$ with Z the center G and dk_* is normalized Haar measure in

K_*. We note that if z is in Z then $kznC = knC$, thus the above integral formula makes sense.

PROOF We first note that since A normalizes N, $G = KNA$ and the map $\varphi: K \times N \times A \to G$ given by $\varphi(k, n, a) = kna$ is a surjective diffeomorphism.

(1) If F is in $C_0(G)$ then $\int_G F(g)dg = \int_{K\times N\times A} F(kna)\, dk\, dn\, da$.

In fact if $S = NA$ then $d_R s = dnda$. (1) now follows from the proof of 7.6.4.

Let dh be invariant measure on C. There is F in $C_0(G/Z)$ so that $f(gC) = \int_C F(gh)dh$ (see 3.6.3). With a suitable normalization of dh we have

$$\int_{G/Z} F(g)dg = \int_{G/C}\int_{C/Z} F(gh)dh_* d\bar{g} = \int_{G/C} f(\bar{x})d\bar{x}.$$

On the other hand

$$\int_{G/Z} F(g)dg = \int_{K_*\times N\times A} F(kna)\, dk_*\, dn\, da = \int_{K_*\times N}\left(\int_A F(kna)da\right)dk_* dn.$$

Now if φ is in $C(K_*)$ then

$$\int_{K_*} \varphi(k_*)dk_* = \int_{K_*}\left(\int_{A_-/Z} \varphi(kh)dh\right)dk_*$$

with the invariant measure dn on A_-/Z suitably normalized. Thus

$$\int_{G/Z} F(g)dg = \int_{K_*\times A^-\times N\times A} F(khna)\, dh\, dn\, da\, dk_*.$$

The map $n \mapsto unu^{-1}$ preserves du since A_-/Z is compact. Thus we have

$$\int_{G/Z} F(g)dg = \int_{K_*\times N}\left(\int_C F(knh)dh\right)dk_* dn = \int_{K_*\times N} f(knC)dk_* du.$$

This proves the Lemma.

7.7.9 *Lemma* Let f be in $C_0(G)$. Let h be in $C \cap M_1'$ (here M_1' has the meaning of 7.5.21). Then $\int_G |f(ghg^{-1})|d\bar{g}$ exists and

$$\int_{G/C} f(ghg^{-1})d\bar{g} = \int_{K_*\times N} |\det(\mathrm{Ad}(h^{-1}) - I)|_{\mathfrak{n}}{}^{-1} f(khnk^{-1})dk_* dn.$$

PROOF Suppose that $\int_{G/C} |f(ghg^{-1})|d\bar{g}$ exists. Then 7.7.8 implies that

$$\int_{G/C} f(ghg^{-1})d\bar{g} = \int_{K_*\times N} f(knhn^{-1}k^{-1})dk_* dn.$$

Now 7.5.22 and 7.5.22(2) (in the proof of 7.5.6) imply that

(1) $$\int_{K_* \times N} f(khnk^{-1})|\det((\mathrm{Ad}(h^{-1}) - I)|_{\mathfrak{n}})|^{-1} dk_* dn$$
$$= \int_{K_* \times N} f(knhn^{-1}k^{-1}) dk_* dn.$$

But the right-hand side of (1) is obviously absolutely convergent. Hence retracing the equalities we find that $\int_{G/C} |f(ghg^{-1})| d\bar{g}$ exists.

7.7.10 *Corollary* Let f be in $C_0(G)$. Then

$$e^{\rho(\log a)} \int_{K_* \times N} f(kmank^{-1}) dk_* dn = D(ma) \int_{G/C} f(gmag^{-1}) d\bar{g}$$

where m is in M, a is in A, and ma is in M_1'. Also,

$$D(ma) = \prod_{\lambda \in \Lambda^+} |\det(e^{-\lambda/2(\log a)} I - e^{\lambda/2(\log a)} \rho_\lambda(m))|$$

where $\rho_\lambda(m) = \mathrm{Ad}(m)|_{\mathfrak{g}_\lambda}$ for m in M.

PROOF

$$|\det((\mathrm{Ad}(ma)^{-1} - I)|_{\mathfrak{n}})$$
$$= \prod_{\lambda \in \Lambda^+} |\det(e^{-\lambda(\log a)} \rho_\lambda(m)^{-1} - I)|^{-1}$$
$$= \prod_{\lambda \in \Lambda^+} e^{(\lambda/2)(\log a) m_\lambda} |\det(e^{-\lambda(\log a)/2} I - e^{\lambda(\log a)/2} \rho_\lambda(m)|^{-1}.$$

The result now follows from 7.7.9.

7.7.11 *Lemma* Let f be in $C_0(G)$. Define for m in M, a in A,

$$F_f(ma) = e^{\rho(\log a)} \int_{K_* \times N} f(kmank^{-1}) dk_* dn.$$

If m^* is in M^* then $F_f(m^* ma\, m^{*-1}) = F_f(ma)$.

PROOF

$$F_f(m^* mam^{*-1})$$
$$= e^{\rho(\mathrm{Ad}(m^*)\log a)} \int_{K_* \times N} f(km^* mam^{*-1} nk^{-1}) dk_* dn.$$

Let $s = m^{*-1}M$ in $W(A)$. Let $N_s = m^{*-1} N m^*$. If φ is in $C_0(N_s)$ then $\int_N \varphi(m^{*-1} nm^*) dn = \int_{N_s} \varphi(n_s) dn_s$ where dn_s is invariant measure on N_s. Hence

$F_f(m^*mam^{*-1}) = e^{s\rho(\log a)} \int_{K_* \times N_s} f(kman_sk^{-1})dk_*dn_s$. Replacing N by N_s in 7.7.10 we find that if D_s is the corresponding D then

$$e^{s\rho(\log a)} \int_{K_* \times N_s} f(kman_sk^{-1})dk_*dn_s = D_s(ma) \int_{G/C} f(gmag^{-1})d\bar{g}$$

for ma in M_1'. The lemma will clearly be proved if we can show that $D_s(ma) = D(ma)$ since M_1' is dense in M_1. Choose $\mathfrak{h}_0$ so that $\mathrm{Ad}(ma)|_{\mathfrak{h}_0} = I$. If Σ is as in 7.5.11(3) for N, then $s^{-1}\Sigma$ is the corresponding system for N_s (here we take a representative for m^* in C^*). If α is in Δ, let $\xi_\alpha(m) = \mathrm{Ad}(m)|_{(\mathfrak{g}_C)_\alpha}$. Then

$$D(ma) = \prod_{\alpha\in\Sigma} |(e^{-\alpha(\log a)/2} - e^{\alpha(\log a)/2}\xi_\alpha(m))|.$$

Thus

$$\begin{aligned} D(ma)^2 &= \prod_{\alpha\in\Sigma} |(e^{-\alpha(\log a)/2} - e^{\alpha(\log a)/2}\xi_\alpha(m))| \prod_{\alpha\in\Sigma} |e^{-\alpha(\log a)/2}\xi_\alpha(m)^{-1} - e^{\alpha(\log a)/2}| \\ &= \prod_{\alpha\subset\Sigma\cup(-\Sigma)} |(e^{-\alpha(\log a)/2} - e^{\alpha(\log a)/2}\xi_\alpha(m)|. \end{aligned}$$

Arguing similarly for $D_s(ma)$ we find that $D_s(ma)^2 = D(ma)^2$ since $s\Sigma \bigcup(-s\Sigma) = \Sigma \bigcup(-\Sigma)$. Thus $D_s(ma) = D(ma)$. This completes the proof of the lemma.

7.8 Integral Formulas for the Adjoint Representation

7.8.1 We retain the notation of 7.7. Let $W = W(\mathfrak{h}_0)$ be as in 7.7.5. Then W acts on the right on G/C as follows: Let m^* be in C^*, $\bar{x} = xC$, x in G, then if $s = m^*C$, $\bar{x}s = xm^*C$. Clearly $\bar{x}s$ is well defined. If $\bar{x}s = \bar{x}$ then $xm^*C = xC$, hence $m^*C = C$. Thus $s = I$. This implies that W acts freely on G/C.

7.8.2 Let $\mathfrak{h}_0$ be as in 7.5. Let $\mathfrak{h}$ be the complexification of $\mathfrak{h}_0$ and let Δ be the root system of $\mathfrak{g}_C$ relative to $\mathfrak{h}$. Let $\mathfrak{h}_0' = \{H \text{ in } \mathfrak{h}_0 | \alpha(H) \neq 0 \text{ for any } \alpha \text{ in } \Delta\}$. Set $\mathfrak{g}_1 = \bigcup_{g\in G} \mathrm{Ad}(g)\mathfrak{h}_0'$. Let Δ^+ be as in 7.5.9.

7.8.3 *Lemma* $\mathfrak{g}_1$ is open in $\mathfrak{g}$. Let $d\bar{x}$ be as in 7.7.7. If f is absolutely integrable on $\mathfrak{g}_1$ then

$$\int_{\mathfrak{g}_1} f(X)dX = \int_{\mathfrak{h}_1} \prod_{\alpha\in\Delta^+} |\alpha(H)|^2 \int_{G/C} f(\mathrm{Ad}(x)H)d\bar{x}dH$$

with dX and dH suitably normalized Euclidean measures on $\mathfrak{g}$ and $\mathfrak{h}_0$.

PROOF Let $\psi(\bar{x}, H) = \mathrm{Ad}(x)H$ for x in G, H in $\mathfrak{h}_0$. ψ is a C^∞ mapping of $G/C \times \mathfrak{h}_0 \to \mathfrak{g}$. We compute ψ_*. Let B be the Killing form of $\mathfrak{g}$. Then B is nondegenerate on $\mathfrak{h}_0$. We may thus identify $T(G/C)_{\bar{e}}$ with $\mathfrak{h}_0^\perp$ relative to B. If f is in $C^\infty(\mathfrak{g})$ and if X is in $\mathfrak{h}_0^\perp$, then

$$\psi_{*(\bar{e},H)}(X, 0)f = \frac{d}{dt}f(\mathrm{Ad}(\exp tX)H)|_{t=0} = \frac{d}{dt}f(e^{\mathrm{ad}tX}H)|_{t=0}$$

$$= \frac{d}{dt}f(H - t[H, X])|_{t=0} = -([H, X]f)(H).$$

Thus

(1) $\psi_{*(\bar{e},H)}(X, 0) = -[H, X]$ for H in $\mathfrak{h}_0$, X in $\mathfrak{h}_0^\perp$.

If H_1 is in $\mathfrak{h}_0$ then clearly $\psi_{*(\bar{e},H)}(0, H_1) = H_1$. Combining this with (1) we have

(2) $\psi_{*(\bar{e},H)}(X, H_1) = H_1 - [H, X]$ for H, H_1 in $\mathfrak{h}_0$, X in $\mathfrak{h}_0^\perp$.

Now $\psi(g\bar{x}, H) = \mathrm{Ad}(g)\psi(\bar{x}, H)$. Thus $\psi_{*(\bar{x},H)} = \mathrm{Ad}(x) \circ \psi_{*(\bar{e},H)}$. Hence

(3) $\psi_{*(\bar{x},H)}(X, H_1) = \mathrm{Ad}(x)(H_1 - [H, X])$ for x in G, H, H_1 in $\mathfrak{h}_0$, X in $\mathfrak{h}_0^\perp$ (X is looked upon as the tangent vector $L_{x_*}X$ at $\bar{x}$ in G/C).

(3) implies the following:

(4) If H is in $\mathfrak{h}_0'$ then $\psi_{*(\bar{x},H)}$ is nonsingular.

(4) clearly implies (by the inverse function theorem) that $\mathfrak{g}_1$ is open in $\mathfrak{g}$.

Let $q = \dim \mathfrak{h}_0$. Let ω in $\wedge^q \mathfrak{h}_0^*$ define dH, the Euclidean volume element of $\mathfrak{h}_0$. We look upon ω as a q-form on $\mathfrak{g}$ using the projection $p\colon \mathfrak{h}_0 \oplus \mathfrak{h}_0^\perp \to \mathfrak{h}_0$. Let η be a nonzero element of $\Lambda^r(\mathfrak{h}_0^\perp)^*$ ($r = \dim \mathfrak{h}_0^\perp$). Then we have seen in 7.6.6 that we may choose η so that η defines $d\bar{x}$ on G/C. We also look at η as an r-form on $\mathfrak{g}$. We take for dX, $\omega \wedge \eta$. There is a C^∞ function $h\colon (G/C) \times \mathfrak{h}_0 \to R$ so that $\psi^* dX = h d\bar{x} dH$.

Suppose that H_1, H_2 are in $\mathfrak{h}_0'$, that x_1, x_2 are in G, and that $\psi(\bar{x}_1, H_1) = \psi(\bar{x}_2, H_2)$. Then $\mathrm{Ad}(x_1)H_1 = \mathrm{Ad}(x_2)H_2$. Hence $\mathrm{Ad}(x_1^{-1}x_2)H_2 = H_1$. Thus since the centralizer of H_i in $\mathfrak{g}$ is $\mathfrak{h}_0$ for $i = 1, 2$, $x_1^{-1}x_2$ is in C^*. Thus there is s in $W(\mathfrak{h}_0)$ so that $\bar{x}_2 = \bar{x}_1 s$, $sH_2 = H_1$. Let w be the order of $W(\mathfrak{h}_0)$. Then

(5) $\psi\colon G/C \times \mathfrak{h}_0' \to \mathfrak{g}_1$ is a w-fold covering map. Hence (see A.4.2.11)

(6) $w \int f(x) dX = \int_{G/C \times \mathfrak{h}_0'} (f \circ \psi)(\bar{x}, H)|h| d\bar{x} dH$.

Using (2) we see that if $H_1, \ldots, H_q$ are in $\mathfrak{h}_0$ and $X_1, \ldots, X_r$ are in $\mathfrak{h}_0^\perp$ and so that $(\omega \wedge \eta)(H_1, \ldots, H_q, X_1, \ldots, X_r) = 1$ then

$$h(\bar{x}, H)$$

$$= (-1)^r(\omega \wedge \eta)(\mathrm{Ad}(x)H_1, \ldots, \mathrm{Ad}(x)H_{\bar{e}}, \mathrm{Ad}(x)[H, X_1], \ldots, \mathrm{Ad}(x)[H, X_r])$$

$$= (-1)^r(\det \mathrm{Ad}(x)) \prod_{\alpha \in \Delta} \alpha(H) = (-1)^{r + r/2} \prod_{\alpha \in \Delta^+} \alpha(H)^2.$$

Thus $|h(\bar{x}, H)| = \prod_{\alpha \in \Delta^+} |\alpha(H)|^2$. Replacing dX by $w \cdot dX$, the lemma now follows.

7.8.4 *Lemma* It is possible to normalize the Euclidean measures dX, dH, and dZ on $\mathfrak{g}$, $\mathfrak{h}_0$, and $\mathfrak{n}$, respectively, so that if f is absolutely integrable on $\mathfrak{g}_1$ then

$$\int_{\mathfrak{g}_1} f(X)dX = \int_{\mathfrak{h}_0} \prod_{\alpha\in\Sigma} |\alpha(H)| \prod_{\alpha\in\Delta^+ - \Sigma} |\alpha(H)|^2 \int_{K_*\times\mathfrak{n}} f(\mathrm{Ad}(k)(H+Z))dk_* dZ\, dH$$

where K_* is K/Z, Z the center of G, dk_* is normalized Haar measure on K_*, and Σ is as in 7.5.11(3).

PROOF 7.8.3 implies that

$$\int_{\mathfrak{g}_1} f(X)dX = \int_{\mathfrak{h}_0} \prod_{\alpha\in\Delta^+} |\alpha(H)|^2 \int_{G/C} f(\mathrm{Ad}(x)H)d\bar{x}\, dH.$$

Applying 7.7.8 we have

$$(1) \qquad \int_{\mathfrak{g}_1} f(X)dX = \int_{\mathfrak{h}_0} \prod_{\alpha\in\Delta^+} |\alpha(H)|^2 \int_{K_*\times N} f(\mathrm{Ad}(k)\mathrm{Ad}(n)H)dk_* dn dH.$$

In the notation of 7.5.15, it is clear that $\mathfrak{h}_0' \subset \mathfrak{m}_1'$. Thus the map

$$\mathfrak{h}_0 \times N \xrightarrow{\psi} \mathfrak{h}_0 + \mathfrak{n}$$

given by $\psi(H, n) = \mathrm{Ad}(n)H$ is a bijection of $\mathfrak{h}_0' \times N$ onto $\mathfrak{h}_0' + \mathfrak{n}$. A computation gives

$$(2) \qquad \psi_{*(H,n)}(X, H_1) = \mathrm{Ad}(n)(H_1 - [H, X])$$

for n in N, H, H_1 in $\mathfrak{h}_0$ and X in $\mathfrak{n}$.

This implies that $\psi^*(dndH)_{(n,H)} = (-1)^{r/2} \prod_{\alpha\in\Sigma} \alpha(H)dndH$. Thus if φ is integrable on $\mathfrak{h}_0 + \mathfrak{n}$ then

$$(3) \qquad \int_{N\times\mathfrak{h}_0} \varphi(\mathrm{Ad}(n)H)dndH = \int_{\mathfrak{h}_0\times\mathfrak{n}} \prod_{\alpha\in\Sigma} |\alpha(H)|^{-1}\varphi(H+Z)dHdZ.$$

The lemma now follows from (1) and (3).

7.8.5 *Lemma* Let $\mathfrak{m}_1$ be as in 7.5.14. The Euclidean measures dY and dZ on $\mathfrak{m}_1$ and $\mathfrak{n}$, respectively, can be normalized so that if f is in $C_0(\mathfrak{g})$ and $f(\mathrm{Ad}(k)X) = f(X)$ for all k in K, X in $\mathfrak{g}$ then

$$\int_{\mathfrak{g}_1} f(X)dX = \int_{\mathfrak{m}_1\times\mathfrak{n}} |\det(\mathrm{ad}\, Y|_{\mathfrak{n}})| f(Y+Z)dYdZ.$$

PROOF 7.5.11 implies that $\Delta^+ - \Sigma$ is a system of positive roots for $\mathfrak{m}_1$ relative to $\mathfrak{h}_0$. Applying 7.8.3 to $\mathfrak{m}_1$ noting that $\bigcup_{m\in M_1} \mathrm{Ad}(m)\mathfrak{h}_0'$ is dense in $\mathfrak{m}_1$

we see that if φ is in $C_0(\mathfrak{m}_1)$ and $\varphi(\mathrm{Ad}(m)Y) = \varphi(Y)$ for all m in M_1, Y in $\mathfrak{m}_1$, then

$$\int_{\mathfrak{m}_1} \varphi(Y)dY = \int_{\mathfrak{h}_0'} \prod_{\alpha\in\Delta^+-\Sigma} |\alpha(H)|^2\varphi(H)dH, \tag{1}$$

with suitable normalizations of dH and dY. Now if H is in $\mathfrak{h}_0$ then $\prod_{\alpha\in\Sigma} |\alpha(H)| = |\det(\mathrm{ad}\, H|_{\mathfrak{n}})|$. We therefore see by (1) that

$$\int_{\mathfrak{n}_1\times\mathfrak{m}} |\det(\mathrm{ad}\ Y|_{\mathfrak{n}})| f(Y+Z)dYdZ \tag{2}$$
$$= \int_{\mathfrak{h}_0\times\mathfrak{n}} \prod_{\alpha\in\Delta^+-\Sigma} |\alpha(H)|^2 \prod_{\alpha\in\Sigma} |\alpha(H)| f(H+Z)\, dH\, dZ.$$

7.8.4 combined with (2) imply the lemma since $\int_{K_*} dk_* = 1$.

7.9 Semisimple Lie Groups with One Conjugacy Class of Cartan Subalgebra

7.9.1 *Definition* Let $\mathfrak{g}$ be a real semisimple Lie algebra. Let $\mathfrak{h}_0$ be a subalgebra of $\mathfrak{g}$. Let $\mathfrak{g}_C$ and $\mathfrak{h}$ be the complexifications of $\mathfrak{g}$ and $\mathfrak{h}_0$, resepctively. Then $\mathfrak{h}_0$ is called a Cartan subalgebra of $\mathfrak{g}$ if $\mathfrak{h}$ is a Cartan subalgebra of $\mathfrak{g}_C$.

7.9.2 Let G be a connected semisimple Lie group. Let $G = KAN$ be an Iwasawa decomposition of G (we use all the notation of 7.3). Let M be the centralizer of A in K. Let θ be the Cartan involution of $\mathfrak{g}$, the Lie algebra of G, corresponding to K. Let $\mathfrak{k}$, $\mathfrak{a}$, $\mathfrak{n}$, $\mathfrak{m}$ be, respectively, the Lie algebras of K, A, N, and M. Let $\bar{\mathfrak{n}} = \theta(\mathfrak{n})$.

7.9.3 *Lemma* Let $\mathfrak{h}^-$ be a maximal Abelian subalgebra of $\mathfrak{m}$. Set $\mathfrak{h}_0 = \mathfrak{h}^- + \mathfrak{a}$. Then $\mathfrak{h}_0$ is a Cartan subalgebra of $\mathfrak{g}$.

PROOF This is just a restatement of 7.5.8.

7.9.4 *Definition* Let $\mathfrak{h}_1$, $\mathfrak{h}_2$ be two Cartan subalgebras of $\mathfrak{g}$. Then $\mathfrak{h}_1$, $\mathfrak{h}_2$ are said to be conjugate, or belong to the same conjugacy class of Cartan subalgebra if there is g in G so that $\mathrm{Ad}(g)\mathfrak{h}_1 = \mathfrak{h}_2$. If $\mathfrak{g}$ has only one conjugacy class of Cartan subalgebra we say that G has one conjugacy class of Cartan subalgebra.

7.9.5 *Lemma* G has one conjugacy class of Cartan subalgebra if and only if $\mathfrak{h}^-$ is maximal Abelian in $\mathfrak{k}$.

PROOF Let $\tilde{\mathfrak{h}}^-$ be a maximal Abelian subalgebra of $\mathfrak{k}$. Let $\tilde{\mathfrak{a}} \subset \mathfrak{p}$ ($\mathfrak{p} = \{x \text{ in } \mathfrak{g} | \theta x = -x\}$) be maximal subject to

(1) $[\tilde{\mathfrak{a}}, \tilde{\mathfrak{a}}] = 0$

(2) $[\tilde{\mathfrak{a}}, \tilde{\mathfrak{h}}^-] = 0$.

Set $\tilde{\mathfrak{h}}_0 = \tilde{\mathfrak{h}}^- + \tilde{\mathfrak{a}}$. We assert that $\tilde{\mathfrak{h}}_0$ is a Cartan subalgebra of $\mathfrak{g}$. Let $\tilde{\mathfrak{h}}$ be the complexification of $\tilde{\mathfrak{h}}_0$ in $\mathfrak{g}_C$. Then arguing in exactly the same way as in 7.5.8 (reversing the roles of $\mathfrak{k}$ and $\mathfrak{p}$) we find that $\tilde{\mathfrak{h}}$ is indeed a Cartan subalgebra of $\mathfrak{g}_C$.

If G has one conjugacy class of Cartan subalgebra then there is g in G so that $\text{Ad}(g)\tilde{\mathfrak{h}}_0 = \mathfrak{h}_0$. Now B, the Killing form of $\mathfrak{g}$, is negative definite on $\mathfrak{k}$ and positive definite on $\mathfrak{p}$. Thus the index of $B|_{\tilde{\mathfrak{h}}_0}$ is dim $\tilde{\mathfrak{h}}^-$ (the index of a symmetric bilinear form on a real vector space is the maximal dimension of a subspace on which the form is negative definite). On the other hand B is G-invariant, thus the index of $B|_{\mathfrak{h}_0}$ is the same as the index of $B|_{\tilde{\mathfrak{h}}_0}$. Hence dim $\mathfrak{h}^-$ = dim $\tilde{\mathfrak{h}}^-$. This clearly implies that $\mathfrak{h}^-$ is maximal Abelian in $\mathfrak{k}$.

Suppose that $\mathfrak{h}^-$ is maximal Abelian in $\mathfrak{k}$. Let $\tilde{\mathfrak{h}}_0$ be a Cartan subalgebra of $\mathfrak{g}$. Let $\tilde{\mathfrak{h}}$ be the complexification of $\tilde{\mathfrak{h}}_0$ in $\mathfrak{g}_C$. Then $\tilde{\mathfrak{h}}$ is a Cartan subalgebra of $\mathfrak{g}_C$. Let Δ be the root system of $\mathfrak{g}_C$ relative to $\tilde{\mathfrak{h}}$. Let $\tilde{\mathfrak{a}} = \{H \text{ in } \tilde{\mathfrak{h}}_0 | \alpha(H) \text{ is in } R$ for all α in $\Delta\}$. Then $\tilde{\mathfrak{a}}$ is a subspace of $\tilde{\mathfrak{h}}_0$. Let $\tilde{\mathfrak{h}}^- = \{H \text{ in } \tilde{\mathfrak{h}}_0 | \alpha(H)$ is pure imaginary for all α in $\Delta\}$. We assert that $\mathfrak{h}_0 = \tilde{\mathfrak{h}}^- \oplus \tilde{\mathfrak{a}}$. Indeed if σ is conjugation relative to $\mathfrak{g}$ in $\mathfrak{g}_C$ then $\sigma\tilde{\mathfrak{h}} \subset \tilde{\mathfrak{h}}$. Thus if α is in Δ then $\alpha \circ \sigma$ is in Δ. Thus if $\tilde{\mathfrak{h}}_R = \{H \text{ in } \tilde{\mathfrak{h}} | \alpha(H) \text{ is in } R \text{ for all } \alpha \text{ in } \Delta\}$ then $\sigma\tilde{\mathfrak{h}}_R = \tilde{\mathfrak{h}}_R$. Thus $\tilde{\mathfrak{h}}_R = \tilde{\mathfrak{h}}_R^- \oplus \tilde{\mathfrak{h}}_R^+$ where $\tilde{\mathfrak{h}}_R^\pm = \{H \text{ in } \mathfrak{h}_R | \sigma H = \pm H\}$. Clearly $\tilde{\mathfrak{h}}_R^+ = \tilde{\mathfrak{a}}$ and $\tilde{\mathfrak{h}}_R^- = (-1)^{\frac{1}{2}}\tilde{\mathfrak{h}}^-$.

3.7.5. says that there is a compact real form $\mathfrak{g}_u$ of $\mathfrak{g}_C$ so that $(-1)^{\frac{1}{2}}\tilde{\mathfrak{h}}_R$ is a maximal Abelian subalgebra of $\mathfrak{g}_u$. Let τ be conjugation in $\mathfrak{g}_C$ relative to $\mathfrak{g}_u$. 3.7.6 says that there is a one-parameter group of automorphism $A(t)$ of $\mathfrak{g}_C$ so that $A(1)\tau A(1)^{-1}$ and σ commute; furthermore $A(t)$ may be chosen so that $A(t)\mathfrak{h} \subset \mathfrak{h}$ (this is implicit in the proof of 3.7.6). We may thus assume that τ and σ commute. But then $\tau\mathfrak{g} = \mathfrak{g}$ and $\tau|_{\mathfrak{g}} = \theta'$ defines a Cartan involution of $\mathfrak{g}$. 3.7.9(3) says that we may assume $\theta' = \theta$ (up to conjugation of $\tilde{\mathfrak{h}}_0$). We therefore see that we may assume $\theta\tilde{\mathfrak{h}}_0 = \tilde{\mathfrak{h}}_0$. Hence $\tilde{\mathfrak{a}} \subset \mathfrak{p}$. By conjugating by an element of K we may assume that $\tilde{\mathfrak{a}} \subset \mathfrak{a}$ (here we use 7.5.5). But $\tilde{\mathfrak{h}}^-$ is an Abelian subalgebra of $\mathfrak{k}$ and dim $\tilde{\mathfrak{h}}_0$ = dim $\mathfrak{h}_0$. Thus since dim $\tilde{\mathfrak{a}} \leqslant$ dim $\mathfrak{a}$ we see that dim $\tilde{\mathfrak{h}}^- \geqq$ dim $\mathfrak{h}^-$. But we have assumed that dim $\mathfrak{h}_0^-$ is maximal. Hence dim $\tilde{\mathfrak{h}}^-$ = dim $\mathfrak{h}^-$. Thus $\mathfrak{a} = \tilde{\mathfrak{a}}$. We therefore see that $\tilde{\mathfrak{h}}^-$ is contained in $\mathfrak{m}$. Now Ad(M) is compact. Hence there is m in M so that $\text{Ad}(m)\tilde{\mathfrak{h}}^- = \mathfrak{h}^-$. Thus $\tilde{\mathfrak{h}}_0$ and $\mathfrak{h}_0$ are conjugate.

7.9.6 *Examples* 1. If G is compact then G has one conjugacy class of Cartan subalgebra.

2. If G is complex semisimple then G has one conjugacy class of Cartan subalgebra.

3. Let on R^{n+1}, $Q(x) = x_1^2 + \cdots + x_n^2 - x_{n+1}^2$ ($x = (x_1, \ldots, x_{n+1})$). Let $SO(n, 1)$ be identity component of the group $G = \{g \text{ in } SL(n+1, R) | Q(g \cdot x) = Q(x)\}$.

It is an exercise (see 7.13.12) to show that if n is even then $SO(n, 1)$ has two conjugacy classes of Cartan subalgebra. If n is odd then $SO(n, 1)$ has one conjugacy class of Cartan subalgebra.

7.9.7 Retaining the notation of 7.9.2 and 7.9.3, let Σ be as in 7.5.11(3).

7.9.8 *Lemma* Suppose that G has one conjugacy class of Cartan subalgebra. If H is in $\mathfrak{h}_0$ then $\prod_{\alpha \in \Sigma} \alpha(H)$ is real and nonnegative.

PROOF Let $\Sigma_1 = \{\alpha \text{ in } \Sigma | \sigma(\mathfrak{g}_C)_\alpha = (\mathfrak{g}_C)_\alpha\}$. We assert that $\Sigma_1 = \phi$. Indeed if $\Sigma_1 \neq \phi$, let α be in Σ_1. Then there is E_α in $(\mathfrak{g}_C)_\alpha$ so that $E_\alpha \neq 0$ and $RE_\alpha = (\mathfrak{g}_C)_\alpha \cap \mathfrak{g}$. If H is in $\mathfrak{h}^-$ then

$$\sigma([H, E_\alpha]) = \overline{\alpha(H)}E = -\alpha(H)E_\alpha \qquad (\alpha(\mathfrak{h}^-) \subset (-1)^{\frac{1}{2}}R).$$

But $[H, E_\alpha]$ is in $\mathfrak{g}$, thus $\sigma([H, E_\alpha]) = [H, E_\alpha]$. Hence $\alpha(\mathfrak{h}^-) = 0$. Let θ be as in 7.8.2. Then $\theta|_{\mathfrak{h}^-} = I$. Thus $\theta\alpha(\mathfrak{h}^-) = 0$. Let $Z = E_\alpha + \theta E_\alpha$. Then $Z \neq 0$ since $\theta\alpha < 0$, $\alpha > 0$. Furthermore Z is in $\mathfrak{k}$ and $[\mathfrak{h}^-, Z] = 0$. Now 7.8.6 implies $\mathfrak{h}^-$ is maximal Abelian. This contradiction implies that $\Sigma_1 = \phi$ as asserted.

We know that if α is in Σ, $\sigma(\mathfrak{g}_C)_\alpha \neq (\mathfrak{g}_C)_\alpha$. Fix for each α in Σ, E_α in $(\mathfrak{g}_C)_\alpha$. Set $X_\alpha = E_\alpha + \sigma E_\alpha$ and $W_\alpha = (-1)^{\frac{1}{2}}(E_\alpha - \sigma E_\alpha)$. Let $\Sigma' \subset \Sigma$ be a subset so that $\Sigma' \cup \sigma\Sigma' = \Sigma$, $\Sigma' \cap \sigma\Sigma' = \phi$. Then $\mathfrak{n} = \sum_{\alpha \in \Sigma} (RX_\alpha + RW_\alpha)$. If H is in $\mathfrak{h}_0$ then $H = H_+ + H_-$ with H_+ in $\mathfrak{a}$, H_- in $\mathfrak{h}^-$. A computation shows that relative to the basis X_α, W_α of $V_\alpha = RX_\alpha + RW_\alpha$,

$$\operatorname{ad} H|_{V_\alpha} = \begin{bmatrix} \alpha(H_+) & -\sqrt{-1}\alpha(H_-) \\ \sqrt{-1}\alpha(H_-) & \alpha(H_+) \end{bmatrix}.$$

Thus $\det(\operatorname{ad} H|_{V_\alpha}) = \alpha(H_+)^2 + (\sqrt{-1}\alpha(H_-))^2 \geqq 0$ (since $\sqrt{-1}\alpha(H_-)$ is *real*). Now $\prod_{\alpha \in \Sigma} \alpha(H) = \prod_{\alpha \in \Sigma'} \det(\operatorname{ad} H|_{V_\alpha}) \geqq 0$. This proves the lemma.

7.9.9 *Corollary* Suppose that G has one conjugacy class of Cartan subalgebra. Let $\mathfrak{m}_1 = \mathfrak{m} \oplus \mathfrak{a}$ ($\mathfrak{m}$, $\mathfrak{a}$ as in 7.9.2). If X is in $\mathfrak{m}_1$ then $\det(\operatorname{ad} X|_{\mathfrak{n}}) \geqq 0$ ($\mathfrak{n}$ is as in 7.9.2).

PROOF If X is in $\mathfrak{m}_1$ then $X = X_1 + H$ with X_1 in $\mathfrak{m}$ and H in $\mathfrak{a}$. Let $\tilde{\mathfrak{h}}^-$ be a maximal Abelian subalgebra of $\mathfrak{m}_1$ containing X_1. Then 7.9.9 follows from 7.9.8 replacing $\mathfrak{h}_0$ with $\tilde{\mathfrak{h}}^- + \mathfrak{a}$.

7.9.10 *Proposition* If G has one conjugacy class of Cartan subalgebra then K is compact and semisimple and M is connected. Furthermore G is Lie isomorphic with a Lie subgroup of $GL(n, C)$ for some n.

PROOF Let $\mathfrak{h}_0$, $\mathfrak{h}$, $\mathfrak{g}_C$, and Δ be as above. Let Δ^+, Σ, and Δ_1 be as in 7.5.11. 7.5.11 says that the complexification of $\mathfrak{m}$, $\mathfrak{m}_C$ is $\sum_{\alpha\in\Delta_1}(\mathfrak{g}_C)_\alpha + \mathfrak{h}^-$. Let for α in Δ, E_α in $(\mathfrak{g}_C)_\alpha$ be chosen, $E_\alpha \neq 0$. Then

$$\mathfrak{k}_C = \mathfrak{h}_C^- + \sum_{\alpha\in\Delta_1}(\mathfrak{g}_C)_\alpha + \sum_{\alpha\in\Sigma} C(E_\alpha + \theta E_\alpha) + \sum_{\alpha\in\Sigma} C(E_{-\alpha} + \theta E_{-\alpha}).$$

If α is in Σ then $\theta\alpha = -\sigma\alpha < 0$. Thus $E_\alpha \neq \sigma E_\alpha$ and we actually have

(1) $$\mathfrak{k}_C = \mathfrak{h}_C^- + \sum_{\alpha\in\Delta_1}(\mathfrak{g}_C)_\alpha + \sum_{\alpha\in\Sigma} C(E_\alpha + \theta E_\alpha).$$

This implies that the root system of $\mathfrak{k}_C$ is $\{\alpha|_{\mathfrak{h}^-} | \alpha \text{ in } \Delta\}$.

The Killing form B of $\mathfrak{g}$ is negative definite on $\mathfrak{k}$. Thus $\mathfrak{k} = \mathfrak{z} \oplus \mathfrak{k}_1$ with $\mathfrak{z}$ the center of $\mathfrak{k}$ and $\mathfrak{k}_1 = [\mathfrak{k}, \mathfrak{k}]$ compact. $\mathfrak{z} \subset \mathfrak{h}^-$ since $\mathfrak{h}^-$ is maximal Abelian in $\mathfrak{k}$ (see 7.9.5). If Z is in $\mathfrak{z}$ then every root of $\mathfrak{k}_{1_C}$ vanishes on Z. But then (1) implies that $\alpha(Z) = 0$ for all α in Δ. Hence $Z = 0$. This implies that $\mathfrak{z} = (0)$. Hence $\mathfrak{k}$ is compact. 3.6.6 now implies that K is compact.

Let T be the maximal torus of K corresponding to $\mathfrak{h}^-$. Then $T \subset M$. Let C be the centralizer of $\mathfrak{h}_0$ in G. 7.6.2 says that $C = A_- A$ where A_- is the centralizer of $\mathfrak{h}_0$ in K. But then A_- centralizes T. On the other hand, T is a maximal torus of K, thus 3.10.8 implies $A_- = T$.

Let m be in M. Then $\mathrm{Ad}(m)$ fixes a regular element X of $\mathfrak{m}$ (see 3.9.2). Let $\tilde{\mathfrak{h}}^-$ be the centralizer of X in $\mathfrak{m}$. Then $\tilde{\mathfrak{h}}^-$ is maximal Abelian in $\mathfrak{m}$ and thus in $\mathfrak{k}$. Let M_0 be the identity component of M. Let m_0 in M_0 be so that $\mathrm{Ad}(m_0)\tilde{\mathfrak{h}}^- = \mathfrak{h}^-$. Then $\mathrm{Ad}(m_0 m m_0^{-1})\,\mathrm{Ad}(m_0)X = \mathrm{Ad}(m_0)X$. Since $\mathrm{Ad}(m_0 m m_0^{-1})$ is in $\mathrm{Ad}(K)$ the proof (1) of 3.9.4 implies that $m_0 m m_0^{-1}$ is in T. Thus m is in $m_0^{-1} T m_0 \subset M_0$. Hence $M = M_0$.

To prove the last statement it is enough to show that if G_C is the simply connected Lie group with Lie algebra $\mathfrak{g}_C$ then the connected subgroup G of G_C with Lie algebra $\mathfrak{g}$ is simply connected. Now, $G = K \times A \times N$ topologically. It is therefore enough to show that K is simply connected. To do this it is enough to show that $\ker(\exp|_{\mathfrak{h}^-})$ is the lattice generated by the elements $(2\pi(-1)^{\frac{1}{2}})(2/\beta(H_\beta))H_\beta$ for β a root of $(\mathfrak{k}, \mathfrak{h}^-)$ (see 4.6.7).

If $\mathfrak{g}_u = \mathfrak{k} + (-1)^{\frac{1}{2}}\mathfrak{p}$ and if G_u is the connected subgroup of G_C corresponding to $\mathfrak{g}_u$ then G_u is simply connected (see 3.6.6). The unit lattice of G_u (that is, $\ker(\exp: \mathfrak{h}^- + (-1)^{\frac{1}{2}}\mathfrak{a} \to G_u)$) is the lattice generated by the elements $2\pi(-1)^{\frac{1}{2}}(2/\alpha(H_\alpha))H_\alpha$, the α running over the elements of Δ (see 4.6.7). This implies that the unit lattice of K is contained in the lattice generated by the elements $2\pi(-1)^{\frac{1}{2}}(2/\alpha(H_\alpha))(H_\alpha + \theta H_\alpha)$. Now the roots of $\mathfrak{k}$ are the elements $\beta = \frac{1}{2}(\alpha + \theta\alpha)$, α in Δ. We may take $H_\beta = \frac{1}{2}(H_\alpha + \theta H_\alpha)$. Thus $\beta(H_\beta) = \frac{1}{2}(\alpha(H_\alpha) + \alpha(\theta H_\alpha))$. This implies that

$$(2/\beta(H_\beta))H_\beta = (2/(\alpha(H_\alpha) + \alpha(\theta H_\alpha))(H_\alpha + \theta H_\alpha).$$

If $\theta H_\alpha = H_\alpha$(i.e., α is an element of $\Delta_1\}$ then $(2/\beta(H_\beta))H_\beta = (2/\alpha(H_\alpha))H_\alpha$. If $\theta H_\alpha \neq H_\alpha$ then we assert that $\alpha(\theta H_\alpha) = 0$. Indeed, if $\alpha(\theta H_\alpha) \neq 0$ then $\alpha + \theta\alpha$ or $\alpha - \theta\alpha$ is in Δ. If $\alpha + \theta\alpha$ is in Δ then 2β is a root of K which is not possible. If $\alpha - \theta\alpha$ is in Δ then $\sigma\mathfrak{g}_{\alpha-\theta\alpha} \subset \mathfrak{g}_{\alpha-\theta\alpha}$ since $\sigma^*\alpha = -\theta\alpha$. Arguing as in the beginning of the proof of 7.9.8 we are led to a contradiction. Hence $\alpha(\theta H_\alpha) = 0$. This implies that $(2/\beta(H_\beta))H_\beta = (2/\alpha(H_\alpha))(H_\alpha + \theta H_\alpha)$. This combined with 4.6.7 completes the proof.

7.10 Differential Operators on a Reductive Lie Algebra

7.10.1 *Definition* Let $\mathfrak{g}$ be a Lie algebra over R. Then $\mathfrak{g}$ is said to be reductive if $\mathfrak{g} = \mathfrak{z} \oplus \mathfrak{g}_1$ with $\mathfrak{z}$ the center of $\mathfrak{g}$ and $\mathfrak{g}_1 = [\mathfrak{g}, \mathfrak{g}]$ is semisimple.

7.10.2 Let $\mathfrak{g}$ be a reductive Lie algebra and let $\mathfrak{g}_C$ be its complexification. Let $\mathfrak{h}_1$ be a Cartan subalgebra of $\mathfrak{g}_1$ and set $\mathfrak{h}_0 = \mathfrak{h}_1 \oplus \mathfrak{z}$. Let $\mathfrak{h}$ be the complexification of $\mathfrak{h}_0$ in $\mathfrak{g}_C$. Then $\mathfrak{g}_C = \mathfrak{h} + \sum_{\alpha\in\Sigma} (\mathfrak{g}_C)_\alpha$ where Δ is the root system of $(\mathfrak{g}_1)_C$ relative to $(\mathfrak{h}_1)_C$.

7.10.3 Let $S(\mathfrak{g}_C)$ be the symmetric algebra of $\mathfrak{g}_C$. We define a map $\lambda: S(\mathfrak{g}_C) \to U(\mathfrak{g}_C)$, the universal enveloping algebra of $\mathfrak{g}_C$, as follows: Let for $X_1, \ldots, X_k$ in $\mathfrak{g}_C$, $\lambda_k(X_1, \ldots, X_k) = (1/k!)\sum_{\sigma\in S_k} X_{\sigma 1} \cdots X_{\sigma k}$, S_k is the symmetric group in k letters. Then λ_k is a symmetric mapping of $\times^k \mathfrak{g}_C \to U(\mathfrak{g}_C)$. Hence λ_k extends to a linear map of $S^k(\mathfrak{g}_C)$ into $U(\mathfrak{g}_C)$. We take $\lambda = \sum_{k=0}^{\infty}\lambda_k$.

7.10.4 *Lemma* λ is a linear isomorphism of $S(\mathfrak{g}_C)$ onto $U(\mathfrak{g}_C)$.

PROOF Let $S_k(\mathfrak{g}_C) = \sum_{j\leq k} S^k(\mathfrak{g}_C)$. Let $U^k(\mathfrak{g}_C)$ be the subspace of $U(\mathfrak{g}_C)$

spanned by products of k or less elements of $\mathfrak{g}_C$. Clearly $\lambda: S_k(\mathfrak{g}_C) \to U^k(\mathfrak{g}_C)$. Let $X_1, \ldots, X_n$ be a basis of $\mathfrak{g}_C$. Let $M = (m_1, \ldots, m_n)$, $m_i \geqq 0$, m_i an integer. Set $(X)^M = X_1^{m_1} \cdot X_2^{m_2} \cdot \cdots \cdot X_n^{m_n}$ where $X \cdot Y$ is the product in $S(\mathfrak{g})$. Set $X^M = X_1^{m_1} X_2^{m_2} \cdots X_n^{m_n}$. Here we use the product XY in $U(\mathfrak{g}_C)$. The proof of 4.2.4 implies that $\lambda((X)^M) \equiv X^M \bmod U^{|M|-1}(\mathfrak{g}_C)$ where $|M| = m_1 + \cdots + m_n$. Thus 4.2.4 implies that λ is surjective. We prove that λ is injective by induction on k. Clearly $\lambda: S_0(\mathfrak{g}_C) \to U^0(\mathfrak{g}_C)$ is injective since $\lambda(1) = 1$. Suppose that we have shown that $\lambda: S_k(\mathfrak{g}_C) \to U^k(\mathfrak{g}_C)$ is injective. Let p be an element of $S_{k+1}(\mathfrak{g}_C)$. Then $p = p_1 + p_2$, p_1 in $S^{k+1}(\mathfrak{g}_C)$ and p_2 in $S_k(\mathfrak{g}_C)$. If $p_1 = 0$ then $\lambda(p_2) = 0$ implies $p_2 = 0$ by the inductive hypothesis. If $p_1 \neq 0$ then $p_1 = \sum_{|M|=k+1} a_M (X)^M$. By the above

$$\lambda(p_1) \equiv \textstyle\sum_{|M|=k+1} a_M X^M \bmod U^k(\mathfrak{g}_C).$$

Now 4.2.9 implies that if $\lambda(p_1) = 0$ then $p_1 = 0$. This proves the lemma.

7.10.5 Let Δ^+ be a choice of positive roots for $(\mathfrak{g}_C, \mathfrak{h})$. Set $\mathfrak{n}^+ = \sum_{\alpha\in\Delta^+} (\mathfrak{g}_C)_\alpha$, $\mathfrak{n}^- = \sum_{\alpha\in\Delta^+} (\mathfrak{g}_C)_{-\alpha}$. Then $\mathfrak{g}_C = \mathfrak{n}^+ \oplus \mathfrak{n}^- \oplus \mathfrak{h}$, a direct sum of Lie algebras.

7.10.6 Since $\mathfrak{h}$ is Abelian we may identify $U(\mathfrak{h})$ with $S(\mathfrak{h})$. Set $\mathscr{V} = \lambda(S(\mathfrak{n}^+ \oplus \mathfrak{n}^-)) \subset U(\mathfrak{g}_C)$.

7.10.7 *Lemma* The map $\mathscr{V} \otimes S(\mathfrak{h}) \to U(\mathfrak{g}_C)$ given by $v \otimes h \mapsto vh$ is a linear bijection.

PROOF Let $\mathscr{V}^k = \lambda(S^k(\mathfrak{n}^+ \oplus \mathfrak{n}^-))$. Let $Z_1, \ldots, Z_r$ be a basis of $\mathfrak{n}^+ \oplus \mathfrak{n}^-$ and let $H_1, \ldots, H_l$ be a basis of $\mathfrak{h}$. Then $Z_1, \ldots, Z_r, H_1, \ldots, H_l$ is a basis of $\mathfrak{g}_C$. Let for $P = (p_1, \ldots, p_r)$, $Q = (q_1, \ldots, q_l)$, $Z^P H^Q = Z_1^{p_1} \cdots Z_r^{p_r} H_1^{q_1} \cdots H_l^{p_l}$. Set $|P| = p_1 + \cdots + p_r$, $|Q| = q_1 + \cdots + q_l$. Then 4.2.9 says that the elements $Z^P H^Q$ with $|P| + |Q| \leqq k$ form a basis of $U^k(\mathfrak{g}_C)$. This clearly implies that $U^k(\mathfrak{g}_C) = \sum_{m+p\leqq k} \mathscr{V}^m \cdot S^p(\mathfrak{h})$. But now

$$\dim \sum_{m+p<k} \mathscr{V}^m \cdot S^p(\mathfrak{h}) \leqq \sum_{m+p<k} \dim S^m(\mathfrak{n}^+ + \mathfrak{n}^-) \dim S^p(\mathfrak{h}) \leqq \dim S_k(\mathfrak{g}_C).$$

But by the above

$$\dim \sum_{m+p<k} \mathscr{V}^m \cdot S^p(\mathfrak{h}) \geqq \dim U^k(\mathfrak{g}_C) = \dim S_k(\mathfrak{g}_C).$$

Thus all inequalities are equalities and the lemma is proved.

7.10.8 If p is in $S(\mathfrak{g}_C)$ define $L_p q = pq$ for q in $S(\mathfrak{g}_C)$. If X is in $\mathfrak{g}$ let d_X be the derivation of $S(\mathfrak{g}_C)$ extending $d_X Y = [X, Y]$ for Y in $\mathfrak{g}_C$.

7.10.9 *Lemma* Let Y in $\mathfrak{g}$ be fixed. Then there is a unique linear map $\Gamma_Y: U((\mathfrak{g}_1)_C) \otimes S(\mathfrak{g}_C) \to S(\mathfrak{g}_C)$, so that

(1) $\Gamma_Y(1 \otimes p) = p$ for p in $S(\mathfrak{g}_C)$,

(2) $\Gamma_Y(X_1 \cdots X_r \otimes p) = (L_{[X_1,Y]} + d_{X_1}) \cdots (L_{[X_r,Y]} + d_{X_r}) \cdot p$ for $X_1, \ldots, X_r$ in $\mathfrak{g}$, p in $S(\mathfrak{g}_C)$.

PROOF Let $Y_1, \ldots, Y_n$ be a basis of $\mathfrak{g}$. Then relative to this basis we have coordinates $t_1, \ldots, t_n$ on $\mathfrak{g}$. That is, $Y = \Sigma\, t_i(Y) Y_i$, Y in $\mathfrak{g}$. Let $D_0(\mathfrak{g})$ be the algebra of all differential operators of the form $\Sigma\, a_I D^I$ where a_I is a complex number, $I = (i_1, \ldots, i_n)$ is an n-tuple of nonnegative integers, and

$$D^I = \frac{\partial^{|I|}}{\partial t_{t_1}^{i_1} \cdots \partial t_{t_n}^{i_n}}.$$

If X is in $\mathfrak{g}_C$ set $\partial(X) = \Sigma\, a_i(\partial/\partial t_i)$ with $X = \Sigma\, a_i Y_i$. Then it is clear that if X is in $\mathfrak{g}$ and f is in $C^\infty(\mathfrak{g})$ then $(\partial(X)f)(Y) = (d/dt)f(Y + tX)|_{t=0}$. $\partial: \mathfrak{g}_C \to D_0(\mathfrak{g})$ is a linear injection and thus extends to a homomorphism $\partial: S(\mathfrak{g}_C) \to D_0(\mathfrak{g})$. Clearly $\partial((Y)^M) = D^M$ (here we use the notation of the proof of 7.10.4). Thus ∂ is an algebra isomorphism of $S(\mathfrak{g}_C)$ onto $D_0(\mathfrak{g})$.

Let G be the identity component of the automorphism group of $\mathfrak{g}_1$. If g is in G let $g(Z + X) = Z + g \cdot X$ for Z in $\mathfrak{z}$, X in $\mathfrak{g}_1$. Then G acts as a group of automorphisms of $\mathfrak{g}$.

If f is in $C^\infty(\mathfrak{g})$, set $f(x\colon X) = f(x \cdot X)$ for x in G and X in $\mathfrak{g}$. Then $(x, X) \mapsto f(x\colon X)$ is a C^∞ function on $G \times \mathfrak{g}$.

We identify $U((\mathfrak{g}_1)_C)$ with the space of all left invariant differential operators (with complex coefficients) on G (see 4.2.7). If Z is in $U((\mathfrak{g}_1)_C)$, f is in $C^\infty(\mathfrak{g})$, set $f(x; Z\colon X) = (Z \cdot f)_X(Y)$ where $f_X(x) = f(x\colon X)$. If p is in $S(\mathfrak{g}_C)$ set $f(x\colon X; \partial(p)) = (\partial(p) f_x)(X)$ where $f_x(X) = f(x\colon X)$. Hence $f(x; Z\colon X; \partial(p))$ makes sense for x in G, Z in $U((\mathfrak{g}_1)_C)$, X in $\mathfrak{g}$, p in $S(\mathfrak{g}_C)$.

(1) $f(x\colon Y; \partial(p)) = f(xY; \partial(x \cdot p))$ where $x \cdot p$ denotes the extension of the action of G on $\mathfrak{g}_C$ to $S(\mathfrak{g}_C)$.

To prove (1) it is enough to consider $p = X^j$ for some X in $\mathfrak{g}$ (indeed, the X^j, X in $\mathfrak{g}$, $j \geqq 0$, j an integer, span $S(\mathfrak{g}_C)$). Then

$$\begin{aligned} f(x\colon Y; \partial(p)) &= \frac{d^j}{dt^j} f(x \cdot (Y + tX))|_{t=0} \\ &= \frac{d^j}{dt^j} f(x \cdot Y + tx \cdot Y)|_{t=0} = (\partial((x \cdot X)^j) f)(x \cdot Y) \\ &= f(x \cdot Y; (x \cdot p)) \end{aligned}$$

as was asserted.

(2) $f(x; X_1 \cdots X_r : Y; \partial(p)) = f(x : Y; \partial(q))$
with $q = (L_{[X_1,Y]} + d_{X_1}) \cdots (L_{[X_r,Y]} + d_{X_r})p$ for $X_1, \ldots, X_r$ in $(\mathfrak{g}_1)_C$, Y in $\mathfrak{g}$, p in $S(\mathfrak{g}_C)$, x in G.

We prove (2) by induction on r. If $r = 1$ and X_1 is in $\mathfrak{g}_1$ then

$$\begin{aligned} f(x; X_1 : Y; \partial(p)) &= \frac{d}{dt} f(x \exp tX_1 : Y; \partial(p))\Big|_{t=0} \\ &= \frac{d}{dt} f(x \exp tX_1 \cdot Y; (x \exp tX_1 \cdot p))\Big|_{t=0} \quad \text{(by (1))} \\ &= \frac{d}{dt} f(x \exp tX_1 \cdot Y; (x \cdot p))\Big|_{t=0} \\ &\quad + \frac{d}{dt} f(x \cdot Y : (x \exp tX_1 \cdot p))|_{t=0} \end{aligned}$$

(here we use the rule for differentiation of bilinear functions). Now $\exp tX_1 \cdot Y$; $Y + t[X_1, Y] + O(t^2)$. Thus

$$\begin{aligned} f(x; X_1 : Y : \partial(p)) &= f(x : Y; \partial(L_{[X_1,Y]}p)) + f(x : Y; \partial(d_{X_1}p)) \\ &= f(x : Y; \partial((L_{[X_1,Y]} + d_{X_1})p)). \end{aligned}$$

This proves (2) for $r = 1$. Suppose the result is true for $1 \leqq j \leqq r$. Set $q' = (L_{[X_2,Y]} + d_{X_2}) \cdots (L_{[Xr,Y]} + d_{X_r})p$. Then the inductive hypothesis implies that $f(x; X_2 \cdots X_r : Y; \partial(p)) = f(x : Y; \partial(q'))$. Now

$$\begin{aligned} f(x; X_1 \cdots X_r : Y; \partial(p)) &= f(x; X_1 : Y; \partial(q')) \\ &= f(x : Y; \partial((L_{[X_1,Y]} + d_{X_1})q')) = f(x : Y; \partial(q)) \end{aligned}$$

as asserted. (2) is now completely proved.

Let $\tilde{\Gamma}_Y : U((\mathfrak{g}_1)_C) \otimes S(\mathfrak{g}_C) \to D_0(\mathfrak{g}_C)$ be defined by $(\tilde{\Gamma}_Y(Z \otimes p)f(Y) = f(e; Z : Y; \partial(p))$. Then there is a unique element $\Gamma_Y(Z \otimes p)$ in $S(\mathfrak{g}_C)$ so that $\partial(\tilde{\Gamma}_Y(Z \otimes p)) = \Gamma_Y(Z \otimes p)$.

(2) implies that Γ_Y has the desired properties.

7.10.10 *Lemma* Let $\pi = \prod_{\alpha \in \Delta^+} \alpha$. If H is in $\mathfrak{h}_0$ and $\pi(H) \neq 0$ then $\Gamma_H : \mathscr{V} \otimes S(\mathfrak{h}) \to S(\mathfrak{g}_C)$ is bijective.

PROOF Let $S^k(\mathfrak{g}_C)$, $S^k(\mathfrak{n}^+ + \mathfrak{n}^-)$, and $S^k(\mathfrak{h})$ denote the homogeneous elements of degree k in the respective symmetric algebras. Let $\mathscr{V}^k = \lambda(S^k(\mathfrak{n}^+ + \mathfrak{n}^-))$ (see 7.10.6 and the proof of 7.10.7). Set (as usual) $S_k(\mathfrak{g}_C) = \sum_{j \leqq k} S^j(\mathfrak{g}_C)$.

(1) $$\sum_{d+e<r} \Gamma_H(\mathscr{V}^d \otimes S^e(\mathfrak{h})) = S_r(\mathfrak{g}_C).$$

We prove (1) by induction on r. If $r = 0$ there is nothing to prove. Clearly if we show

$$(1)' \qquad \sum_{d+e<r} \Gamma_H(\mathscr{V}^d \otimes S^e(\mathfrak{g})) + S_{r-1}(\mathfrak{g}_C) \supset S^r(\mathfrak{g}_C)$$

then (1) will follow by induction.

$\mathfrak{g}_C = \mathfrak{n}^+ + \mathfrak{n}^- + \mathfrak{h}$. Thus $S^r(\mathfrak{g}_C)$ is spanned by elements of the form $Z_1 \cdots Z_d H_1 \cdots H_e$, $d + e = r$, Z_i in $\mathfrak{n}^+ + \mathfrak{n}^-$, $i = 1, \ldots, d$, H_i in $\mathfrak{h}$, $i = 1, \ldots, e$. Furthermore $|\det(\text{ad } H|_{\mathfrak{n}^+ + \mathfrak{n}^-})| = |\pi(H)|^2 \neq 0$. There are therefore elements $Y_1, \ldots, Y_d$ in $\mathfrak{n}^+ + \mathfrak{n}^-$ so that $\text{ad } H \cdot Y_i = -Z_i$, $1 \leqq i \leqq d$. Now

$$\begin{aligned}\Gamma_H(Y_1 \cdots Y_d \otimes H_1 \cdots H_e) &= (L_{Z_1} + d_{Y_1}) \cdots (L_{Z_d} + d_{Y_r})(H_1 \cdots H_e)\\ &\equiv Z_1 \cdots Z_d H_1 \cdots H_e \text{ mod } S_{r-1}(\mathfrak{g}_C).\end{aligned}$$

Also $Y_1 \cdot Y_2 \cdots Y_d - \lambda(Y_1 \cdot Y_2 \cdots Y_d)$ is in $\lambda(S_{d-1}(\mathfrak{g}_C))$ and

$$\Gamma_H(\lambda(S_{d-1}(\mathfrak{g}_C) \otimes H_1 \cdots H_e) \subset S_{d+e-1}(\mathfrak{g}_C) = S_{r-1}(\mathfrak{g}_C).$$

Thus (1)′ follows.

Now $\dim \lambda(S^d(\mathfrak{n}^+ + \mathfrak{n}^-)) = \dim S^d(\mathfrak{n}^+ + \mathfrak{n}^-)$ and

$$S_r(\mathfrak{g}_C) = \sum_{d+e \leqslant r} S^d(\mathfrak{n}^+ + \mathfrak{n}^-) \otimes S^e(\mathfrak{h}).$$

Thus $\dim \sum_{d+e \leqq r} \lambda(S^d(\mathfrak{n}^+ + \mathfrak{n}^-)) \otimes S^e(\mathfrak{h}) = \dim S_r(\mathfrak{g}_C)$. Thus (1) implies that Γ_H is bijective. The lemma is thus completely proved.

7.10.11 *Definition* Let V_1, V_2 be vector spaces over R and C, respectively, with $\dim V_1 < \infty$. Let $f\colon V_1 \to V_2$ be a mapping. Then f is said to be a polynomial mapping if

(1) $f(V_1)$ is contained in a finite dimensional subspace of V_2.

(2) If μ is a linear form on V_2 then $\mu \circ f$ is a polynomial mapping of V_1 to C.

7.10.12 *Lemma* If p is in $S(\mathfrak{g}_C)$ then there exists a polynomial mapping $\gamma_p\colon \mathfrak{h}_0 \to \mathscr{V} \otimes S(\mathfrak{h})$ and $r_p \geqq 0$, an integer, so that $\Gamma_H(\gamma_p(H)) = \pi(H)^{r_p} p$.

PROOF If $p = 0$ take $\gamma_p \equiv 0$. Suppose that for all p in $S_{r-1}(\mathfrak{g}_C)$, γ_p and r_p have been found. Suppose that $p = Z_1 \cdots Z_d H_1 \cdots H_e$, $Z_1, \ldots, Z_d$ in $\mathfrak{n}^+ + \mathfrak{n}^-$ and $H_1, \ldots, H_e$ in $\mathfrak{h}$.

Set $\det((tI - \text{ad } H)|_{\mathfrak{n}^+ + \mathfrak{n}^-}) = \Sigma\, (-t)^{q-i} \sigma_i(H)$ where $q = \dim(\mathfrak{n}^+ + \mathfrak{n}^-)$. Then $\sigma_q(H) = \prod_{\alpha \in \Delta} \alpha(H) = \varepsilon \pi(H)^2$ with $\varepsilon = (-1)^{q/2}$. Set

$$B(H) = (-1)^{q+1}\, \varepsilon \sum_{i=1}^{q-1} (-1)^i \sigma_i(H)(\text{ad } H)^{q-1-i}.$$

Then $B(H) \operatorname{ad} H|_{\mathfrak{n}^+ + \mathfrak{n}^-} = \pi(H)^2 I$. Set $Y_i = B(H)Z_i$ and $q(H) = \lambda(Y_1 \cdots Y_d)$. Then

$$\Gamma_H(q(H) \otimes H_1 \cdots H_e) \equiv \pi(H)^{2d} Z_1 \cdots Z_d H_1 \cdots H_e \bmod S_{r-1}(\mathfrak{g}_C).$$

If $b_0(H) = \pi(H)^{2d} p - \Gamma_H(q(H) \otimes H_1 \cdots H_e)$ then we have just shown that $b_0(H)$ is in $S_{r-1}(\mathfrak{g}_C)$. Let $b_1, \ldots, b_N$ be a basis of $S_{r-1}(\mathfrak{g}_C)$. Then $b_0(H) = \sum_{i=1}^N a_i(H) b_i$. This defines $a_i(H)$ and tracing back through the definition of $q(H)$ and Γ_H we see that the $a_i\colon \mathfrak{h}_0 \to C$ are polynomial mappings. Now the inductive hypothesis implies that there are for each $i = 1, \ldots, N$, γ_{b_i} and r_i so that $\Gamma_H(\gamma_{b_i}(H)) = \pi(H)^{r_i} b_i$. Let $s = \max_{1 \leq i \leq N}(2d, r_i)$ and set

$$\gamma_p(H) = \pi(H)^{s-2d} q(H) \otimes H_1 \cdots H_e + \sum_{i=1}^{N} q_i(H) \pi(H)^{s-r_i} \gamma_{b_i}(H).$$

Then $\gamma_p\colon \mathfrak{h}_0 \to \mathscr{V} \otimes S(\mathfrak{h})$ is a polynomial map and $\Gamma_H(\gamma_p(H)) = \pi(H)^s p$.

If p is in $S^r(\mathfrak{g}_C)$ then $p = \sum_{i=1}^N a_i p_i$ with p_i of the form $Z_1 \cdots Z_d H_1 \cdots H_e$, $d + e = r$. Let for each p_i, γ_{p_i} and r_i be defined as above so that $\Gamma_H(\gamma_{p_i}(H)) = \pi(H)^{r_i} p_i$. Let $s = \max\{r_1, \ldots, r_j\}$. Set $\gamma_p(H) = \Sigma\, a_i \pi(H)^{s-r_i} \gamma_{p_i}$. Then $\Gamma_H(\gamma_p(H)) = \pi(H)^s p$. This proves the lemma.

7.10.13 Let B be a $\mathfrak{g}$-invariant symmetric, nondegenerate, bilinear form on $\mathfrak{g} \times \mathfrak{g}$. Such B exist since $\mathfrak{g}$ is reductive. Indeed, we may take B to be any inner product on $\mathfrak{z}$ and B to be the Killing form of $\mathfrak{g}$ on $\mathfrak{g}_1$. We also use the symbol B for the complex bilinear extension of B to $\mathfrak{g}_C$. B induces a linear isomorphism of $\mathfrak{g}_C \to \mathfrak{g}_C^*$ $(X \mapsto B(X, \cdot))$. We may thus identify $S(\mathfrak{g}_C)$ with $S(\mathfrak{g}_C^*)$, the algebra of a complex valued polynomial functions on $\mathfrak{g}$ (here we identify, using B, $\mathfrak{g}_C^*$ with the complex valued linear forms on $\mathfrak{g}$). Let $\mathscr{D}(\mathfrak{g})$ be the algebra generated by $S(\mathfrak{g}_C)$ and $\partial(S(\mathfrak{g}_C))$ (here we look at functions as differential operators). Then in the notation of the proof of 7.10.9 if D is in $\mathscr{D}(\mathfrak{g})$, $D = \Sigma\, a_M D^M$ with $a_M\colon \mathfrak{g} \to C$ a polynomial function. If X is in $\mathfrak{g}$ let $D_X = \Sigma\, a_M(X) D^M$ in $\partial(S(\mathfrak{g}_C))$.

7.10.14 *Lemma* Set $\mathfrak{h}_0' = \{H \text{ in } \mathfrak{h}_0 | \pi(H) \neq 0\}$. Let H be in $\mathfrak{h}_0$ and let p be in $S(\mathfrak{g}_C)$. There is a unique element $\beta_H(p)$ in $S(\mathfrak{h})$ so that $p - \beta_H(p)$ is in $\Gamma_H(\mathscr{V}' \otimes S(\mathfrak{h}))$ where $\mathscr{V}' = \lambda(\sum_{r=1}^{\infty} S^r(\mathfrak{n}^+ + \mathfrak{n}^-))$.

PROOF This follows directly from 7.10.10.

7.10.15 Let H be in $\mathfrak{h}_0'$. Define $\delta_H'(\partial(p)) = \partial(\beta_H(p))$ for p in $S(\mathfrak{g}_C)$ (here $\beta_H(p)$ is in $S(\mathfrak{h})$ and $\partial(\beta_H(p))$ is thus a constant coefficient differential operator on $\mathfrak{h}_0$). We extend δ_H' to $\mathscr{D}(\mathfrak{g})$ by setting $\delta_H'(D) = \delta_H'(D_H)$.

7.10.16 *Lemma* If D is in $\mathscr{D}(\mathfrak{g})$ then there exists a differential operator $\delta'(D)$ on $\mathfrak{h}_0'$ so that if H is in $\mathfrak{h}_0'$ then $\delta'(D)_H = \delta_H'(D)$.

PROOF 7.10.12 implies that there is a polynomial mapping γ of $\mathfrak{h}_0$ into $\partial(S(\mathfrak{h}))$ and a nonnegative integer m so that $\delta_H'(D) = \pi(H)^{-m}\gamma(H)$. Thus if f is in $C^\infty(\mathfrak{h}_0')$ and if we define for H in $\mathfrak{h}_0'$, $(\delta'(D)f)(H) = (\delta_H'(D)f)(H)$ then $\delta'(D)$ is a differential operator on $\mathfrak{h}_0'$.

7.10.17 *Lemma* Let f be in $C^\infty(\mathfrak{g})$ and suppose that $f(xH) = f(H)$ for x in G, H in $\mathfrak{h}_0$ (see the proof of 7.9.9 for the definition of G). If D is in $\mathscr{D}(\mathfrak{g})$ then

$$(Df)(H) = (\delta'(D)(f|_{\mathfrak{h}_0'}))(H)$$

for H in $\mathfrak{h}_0'$.

PROOF Let H be in $\mathfrak{h}_0$. Let p be in $S(\mathfrak{g}_C)$ so that $D_H = \partial(p)$. Then we know that

$$p = \Gamma_H(1 \otimes \beta_H(p)) + \sum_{i=1}^{r} s_i \otimes h_i$$

with s_i in $\mathscr{V}'$, h_i in $S(\mathfrak{h})$ (see 7.10.14). If x is in G then (in the notation of the proof of 7.10.9)

$$f(x\colon H; \partial(p)) = f(x\colon H; \partial(\beta_H(p)) + \sum_{i=1}^{r} f(x; s_i\colon H; \partial(h_i))$$

(see (2) of the proof of 7.10.9). But by assumption on f, $f(x; Z\colon H; \partial(h_i)) = 0$ for Z in $\mathscr{V}'$. Thus $f(x\colon H; \partial(p)) = f(x\colon H; \partial(\beta_H(p))$. Now $(D_H f)(H) = (\partial(p)f)(H)$ and $\partial(\beta_H(p)) = \delta_H'(p) = \delta_H'(D_H) = \delta_H'(D)$. Thus $(Df)(H) = (\delta'(D)(f|_{\mathfrak{h}_0'}))(H)$ as was asserted.

7.10.18 Let for f in $C^\infty(\mathfrak{g})$, x in G, X in $\mathfrak{g}$, $(\rho(x)f)(X) = f(xX)$. Let $\mathscr{T}(\mathfrak{g}) = \{D \text{ in } \mathscr{D}(\mathfrak{g}) \mid D \circ \rho(x) = \rho(x) \circ D \text{ for all } x \text{ in } G\}$. $\mathscr{T}(\mathfrak{g})$ is clearly a subalgebra of $\mathscr{D}(\mathfrak{g})$.

7.10.19 *Lemma* $\delta' = \mathscr{T}(\mathfrak{g}) \to \operatorname{diff}(\mathfrak{h}_0')$ ($\operatorname{diff}(\mathfrak{h}_0')$ is the algebra of all differential operators on $\mathfrak{h}_0'$) is an algebra homomorphism.

PROOF Let A be the centralizer of $\mathfrak{h}_0$ in G. That is, $A = (g \text{ in } G | g|_{\mathfrak{h}_0} = I\}$.

Then A is a closed subgroup of G. Arguing as in the proof of 7.5.3 we find that the map $\psi: G/A \times \mathfrak{h}_0' \to \mathfrak{g}$ given by $\psi(\bar{x}, H) = x \cdot H$ ($\bar{x} = xA$), x in G, H in $\mathfrak{h}_0'$ is everywhere regular. Fix H_0 in $\mathfrak{h}_0'$. Then there are open neighborhoods U and V of $\bar{I}$ (identity coset) and H_0 in G/A and $\mathfrak{h}_0'$ so that

$$\psi: V \times U \to \psi(V \times U) = W$$

is a diffeomorphism onto W, an open subset of $\mathfrak{g}$. Clearly W is a neighborhood of H_0. Let g be in $C^\infty(U)$. Define $\tilde{g}$ in $C^\infty(W)$ by $\tilde{g}(\psi(\bar{x}, H)) = g(H)$ for $\bar{x}$ in V, H in U. Then $\tilde{g}(xH) = \tilde{g}(H)$ for all x in $\tilde{V} = VA$ and H in U.

Let D_1, D_2 be in $\mathscr{T}(\mathfrak{g})$. Set $F = D_2\tilde{g}$. Since 7.9.17 is equally applicable to locally defined functions with the appropriate local invariance properties we see that since D_2 is in $\mathscr{T}(\mathfrak{g})$, $(D_1F)(H) = (\delta'(D_1)F|_U)(H)$. But $(D_2\tilde{g})(H) = \delta'(D_2)(g|_U)(H)$. Thus $(D_1D_2\tilde{g})(H) = \delta'(D_1)\delta'(D_2)(\tilde{g}|_U)(H)$. But $(D_1D_2\tilde{g})(H) = (\delta'(D_1D_2)(g|_U)(H)$ and since $\tilde{g}|_U = g$, an arbitrary C^∞ function on U, we see that $\delta'(D_1D_2)$ and $\delta'(D_1)\delta'(D_2)$ have the same local expression at H_0. Since H_0 in $\mathfrak{h}_0$ is arbitrary this says that $\delta'(D_1D_2) = \delta'(D_1)\delta'(D_2)$ as asserted.

7.10.20 If p is in $S(\mathfrak{g}_C)$ then using B (see 7.10.13) we may identify p with a polynomial function on $\mathfrak{g}$. Thus $\bar{p} = p|_{\mathfrak{h}_0}$ is a polynomial function on $\mathfrak{h}_0$. Noting that $B|_{\mathfrak{h}_0 \times \mathfrak{h}_0}$ is nondegenerate we may identify $\bar{p}$ with an element of $S(\mathfrak{h})$. Thus using B we have a map $p \mapsto \bar{p}$ of $S(\mathfrak{g}_C)$ into $S(\mathfrak{h})$.

7.10.21 Let $I(\mathfrak{g})$ be the set of all p in $S(\mathfrak{g}_C)$ so that $xp = p$ for all x in G. Then clearly if $D_pf = pf$ for p in $I(\mathfrak{g}_C)$, f in $C^\infty(\mathfrak{g})$, then D_p is in $\mathscr{T}(\mathfrak{g})$. Furthermore it is clear that if p is in $I(\mathfrak{g})$ then $\delta'(D_p) = \bar{p}$.

7.10.22 The purpose of this section is to derive a formula for $\delta'(\partial(p))$ for p in $I(\mathfrak{g})$ in terms of $\bar{p}$.

7.10.23 Let $W(\Delta)$ be the group of linear maps of $\mathfrak{h} \to \mathfrak{h}$ generated by the s_α, α in Δ, where $s_\alpha(H) = H - 2\{\alpha(H)/\alpha(H_\alpha)\}H_\alpha$ where $B(H_\alpha, H) = \alpha(H)$, H_α in $\mathfrak{h}$. The action of $W = W(\Delta)$ extends to $S(\mathfrak{h})$ in the canonical fashion. Let us denote by $s \cdot p$, s in W, p in $S(\mathfrak{h})$ the corresponding action. Let

$$I(\mathfrak{h}) = \{p \text{ in } S(\mathfrak{h}) | sp = p\}.$$

7.10.24 *Lemma* If q is in $S(\mathfrak{h})$ and $sq = \det(s)q$ then $q = \pi q_0$ with q_0 in $I(\mathfrak{h})$.

PROOF Let α be a simple root in Δ^+. Then $s_\alpha\Delta^+ = (\Delta^+ - \{\alpha\}) \cup \{-\alpha\}$

(see 3.10.10). Thus $s_\alpha \pi = (\prod_{\beta \in \Delta^+, \beta \neq \alpha} \beta)(-\alpha) = -\pi$. But det $s_\alpha = -1$ and the s_α, α simple, generate W. Thus $s \cdot \pi = \det(s)\pi$ for all s in W.

If α is in Δ then set $P_\alpha = \{H \text{ in } \mathfrak{h} | \alpha(H) = 0\}$. If p is in $S(\mathfrak{h})$ then $(s_\alpha p - p)(H) = p(s_\alpha H) - p(H) = p(H) - p(H) = 0$ if H is in P_α. Thus $s_\alpha p - p$ vanishes on P_α. Thus $s_\alpha p - p = \alpha \cdot q$ with q in $S(\mathfrak{h})$. Thus if q is as in the statement of the lemma, $q = -\frac{1}{2}(s_\alpha q - q)$ and hence q is divisible by α for all α in Δ^+. But the P_α, in Δ^+, are all distinct and $S(\mathfrak{h})$ clearly has unique factorization. Thus q is divisible by π. Clearly q/π is in $I(\mathfrak{h})$.

7.10.25 *Corollary* Let p be in $I(\mathfrak{h})$ and homogeneous of degree $k > 0$. Then $\partial(p)\pi = 0$.

PROOF Let s be in W. Then $s \cdot \partial(p)\pi = \partial(sp)s\pi = \det(s)\partial(p)\pi$. But then 7.10.24 implies that $\partial(p)\pi$ is divisible by π. This implies that $\partial(p)\pi = 0$ since $\deg(\partial(p)\pi) < \deg \pi$.

7.10.26 *Lemma* Let for X in $\mathfrak{g}$, $\Omega_0(X) = B(X, X)$. Then Ω_0 is in $I(\mathfrak{g})$. If $\Omega = \partial(\Omega_0)$ then $(\pi(H)\delta'(\Omega)f)(H) = (\partial(\overline{\Omega}_0)\pi f)(H)$ for H in $\mathfrak{h}_0'$ and f in $C^\infty(\mathfrak{h}_0')$.

PROOF Let $X_1, \ldots, X_n$ be a basis of $\mathfrak{g}$ and let $Y_1, \ldots, Y_n$ be so that $B(X_i, Y_j) = \delta_{ij}$. Then a straightforward computation shows that our assumptions say that $\Omega_0 = \Sigma X_i Y_i$. If $Z_1, \ldots, Z_n$ is a basis of $\mathfrak{g}_C$ and if $W_1, \ldots, W_n$ in $\mathfrak{g}_C$ are such that $B(Z_i, W_j) = \delta_{ij}$ then again $\Omega_0 = \Sigma Z_i W_i$.

Now $\mathfrak{g}_C = \mathfrak{h} + \sum_{\alpha \in \Delta} (\mathfrak{g}_C)_\alpha$. For each α in Δ^+ let $E_\alpha \neq 0$ in $(\mathfrak{g}_C)_\alpha$ be chosen. Since B is $\mathfrak{g}$-invariant we see $B((\mathfrak{g}_C)_\alpha, (\mathfrak{g}_C)_\beta) = 0$ if $\alpha + \beta \neq 0$. Let for α in Δ^+, $E_{-\alpha}$ in $(\mathfrak{g}_C)_{-\alpha}$ be chosen so that $B(E_\alpha, E_{-\alpha}) = 1$. Now if H is in $\mathfrak{h}$, $B([E_\alpha, E_{-\alpha}], H) = B(E_{-\alpha}, [H, E_\alpha]) = \alpha(H)B(E_{-\alpha}, E_\alpha) = \alpha(H)$. Thus $[E_\alpha, E_{-\alpha}] = H_\alpha$.

Let $H_1, \ldots, H_l$ be a basis of $\mathfrak{h}$ so that $B(H_i, H_j) = \delta_{ij}$. Then $\Omega_0 = \sum_{i=0}^{l} H_i^2 + 2\sum_{\alpha \in \Delta^+} E_\alpha E_{-\alpha}$. Now $\overline{\Omega}_0 = \sum_{i=1}^{l} H_i^2$. Thus $\Omega_0 = \overline{\Omega}_0 + 2\sum_{\alpha \in \Delta^+} E_\alpha E_{-\alpha}$. ($\overline{\Omega}_0$ is extended to $\mathfrak{g}$ by using the formula for $\overline{\Omega}_0$, that is, $\Omega_0(X) = \sum_{i=1}^{l} B(H_i, X)^2$).

If H_0 is in $\mathfrak{h}_0'$ and if α is in Δ^+ then

(1)
$$\begin{aligned} \Gamma_{H_0}(E_\alpha \cdot E_{-\alpha} \otimes 1) &= (L_{[E_\alpha, H_0]} + d_{E_\alpha})([E_{-\alpha}, H_0]) \\ &= -\alpha(H_0)^2 E_\alpha E_{-\alpha} + \alpha(H_0)H_\alpha. \end{aligned}$$

Here we are using the product $X \cdot Y$ in $U(\mathfrak{g}_C)$ and XY in $S(\mathfrak{g}_C)$. In $U(\mathfrak{g}_C)$, $E_\alpha \cdot E_{-\alpha} - E_{-\alpha} \cdot E_\alpha = H_\alpha$. Thus

(2)
$$E_\alpha \cdot E_{-\alpha} = \frac{1}{2}(E_\alpha \cdot E_{-\alpha} + E_{-\alpha} \cdot E_\alpha + H_\alpha)$$

in $U(\mathfrak{g}_C)$. (2) implies that $E_\alpha \cdot E_{-\alpha} = \lambda(E_\alpha E_{-\alpha}) + \frac{1}{2}H_\alpha$. Using (1) we see that

(3) $$\Gamma_{H_0}\Big(-\sum_{\alpha\in\Delta+} 2\alpha(H_0)^{-2}(E_\alpha \cdot E_{-\alpha}) \otimes 1 + 1 \otimes \overline{\Omega}_0 + 2\sum_{\alpha\in\Delta+} \alpha(H_0)^{-1}(1 \otimes H_\alpha)\Big)$$
$$= \overline{\Omega}_0 + 2\sum_\alpha E_\alpha E_{-\alpha} = \Omega_0.$$

Using (3) and (2) we compute $\beta_{H_0}(\Omega_0)$.

$$-\sum_{\alpha\in\Delta+} 2\alpha(H_0)^{-2}(E_\alpha \cdot E_{-\alpha}) \otimes 1$$
$$= -\sum_{\alpha\in\Delta+} 2\alpha(H_0)^{-2}\lambda(E_\alpha E_{-\alpha}) \otimes 1 - \sum_{\alpha\in\Delta+} \alpha(H_0)^{-2}H_\alpha \otimes 1.$$

Now

$$\Gamma_{H_0}(H_\alpha \otimes 1) = (L_{[H_\alpha, H_0]} + d_{H_\alpha})1 = 0.$$

Thus

$$\beta_{H_0}(\Omega_0) = \overline{\Omega}_0 + 2\sum_{\alpha\in\Delta+} \alpha(H_0)^{-1}H_\alpha.$$

We therefore see that

(4) $$\delta'(\Omega) = \partial(\overline{\Omega}_0) + 2\sum_{\alpha\in\Delta+} \alpha^{-1}\partial(H_\alpha).$$

The lemma will be proved if we can show

(5) $$\pi^{-1}\partial(\overline{\Omega}_0)D_\pi = \partial(\overline{\Omega}_0) + 2\sum_{\alpha\in\Delta+} \alpha^{-1}\partial(H_\alpha) \text{ on } \mathfrak{h}_0'.$$

Now $\pi^{-1}\partial(\overline{\Omega}_0)D_\pi 1 = \pi^{-1}(\partial(\overline{\Omega}_0)\pi) = 0$. Thus $\pi^{-1}\partial(\overline{\Omega}_0)D_\pi = \partial(\overline{\Omega}_0) + D_1$ with D_1 a first-order operator on $\mathfrak{h}_0'$. To compute D_1 we note that if f is in $S^1(\mathfrak{h}_0)$, then $\partial(\overline{\Omega}_0)f = 0$. Thus $\pi^{-1}\partial(\overline{\Omega}_0)D_\pi f = D_1 f$. Let $f = \mu_{H_0}$ where $\mu_{H_0}(H) = B(H_0, H)$.

Now $\partial(\Omega_0)\pi = 0$ and $\partial(\overline{\Omega}_0)\mu_{H_0} = 0$. Thus

$$\partial(\Omega_0)\pi\mu_{H_0} = 2\sum_{i=1} (\partial(H_i)\pi)\partial(H_i)\mu_{H_0}$$
$$= 2\sum_{i=1}^{l} (\partial(H_i)\pi)B(H_0, H_i) = 2\partial(H_0)\pi = 2\sum_{\alpha\in\Delta+} \alpha(H_0)\alpha^{-1}\pi.$$

Thus

$$\pi^{-1}\partial(\overline{\Omega})\pi\mu_{H_0} = 2\sum_{\alpha\in\Delta+} \alpha^{-1}\alpha(H_0) = 2\sum_{\alpha\in\Delta+} \alpha^{-1}\partial(H_\alpha)\mu_{H_0}.$$

Thus $D_1 = 2\sum_{\alpha\in\Delta+} \alpha^{-1}\partial(H_\alpha)$ as was asserted.

7.10.27 *Theorem* Let p be in $I(\mathfrak{g})$. Then $\delta'(\partial(p)) = \pi^{-1}\partial(\bar{p})D_\pi$ as a differential operator on $\mathfrak{h}_0'$.

PROOF Let for D in $\mathscr{D}(\mathfrak{g})$, $2\mu(D) = \Omega D - D\Omega$.

(1) $\mu(D_X) = \partial(X)$ for X in $\mathfrak{g}_C$. Here $(D_X f)(Y) = B(X, Y)f(Y)$ for f in $C^\infty(\mathfrak{g})$, Y in $\mathfrak{g}$.

In fact, $(\Omega D_X - D_X\Omega)f = \Omega B(X, \cdot)f - B(X, \cdot)\Omega f$. Let $X_1, \ldots, X_n$ be a basis of $\mathfrak{g}$, and $Y_1, \ldots, Y_n$ in $\mathfrak{g}$ so that $B(X_i, Y_j) = \delta_{ij}$. Then $\Omega = \Sigma\, \partial(X_i)\partial(Y_i)$. Hence

$$\begin{aligned}\Omega B(X, \cdot)f &= \Sigma\, \partial(X_i)B(X, Y_i)f + \Sigma\, \partial(X_i)B(X, \cdot)\partial(Y_i)f \\ &= \Sigma\, B(X, Y_i)\partial(X_i)f + \Sigma\, B(X, X_i)\partial(Y_i)f + B(X, \cdot)\Omega f \\ &= 2\partial(X)f + B(X, \cdot)\Omega f.\end{aligned}$$

Hence (1).

(2) If p is in $S^k(\mathfrak{g}_C)$ and $k \geqq 0$ then $\mu(\partial(p)) = 0$.

This follows from the fact that $D_0(\mathfrak{g})$ is commutative and $\partial(p)$, Ω are in $D_0(\mathfrak{g})$.

(3) If D_1, D_2 are in $\mathscr{D}(\mathfrak{g})$ then $\mu^m(D_1D_2) = \Sigma\binom{m}{k}\mu^{m-k}(D_1)\mu^k(D_2)$.

This follows from the observation that μ is a derivation of $\mathscr{D}(\mathfrak{g})$.

(4) If $X_1, \ldots, X_m$ are in $\mathfrak{g}_C$ then $\mu^m(D_{X_1}D_{X_2}\cdots D_{X_m}) = m!\partial(X_1)\cdots\partial(X_m)$.

We prove (4) by induction on m. If $m = 1$, (4) is the same as (1). Now

$$\mu^m(D_{X_1}\cdots D_{X_m}) = \Sigma\binom{m}{k}\mu^{m-k}(D_{X_1}\cdots D_{X_{m-1}})\mu^k(D_{X_m})$$

by (3). But $\mu^k(D_{X_m}) = 0$ for $k \geqq 2$ by (2). Thus

$$\mu^m(D_{X_1}\cdots D_{X_m}) = \mu^m(D_{X_1}\cdots D_{X_{m-1}})D_{X_m} + m\mu^{m-1}(D_{X_1}\cdots D_{X_{m-1}})\mu(D_{X_m}).$$

But $\mu^m(D_{X_1}\cdots D_{X_{m-1}}) = 0$ by (2). Thus

$$\mu^m(D_{X_1}\cdots D_{X_m}) = m\mu^{m-1}(D_{X_1}\cdots D_{X_{m-1}})\partial(X_m)$$

by the above and (1). (4) now follows by induction.

(5) If p is in $S^m(\mathfrak{g}_C)$ then $\mu^m(D_p) = m!\partial(p)$.

This follows from (4) (note that $D_p f = pf$).

Let diff($\mathfrak{h}_0'$) be the algebra of all differential operators on $\mathfrak{h}_0'$. If ξ is in diff($\mathfrak{h}_0$) let $\bar{\mu}(\xi) = \frac{1}{2}(\delta'(\Omega)\xi - \xi\delta'(\Omega))$. Then $\bar{\mu}$ is a derivation of diff($\mathfrak{h}_0'$). 7.10.19 implies that if D is in $\mathscr{T}(\mathfrak{g})$ then $\delta'(\mu(D)) = \bar{\mu}\delta'(D)$. Since $\mu\mathscr{T}(\mathfrak{g}) = \mathscr{T}(\mathfrak{g})$ we find that

(6) if p is in $S^m(\mathfrak{g}_C) \cap I(\mathfrak{g})$ then $m!\delta'(\partial(p)) = \delta'(\mu^m D_p) = \bar{\mu}^m\delta'(D_p) = \bar{\mu}^m D_{\bar{p}}$.

On the other hand

$$\begin{aligned}\bar{\mu}\xi &= \frac{1}{2}(\delta'(\Omega)\xi - \xi\delta'(\Omega)) \\ &= \frac{1}{2}(\pi^{-1}\partial(\overline{\Omega}_0)\pi\xi - \xi\pi^{-1}\partial(\overline{\Omega}_0)\pi) = \frac{1}{2}\pi^{-1}(\partial(\overline{\Omega}_0)\xi - \xi\partial(\overline{\Omega}_0))D_\pi.\end{aligned}$$

Replacing $\mathfrak{g}$ by $\mathfrak{h}_0$ in (1)–(5) we find that if $\nu(\xi) = \frac{1}{2}\{\partial(\overline{\Omega}_0)\xi - \xi\partial(\overline{\Omega}_0)\}$ then
(7) $\nu^m(D_{\bar{p}}) = m!\partial(\bar{p})$ for p in $S^m(\mathfrak{g}_C)$.
We therefore see that $\bar{\mu}^m(D_{\bar{p}}) = \pi^{-1}(\nu^m(D_{\bar{p}}))D_\pi = m!\pi^{-1}\partial(\bar{p})D_\pi$.
Combining (6) with (7) we have

$$m!\,\delta'(\partial(p)) = m!\pi^{-1}\partial(\bar{p})D_\pi.$$

This proves the theorem.

7.11 A Formula for Semisimple Lie Groups with One Conjugacy Class of Cartan Subalgebra

7.11.1 We return to the notation of 7.8 and 7.9. Let G be a connected semisimple Lie group and let $\mathfrak{g}$ be the Lie algebra of G. Let $\mathfrak{g} = \mathfrak{k} \oplus \mathfrak{a} \oplus \mathfrak{n}$ and $G = KAN$ be Iwasawa decompositions of $\mathfrak{g}$ and G (see 7.4.3). Let $\mathfrak{m}$ and M be, respectively, the centralizers of $\mathfrak{a}$ in $\mathfrak{k}$ and K. Set $\mathfrak{m}_1 = \mathfrak{m} \oplus \mathfrak{a}$.

7.11.2 Let $S(\mathfrak{g}_C)$ be the symmetric algebra on $\mathfrak{g}_C$. Then as in 7.10 we can look at $S(\mathfrak{g}_C)$ as $D_0(\mathfrak{g})$, the space of all constant coefficient differential operators on $\mathfrak{g}$. If p is in $S(\mathfrak{g}_C)$ then let $\partial(p)$ be the corresponding constant coefficient differential operator.

7.11.3 Let B be the Killing form of $\mathfrak{g}$. Then using B we can (as in 7.10) identify $S(\mathfrak{g}_C)$ with the space of all complex valued polynomial functions on $\mathfrak{g}$.

7.11.4 Let f be in $C_0^\infty(\mathfrak{g})$; we define

(1) $$\hat{f}(Y) = \frac{1}{(2\pi)^{n/2}} \int_{\mathfrak{g}} f(X) e^{(-1)^{1/2}B(X,Y)}\,dX$$

for Y in $\mathfrak{g}$ and $n = \dim \mathfrak{g}$.
The Fourier inversion theorem (see 5.A.5) says

(2) $$f(X) = \frac{1}{(2\pi)^{n/2}} \int_{\mathfrak{g}} \hat{f}(Y) e^{-(-1)^{1/2}B(X,Y)}\,dY.$$

7.11.5 *Lemma* Let f be in $C_0^\infty(\mathfrak{g})$. Let $p\colon \mathfrak{g} \to C$ be a polynomial function. Then

$$(\partial(p)f)^{\wedge}(Y) = p(-(-1)^{1/2}Y)\hat{f}(Y)$$

for Y in $\mathfrak{g}$ (here p is extended to $\mathfrak{g}_C$ in the obvious fashion).

PROOF Let λ be a real valued linear form on $\mathfrak{g}$. Let X_λ in $\mathfrak{g}$ be defined by $B(X_\lambda, X) = \lambda(X)$ for all X in $\mathfrak{g}$. Then the map $\lambda \mapsto X_\lambda$ is in fact the identification we have been assuming between $S(\mathfrak{g}_C)$ and the algebra of all polynomial maps of $\mathfrak{g}$ to C. Now $(\partial(\lambda)f)(X) = (d/dt)f(X + tX_\lambda)|_{t=0}$. Integrating by parts we find

$$\begin{aligned}(\partial(\lambda)f)^{\wedge}(Y) &= \frac{1}{(2\pi)^{n/2}}\int_{\mathfrak{g}} (\partial(\lambda)f)(X)e^{(-1)^{1/2}B(X,Y)}dY \\ &= \frac{1}{(2\pi)^{n/2}}\int_{\mathfrak{g}} \frac{d}{dt}f(X + tX_\lambda)\bigg|_{t=0} e^{(-1)^{1/2}B(X,Y)}dY \\ &= -\frac{1}{(2\pi)^{n/2}}\int_{\mathfrak{g}} f(X)\frac{d}{dt}e^{(-1)^{1/2}B(X+tX_\lambda,Y)}\bigg|_{t=0} dY \\ &= -\lambda((-1)^{1/2}Y)\hat{f}(Y).\end{aligned}$$

The result is thus true for linear forms. Since every element of $S(\mathfrak{g}_C)$ is a linear combination of products of linear forms the lemma follows.

7.11.6 Let for X in $\mathfrak{m}_1$, $\tilde{\omega}(X) = \det(\operatorname{ad} X|_{\mathfrak{n}})$. Then applying the techniques of 7.10 to the pair $\mathfrak{m}_1$, $B|_{\mathfrak{m}_1 \times \mathfrak{m}_1}$ we may associate to $\tilde{\omega}$ a differential operator $\partial(\tilde{\omega})$ in $D_0(\mathfrak{m}_1)$.

7.11.7 *Lemma* Assume that G has one conjugacy class of Cartan subalgebra. There is a real constant $c \neq 0$ so that if F is in $C_0^\infty(\mathfrak{g})$ and $F(\operatorname{Ad}(k)X) = F(X)$ for all k in K then

$$cF(0) = \lim_{\substack{Y\to 0\\ Y\in\mathfrak{m}_1}} \partial(\tilde{\omega})\int_{\mathfrak{n}} F(Y + Z)dZ.$$

PROOF $\mathfrak{g} = \bar{\mathfrak{n}} \oplus \mathfrak{m}_1 \oplus \mathfrak{n}$ (see 7.3.5). Also, it is easily seen that $B(\mathfrak{n}, \mathfrak{n}) = B(\bar{\mathfrak{n}}, \bar{\mathfrak{n}}) = B(\mathfrak{n}, \mathfrak{m}_1) = B(\bar{\mathfrak{n}}, \mathfrak{m}_1) = 0$. Also, B defines a nondegenerate pairing of $\bar{\mathfrak{n}}$ with $\mathfrak{n}$ and $\mathfrak{m}_1$ with itself. Extend $\tilde{\omega}$ to $\mathfrak{g}$ by setting $\tilde{\omega}(X + Y + Z) = \tilde{\omega}(Y)$ for X in $\bar{\mathfrak{n}}$, Y in $\mathfrak{m}_1$, Z in $\mathfrak{n}$.

$$(1) \qquad \int_{\mathfrak{m}_1\times\mathfrak{n}} \tilde{\omega}(Y)\hat{F}(Y + Z)dYdZ = \varepsilon\int_{\mathfrak{m}_1\times\mathfrak{n}} (\partial(\tilde{\omega})F)^{\wedge}(Y + Z)dYdZ$$

where $\varepsilon = \pm 1$ and dY and dZ are Euclidean measures on $\mathfrak{m}_1$ and $\mathfrak{n}$ respectively.

Equation (1) follows 7.11.5 since

$$(\partial(\tilde{\omega})F)^\wedge(Y + Z) = \tilde{\omega}(-(-1)^{1/2}(Y + Z))F^\wedge(Y + Z)$$

by 7.10.5. But $\tilde{\omega}(-(-1)^{1/2}(Y + Z)) = (-(-1)^{1/2})^r\tilde{\omega}(Y)$ where $r = \dim \mathfrak{n}$. The proof of 7.9.8 implies that r is even. This proves (1).

$$\int_{\mathfrak{m}_1 \times \mathfrak{n}} (\partial(\tilde{\omega})F)^\wedge(Y + Z)dYdZ$$

$$= \frac{1}{(2\pi)^{n/2}} \int_{\mathfrak{m}_1 \times \mathfrak{n}} \int_{\bar{\mathfrak{n}} \times \mathfrak{m}_1 \times \mathfrak{n}} (\partial(\tilde{\omega})F)(W + Y_1 + Z_1) \cdot e^{(-1)^{1/2}B(W+Z, X+Y_1+Z_1)}dWdY_1dZ_1dYdZ.$$

Here dW, dY, dZ are Euclidean measures on $\bar{\mathfrak{n}}$, $\mathfrak{m}_1$, and $\mathfrak{n}$ normalized so that $dWdYdZ = dX$, the Euclidean measure on $\mathfrak{g}$.

Since $\partial(\tilde{\omega})F$ has compact support, we may interchange orders of integration and find (using $B(Y + Z, W + Y_1 + Z_1) = B(Y, Y_1) + B(Z, W)$) that

(2)
$$\int_{\mathfrak{m}_1 \times \mathfrak{n}} (\partial(\tilde{\omega})F)^\wedge(Y + Z)dYdZ$$

$$= \frac{1}{(2\pi)^{n/2}} \int_{\mathfrak{m}_1} \int_{\mathfrak{n}} \int_{\mathfrak{m}_1} \int_{\mathfrak{n}} \int_{\bar{\mathfrak{n}}} (\partial(\tilde{\omega})F)(W + Y_1 + Z_1) \cdot e^{(-1)^{1/2}\{B(Y_1,Y)+B(Z,W)\}}dZ_1dYdY_1dZdW.$$

The Fourier inversion formula implies that if φ is in $C_0^\infty(\mathfrak{m}_1)$, then

(3)
$$\int_{\mathfrak{m}_1} \int_{\mathfrak{m}_1} \varphi(Y_1)e^{(-1)^{1/2}B(Y,Y_1)}\,dY\,dY_1 = (2\pi)^q\varphi(0)$$

where $q = \dim \mathfrak{m}_1$.

Also if ψ is in $C_0^\infty(\bar{\mathfrak{n}})$ we find

(4)
$$\int_{\mathfrak{n}} \int_{\mathfrak{n}} \psi(W)e^{(-1)^{1/2}B(W,Z)}dWdZ = (2\pi)^r\psi(0)$$

where $r = \dim \mathfrak{n} = \dim \bar{\mathfrak{n}}$.

Combining (2), (3), and (4) we find

(5)
$$\int_{\mathfrak{m}_1 \times \mathfrak{n}} (\partial(\tilde{\omega})F)^\wedge(Y + Z)dYdZ = (2\pi)^{q+r-(n/2)} \int_{\mathfrak{n}} (\partial(\tilde{\omega})F)(Z)dZ.$$

Now,

$$\int_{\mathfrak{n}} (\partial(\tilde{\omega})F)(Z)dZ = \lim_{\substack{Y\to 0\\ Y\in\mathfrak{m}_1}} \int_{\mathfrak{n}} (\partial(\tilde{\omega})F)(Y + Z)dZ$$

$$= \lim_{\substack{Y\to 0\\ Y\in\mathfrak{m}_1}} \partial(\tilde{\omega}) \int_{\mathfrak{n}} F(Y + Z)dZ.$$

Combining this with (5) we find

$$(6)\qquad \int_{\mathfrak{m}_1\times\mathfrak{n}} (\partial(\tilde{\omega})F)^\wedge(Y+Z)\,dY\,dZ$$
$$= (2\pi)^{q+r-(n/2)} \lim_{\substack{Y\to 0\\ Y\in\mathfrak{m}_1}} \partial(\tilde{\omega}) \int_{\mathfrak{n}} F(Y+Z)\,dZ.$$

Combining (6) with (1) we find that if $c_1 = (2\pi)^{(n/2)-q-r}\varepsilon$ then

$$(7)\qquad c_1 \int_{\mathfrak{m}_1\times\mathfrak{n}} \tilde{\omega}(Y)\hat{F}(Y+Z)\,dY\,dZ = \lim_{\substack{Y\to 0\\ Y\in\mathfrak{m}_1}} \partial(\tilde{\omega}) \int_{\mathfrak{n}} F(Y+Z)\,dZ.$$

Now 7.8.5 combined with 7.9.9 implies that

$$(8)\qquad \int_{\mathfrak{m}_1\times\mathfrak{n}} \tilde{\omega}(Y)\hat{F}(Y+Z)\,dY\,dZ = \int_{\mathfrak{g}} \hat{F}(X)\,dX = (2\pi)^{n/2}F(0).$$

Equations (7) and (8) imply the lemma with $c = (2\pi)^{n/2}c_1 = \varepsilon(2\pi)^r$.

7.11.8 *Lemma* Suppose that G has one conjugacy class of Cartan subalgebra. Then there is a real constant $c \neq 0$ so that if f is in $C_0^\infty(G)$ then

$$cf(e) = \lim_{\substack{Y\to 0\\ Y\in\mathfrak{m}_1}} \partial(\tilde{\omega})\delta_-(Y)F_f(\exp Y)$$

where F_f is defined as in 7.7.11 and $\delta_-(Y) = |\det((I - e^{-\operatorname{ad}Y}|_{\mathfrak{m}_1})/\operatorname{ad}Y|_{\mathfrak{m}_1})|^{1/2}$.

PROOF Set $f_1(x) = \int_{K_*} f(kxk^{-1})\,dk_*$ for x in G. Since $K_* = \operatorname{Ad}_G(K)$ is compact, f_1 is in $C_0^\infty(G)$. Let U be a neighborhood of 0 in $\mathfrak{m}_1$ as in 7.5.23. Let V be a neighborhood of 0 in U so that $\bar{V} \subset U$ and $\bar{V}$ is compact.

Let $F_1(X) = f_1(\exp X)$. Then F_1 is in $C^\infty(\mathfrak{g})$ and $F_1(\operatorname{Ad}(k)X) = F_1(X)$ for all X in $\mathfrak{g}$ and k in K. Set $V_1 = (\bar{V}+\mathfrak{n}) \cap \operatorname{supp} F_1$. Set $E = \bigcup_{k\in K} \operatorname{Ad}(k)V_1$. If X is in V_1 then exp is regular at X. Thus exp is regular at each X in E.

(1) V_1 is compact.

In fact, $\exp\colon U + \mathfrak{n} \to M_1N$ is a diffeomorphism into and $V_1 = (f_1|_{\exp(\bar{V}+\mathfrak{n})})^{-1}(\operatorname{supp} f_1)$. But $\exp(\bar{V}+\mathfrak{n})$ is closed in G (see 7.5.24). Thus V_1 is compact.

Now $\operatorname{Ad}(K)$ is compact so (1) implies

(2) E is a compact subset of $\mathfrak{g}$.

Let V_2 be an open subset of $\mathfrak{g}$ containing E and so that $\bar{V}_2$ is compact and if X is in $\bar{V}_2$, then exp is regular at X. (This is possible by (2) and the fact that exp is regular at each X in E).

Set $V_3 = \bigcup_{k\in K} \operatorname{Ad}(k)\bar{V}_2$. Then V_3 is compact and exp is regular at each X in V_3. Finally let V_4 be an open subset of $\mathfrak{g}$ containing V_3 and so that $\bar{V}_4$ is compact and exp is regular at each X in $\bar{V}_4$. Let ϕ be an element of $C_0^\infty(\mathfrak{g})$ so that $\phi \equiv 1$ on $\bar{V}_3$ and $\phi \equiv 0$ on $\mathfrak{g} - V_4$.

If X is in $\mathfrak{g}$ set $T(X) = ((I - e^{-\operatorname{ad} X})/\operatorname{ad} X)$. If X is in V_4 then $\det(T(X)) \neq 0$. Thus if $h(X) = |\det(T(X))|^{1/2}$ then h is C^∞ on V_4. Set $F_2(X) = F_1(X)\phi(X)h(X)$. Then F_2 is in $C_0^\infty(\mathfrak{g})$. Finally, set $F(X) = \int_{K_*} F_2(\operatorname{Ad}(k)X)dk_*$. Then F is in $C_0^\infty(\mathfrak{g})$ and $F(\operatorname{Ad}(k)X) = F(X)$ for all k in K.

If X is in V_3 and u is in K then $F_2(\operatorname{Ad}(u)X) = F_1(\operatorname{Ad}(u)X)h(\operatorname{Ad}(u)X) = F_1(X)h(X)$. Thus $F(X) = F_2(X) = F_1(X)h(X)$ for X in V_3. If X is in $\bar{V} + \mathfrak{n}$ but X is not in V_3, then X is not in supp F_1. Since $\operatorname{Ad}(K)$ supp F_1 = supp F_1, we see that $\operatorname{Ad}(K)X \cap \operatorname{supp} F_1 = \phi$. Thus $F(X) = \int_{K_*} F_2(\operatorname{ad}(k)X)dk_* = 0 = F_1(X)h(X)$. We therefore have

(3) $F|_{\bar{V}+\mathfrak{n}} = F_1 h|_{\bar{V}+\mathfrak{n}}$.

Lemma 7.11.7 applies to F and we find that

$$cF(0) = \lim_{\substack{Y\to 0\\ Y\in\mathfrak{m}_1}} \partial(\tilde{\omega})\int_{\mathfrak{n}} F(Y+Z)dZ \tag{4}$$

$$= \lim_{\substack{Y\to 0\\ Y\in\mathfrak{m}_1}} \partial(\tilde{\omega})\int_{\mathfrak{n}} F_1(Y+Z)h(Y+Z)dZ.$$

Fix Y in $\mathfrak{m}_1$. Then we may choose $\mathfrak{h}^- \subset \mathfrak{m}$ so that Y is in $\mathfrak{h}^- \oplus \mathfrak{a} = \mathfrak{h}_0$ (see 7.5.6). We now use the notation of Lemma 7.5.11. Let Z be in $\mathfrak{n}$. Then $\operatorname{ad}(Y+Z)$ can be put in upper triangular form as an endomorphism of $\mathfrak{g}_C$ with diagonal entries 0 or $\alpha(Y)$ for α in Δ. Thus $((I - e^{-\operatorname{ad}(Y+Z)})/\operatorname{ad}(Y+Z))$ is in upper triangular form with diagonal entries 1 or $((1 - e^{-\alpha(Y)})/\alpha(Y))$ for α in Δ. We therefore have

$$\det\left(\frac{I - e^{-\operatorname{ad}(Y+Z)}}{\operatorname{ad}(Y+Z)}\right) = (-1)^s \prod_{\alpha\in\Delta^+} \frac{1 - e^{-\alpha(Y)}}{\alpha(Y)} \prod_{\alpha\in\Delta^+} \frac{1 - e^{\alpha(Y)}}{\alpha(Y)} \tag{5}$$

with s the number of elements in Δ^+.

Now

$$\prod_{\alpha\in\Delta^+} 1 - e^{-\alpha(Y)} = e^{-(1/2)\Sigma_{\alpha\in\Delta^+}\alpha} \prod_{\alpha\in\Delta^+} \{e^{\alpha(Y)/2} - e^{-\alpha(Y)/2}\}$$

and

$$\prod_{\alpha\in\Delta^+} 1 - e^{\alpha(Y)} = (-1)^s e^{(1/2)\Sigma_{\alpha\in\Delta^+}\alpha} \prod_{\alpha\in\Delta^+} \{e^{\alpha(Y)/2} - e^{-\alpha(Y)/2}\}.$$

This combined with (5) gives

$$h(Y+Z) = \left|\prod_{\alpha\in\Delta^+} \frac{e^{\alpha(Y)/2} - e^{-\alpha(Y)/2}}{\alpha(Y)}\right|. \tag{6}$$

Let $\mathfrak{m}_1'$ be as in 7.5.21. Set $\bar{V}' = \bar{V} \cap \mathfrak{m}_1'$. If Y is in $\bar{V}'$ then setting $\phi(Z) =$

$\exp(-Y)\exp(Y + Z)$ we see that 7.5.15 combined with 7.5.22 implies that ϕ is a diffeomorphism of $\mathfrak{n}$ onto N. A direct computation shows that

$$(7) \qquad \phi^* dn = \left|\det\left(\frac{I - e^{-\operatorname{ad}(Y+Z)}|_{\mathfrak{n}}}{\operatorname{ad}(Y+Z)|_{\mathfrak{n}}}\right)\right| dZ$$

where dn is normalized so that there are no constants in this formula.

Now

$$\left|\det\left(\frac{I - e^{-\operatorname{ad}(Y+Z)}|_{\mathfrak{n}}}{\operatorname{ad}(Y+Z)|_{\mathfrak{n}}}\right)\right|$$
$$= |\prod_{\alpha\in\Delta^+-\Sigma} \alpha(Y)|e^{\rho(H)}| \prod_{\alpha\in\Delta^+-\Sigma} e^{\alpha(Y)/2} - e^{-\alpha(Y)/2}|^{-1} h(Y+Z)$$

using (6). Here $Y = Y_1 + H$ with Y_1 in $\mathfrak{m}$, H in $\mathfrak{a}$ and $\rho = \frac{1}{2}\sum_{\alpha\in\Sigma} \alpha$. This shows that if Y is in $\overline{V}'$ then

$$(8) \qquad h(Y+Z)\int_{\mathfrak{n}} F_1(Y+Z)dZ = \frac{e^{\rho(H)}}{|\prod_{\alpha\in\Delta^+-\Sigma} \alpha(Y)|} \cdot$$
$$\cdot|(\prod_{\alpha\in\Delta^+-\Sigma} e^{\alpha(Y)/2} - e^{-\alpha(Y)/2})| \int_N f_1(\exp(Y)n)\, dn$$

with the notation as above.

Noting that

$$\left|\frac{\prod_{\alpha\in\Delta^+-\Sigma})e^{\alpha(Y)/2} - e^{-\alpha(Y)/2})}{\prod_{\alpha\in\Delta^+-\Sigma}\alpha(Y)}\right| = \delta_-(Y)$$

(see the statement of the lemma) and that

$$e^{\rho(H)}\int_N f_1(\exp(Y)n)dn = F_f(\exp Y)$$

we have shown using (8), (4), and (3) that

$$(9) \qquad cF(0) = \lim_{\substack{Y\to 0\\ Y\in\mathfrak{m}_1}} \partial(\varpi)\delta_-(Y)F_f(\exp Y).$$

But $F(0) = F_1(0)h(0) = F_1(0) = f_1(e) = f(e)$. Thus (9) proves the lemma.

7.12 The Fourier Expansion of F_f

7.12.1 Let G be a connected semisimple Lie group. Let K, A, N, $\mathfrak{k}$, $\mathfrak{a}$, $\mathfrak{n}$, M, and $\mathfrak{m}$ be as in 7.11.1.

7.12.2 We assume that G has finite center. Thus K and M are compact. Let $\hat{M}$ be the set of all equivalence classes of irreducible finite dimensional unitary representations of M. If ξ is in $\hat{M}$, let χ_ξ be its character. Let $\mathfrak{a}^*$ be the set of all real valued linear functionals on $\mathfrak{a}$.

7.12.3 *Lemma* Let for φ in $C_0^\infty(MA)$,

$$\hat{\varphi}(\xi, \nu) = \frac{1}{(2\pi)^{l/2}} \int_{M \times A} \varphi(ma)\chi_\xi(m)e^{(-1)^{1/2}\nu(\log a)}\, dm\, da$$

where $l = \dim A$, ξ is in $\hat{M}$, ν is in $\mathfrak{a}^*$.

If φ is in $C_0^\infty(MA)$ and if $\varphi(m_0 m m_0^{-1} a) = \varphi(ma)$ for m_0, m in M, a in A then

$$(1) \qquad \varphi(ma) = \frac{1}{(2\pi)^{l/2}} \sum_{\xi \in \hat{M}} \int_{\mathfrak{a}^*} \hat{\varphi}(\xi, \nu)e^{-(-1)^{1/2}\nu(\log a)}\overline{\chi_\xi(m)}\, d\nu$$

where $d\nu$ is the Euclidean volume element on $\mathfrak{a}^*$ gotten by taking $\log^* da$ on $\mathfrak{a}$ and identifying $\mathfrak{a}$ with $\mathfrak{a}^*$. The convergence in (1) is in the sense of H^s for all s (see 5.7.4).

PROOF MA is Lie isomorphic with the product group $M \times A$. A is isomorphic under log with $\mathfrak{a}$. The lemma is now just a combination of the Fourier integral theorem (5.A.5), 2.9.6, and 5.7.10.

7.12.4 *Definition* Let F_f be as in 7.7.11, for f in $C_0^\infty(G)$. Define for ξ in $\hat{M}$, ν in $\mathfrak{a}^*$,

$$(1) \qquad \Theta_{\xi,\nu}(f) = \hat{F_f}(\xi, \nu).$$

It is clear that $\Theta_{\xi,\nu}$ defines a measure on G (see Appendix 7).

7.12.5 *Lemma* If f is in $C_0^\infty(G)$, then

$$F_f(ma) = \frac{1}{(2\pi)^{l/2}} \sum_{\xi \in \hat{M}} \int_{\mathfrak{a}^*} \Theta_{\xi,\nu}(f)\overline{\chi_\xi(m)}e^{-(-1)^{1/2}\nu(\log a)}\, d\nu$$

with convergence in the sense of 7.12.3.

PROOF This is a direct consequence of 7.12.3 and the definition of $\Theta_{\xi,\nu}$ in 7.12.4.

7.12.6 We will see in Chapter 8 that the $\Theta_{\xi,\nu}$ play the roles of characters of certain unitary representations of G. Theorem 7.12.9 below will then play the role of the Peter–Weyl theorem for the groups in question.

7.12.7 Suppose that G is a connected semisimple Lie group with one conjugacy class of Cartan subalgebra. Then 7.9.10 implies that G has finite center and that M is connected. Furthermore if C and A_- are as in 7.7.1 and 7.7.2, then A_- is a maximal torus of M and $C = A_-A$. Also $\Delta_1^+ = \Delta^+ - \Sigma$ (see 7.9 for notation) is a system of positive roots for M relative to A_-. If ξ is in $\hat{M}$ then the theorem of the highest weight says that ξ is completely determined by its highest weight Λ_ξ, a linear form on $\mathfrak{h}^-$ the Lie algebra of A_-.

7.12.8 If λ is a complex valued (real) linear form on $\mathfrak{h}^-$ and if μ is a complex valued linear form on $\mathfrak{a}$ we look upon $\lambda + \mu$ as a complex valued linear form on $\mathfrak{h}^- + \mathfrak{a} = \mathfrak{h}_0$ as follows: If H_1 is in $\mathfrak{h}^-$, H_2 is in $\mathfrak{a}$, then $(\lambda + \mu)(H_1 + H_2) = \lambda(H_1) + \mu(H_2)$.

7.12.9 *Theorem* Let G be a semisimple Lie group with one conjugacy class of Cartan subalgebra. Then there is a constant $c \neq 0$ so that if f is in $C_0^\infty(G)$ then

$$cf(e) = \sum_{\xi \in \hat{M}} \int_{\mathfrak{a}^*} \Theta_{\xi,v}(f) m(\xi, v)\, dv$$

with $m(\xi, v) = d(\xi)\prod_{\alpha\in\Sigma} (\Lambda_\xi + \rho_M + (-1)^{1/2}v)(H_\alpha)$; here $d(\xi)$ is the dimension of any element of ξ, Λ_ξ is the highest weight of ξ, and $\rho_M = \frac{1}{2}\sum_{\alpha\in\Delta_1^+}\alpha$.

PROOF In the notation of 7.11.8 there is a real nonzero constant c_1 so that

$$(1) \qquad c_1 f(e) = \lim_{\substack{Y\to 0\\ Y\in\mathfrak{m}_1}} \partial(\tilde{\omega})\delta_-(Y)F_f(\exp Y).$$

Now 7.12.5 says that if $Y = Y_1 + H$, Y_1 in $\mathfrak{m}$, H in $\mathfrak{a}$, then

$$(2) \qquad F_f(\exp Y)) = \frac{1}{(2\pi)^{l/2}} \sum_{\xi \in M} \int_{\mathfrak{a}^*} \Theta_{\xi,v}(f)\overline{\chi_\xi(\exp Y_1)} e^{-(-1)^{1/2}v(H)}\, dv.$$

Combining (1) and (2) and the convergence statement of 7.12.5 we have

$$(3) \qquad c_1 f(e) = \frac{1}{(2\pi)^{l/2}} \sum_{\xi \in \hat{M}} \int_{\mathfrak{a}^*} \Theta_{\xi,v}(f) m(\xi, v)\, dv$$

with

$$m(\xi, \nu) = \lim_{\substack{Y_1+H\to 0\\ Y_1\in\mathfrak{m}\\ H\in\mathfrak{a}}} \partial(\tilde{\omega})\delta_-(Y_1 + H)\overline{\chi_\xi(\exp Y_1)}e^{-(-1)^{1/2}\nu(H)}.$$

To complete the proof of the theorem we need only compute $m(\xi, \nu)$.

Let W_M be the Weyl group of M relative to A_-. Let $\pi_- = \prod_{\alpha\in\Delta_1^+}\alpha$. Set $\tilde{\omega}_1 = \tilde{\omega}|_{\mathfrak{h}_0}$. Also set

$$g(Y_1 + H) = \delta_-(Y_1 + H)\overline{\chi_\xi(\exp Y_1)}e^{-(-1)^{1/2}\nu(H)}.$$

We note that $\tilde{\omega}$ is in $I(\mathfrak{m}_1)$ (see 7.10.21) and that g is an invariant C^∞ function on $\mathfrak{m}_1$, that is, $g(\mathrm{Ad}(m)X) = g(X)$ for m in M_1, X in $\mathfrak{m}_1$. We may thus apply 7.10.17 and 7.10.27 to $\tilde{\omega}$, g and find that if H is in $\tilde{\mathfrak{h}}_0 = \{H \text{ in } \mathfrak{h}_0 | \pi_-(H) \neq 0\}$, then

$$(\partial(\tilde{\omega})g)(H) = \pi_-(H)^{-1}(\partial(\tilde{\omega}_1)(\pi_- g))(H)(\tilde{\omega}_1 = \tilde{\omega}|_{\mathfrak{h}_0}). \tag{4}$$

Let H be in $\tilde{\mathfrak{h}}_0$. Then $H = H_- + H_+$, H_- in $\mathfrak{h}^-$, H_+ in $\mathfrak{a}$.

$$\prod_{\alpha\in\Delta_1+}(e^{\alpha(H_-)/2} - e^{-\alpha(H)/2})\chi_\xi(H_-) = \sum_{s\in W_M}\det(s)\exp(s(\Lambda_\xi + \rho_M))(H_-). \tag{5}$$

(see 4.9.6).

Also a computation of $\delta_-(H)$ gives (see the last part of the proof of 7.11.8)

$$\delta_-(H) = \frac{|\prod_{\alpha\in\Delta_1+} e^{\alpha(H_-)/2} - e^{-\alpha(H_-)/2}|}{|\prod_{\alpha\in\Delta_1^+}\alpha(H_-)|}. \tag{6}$$

Let now $\mathfrak{h}_R^- = (-1)^{1/2}\mathfrak{h}^-$. If α is in Δ_1 then $\alpha(\mathfrak{h}_R^-) \subset R$. Let X be in $\mathfrak{h}_R^-$ so that $\pi_-(X) \neq 0$, then

$$\frac{e^{\alpha((-1)^{1/2}X)/2} - e^{-\alpha((-1)^{1/2}X)/2}}{\alpha((-1)^{1/2}X)} = 2(-1)^{1/2}\frac{\sin(\alpha(X)/2)}{(-1)^{1/2}\alpha(X)} = \frac{\sin(\alpha(X)/2)}{\alpha(X)/2}. \tag{7}$$

Thus if X is sufficiently near 0, (7) implies that

$$\frac{e^{\alpha((-1)^{1/2}X)/2} - e^{-\alpha((-1)^{1/2}X)/2}}{\alpha((-1)^{1/2}X)} > 0.$$

We therefore see that there is a neighborhood of 0, U in $\mathfrak{h}_0$ so that if H is in $U \cap \tilde{\mathfrak{h}}_0$ then

$$\delta_-(H) = \frac{\prod_{\alpha\in\Delta_1^+}(e^{\alpha(H_-)/2} - e^{-\alpha(H_-)/2})}{\pi_-(H_-)}. \tag{8}$$

Combining (8) with (5) we find

(9) If H is in $U \cap \tilde{\mathfrak{h}}_0$ then

$$\delta_-(H)\overline{\chi_\xi(H)}e^{-(-1)^{1/2}\nu(H_+)}$$
$$= \pm\pi_-(H)^{-1}\sum_{s\in W_M}\det(s)\exp(-\{s(\Lambda_\xi+\rho_M)+(-1)^{1/2}\nu\}(H)).$$

Now $\partial(\tilde{\omega}_1) = \prod_{\alpha\in\Sigma}\partial(H_\alpha)$. Thus we find using (9) and (4) that if H is in $U \cap \tilde{\mathfrak{h}}_0$ then

$$\text{(10)}\quad \pm(\partial(\tilde{\omega})g)(H)$$
$$= \pi_-(H)^{-1}\sum_{s\in W_M}\det(s)\Big(\prod_{\alpha\in\Sigma}\partial(H_\alpha)\Big)$$
$$\cdot e^{(-\{s(\Lambda_\xi+\rho_M)+(-1)^{1/2}\nu\})}(H)$$
$$= \pi_-(H)^{-1}\sum_{s\in W_M}\det(s)\prod_{\alpha\in\Sigma}(-(s(\Lambda_\xi+\rho_M)+(-1)^{1/2}\nu)(H_\alpha))$$
$$\cdot e^{-\{s(\Lambda_\xi+\rho_M)+(-1)^{1/2}\nu\}}(H).$$

But Σ is of even order. Furthermore $W_M \subset W(\mathfrak{h}_0)$, the Weyl group of Δ. If H is in $\mathfrak{a}$ then $sH = H$ for all s in W_M. Also, $s\Sigma = \Sigma$ for all $s \in W_M$; this implies that

$$\text{(11)}\quad \prod_{\alpha\in\Sigma}(-(s(\Lambda_\xi+\rho_M)+(-1)^{1/2}\nu)(H_\alpha))$$
$$= \prod_{\alpha\in\Sigma}(\Lambda_\xi+\rho_M+(-1)^{1/2}\nu)(H_\alpha)$$

for all s in W_M.

Combining (10) with (11) we have for H in $U \cap \tilde{\mathfrak{h}}_0$

$$\text{(12)}\quad \pm(\partial(\tilde{\omega})g)(H) = \prod_{\alpha\in\Sigma}(\Lambda_\xi+\rho_M+(-1)^{1/2}\nu)(H_\alpha)e^{-(-1)^{1/2}\nu(H_+)}$$
$$\cdot\pi_-(H_-)^{-1}\sum_{s\in W_M}\det(s)e^{-s(\Lambda_\xi+\rho_M)(H_-)}.$$

Now

$$\lim_{H\to 0}\frac{\prod_{\alpha\in\Delta_1^+}e^{\alpha(H_-)/2}-e^{-\alpha(H_-)/2}}{\pi_-(H_-)} = 1$$

according to (7). Thus

$$\lim_{H\to 0}\pi_-(H_-)^{-1}\sum_{s\in W_M}\det(s)e^{-s(\Lambda_\xi+\rho_M)(H_-)} = \pm\overline{\chi_\xi(e)} = \pm d(\xi).$$

This gives

$$\text{(13)}\quad \lim_{H\to 0}(\partial(\tilde{\omega})g)(H) = d(\xi)\prod_{\alpha\in\Sigma}(\Lambda_\xi+\rho_M+(-1)^{1/2}\nu)(H_\alpha).$$

The theorem is now completely proved.

7.13 Exercises

7.13.1 Let $G = SL(n, R)$ (resp. $SL(n, C)$). Define for $g = (g_{ij})$ in G, $\Delta_k(g) = \det((g_{ij})_{i,j \leqslant k})$. Show that if $K = SO(n)$ (resp. $SU(n)$),

$$A = \left\{ \begin{bmatrix} a_1 & & 0 \\ & \ddots & \\ 0 & & a_n \end{bmatrix} \middle| a_i > 0,\ a_1 \cdots a_n = 1 \right\}, \qquad N = \left\{ n = \begin{bmatrix} 1 & & * \\ & \ddots & \\ 0 & & 1 \end{bmatrix} \middle| n \text{ in } G \right\}.$$

Then $G = KAN$ is an Iwasawa decomposition of G. Show that $\bar{N}MAN = g$ G, $\Delta_k(g) \neq 0$, for $k = 1, \ldots, n$.

7.13.2 Let G be as in 7.13.1; derive 7.5.18 for G using 7.13.1.

7.13.3 Let G be a connected semisimple Lie group. Use the techniques of the proof of 7.4.3 and 7.13.1 rather than 7.5.18 to prove 7.5.20.

7.13.4 Let $G = SU(2)$ and let $\mathfrak{g}$ be its Lie algebra. Let

$$H = \begin{bmatrix} i & 0 \\ 0 & -i \end{bmatrix}.$$

Then RH is a maximal Abelian subalgebra of $\mathfrak{g}$. Show that

$$t \int_{SU(2)} e^{\operatorname{tr}(tgHg^{-1}H)} dg = \frac{1}{4}(e^{2t} - e^{-2t}).$$

(Hint: Let $f(X) = \int_{SU(2)} e^{\operatorname{tr}(gXg^{-1}H)} dg$. Let X_1, X_2, X_3 be in $\mathfrak{g}$ so that $X_1 = H$, $\operatorname{tr} X_i X_j = -2\delta_{ij}$. Let $\Omega = \Sigma X_i^2$. Compute $\partial(\Omega) f$ and use 7.10.27).

7.13.5 Let (π, C^3) be the irreducible representation of $SU(2)$ on C^3 (show that there is exactly one). Show that there is a nonzero constant c so that if f is in $C_0^\infty(C^3)$ then

$$\lim_{z_2 \to 0} \frac{\partial^2}{\partial z_2 \partial \bar{z}_2} \int_C \left(\int_{SU(2)} f(\pi(g)(0, z_2, z_3)) dg \right) dz_3 = cf(0).$$

Compute c. (Hint: Use 7.11.7).

7.13.6 Let $G = U(n)$. Let $\mathfrak{g}$ be its Lie algebra and let $\mathfrak{h}$ be the maximal

Abelain subalgebra consisting of the diagonal matrices. If X is a diagonal matrix with entries $z_1, \ldots, z_n$ set $X = [z_1, \ldots, z_n]$. Let T be the maximal torus in G corresponding to $\mathfrak{h}$ and let $W(T)$ be the corresponding Weyl group. Show that $W(T)$ consists of all s such that if $H = [it_1, \ldots, it_n]$ then $sH = [it_{\sigma_1}, \ldots, it_{\sigma_n}]$ where σ is a permutation on n letters depending only on s.

7.13.7 Let the notation be as in 7.13.6. Let $I(\mathfrak{g})$ be as in 7.10.21. Let $p_1, \ldots, p_n$ be defined by $\det(X - tI) = \pm t^n + \sum_{j=1}^n p_j(X)t^{n-j}$. Show that if q is in $I(\mathfrak{g})$ then $q = q_1(p_1, \ldots, p_n)$ with q_1 a polynomial in n variables. (Hint: $P_j|_{\mathfrak{h}}$ is a constant multiple of the polynomial $\sigma_j([it_1, \ldots, it_n]) = \sum_{1 \leqq i_1 < \cdots < i_j} t_{i_1} \cdots t_{i_j}$.)

7.13.8 (7.13.6 *continued*) Suppose that f: $\mathfrak{g} \to C$ is analytic (in the real sense) and that (in the notation of 7.13.7) $\partial(\sigma_j)f = \mu_j f, j = 1, \ldots, n$. Suppose that there is H_0 in $\mathfrak{h}$ so that $H_0 = [it_1, \ldots, it_n]$, $\prod_{i<j}(t_i - t_j) \neq 0$, and $\sigma_j(H_0) = \mu_j$, $j = 1, \ldots, n$. Show that $f(H) = \sum_{s \in W} c_s e^{\operatorname{tr}(H(sH_0))}$. (Hint: Prove the result by induction noting that the restriction of the σ_j to the "$\mathfrak{h}$" corresponding to $U(n-1)$ are the corresponding "σ_j".)

7.13.9 Use 7.13.8 to derive a generalization of 7.13.4 to $U(n)$.

7.13.10 Use the results of 7.13.9 and 7.8.3 to analyze the Fourier transform on the Lie algebra of $U(n)$ using techniques of 7.11.

7.13.11 Write out the formula 7.12.9 as explicitly as you can for $G = SL(2, C)$. (That is, write out all decompositions and integral formulas using explicit coordinates, etc.)

7.13.12 Let G be a semisimple Lie group with Lie algebra $\mathfrak{g}$. Let θ be a Cartan involution for $\mathfrak{g}$. Show that if $\mathfrak{h}$ is a Cartan subalgebra of $\mathfrak{g}$ then there is a g in G so that $\theta(\operatorname{Ad}(g)\mathfrak{h}) = \operatorname{Ad}(g)\mathfrak{h}$.

7.13.13 Let G be a semisimple Lie group with finite center. Let $G = KAN$ be an Iwasawa decomposition of G. Let M be the centralizer of A in K. Use 7.13.12 to show that if $\dim A = 1$ then G has $2^{(\operatorname{rank} K - \operatorname{rank} M)}$ conjugacy classes of Cartan subalgebra.

7.14 Notes

7.14.1 7.2.5 is due to Cartan [1]. The proof we use of this result is due to Mostow [1].

7.14.2 The results of 7.3 and 7.4 are due to Iwasawa [1].

7.14.3 Lemma 7.5.15 is taken from Harish-Chandra [3].

7.14.4 7.5.18 is the so-called Bruhat lemma originally proved by Bruhat for the classical groups and by Harish-Chandra in the general case. See Bruhat [1] and Harish-Chandra [5].

7.14.5 7.5.20 was originally proved by Gelfand and Naimark [1] for the classical groups. We develop their original proof in 7.13.2 and 7.13.3.

7.14.6 7.5.22 is a result of Harish-Chandra [3]. Harish-Chandra's proof of this lemma (Harish-Chandra [3], see also Helgason [1], p. 232, Theorem 4.7) is quite complicated. 7.5.22 is the key to "F_f" and the original proof of 7.5.22 was the main reason for the seeming difficulty of "F_f".

7.14.7 The integral formulas of 7.6 are due to Harish-Chandra [3] and [8]. The trick of using all the "built-in invariances" to compute integral formulas is taken from Harish-Chandra's recent lectures on harmonic analysis on p-adic groups.

7.14.8 The integral formulas of 7.7 and 7.8 are slight variations on the theme of Harish-Chandra [3].

7.14.9 The notion F_f (7.7.11) is due to Harish-Chandra [8] and its restriction to A plays a critical role in the analysis of spherical functions (see Helgason [1, 3]).

7.14.10 Section 7.10 is taken almost verbatum out of Harish-Chandra [6].

7.14.11 7.11.8 generalizes a formula of Harish-Chandra [3] for complex semisimple Lie groups. The techniques of the proof are modifications of Harish-Chandra's proof of this result in the complex case.

7.14.12 7.12.9 is essentially the Plancherel theorem for the groups in question. This formula was derived in the complex case by Harish-Chandra in [3].

CHAPTER 8

Representations of Semisimple Lie Groups

8.1 Introduction

In this chapter the basic (infinite dimensional) representation theory of semisimple Lie groups is developed. In Sections 8.3 and 8.4 the principal series of representations is defined. These representations are used as models for the results on K-finite representations. In Section 8.5, we study the finite dimensional representations of a semisimple Lie group as subrepresentations of the principal series. We use the finite dimensional theory to study the principal series. Indeed, 8.5.15 says in essence that principal series representations are "limits" of finite dimensional representations.

In Section 8.6, the character theory of K-finite representations is developed. 8.6.9 takes the place of the orthogonality relations in the compact theory. The character is defined in 8.6.19 and in 8.6.21 it is shown (following Harish-Chandra) that characters do indeed determine unitary representations.

In Section 8.7 a more refined notion of character is developed. We use the results of Chapter 5 to prove that this character exists subject to some growth properties of the dimensions of the K subrepresentations.

In Section 8.8 the character of a principal series representation is computed to be a "Fourier transform" in the compact Abelian sense of F_f (see 7.7.11 and 7.12). This relates the analysis of Chapter 7 to representation theory. The character of a unitary principal series representation predicts certain intertwining operators. Sections 8.9, 8.10, 8.11, and 8.14 are concerned with finding in an explicit manner these intertwining operators. In the process we develop the Kunze–Stein, Schiffmann, Knapp–Stein, Helgason theory of intertwining operators, using heavily a celebrated technique of Gindinkin-Karpelevic. Our technique of proving the meromorphic continuation of the intertwining operators is probably the hardest way. We, however, prove much more. Indeed, we give explicit formulas for the matrix entries of the intertwining operators in terms of gamma functions. We feel that 8.11.3 (a result of Kostant) has independent interest.

The results of Section 8.12 foreshadow the Harish-Chandra "philosophy of cusp forms." 8.12.6 should have some relation with "S-matrix" theory.

The purpose of Section 8.13 is to prove that "almost all" principal series representations are irreducible.

In Section 8.14 using the results of Section 8.13 we give essentially Kunze–Stein normalization of the intertwining operators.

In Section 8.15, we make the link with the results of Chapter 7 (7.12 in particular) and give the Plancherel theorem for semisimple Lie groups with one conjugacy class of Cartan subalgebra. We then relate the norm of the intertwining operators with the Plancherel measure (following Knapp and Stein) for the Lorentz groups, $SO(2n + 1, 1)$, Spin$(2n + 1, 1)$. The most general form of this result (for arbitrary cuspidal representations of arbitrary semisimple Lie groups with finite center) is called the Mass–Selberg relations by Harish-Chandra. The Mass–Selberg relations have been used by Harish-Chandra to give explicitly the Plancherel measure for semisimple Lie groups with finite centers.

8.2 Finite Dimensional Unitary Representations

8.2.1 Let G be a connected Lie group. Let (π, H) be a finite dimensional unitary representation of G. We may identify H with C^n. Then $\pi: G \to U(n)$ is a Lie homomorphism. Let $\mathfrak{g}$ be the Lie algebra of G. π induces a homomorphism of $\mathfrak{g}$, which we also denote by π. If X, Y are in $\mathfrak{g}$ define $(X, Y) = -\operatorname{tr}\pi(X)\pi(Y)$. Then $(\ , \ )$ is positive semidefinite and $\mathfrak{g}$-invariant.

8.2.2 *Lemma* Let $\mathfrak{g}_\pi = \{X \text{ in } \mathfrak{g}_i | (X, Y) = 0 \text{ for all } Y \text{ in } \mathfrak{g}\}$. Then $\mathfrak{g}_\pi = \ker \pi$.

PROOF If $\pi(X) = 0$ then $(X, Y) = 0$ for all Y in $\mathfrak{g}$. If $(X, Y) = 0$ for all Y in $\mathfrak{g}$ then $(X, X) = 0$. But $\pi(X)$ is skew Hermitian and thus if $\operatorname{tr}\pi(X)\pi(X) = 0$, then $\pi(X) = 0$.

8.2.3 8.2.2 says that (,) induces a positive definite inner product on $\mathfrak{g}/\mathfrak{g}_\pi$ which is $\mathfrak{g}$-invariant.

8.2.4 *Lemma* $\mathfrak{g}/\mathfrak{g}_\pi = \mathfrak{a} \oplus \mathfrak{g}_1$ with $\mathfrak{a}$ the center of $\mathfrak{g}/\mathfrak{g}_\pi$ and $\mathfrak{g}_1$ a compact ideal.

PROOF (,) is positive definite on $\mathfrak{g}/\mathfrak{g}_\pi$, hence the Killing form of $\mathfrak{g}/\mathfrak{g}_\pi$ is negative semidefinite. If $\mathfrak{g}/\mathfrak{g}_\pi$ has no ideals then either $\mathfrak{g}/\mathfrak{g}_\pi$ is Abelian or $\mathfrak{g}/\mathfrak{g}_\pi$ is compact and the lemma is true in this case. If $\mathfrak{g}/\mathfrak{g}_\pi$ has an ideal $\mathfrak{U}$, then $\mathfrak{U}^\perp = \{Z \text{ in } \mathfrak{g}/\mathfrak{g}_{\pi;}|(\mathfrak{U}, Z) = 0\}$ is an ideal in $\mathfrak{g}/\mathfrak{g}_\pi$ and thus $\mathfrak{g}/\mathfrak{g}_\pi = \mathfrak{U} \oplus \mathfrak{U}^\perp$. We may continue splitting $\mathfrak{U}$ and $\mathfrak{U}^\perp$ until we have the lemma.

8.2.5 Let $\mathfrak{g}_0 = \bigcap \mathfrak{g}_\pi$ the intersection running over all finite dimensional unitary representations of G.

8.2.6 *Lemma* There is a finite dimensional unitary representation of G, (π, H), so that $\mathfrak{g}_\pi = \mathfrak{g}_0$.

PROOF Let $X_1, \ldots, X_n$ be a basis of $\mathfrak{g}/\mathfrak{g}_0$. Then for each X_i there is (π_i, H_i), a finite dimensional unitary representation of G so that $\pi_i(\tilde{X}_i) \neq 0$ where $\tilde{X}_i$ is in $\tilde{X}_i + \mathfrak{g}_0 = X_i$. Let $\pi = \pi_1 \oplus \cdots \oplus \pi_n$.

8.2.7 *Proposition* Let G be a connected Lie group. Then there is a closed normal subgroup G_0 of G so that the Lie algebra of G/G_0 is of the form $\mathfrak{a} \oplus \mathfrak{g}_1$ with $\mathfrak{a}$ the center and $\mathfrak{g}_1$ a compact ideal and such that if (π, H) is a finite dimensional unitary representation of G then $\ker \pi \supset G_0$.

PROOF G_0 can be taken to be the intersection of all kernels of finite dimensional unitary representations of G. The result now follows from 8.2.6 and 8.2.4.

8.2.8 *Corollary* If G is a connected semisimple Lie group whose Lie algebra has no compact ideals then the only finite dimensional unitary representation of G is the trivial one-dimensional representation (i.e., $G = G_0$ in 8.2.7).

8.2.9 If G is a compact Lie group, then 2.8.5 says that the space of matrix elements of finite dimensional representations of G is dense in $C(G)$ relative to the uniform topology. 8.2.7 says that there is no analog of this result for general noncompact Lie groups. However, there is an analog of the statement $L^2(G) = \sum_{\gamma \in \hat{G}} V_\gamma \otimes V_\gamma^*$ for noncompact Lie groups, if $\hat{G}$ is taken to be the set of all equivalence classes of irreducible unitary representations of G. One of the main tasks of this chapter is to find the analog of the L^2 Peter–Weyl theorem for certain semisimple Lie groups.

8.3 The Principal Series

8.3.1 Let G be a connected semisimple Lie group with finite center. Let $G = KAN$ be an Iwasawa decomposition of G (see 7.4.3). Let M be the centralizer of A in K and set $B = MAN$.

8.3.2 Let $\mathfrak{a}$ be the Lie algebra of A and let $\mathfrak{a}^*$ and $\mathfrak{a}_C^*$ be, respectively, the spaces of real valued and complex valued real linear forms on $\mathfrak{a}$. If $\nu \in \mathfrak{a}_C^*$ we define a representation of A on C by

$$a \mapsto a^\nu = e^{\nu(\log a)}$$

where $\log: A \to \mathfrak{a}$ is the inverse to $\exp: \mathfrak{a} \to A$ (see 7.4.3).

8.3.3 Let $\hat{M}$ be the set of all equivalence classes of finite dimensional irreducible unitary representations of M. If ξ is in $\hat{M}$ let (ξ, H_ξ) in ξ be fixed (we abuse notation and use the same symbol for ξ in $\hat{M}$ and the action of ξ on H_ξ).

8.3.4 If ξ is in $\hat{M}$, ν is in $\mathfrak{a}_C^*$, we define a representation $\mu_{\xi,\nu}$ of B on H_ξ by the formula

(1) $$\mu_{\xi,\nu}(man) = \xi(m)a^{\sqrt{-1}\nu}, \ m \text{ in } M,\ a \text{ in } A,\ n \text{ in } N.$$

We note that $\mu_{\xi,\nu}$ is unitary if and only if ν is in $\mathfrak{a}^*$.

8.3.5 Let $E_{\xi,\nu}$ be the homogeneous vector bundle $G\times_{\mu_{\gamma,\xi}} H_\xi$ over G/B (see 5.2.2). 7.4.3 says that $G = KAN$, thus $G/B = K/M$ as a homogeneous space of K. Since G has finite center, K is compact and thus G/B is a compact homogeneous space of G. Furthermore since $G/B = K/M$ there is a unique K-invariant volume element dx on G/B so that $\int_{K/M} dx = 1$.

8.3.6 If g is in G then g induces a diffeomorphism of K/M given by $gkM = gkMAN$ in $G/B = K/M$. Let $g^*dx = C(g, x)dx$. Then 7.6.6 says $C(g, x) = e^{-2\rho(H(gx))}$ where $H(kan) = \log a$ for k in K, a in A, n in N, and $\rho(H) = \frac{1}{2}\mathrm{tr}(\mathrm{ad}\, H|_{\mathfrak{n}})$ for H in $\mathfrak{a}$ (here $\mathfrak{n}$ is the Lie algebra of N).

8.3.7 As a vector bundle over K/M, $E_{\xi,\nu} = K \times_{\xi} H_{\xi}$. Since ξ is unitary, $E_{\xi,\nu}$ has a K-invariant unitary structure $\langle\ ,\ \rangle$ (see 5.2.7). If ν is in $\mathfrak{a}^*$ then $\mu_{\xi,\nu}$ is unitary, hence $\langle\ ,\ \rangle$ may be taken to be G-invariant.

8.3.8 Since K/M is compact, $(E_{\xi,\nu}, \langle\ ,\ \rangle)$ is G-admissible in the sense of 2.4.4. 2.4.6 implies that $(\pi_{\xi,\nu}, L^2(E_{\xi,\nu}, dx))$ defines a representation of G where $(\pi_{\xi,\nu}(g)f)(x) = e^{-\rho(H(g^{-1}k))}gf(g^{-1}x)$ for $x = kB$, k in K, g in G, and f in $\Gamma(E_{\xi,\nu})$. We set $H^{\xi,\nu} = L^2(E_{\xi,\nu}, dx)$. Then $(\pi_{\xi,\nu}, H^{\xi,\nu})$ defines a unitary representation of G if ν is in $\mathfrak{a}^*$ (see 2.4.6).

8.3.9 *Definition* The series of unitary representations $(\pi_{\xi,\nu}, H^{\xi,\nu})$ for ξ in $\hat{M}$, ν in $\mathfrak{a}^*$ is called the principal series of representations of G. The series of representations $(\pi_{\xi,\nu}, H^{\xi,\nu})$ for ξ in $\hat{M}$, ν in $\mathfrak{a}_C^*$ is called the analytic continuation of the principal series or the nonunitary principal series.

8.3.10 As a Hilbert space $H^{\xi,\nu} = L^2(E_{\xi}, dx)$ and we may thus look at $\pi_{\xi,\nu}$ as a representation of G on $L^2(E_{\xi}, dx)$. That is, the $\pi_{\xi,\nu}$ for $\nu \in \mathfrak{a}_C^*$ and ξ in M fixed all act on the same Hilbert space.

8.3.11 *Lemma* If f_1, f_2 are in $L^2(E_{\xi}, dx)$ let $(f_1, f_2) =$

$$\int_{K/M} \langle f_1(x), f_2(x)\rangle_x \, dx$$

be the inner product on $L^2(E_{\xi}, dx)$. Then $(\pi_{\xi,\nu}(g)f_1, \pi_{\xi,\bar{\nu}}(g)f_2) = (f_1, f_2)$ for all g in G, ν in $\mathfrak{a}_C^*$.

PROOF If v is in $(E_{\xi,\nu})_{K/M}$ then $v = k \cdot v_0$ with v_0 in $(E_{\xi,\nu})_{eM}$ and $g \cdot v = gkv_0 = k(gk)\exp(H(gk))n(gk)v_0$ where if x is in G,

$$x = k(x)\exp(H(x))n(x)$$

with $k(x)$ in K, $H(x)$ in $\mathfrak{a}$, $n(x)$ in N. Now $\exp(H(gk))n(gk)v_0 = e^{\sqrt{-1}\nu(H(gk))}v_0$. Thus $g \cdot v = e^{\sqrt{-1}\nu(H(gk))}k(gk)v_0$. Hence if we denote the action of G on $E_{\xi,\nu}$ by τ_ν then

$$\langle(\pi_{\xi,\nu}(g)f_1)(x), (\pi_{\xi,\bar\nu}(g)f_2)(x)\rangle_x$$
$$= e^{-2\rho(H(g^{-1}k))}\langle(\tau_\nu(g)f(g^{-1}x), \tau_\nu(g)f(g^{-1}x)\rangle_x$$
$$= e^{-2\rho(H(g^{-1}k))}\langle e^{\sqrt{-1}\nu(H(gk))}f_1(g^{-1}x), e^{-\sqrt{-1}\nu(H(gk))}f_2(g^{-1}x)\rangle_{g^{-1}x}.$$

Hence

$$(\pi_{\xi,\nu}(g)f_1, \pi_{\xi,\bar\nu}(g)f_2) = \int_{K/M} e^{-2\rho(H(g^{-1}k))}\langle f(g^{-1}x), f_2(g^{-1}x)\rangle_{g^{-1}x}\,dx$$
$$= \int_{K/M} \langle f_1(x), f_2(x)\rangle\,dx \text{ by 7.6.6.}$$

8.4 Other Realizations of the Principal Series

8.4.1 We retain the notation of 8.3. Let ξ be in $\hat{M}$ and ν in $\mathfrak{a}_{\mathbb{C}}^*$. Let $C_{\xi,\nu}(G)$ be the space of all continuous functions f from G to H_ξ so that

(1) $$f(gman) = \xi(m)^{-1}e^{-\nu(\log a)}f(g).$$

Let for f in $C_{\xi,\nu}(G)$. $\|f\|^2 = \int_{K/M}\|f(x)\|^2\,dx$ where $\|f(kM)\|^2 = \|f(k)\|^2$ is well defined since ξ is unitary.

Let $(\tilde\pi_{\xi,\nu}(g)f)(x) = f(g^{-1}x)$ for g, x in G.

8.4.2 *Lemma* $(\tilde\pi_{\xi,\nu}, C_{\xi,\nu}(G))$ extends to a representation of G on $\tilde{H}^{\xi,\nu}$, the Hilbert space completion of $C_{\xi,\nu}$. Furthermore if ν is in $\mathfrak{a}_{\mathbb{C}}^*$ and f is in $\Gamma(E_{\xi,\nu})$ define $\tilde{f}(g) = e^{-\rho(H(g))}g^{-1}f(gB)$ (here we identify $(E_{\xi,\nu})_{eB}$ with H_ξ). Then $\tilde{f}$ is in $C_{\xi,\rho+\sqrt{-1}\nu}(G)$ and if $A_{\xi,\nu}(f) = \tilde{f}$ then $A_{\xi,\nu}$ extends to an isometry of the Hilbert spaces $H^{\xi,\nu}$ and $H^{\xi,\rho+\sqrt{-1}\nu}$ such that $A_{\xi,\nu}\circ\pi_{\xi,\nu}(g) = \tilde\pi_{\xi,\rho+\sqrt{-1}\nu}(g)A_{\xi,\nu}$. In particular, $(\pi_{\xi,\nu}, H^{\xi,\nu})$ and $(\tilde\pi_{\xi,\rho+\sqrt{-1}\nu}, \tilde{H}^{\xi,\rho+\sqrt{-1}\nu})$ are equivalent representations of G.

PROOF Let f be in $\Gamma(E_{\xi,\nu})$. Set $f_1(g) = (\pi_{\xi,\nu}(g)^{-1}f)(eK)$. Then $(\pi_{\xi,\nu}(g)^{-1}f)(e) = e^{-\rho(H(g))}gf(g^{-1}x) = \tilde{f}(g)$. Now

$$f(gman) = e^{-\rho(H(gman))}n^{-1}a^{-1}m^{-1}g^{-1}f(gB)$$
$$= e^{-\rho(H(g)+(\log a))}e^{-\sqrt{-1}\nu(\log a)}\xi(m^{-1})g^{-1}f(gB).$$

Since $g^{-1}f(gB)$ is in $(E_{\xi,\nu})_{eB} = H_\xi$ and if v is in $(E_{\xi,\nu})_{eB}$ then $n\cdot v = v$ for n in N, $m\cdot v = \xi(m)v$, and $a\cdot v = e^{\sqrt{-1}\nu(\log a)}v$. Thus $A_{\xi,\nu}f$ is in $C_{\xi,\rho+\sqrt{-1}\nu}(G)$. Since $(\pi_{\xi,\nu}(g)^{-1}f)(e) = f(g)$ we see that $A_{\xi,\nu}\pi_{\xi,\nu}(g) = \tilde\pi_{\xi,\rho+\sqrt{-1}\nu}(g)A_{\xi,\nu}$ for g in G. 5.3.4 implies that $A_{\xi,\nu}$ extends to an isometry of Hilbert spaces which is bijective. The lemma now follows.

8.4.3 The realization $(\tilde{\pi}_{\xi,\lambda}, \tilde{H}^{\xi,\lambda})$ of the principal series is useful for the purposes of proving irreducibility. We use this model to find yet another model.

8.4.4 7.5.20 implies that if f is in $C_{\xi,\rho+\sqrt{-1}\nu}(G)$ then $f|_{\bar{N}}$ completely determines f (here we use the notation of 7.5.20). Furthermore using 7.6.8 we see that $\|f\|^2 = \int_{\bar{N}} \|f(k(\bar{n}))\|^2 e^{-2\rho(H(\bar{n}))} d\bar{n}$ where $d\bar{n}$ is Haar measure on $\bar{N}$ normalized so that $\int_{\bar{N}} e^{-2\rho(H(\bar{n}))} d\bar{n} = 1$.

8.4.5 If g is in $\bar{N}MAN$ let $\bar{n}(g)$, $m(g)$, $a(g)$, $n(g)$ be so that $\bar{n}(g)$ is in $\bar{N}$, $m(g)$ is in M, $a(g)$ is in A, $n(g)$ is in N, and $g = \bar{n}(g)m(g)a(g)n(g)$. Then n, m, a, m are defined on an open dense subset of G (see 7.5.20).

8.4.6 Let $C_0(\bar{N}, H_\xi)$ be the space of all continuous compactly supported functions $f: \bar{N} \to H_\xi$. Let $L^2(\bar{N};\ \xi)$ be the Hilbert space completion of $C_0(\bar{N}, H_\xi)$ relative to the inner product $(f_1, f_2) = \int_{\bar{N}} \langle f_1(\bar{n}), f_2(\bar{n})\rangle\, d\bar{n}$.

8.4.7 *Lemma* If f is in $\tilde{H}^{\xi,\rho+\sqrt{-1}\nu}$ and ν is in $\mathfrak{a}^*$ set $(B_{\xi,\nu}f)(\bar{n}) = f(\bar{n})$. Then $B_{\xi,\nu}f$ is in $L^2(\bar{N};\ H_\xi)$ and $B_{\xi,\nu}: \tilde{H}^{\xi,\rho+\sqrt{-1}\nu} \to L^2(\bar{N};\ H_\xi)$ is an isometry of Hilbert spaces. Furthermore

(1) $$(B_{\xi,\nu}\tilde{\pi}_{\xi,\rho+\sqrt{-1}\nu}(g)f)(\bar{n}) = \xi(m(g^{-1}\bar{n}))^{-1} e^{-(\rho+\sqrt{-1}\nu)(\log a(g^{-1}\bar{n}))}(B_{\xi,\nu}f)(\bar{n}(g^{-1}\bar{n}))$$

for almost all g in G, $\bar{n}$ in $\bar{N}$.

PROOF Let f be in $\tilde{H}^{\xi,\rho+\sqrt{-1}\nu}$. Then $f(\bar{n}) = f(k(\bar{n})\exp(H(\bar{n}))n)$ with n in N (if g is in G, $g = k(g)\exp(H(g))n$ with $k(g)$ in K, $H(g)$ in $\mathfrak{a}$). Thus $f(\bar{n}) = e^{-(\rho+\sqrt{-1}\nu)(H(\bar{n}))} f(k(\bar{n}))$. Thus $B_{\xi,\nu}(f)(\bar{n}) = e^{-(\rho+\sqrt{-1}\nu)(H(\bar{n}))} f(k(\bar{n}))$. Hence

$$\int_{\bar{N}} \langle B_{\xi,\nu}(f)(\bar{n}), B_{\xi,\nu}(f)(\bar{n})\rangle d\bar{n} = \int_{\bar{N}} e^{-2\rho(H(\bar{n}))} \langle f(k(\bar{n})), f(k(\bar{n}))\rangle\, d\bar{n} = \int_{K/M} \langle f(x), f(x)\rangle\, dx,$$

by 7.6.8. Thus $\|B_{\xi,\nu}(f)\|^2 = \|f\|^2$. If f is in $L^2(\bar{N};\ H_\xi)$ let $f_1(\bar{n}man) = \xi(m)^{-1} e^{(\rho+\sqrt{-1}\nu)(\log a)} f(\bar{n})$. Then $B_{\xi,\nu}f_1 = f$. Thus $B_{\xi,\nu}$ is an isometry of the Hilbert spaces $\tilde{H}^{\xi,\rho+\sqrt{-1}\nu}$ and $L^2(\bar{N};\ H_\xi)$.

$$\begin{aligned}(B_{\xi,\nu}\tilde{\pi}_{\xi,\rho+i\nu}(g)f)(\bar{n}) &= f(g^{-1}\bar{n}) \\ &= f(\bar{n}(g^{-1}\bar{n})m(g^{-1}\bar{n})a(g^{-1}\bar{n})n(g^{-1}\bar{n})) \\ &= \xi(m(g^{-1}\bar{n}))^{-1}e^{-(\rho+\sqrt{-1}\nu)(\log a(g^{-1}\bar{n}))}f(\bar{n}(g^{-1}\bar{n})) \\ &= \xi(m(g^{-1}\bar{n}))^{-1}e^{-(\rho+\sqrt{-1}\nu)(\log(a(g^{-1}\bar{n}))}(B_{\xi,\nu}f)(\bar{n}(g^{-1}\bar{n})).\end{aligned}$$

8.4.8 Let ξ be in $\hat{M}$, ν in $\mathfrak{a}^*$, then 8.4.7 implies that if we define for f in $L^2(\bar{N}; H_\xi)$ g in G

$$(\bar{\pi}_{\xi,\nu}(g)f)(\bar{n}) = \xi(m(g^{-1}\bar{n}))^{-1}e^{-(\rho+\sqrt{-1}\nu)(\log a(g^{-1}\bar{n}))}f(\bar{n}(g^{-1}\bar{n}))$$

then $(\bar{\pi}_{\xi,\nu}, L^2(\bar{N}, H_\xi))$ is a unitary representation unitarily equivalent with the principal series representation $(\pi_{\xi,\nu}, H^{\xi,\nu})$.

8.4.9 *Example 1 SL(2, C)* Let $G = SL(2, C)$, the group of all 2×2 complex matrices g so that $\det(g) = 1$. We may take

$$\bar{N} = \left\{\begin{bmatrix}1 & 0\\ z & 1\end{bmatrix}\middle|\, z \text{ in } C\right\}, \qquad M = \left\{\begin{bmatrix}e^{i\theta} & 0\\ 0 & e^{-i\theta}\end{bmatrix}\middle|\, \theta \text{ in } R\right\},$$

$$A = \left\{\begin{bmatrix}a & 0\\ 0 & a^{-1}\end{bmatrix}\middle|\, a > 0,\ a \text{ in } R\right\}, \qquad N = \left\{\begin{bmatrix}1 & z\\ 0 & 1\end{bmatrix}\middle|\, z \text{ in } C\right\}.$$

A computation shows that g is in $\bar{N}MAN$ if and only if $g_{11} \neq 0$ and

$$\bar{n}(g) = \begin{bmatrix}1 & 0\\ \dfrac{g_{21}}{g_{11}} & 1\end{bmatrix}, \qquad m(g) = \begin{pmatrix}\dfrac{g_{11}}{|g_{11}|} & 0\\ 0 & \dfrac{|g_{11}|}{g_{11}}\end{pmatrix},$$

$$a(g) = \begin{bmatrix}|g_{11}| & 0\\ 0 & |g_{11}|^{-1}\end{bmatrix}, \qquad n(g) = \begin{bmatrix}1 & \dfrac{g_{12}}{g_{11}}\\ 0 & 1\end{bmatrix}$$

with

$$g = \begin{bmatrix}g_{11} & g_{12}\\ g_{21} & g_{22}\end{bmatrix}.$$

$\hat{M}$ is parametrized by the integers with

$$\xi_k\left(\begin{bmatrix}e^{i\theta} & 0\\ 0 & e^{-i\theta}\end{bmatrix}\right) = e^{ik\theta}.$$

Thus

$$\mu_{\xi_{k,\nu}}\left(\begin{bmatrix} e^{i\theta} & 0 \\ 0 & e^{-i\theta}\end{bmatrix}\begin{bmatrix} a & 0 \\ 0 & a^{-1}\end{bmatrix}\right) = e^{ik\theta}a^{\sqrt{-1}\nu}, \quad \text{in } C.$$

Also

$$\begin{bmatrix} a & 0 \\ 0 & a^{-1}\end{bmatrix}^{\rho} = a^2.$$

Indeed

$$\mathfrak{n} = \left\{\begin{bmatrix} 0 & z \\ 0 & 0\end{bmatrix}\middle| z \text{ in } C\right\}.$$

$$\mathfrak{a} = RH \text{ with } H = \begin{bmatrix} 1 & 0 \\ 0 & -1\end{bmatrix}.$$

$$\left[H, \begin{bmatrix} 0 & z \\ 0 & 0\end{bmatrix}\right] = \begin{bmatrix} 0 & 2z \\ 0 & 0\end{bmatrix}.$$

Thus $\operatorname{ad} H|_{\mathfrak{n}} = 2I$. Thus $\operatorname{tr}(\operatorname{ad} H|_{\mathfrak{n}}) = 2 \dim_R \mathfrak{n} = 4$. Thus $\rho(H) = 2$. Hence $(e^{tH})^{\rho} = e^{2t}$. Let $\bar{\pi}_{\xi_{k,\nu}} = \bar{\pi}_{k,\nu}$. As a Lie group $\bar{N}$ is Lie isomorphic with the additive group of complex numbers. Using the formula of 8.4.8 we find that if f is in $L^2(C)$

$$(\bar{\pi}_{\nu,\nu}(g)f)(z) = \left(\frac{g_{22} - g_{12}z}{|g_{22} - g_{12}z|}\right)^{-n} |g_{22} - g_{12}z|^{-2-\sqrt{-1}\nu} f\left(\frac{g_{11}z - g_{21}}{-g_{12}z + g_{22}}\right)$$

for

$$g = \begin{bmatrix} g_{11} & g_{12} \\ g_{21} & g_{22}\end{bmatrix}$$

in G and z in C.

Thus the representations $\bar{\pi}_{n,\nu}$ coincide with the representations of Gelfand *et al.* of [1], p. 145.

8.4.10 *Example 2. SL(2, R)* Let G be the group of all 2×2 real matrices of determinant 1. Then proceeding as in 8.4.9 we find that

$$\bar{N} = \left\{\begin{bmatrix} 1 & 0 \\ t & 1\end{bmatrix}\middle| t \text{ in } R\right\}, \qquad M = \left[\begin{matrix} \varepsilon & 0 \\ 0 & \varepsilon\end{matrix}\right]\left| \varepsilon = \pm 1\right\},$$

$$A = \left\{\begin{bmatrix} a & 0 \\ 0 & a^{-1}\end{bmatrix}\middle| a > 0 \text{ in } R\right\}, \qquad N = \left\{\begin{bmatrix} 1 & t \\ 0 & 1\end{bmatrix}\middle| t \text{ in } R\right\}.$$

In this case

$$\left(\begin{bmatrix} a & 0 \\ 0 & a^{-1} \end{bmatrix}\right)^{\rho} = a. \ \hat{M} = Z_2 = \{0, 1\}.$$

The principal series is realized on $L^2(R)$ and

$$(\bar{\pi}_{\delta,\nu}(g)f)(x) =$$

$$\left(\frac{g_{22} - g_{12}}{|g_{22} - g_{12}|}\right)^{\delta} |g_{22} - g_{12}|^{-1-\sqrt{-1}\nu} f\left(\frac{g_{11} - g_{21}x}{-g_{12} + g_{22}x}\right), \ \delta = 0 \text{ or } 1.$$

Compare with Gelfand *et al.* [1], p. 393.

8.5 Finite Dimensional Subrepresentations of the Nonunitary Principal Series

8.5.1 In this section we retain the notation of 8.4. We will use the realization $\{\tilde{\pi}_{\xi,\nu}\}$ of the principal series as explained in 8.4.

8.5.2 Let $\bar{N}$, M, A, N be as in 7.4.20. If (π, V) is a finite dimensional representation of G set

$$V^{\bar{N}} = \{v \text{ in } V | \pi(\bar{n})v = v \text{ for all } \bar{n} \text{ in } \bar{N}\}$$

and

$$V^{N} = \{\mathrm{v} \text{ in } V | \pi(\mathrm{n})\mathrm{v} = \mathrm{v} \text{ for all n in N}\}.$$

8.5.3 *Lemma* Let (π, V) be an irreducible finite dimensional representation of G (we do not in this lemma assume that G has finite center). Let (π^*, V^*) be the contragradient representation. Then $V^{\bar{N}}$ and V^{*^N} are MA-invariant and irreducible representations of MA. Furthermore as representations of MA, $V^{\bar{N}}$ is equivalent with $(V^{*^N})^*$.

PROOF Let $i: V \to (V^{*^N})^*$ be defined by $i(v)(\lambda) = \lambda(v)$ for v in V, λ in V^{*^N}. Clearly ker i is invariant under MAN. Suppose that v is in ker $i \cap V^{\bar{N}}$. If n is in N, m is in M, a is in A, n is in N, then

$$\lambda(\pi(nma\bar{n})v) = (\pi^*(nma)^{-1}\lambda)(\pi(\bar{n})v) = i(v)(\pi^*(ma)^{-1}\pi^*(n)^{-1}\lambda) = 0.$$

Thus $(\pi(g)^*\lambda)(v) = 0$ for all g in G ($NMA\bar{N}$ is dense in G by 7.5.20). But (π^*, V^*) is irreducible. Hence $v = 0$. Thus i: $V^{\bar{N}} \to (V^{*^N})^*$ is injective. Interchanging the roles of V and V^*, N and $\bar{N}$, we find that there is also an injective map j: $V^{*^N} \to (V^{\bar{N}})^*$. Thus $\dim V^{\bar{N}} = \dim(V^{*^N})^*$. Hence i is a bijection from $V^{\bar{N}}$ to $(V^*)^{N*}$.

Let m be in MAN, then

$$i(\pi(m)v)(\lambda) = \lambda(\pi(m)v) = (\pi^*(m)^{-1}\lambda)(v) = (\sigma^*(m)i(v))(\lambda)$$

where (σ, V^{*^N}) is the representation of MA given by $\sigma(m)\lambda = \pi^*(m)\lambda$ for m in MA, λ in V^{*^N}. Thus the representations $V^{\bar{N}}$ and $(V^{*^N})^*$ of MA are equivalent.

Suppose that as an MA representation $V^{\bar{N}}$ is reducible. Let $U \subset V^{\bar{N}}$ be a nonzero invariant, proper subspace of $V^{\bar{N}}$. Let v be a nonzero element of U. Since $NMA\bar{N}$ is dense in $V^{\bar{N}}$ we see that $\pi(NMA)v$ spans V. Hence $\pi(N)U$ spans V. This implies that since $U \neq V^{\bar{N}}$, $V^{\bar{N}} \cap V_N \neq (0)$ where $V_N = \{\pi(n)w - w | n \text{ in } N, w \text{ in } V\}$. But $V_N \subset \ker i$. This is a contradiction.

8.5.4 Let (π, V) be a finite dimensional irreducible representation of G. Then $V^{\bar{N}}$ is an irreducible MA representation according to 8.5.3. If m is in MA and a is in A, then $ma = am$ by the definition of M. Thus if a is in A, A, $\pi(a)|_{V^{\bar{N}}} = c(a)I$ with $c(a)$ in C. Since $c(a_1a_2) = c(a_1)c(a_2)$ we see that there is ν in $\mathfrak{a}_C^*$ so that $c(a) = a^\nu$ (see 8.3.2). Since the roots of $\mathfrak{g}_C$ are real on $\mathfrak{a}$ we see that ν is in $\mathfrak{a}^*$. In fact, if Λ is the lowest weight of V as a representation of $\mathfrak{g}_C$, $\nu = \Lambda|_{\mathfrak{a}}$.

8.5.5 *Lemma* Let ξ be in $\hat{M}$, ν in $\mathfrak{a}_C^*$. $\tilde{H}^{\xi,\nu}$ contains at most one nonzero invariant finite dimensional subspace.

PROOF Let $(0) \neq V \subset \tilde{H}^{\xi,\nu}$ be an invariant finite dimensional subspace. Let f be in $V^{\bar{N}}$. Then $f(\bar{n}man) = \xi(m)^{-1}a^{-\nu}f(e)$ for $\bar{n}$ in $\bar{N}$, m in M, a in A, and n in N. Thus the map ε: $V^{\bar{N}} \to H_\xi$ given by $\varepsilon(f) = f(e)$ sets up an equivalence between $V^{\bar{N}}$ as an M representation and (ξ, H_ξ). Thus $V^{\bar{N}}$ is irreducible. This combined with 8.5.3 proves the result.

8.5.6 If there is a nonzero finite dimensional subrepresentation of $\tilde{H}^{\xi,\nu}$ denote it $V_{\xi,\nu}$. If $\tilde{H}^{\xi,\nu}$ has no nonzero, finite dimensional, invariant subspaces set $V_{\xi,\nu} = (0)$.

8.5.7 *Lemma* Let (π, V) be a finite dimensional, irreducible representation of G. Let ξ be the action of M on $V^{\bar{N}}$ and let ν in $\mathfrak{a}^*$ be so that $\pi(a)|_{V^{\bar{N}}} = a^\nu I$. Then (π, V) is equivalent with $V_{\xi,\nu}$.

PROOF 8.5.3 implies that $V^{\bar{N}}$ is equivalent with $(V^{*^N})^*$ as an MA representation. We thus identify H_ξ with $(V^{*^N})^*$. If v is in V and g in G we define $\alpha(v)(g)$ to $(V^{*^N})^*$ by the formula $(\alpha(v)(g))(\lambda) = \lambda(\pi(g^{-1})v)$. Then

$$\begin{aligned}(\alpha(v)(gman))(\lambda) &= \lambda(\pi(n)^{-1}\pi(m)^{-1}\pi(a)^{-1}\pi(g^{-1})v)\\ &= \lambda(\pi(m)^{-1}\pi(a)^{-1}\pi(g)^{-1}v) = (\pi^*(ma)\lambda)\pi(g^{-1})v)\\ &= (\xi(m)^{-1}(a^{-\nu}\alpha(v)(g))(\lambda).\end{aligned}$$

(Here ξ is the contragradient action to π^* on V^{*^N}.) Thus $\alpha(v)$ is in $\tilde{H}^{\xi,\nu}$. Now

$$\begin{aligned}((\alpha(\pi(g)v)(x)(\lambda) &= \lambda(\pi(x)^{-1}\pi(g)v)\\ &= \lambda(\pi(g^{-1}x)^{-1}v) = (\alpha(v)(g^{-1}x))(\lambda).\end{aligned}$$

Thus $\alpha(\pi(g)v) = \tilde{\pi}_{\xi,\nu}(g)(\alpha(v))$. Now $\alpha(v)(e) = i(v)$ in the notation of the proof of 8.5.3. Thus $\alpha \neq 0$. Since (π, V) is irreducible (π, V) is equivalent with $\alpha(V)$. The result now follows from 8.5.5.

8.5.8 *Lemma* Let 1 denote the trivial representation of M. Let (π, V) be a finite dimensional irreducible representation of the universal covering group of G. Let Λ be the lowest weight of (π, V) (here we use the order on the weights corresponding to 7.5.9). Set $\lambda = 2\Lambda|_{\mathfrak{a}}$. Then $V_{1,\lambda} \neq (0)$. Furthermore, if ν is in $\mathfrak{a}_C^*$ we set $1_\nu(g) = e^{-\nu(H(g))}$, then 1_λ is in $V_{1,\lambda}$.

PROOF Let $\tilde{G}$ be the universal covering group of G. Let $\tilde{K}$, $\tilde{A}$, $\tilde{N}$, and $\bar{N}^\sim$ be the subgroups of G corresponding to K, A, N, and $\bar{N}$. Let η be the covering map of $\tilde{G}$ onto G. Let $\tilde{M}$ be the centralizer of $\tilde{A}$ in $\tilde{K}$.

Let π also denote the corresponding representation of $\mathfrak{g}$. Then $\pi(\tilde{G})$ is the connected subgroup of $GL(V)$ corresponding to $\pi(\mathfrak{g})$. Thus $\pi(\tilde{G})$ is contained in a complex semisimple Lie group $G_{\mathbb{C}}$, the subgroup of $GL(V)$ corresponding to $\pi(\mathfrak{g})_{\mathbb{C}}$. Let G_u be the subgroup of $G_{\mathbb{C}}$ corresponding to $\pi(\mathfrak{k}) + (-1)^{1/2}\pi(\mathfrak{p})$. Then 3.6.6 implies that G_u is compact. Now the center of $\pi(\tilde{G})$ is contained in the center of G_u which is finite. Thus $\pi(\tilde{K})$ is a finite covering of a compact Lie group. This implies that $\pi(\tilde{K})$ is compact. Hence $\pi(\tilde{M})$ is compact.

$V^{\bar{N}}$ may be identified with $(V^{*^N})^*$ as a representation of M (see 8.5.3). Furthermore since $\pi(\tilde{M})$ is compact we may put an $\tilde{M}$-invariant inner product on $H_\xi = (V^{*N})^*$, $\langle\ ,\ \rangle$. Let α be as in 8.5.7. Define

$$\beta(v \otimes w)(g) = \langle \alpha(v)(g), \alpha(w)(g)\rangle.$$

Then β is a *real* linear mapping of $V \otimes V$ into $C(\tilde{G})$ (the continuous functions on $\tilde{G}$). Let z be in ker η. Then z is in $\tilde{M}$. Furthermore

$$\begin{aligned}\beta(v \otimes w)(gz) &= \langle \alpha(v)(gz), \alpha(w)(gz)\rangle\\ &= \langle \xi(z)^{-1}\alpha(v)(g), \xi(z)^{-1}\alpha(w)(g)\rangle = \langle \alpha(v)(g), \alpha(w)(g)\rangle.\end{aligned}$$

Thus $\beta\colon V \otimes V \to C(G)$.

Now

$$\begin{aligned}(\beta(v \otimes w))(gman) &= \langle \alpha(v)(gman), \alpha(w)(gman)\rangle \\ &= e^{-2\Lambda(\log a)}\langle \xi(m)^{-1}\alpha(v)(g), \xi(m)^{-1}\alpha(w)(g)\rangle \\ &= e^{-2\Lambda(\log a)}\beta(v \otimes w)(g)\end{aligned}$$

for v, w in V, g in G, m in M, a in A, n in N. Thus $\beta(V \otimes V) \subset H^{1,\lambda}$ with $\lambda = 2\Lambda|_{\mathfrak{a}}$. Since $\beta(V \otimes V)$ is a G-invariant finite dimensional subspace of $H^{1,\lambda}$ we see that $\beta(V \otimes V) \subset V_{1,\lambda}$.

Furthermore since $\pi(\tilde{K})$ is compact there is a K-invariant inner product on V, $(\ ,\ )$. Let $v_1, \ldots, v_n$ be an orthonormal basis of V relative to K. Then

$$\sum_{i=1}^{n} \beta(v_i \otimes v_i)(g) = \sum_{i=1}^{n} \langle \alpha(v_i)(g), \alpha(v_i)(g)\rangle.$$

Thus since $\alpha \neq 0$, $\sum_{i=1}^{r} \beta(v_i \otimes v_i)(g) \neq 0$ for all g. Hence $V_{1,\lambda} \neq (0)$. If k is in K then

$$\tilde{\pi}_{1,\lambda}(k) \sum_{i=1}^{n} \beta(v_i \otimes v_i) = \sum_{i=1}^{n} \beta(v_i \otimes v_i).$$

Thus

$$\sum_{i=1}^{n} \beta(v_i \otimes v_i)(kan) = e^{-2\Lambda(\log a)} \sum_{i=1}^{n} \|\alpha(v_i)(e)\|^2.$$

Hence $\sum_{i=1}^{r} \beta(v_i \otimes v_i) = c1_\lambda$ with $c \neq 0$. The lemma is now completely proved.

8.5.9 *Definition* If ξ is in $\hat{M}$ then ξ is said to be extendable if there is a λ in $\mathfrak{a}^*$ so that $V_{\xi,\lambda} \neq (0)$.

8.5.10 8.5.7 says that ξ is extendable if and only if there is an irreducible finite dimensional representation (π, V) of G so that V^N is equivalent with ξ as an M representation.

8.5.11 *Theorem* Suppose that G is Lie isomorphic with a Lie subgroup of $GL(n, R)$ for some n. If ξ is in $\hat{M}$ then ξ is extendable.

PROOF We may assume that the Lie algebra $\mathfrak{g}$ of G is a Lie subalgebra of $M_n(R)$ and that G is the connected subgroup of $GL(n, R)$ corresponding to $\mathfrak{g}$. Let $\mathfrak{g}_C$ be the complex linear span of $\mathfrak{g}$ in $M_n(C)$. Let G_C be the connected subgroup of $GL(n, C)$ corresponding to $\mathfrak{g}_C$. Then G is the connected subgroup of G_C corresponding to $\mathfrak{g}$. Let θ be the Cartan involution of $\mathfrak{g}$ corresponding to $\mathfrak{k}$. Let $\mathfrak{p} = \{X \text{ in } \mathfrak{g} | \theta X = -X\}$. Set $\mathfrak{g}_u = \mathfrak{k} \oplus (-1)^{1/2}\mathfrak{p}$. Then $\mathfrak{g}_u$ is a compact form of $\mathfrak{g}_C$ (see 3.7.8, 3.7.9). Let G_u be the connected subgroup of G_C corresponding to $\mathfrak{g}_u$. Then $\mathfrak{g}_u$ is a compact, semisimple Lie group (see 3.6.6). Let $\mathfrak{m}$ be the Lie algebra of M and let $\mathfrak{h}^-$ be a maximal Abelian subalgebra of $\mathfrak{m}$. Set $\mathfrak{h}_0 = \mathfrak{h}^- \oplus \mathfrak{a}$. Then $\mathfrak{h}_* = \mathfrak{h}^- \oplus (-1)^{1/2}\mathfrak{a}$ is a maximal Abelian subalgebra of $\mathfrak{g}_u$. Let T_1 be the maximal torus of M corresponding to $\mathfrak{h}^-$ and let $T_2 = \exp((-1)^{1/2}\mathfrak{a})$. Then $T_1 T_2 = T$ is the maximal torus of G_u corresponding to $\mathfrak{h}_*$. Let m be an element of M. Then there is a regular element X of $\mathfrak{m}$ so that $\text{Ad}(m)X = X$ (see 3.9.2). Let M_0 be the identity component of M. Then there is m_1 in M_0 so that $\text{Ad}(m_1)X$ is in $\mathfrak{h}^-$. Thus $\text{Ad}(m_1 m m_1^{-1})$ leaves fixed a regular element of $\mathfrak{h}^-$. But then $\text{Ad}(m_1 m m_1^{-1})$ leaves fixed a regular element of $\mathfrak{h}_*$ since $\text{Ad}(m_1 m m_1^{-1})|_{\mathfrak{a}} = I$. This implies that $m_1 m m_1^{-1}$ is in T (see the proof of 3.9.4(1)). This implies

(1) If G is a linear Lie group and if M_0 is the identity component of M then $M = M_0 F$ with $F = \exp((-1)^{1/2}\mathfrak{a}) \cap K$.

Let now E be the space of continuous functions spanned by the matrix elements of the representations of M of the form V^N with (π, V) an irreducible finite dimensional representation of G. We assert that E is an algebra under multiplication. In fact, let V and W be irreducible finite dimensional representations of G. Then $(V \otimes W)^N \supset V^N \otimes W^N$. Since the matrix elements of $V^N \otimes W^N$ are just the set of all products of matrix elements of V^N and W^N, E is an algebra.

We next assert that if f is in E then $\bar{f}$ is in E. Indeed, suppose that ξ is in $\hat{M}$ and $V_{\xi,\lambda} \neq (0)$ for some λ in $\mathfrak{a}^*$, let $v_1, \ldots, v_n$ an orthonormal basis of H_ξ, if $v = \sum a_i v_i$ is in H set $\bar{v} = \sum \bar{a}_i v_i$. Let $\overline{\tilde{H}^{\xi,\lambda}} = \{\bar{f} | f \text{ in } \tilde{H}^{\xi,\lambda}\}$. Here $\bar{f}(g) = \overline{f(g)}$. Then $f(gman) = e^{-\lambda(\log g)}\xi(m)^{-1}f(g)$. Let $\overline{\xi(m)}$ be the matrix relative to the basis $v_1, \ldots, v_n$ obtained by conjugating the matrix elements of $\xi(m)$ relative to $v_1, \ldots, v_n$. Then set $\bar{\xi}(m) = \overline{\xi(m)}$. We then have $\overline{\tilde{H}^{\xi,\lambda}} = H^{\bar{\xi},\lambda}$. Clearly $\bar{V}_{\xi,\lambda} \subset \tilde{H}^{\bar{\xi},\lambda}$ is a finite dimensional invariant space and the matrix elements of $\bar{V}^N$ are the complex conjugates of the matrix elements of ξ. Thus E is indeed closed under complex conjugation.

We show that if m is in M there is f in E so that $0 \neq f(m) \neq f(e)$. Suppose m is in M and $f(m) = f(e)$ for all f in E. This implies that if (π, V) is an irreducible representation of G then $\pi(m)|_{V^N} = I$. But if m_1 is in M, then $\pi(m_1 m m_1^{-1})|_{V^N} = I$ for all (π, V) as above. Thus by the proof of (1) above, we may assume that m is in T and $\pi(m)|_{V^N} = I$ for all (π, V) irreducible finite dimensional representations of G_C (hence G_u). But then if Λ is the lowest weight of (π, V), $\Lambda(m) = 1$. This implies by the theorem of the highest weight that $\Lambda(m) = 1$ for all Λ in $\hat{T}$. But then $m = e$.

We therefore know that E is an algebra over C closed under complex conjugation and E separates the points of M such that for any m in M there is f in E so that $f(m) \neq 0$. The Stone–Weierstrauss theorem implies that E is dense in $C(M)$ (see e.g., Lang [1], p. 51). But this implies that every element of $\hat{M}$ appears as a subrepresentation of the left regular action of M on E. This clearly implies the theorem.

8.5.12 Before proceeding we make a simple, but crucial observation. Let λ be in $\mathfrak{a}^*$ so that $V_{1,\lambda} \neq (0)$. Let ξ be in $\hat{M}$ and let μ be in $\mathfrak{a}^*$ so that $V_{\xi,\mu} \neq (0)$. We note that if f is in $\tilde{H}^{1,\nu}$, g is in $\tilde{H}^{\xi,\mu}$, then $(f \cdot g)(x) = f(x)g(x)$ defines an element of $\tilde{H}^{\xi,\mu+\nu}$. The observation is

$$V_{1,\lambda} \cdot V_{\xi,\mu} = V_{\xi,\lambda+\mu}. \tag{1}$$

To see this, we note that $\dim(V_{1,\lambda} \cdot V_{\xi,\mu}) \leqslant \dim(V_{1,\lambda} \otimes V_{\xi,\mu}) < \infty$ and that since the elements of $V_{1,\lambda}$ and $V_{\xi,\mu}$ consist of analytic functions if f is in $V_{1,\lambda}$, $f \neq 0$, then $f \cdot g \neq 0$ for g in $V_{\xi,\mu}$, $g \neq 0$. Thus $V_{1,\lambda} \cdot V_{\xi,\mu} \neq 0$. The observation now follows from 8.5.5.

8.5.13 *Lemma* Let $\rho(H) = \frac{1}{2}\operatorname{tr}(\operatorname{ad} H|_{\mathfrak{n}})$. Then $V_{1,-2\rho} \neq (0)$ and $1_{-2\rho}$ is in $V_{1,-2\rho}$.

PROOF Let Δ^+ be the system of positive roots of $(\mathfrak{g}_{\mathbb{C}}, \mathfrak{h})$ we have been using throughout this section (see 7.5.9). Let $\tilde{\rho} = \frac{1}{2}\sum_{\alpha\in\Delta^+}\alpha$. Let Δ_1^+ be the positive roots of M (see 7.5.11) and let $\Sigma = \Delta^+ - \Delta_1^+$. Then 7.5.11 implies that $\rho = \tilde{\rho}|_{\mathfrak{n}}$. Since $\tilde{\rho}$ is dominant integral the theorem of the highest weight implies that there is an irreducible representation of G with lowest weight $-\tilde{\rho}$. Thus 8.5.8 implies that $V_{1,-2\rho} \neq (0)$ since $-2\tilde{\rho}|_{\mathfrak{n}} = -2\rho$ and that $1_{-2\rho}$ is in $V_{1,-2\rho}$.

8.5.14 *Proposition* Let $\hat{K}_0 = \{\gamma \text{ in } \hat{K} | \text{if } (\pi_\gamma, V_\gamma) \text{ is in } \gamma \text{ then } V_\gamma^M \neq (0)\}$. ($V_\gamma^M = \{v \text{ in } V_\gamma | \pi_\gamma(m)v = v \text{ for all } m \text{ in } M\}$). If $k \geqslant 0$, k an integer, then $V_{1,2k\rho} = \sum_{\gamma\in\hat{K}} m_k(\gamma)V_\gamma$ as a K representation.

(1) $m_k(\gamma) \geqslant m_s(\gamma)$ if $k \geqslant s$.

(2) $\lim_{k\to\infty} m_k(\gamma) = \dim V_\gamma^M$ for each γ in $\hat{K}$.

PROOF We first note that if X is the smallest closed invariant subspace of $\tilde{H}^{1,2\rho}$ containing $1_{2\rho}$ (see 8.5.14 for the definition of $1_{2\rho}$), then $X = \tilde{H}^{1,2\rho}$. To see this we argue as follows: As a representation of K, $H^{1,2\rho}$ is just the left action of K on $L^2(K/M)$. If $X \neq H^{1,2\rho}$ then there would be an f in $C(K/M)$ so that

$$\int_{K/M} 1_{2\rho}(g^{-1}k)\overline{f(k)}d(k_M) = 0 \tag{1}$$

for all g in G.

Now $1_{2\rho}(g^{-1}k) = e^{-2\rho(H(g^{-1}k))}$. Thus 7.6.6 combined with (1) implies that

$$\int_{K/M} \overline{f(g\cdot x)}\, dx = 0 \tag{2}$$

for all g in G. $g \cdot x$ is the action of G on $K/M = G/MAN$, we therefore extend f to G as an element of $C(G/B)$. But then 7.6.8 implies that

$$\int_{\bar{N}} \overline{f(gk(\bar{n}))}\, e^{-2\rho(H(\bar{n}))}\, d\bar{n} = 0 \tag{3}$$

for all g in G.

Now $\bar{n} = k(\bar{n})\exp(H(\bar{n}))n(\bar{n})$ with $k(\bar{n})$ in K, $n(\bar{n})$ in N. Thus $f(gk(\bar{n})) = f(g\bar{n})$. Hence

$$\int_{\bar{N}} \overline{f(g\bar{n})} e^{-2\rho(H(\bar{n}))}\, d\bar{n} = 0. \tag{4}$$

Let H in $\mathfrak{a}$ be so that $\lambda(H) > 0$ for all λ in Λ^+. Set $a_t = \exp tH$. Then (4) says that $\int_{\bar{N}} \overline{f(ga_t\bar{n})} e^{-2\rho(H(\bar{n}))}\, d\bar{n} = 0$ for all g and all t. Now $f(ga_t\bar{n}) = f(ga_t\bar{n}a_t^{-1})$ ($f(gman) = f(g)$, m in M, a in A, n in N). Hence

$$\lim_{t\to\infty} \int_{\bar{N}} \overline{f(ga_t\bar{n}a_t^{-1})} e^{-2\rho(H(\bar{n}))}\, d\bar{n} = 0. \tag{5}$$

The Lebesque dominated convergence theorem (here we use the fact that f is bounded) implies that since $\lim_{t\to\infty} f(ga_t\bar{n}a_t^{-1}) = f(g)$

$$0 = \overline{f(g)} \int_{\bar{N}} e^{-2\rho(H(\bar{n}))}\, d\bar{n} = \overline{f(g)}. \tag{6}$$

Thus $f = 0$. This proves that $X = H^{1,2\rho}$.

Suppose now k is in K and that $\tilde{\pi}_{1,-2\rho}(k)|_{V_{1,-2\rho}} = I$. Then

$$\tilde{\pi}_{1,-2\rho}(k)\tilde{\pi}_{1,-2\rho}(g)1_{-2\rho} = \pi_{1,-2\rho}(g)1_{-2\rho}$$

for all $g \in G$. But

$$\begin{aligned}(\tilde{\pi}_{1,-2\rho}(k)\tilde{\pi}_{1,-2\rho}(g)1_{-2\rho})(x) &= (\tilde{\pi}_{1,-2\rho}(g)1_{-2\rho})(k^{-1}x)\\ &= 1_{-2\rho}(g^{-1}k^{-1}x) = e^{2\rho(H(g^{-1}k^{-1}x))}.\end{aligned}$$

Thus we see that $e^{2\rho(H(g^{-1}k^{-1}x))} = e^{2\rho(H(g^{-1}x))}$ for all x, g in G. But then $e^{-2\rho(H(g^{-1}k^{-1}x))} = e^{-2\rho(H(g^{-1}x))}$ for all x, g in G. This implies that $\tilde{\pi}_{1,2\rho}(k) = I$. But $H^{1,2\rho}$ is equivalent with $L^2(K/M)$ as a K representation. Thus k must be in M.

This implies that the elements of $V_{1,-2\rho}$ separate the points of K/M. If f

is in $V_{1,-2\rho}$ then $\bar{f}$ is in $V_{1,-2\rho}$ since $\overline{1_{-2\rho}} = 1_{-2\rho}$. Also

$$V_{1,-2k\rho} = V_{1,-2\rho} \cdots V_{1,-2\rho}$$

a k-fold product. Thus $\sum_{k=0}^{\infty} V_{1,-2k\rho}$ is an algebra under multiplication that separates points of K/M and is closed under complex conjugation. The Stone–Weierstrauss theorem combined with 5.3.6 implies that $\sum_{k=0}^{\infty} V_{1,-2k} = \sum_{\gamma\in K_0} \dim(V_\gamma^M)V_\gamma$ as a K representation. We now note that $V_{1,-2(k+1)\rho} \supset \sum_{j=1}^{k} V_{1,-2j\rho}$ as a K representation. Indeed,

$$\begin{aligned} V_{1,-2(k+1)\rho} \supset (1_{-2\rho})^{k-j}(V_{1,-2\rho})^j &= 1_{-2(k-j)\rho}\cdot(V_{1,-2\rho})^j \\ &= 1_{-2(k-j)\rho}\cdot V_{1,-2j\rho} \end{aligned}$$

by 8.5.12. This completes the proof of the proposition.

8.5.15 *Corollary* Let ξ be in M and λ in $\mathfrak{a}^*$ so that $V_{\xi,\lambda} \neq (0)$. Let $\hat{K}_\xi = \{\gamma \text{ in } \hat{K} | \mathrm{Hom}_M(V_\gamma, H_\xi) \neq (0)\}$. Set $m(\xi, \gamma) = \dim \mathrm{Hom}_M(V_\gamma, H_\xi)$. If $k \geqslant 0$, k an integer, then $V_{\xi,\lambda-2k\rho} \neq (0)$ and if $V_{\xi,\lambda-2k\rho} = \sum_{\gamma\in\hat{K}} m_k(\gamma)V_\gamma$ as a K representation then

(1) $m_k(\gamma) \geqslant m_l(\gamma)$ if $k \geqslant l$, γ in $\hat{K}_\xi$

(2) $\lim_{k\to\infty} m_k(\gamma) = m(\xi, \gamma)$ for each γ in $\hat{K}_\xi$.

PROOF Let $C(K; \xi)$ be the space of all $f: K \to H_\xi$, $f(km) = \xi(m)^{-1}f(k)$, k in K, m in M. We assert that since

$$C(K/M) = C(K; I), C(K/M)\cdot V_{\xi,\lambda}|_K = C(K; \xi).$$

In fact, since $V_{\xi,\lambda}^{\bar{N}}$ is equivalent with (ξ, H_ξ) as an M representation we see that if $f_1, \ldots, f_n$ is a basis of $V_{\xi,\lambda}|_K$ then $\{f_1(e), \ldots, f_n(e)\}$ is a basis of H_ξ. There is thus a neighborhood U of e in K so that if k is in U, $\{f_1(k), \ldots, f_n(k)\}$ is a basis of H_ξ. Now if k is in K then $\{\tilde{\pi}_{\xi,\lambda}(k)f_1(u), \ldots, (\tilde{\pi}_{\xi,\lambda}(k)f_e)(u)\}$ is a basis of H_ξ for u in kU. There is thus a finite cover $U_1, \ldots, U_r$ of K and a collection $\{f_1^i, \ldots, f_n^i\}$ of elements of $V_{\xi,\lambda}$ for each i so that $f_1^i(k), \ldots, f_n^i(k)$ is a basis of H_ξ for k in U_i. Let now f be in $C(K; \xi)$. Then $f|_{U_i} = \sum \varphi_j^i f_j^i$ with φ_j^i in $C(K; 1)$. Let $\psi_1, \ldots, \psi_r$ be a partition of unity subordinate to $U_1, \ldots, U_r$. Then $f = \sum_{i,j}(\psi_i\varphi_j)f_j^i$. Thus f is in $C(K; 1)\cdot V_{\xi,\lambda}|_K$. Let X be the space of K-finite elements of $C(K; 1)$ (f in X implies $K\cdot f$ spans a finite dimensional space). Then $X\cdot V_{\xi,\lambda}|_K$ is uniformly dense in $C(K; \xi)$ and is therefore dense in $L^2(K; \xi)$. But if X_ξ is the set of all K-finite elements of $C(K; \xi)$ and f is in X_ξ then 8.5.1 implies f is in $V_{1,-2k\rho}|_K \cdot V_{\xi,\lambda}|_K$ for some $k \geqslant 0$. Now 8.5.12(1) says $V_{1,-2k\rho}|_K \cdot V_{\xi,\lambda}|_K = V_{\xi,\lambda-2k\rho}|_K$. Arguing as in 8.5.14 we find that as a K representation $V_{\xi,\lambda-2k\rho}|_K \supset \sum_{j<k-1} V_{\xi,\lambda-2j\rho}|_K$. This implies the corollary.

8.6 The Character of K-Finite Representation

8.6.1 Let G be a Lie group and let K be a compact subgroup of G.

8.6.2 *Definition* Let (π, H) be a representation of G on the Hilbert space $(H, \langle\ ,\ \rangle)$. We say that (π, H) is K-finite if

(1) $(\pi|_K, H)$ is a unitary representation.

(2) As a representation of K, $H = \sum_{\gamma\in\hat{K}} H_\gamma$ (unitary direct sum) where $H_\gamma = m(\gamma)V_\gamma$, (π_γ, V_γ) an element of $\gamma \in \hat{K}$, and $m(\gamma) < \infty$.

(3) If γ is in $\hat{K}$ and v is in H_γ then the map $G \to H$ given by $g \to \pi(g)v$ is real analytic.

8.6.3 Let (π, H) be a K-finite representation of G. Let $H_F = \{v \text{ in } H | \pi(K)v$ is contained in a finite dimensional subspace$\}$. If X is in $\mathfrak{g}$ (the Lie algebra of G) and v is in H_F, define $\pi(X)v = (d/dt)\pi(\exp tX)v|_{t=0}$. (3) says that this makes sense. Now, if k is in K, X in $\mathfrak{g}$,

$$\pi(k)\pi(X)v = \frac{d}{dt}\pi(k)\pi(\exp tX)v\bigg|_{t=0} = \frac{d}{dt}\pi(k \exp tX)v\bigg|_{t=0}$$

$$= \frac{d}{dt}\pi(k \exp tXk^{-1})\pi(k)v\bigg|_{t=0} = \pi(\mathrm{Ad}(k)X)\pi(k)v.$$

This implies that the map $\mathfrak{g} \otimes H_F \to H$ given by $(X, v) \mapsto \pi(X)v$ is a K representation homomorphism. Hence if v is in H_F, X is in $\mathfrak{g}$, then $\pi(X)v$ is in H_F. Using (3) again we now find $\pi(X)\pi(Y)v - \pi(Y)\pi(X)v = \pi([X, Y])v$ for X, Y in $\mathfrak{g}$, v in H_F.

8.6.4 It can be shown (Harish-Chandra [1]) that 8.6.2(3) is redundant.

8.6.5 *Lemma* Let G be a semisimple Lie group with finite center. Let K be a maximal compact subgroup of G. Then every nonunitary principal series representation of G is K-finite.

PROOF We use the notation of 8.5. Then as a representation of of K, $(\tilde{\pi}_{\xi,\nu}, \tilde{H}^{\xi\nu})$ is just (π_ξ, H^ξ) which is unitary. We have seen that $H^\xi = \sum m(\xi, \gamma)V_\gamma$ with dim $m(\xi, \gamma) = \dim \mathrm{Hom}_M(V_\gamma, H_\xi) \leqslant d(\gamma)$. Thus (1) and (2) of 8.6.2 are true. If f is in $\tilde{H}_\gamma^{\xi,\nu}$ then $f(kan) = e^{-\nu((\log a)}f(k)$. But $f|_K$ transforms as a multiple of an irreducible representation of K. Thus $f|_K$ is real analytic, hence f is a real analytic function on G. Thus for t sufficiently small $f(\exp\text{-}tX \cdot g) = \sum t^n/n!(d^r/dt^r)f(\exp(-tX)g)|_{t=0}$. Thus (3) is satisfied.

8.6.6 *Definition* If (π, H) is a K-finite representation of G then (π, H) is said to be infinitesimally irreducible if (π, H_F) is an irreducible representation of $U(\mathfrak{g}_C)$ (see 8.6.3).

If (π_1, H^1), (π_2, H^2) are K-finite representations then π_1, π_2 are said to be infinitesimally equivalent if there is $T: H_F^1 \to H_F^2$ such that T is a bijection and $T \circ \pi_1(X) = \pi_2(X) \circ T$ for X in $\mathfrak{g}_C$.

8.6.7 *Lemma* Let (π_1, H^1) and (π_2, H^2) be K-finite unitary representations of G. Suppose that (π_1, H^1) is irreducible and that π_1 is infinitesimally equivalent with π_2. Then π_1, π_2 are unitarily equivalent.

PROOF Let $T: H_F^1 \to H_F^2$ be so that T is bijective and $T\pi_1(X) = \pi_2(X)T$ for all X in $U(\mathfrak{g}_C)$. Let $\mathfrak{k}$ be the Lie algebra of K. Then $U(\mathfrak{k})$ is a subalgebra of $U(\mathfrak{g}_C)$.

(1) $$T\pi_1(k) = \pi_2(k)T \text{ for all } k \text{ in } K.$$

In fact, if v is in H_F^1 let W_1 be the smallest subspace of H_F^1 which is K-invariant and contains v. Let W_2 be the smallest K-invariant subspace of H_F^2 containing Tv. Then since K is connected, W_1 is the smallest $U(\mathfrak{k})$-invariant subspace of H_F^1 containing v and W_2 is the smallest $U(\mathfrak{k})$-invariant subspace of H_F^2 containing Tv. Thus $TW_1 = W_2$. Since $\dim W_1 < \infty$ and $\dim W_2 < \infty$ we see that T is a continuous map of W_1 to W_2. Thus

$$T\pi_1(\exp X)v = T\sum_{k=0}^{\infty}\left(\frac{1}{k!}\right)\pi_1(X)^k v$$

$$= \sum_{k=0}^{\infty}\frac{1}{k!}T\pi_1(X)^k v = \sum_{k=0}^{\infty}\frac{1}{k!}\pi_2(X)^k Tv = \pi_2(\exp X)Tv.$$

Hence the connectedness of K implies that $T\pi_1(k)v = \pi_2(k)Tv$ for all k in K.

(2) $$TH_\gamma^1 \subset H_\gamma^2.$$

Indeed, let $E_\gamma^i: H^i \to H_\gamma^i$ be the orthogonal projection $i = 1, 2$. Then $E_\gamma^i v = d(\gamma)\int_K \overline{\chi_\gamma(k)}\pi_i(k)v\,dk$ for v in H^i (see 8.6.15). If v is in H_γ^1 then

$$Tv = TE_\gamma^1 v = \int Td(\gamma)\int_K \overline{\chi_\gamma(k)}\pi_1(k)v\,dk = d(\gamma)\int_K \overline{\chi_\gamma(k)}T\pi_1(k)v\,dk,$$

since the integral in question is a limit of linear combinations of elements of H_γ^1. Thus $Tv = d(\gamma)\int_K \overline{\chi_\gamma(k)}\pi_2(k)Tv\,dk = E_\gamma^2\,Tv$. This clearly proves (2).

Let now $T^*: H_F^2 \to H^1$ be defined by $\langle Tv, w\rangle = \langle v, T^*w\rangle$ for v in H_F^1, W in H_F^2. This defines T^* since H_F^1 and H_F^2 are dense in H^1 and H^2, respectively. Now $TE_\gamma^1 = E_\gamma^2 T$, thus $T^*E_\gamma^2 = E_\gamma^1 T^*$. Thus $T^*: H_\gamma^2 \to H_\gamma^1$. Furthermore if X is in $\mathfrak{g}$, v, w are in H_F^i then

$$\langle \pi_i(X)v, w\rangle = \frac{d}{dt}\langle \pi_i(\exp tX)v, w\rangle\bigg|_{t=0} = \frac{d}{dt}\langle v, \pi_i(\exp -tX)w\rangle\bigg|_{t=0}$$
$$= \langle v, -\pi_i(X)w\rangle.$$

Thus $\pi_i(X)^* = -\pi_i(X)$. Hence

$$T^*\pi_2(X) = -(\pi_2(X)T)^* = -(T\pi_1(X))^* = \pi_1(X)T^*.$$

Now $T^*T\pi_1(X) = T^*\pi_2(X)T = \pi_1(X)T^*T$.

(3) (π_1, H^1) is infinitesimally irreducible. In fact if $W \subset H_F^1$ is a $U(\mathfrak{g})$-invariant subspace then $\overline{W} \subset H^1$ is G-invariant. Arguing as above W is K-invariant hence $W = \sum H_\gamma^1 \cap W$. Thus if $E_\gamma^1\overline{W} \neq H_\gamma^1$ for some γ, $E_\gamma^1 W \neq H_\gamma^1$ for some γ. Thus if $W \neq H_F^1$, $\overline{W} \neq H^1$.

Now $T^*T: H_\gamma^1 \to H_\gamma^1$ for each γ and T^*T is self-adjoint. Let γ be so that $H_\gamma^1 \neq (0)$. Then T^*T has an eigenvalue on H_γ^1, $\lambda \neq 0$. Let $H_\lambda^1 = \{v \text{ in } H_F^1 | T^*Tv = v\}$. Then $H_\lambda^1 \neq (0)$ and H_λ^1 is $U(\mathfrak{g})$-invariant. Thus $H_\lambda^1 = H_F^1$. This implies that $T^*T = \lambda I$ on H_F^1. Let $A = (1/\sqrt{\lambda})T$. Then $A^*A = I$. Thus A defines unitary equivalence.

8.6.8 *Definition* Let (π, H) be a K-finite representation of G. Let $E_\gamma: H \to H_\gamma$ be the orthogonal projection. Let for x in G, $\phi_\gamma^\pi(x) = \mathrm{tr}(E_\gamma\pi(x)E_\gamma|_{H_\gamma})$. Then ϕ_γ^π is called the γ-spherical function of π. We note that ϕ_γ^π is a real analytic function on G (see 8.7.2(3)).

8.6.9 *Lemma* Let $U(\mathfrak{g}_C)$ act on $C^\infty(G)$ by the usual action as left invariant differential operators. Suppose $H_1, \ldots, H_q$ are $U(\mathfrak{g}_C)$-invariant subspaces of $C^\infty(G)$ so that

(1) $H_i = H_i^1 \oplus \cdots \oplus H_i^q$, (direct sum), H_i^j, $U(\mathfrak{g}_C)$-invariant and irreducible and H_i^j equivalent to H_i^1 as a representation of $U(\mathfrak{g}_C)$.

(2) H_i^1 is not equivalent with H_j^1 for $i \neq j$. Then the sum $H_1 + \cdots + H_q$ is direct.

PROOF We prove the result by induction on q. If $q = 1$ there is nothing to prove. Suppose that the result is true for $q - 1$. Suppose that the sum $H_1 + \cdots + H_q$ is not direct. Then we must have

$$H_q \cap (H_1 + \cdots + H_{q-1}) \neq (0).$$

If ϕ is in H_i then $\phi = \phi^1 + \cdots + \phi^{q_i}$ with ϕ^j in H_i^j. Set $P_i^j\phi = \phi^j$. If ϕ is in $H_1 + \cdots + H_{q-1}$ the inductive hypothesis allows us to define $P_i^j\phi$ in H_i^j and $\phi = \sum_{i=1,j}^{q-1} P_i^j\phi$. Now $H_q \cap (H_1 + \cdots + H_{q-1}) \neq (0)$. There is thus a $P_i^j\phi$ so that $P_i^j(H_q \cap (H_1 + \cdots + H_{q-1})) \neq (0)$. For simplicity we may assume $i = 1, j = 1$. If $\ker P_1^1|_{H_q \cap (H_1 + \cdots + H_{q-1})} = (0)$ then

$$H_q \cap (H_1 + \cdots + H_{q-1})$$

is equivalent with H_1^1. There is j so that $P_q^j(H_q \cap (H_1 + \cdots + H_{q-1})) \neq (0)$. Thus H_q^j is equivalent with H_1^1. This is a contradiction. Thus

$$\ker P_1^1|_{H_q \cap (H_1 + \cdots + H_{q-1})} \neq (0).$$

Let $H_{1,1} = \operatorname{Ker} P_1^1|_{H_1 \cap (H_1 + \cdots + H_{q-1})}$. Then arguing in the same way we find $\ker P_1^2|_{H_{1,1}} \neq (0)$. Continuing in the same manner we find that there is a nonzero subspace $V_1 \subset H_q \cap (H_1 + \cdots + H_{q-1})$ so that $P_1^j V_1 = 0$ for all $j = 1, \ldots, q_1$. But then $V_1 \subset H_q \cap (H_2 + \cdots + H_{q-1})$. Applying the inductive hypothesis to $H_2, \ldots, H_q$ we find a contradiction.

8.6.10 *Proposition* Let $(\pi_1, H^1), \ldots, (\pi_n, H^n)$ be K-finite, mutually infinitesimally inequivalent and infinitesimally irreducible representations. If $\gamma_1, \ldots, \gamma_n$ are in K and such that $\phi_{\gamma_i}^{\pi_i} \neq 0$, $i = 1, \ldots, n$, then $\{\phi_{\gamma_1}^{\pi_1}, \ldots, \phi_{\gamma_n}^{\pi_n}\}$ is a linearly independent set of functions.

PROOF Let (π, H) be a K-finite representation of G. Let for w, v in H_F, $A_v(w)(x) = \langle \pi(x)w, v\rangle$. Then $A_v: H_F \to C^\infty(G)$ and $A_v(\pi(X)w) = X \cdot A_v(w)$ for X in $\mathfrak{g}$.

Let now $v_1^i, \ldots, v_{q_i}^i$ be a basis of $H_{\gamma_i}^i$, $i = 1, \ldots, n$. Then

$$\phi_{\gamma_i}^{\pi_i}(x) = \sum_{j=1}^{q_i} \langle \pi_i(x)v_j^i, v_j^i\rangle = \sum_{j=1}^{q_i} A_{v_j^i}^{\pi_i}(v_j^i)(x).$$

Thus $\phi_{\gamma_i}^{\pi_i}$ is in $\sum_{q=1}^{q_i} A_{\gamma_j^i}^{\pi_i}(H_F^i)$. Let $V_i = \sum_{q=1}^{q_i} A_{v_j^i}^{\pi_i}(H_F^i)$. Let $v_{i_1}^i, \ldots, v_{i_k}^i$ be so that

$$(*) V_i = \sum_{j=1}^{k} A_{v_{i_j}^i}^{\pi_i}(H_F^i)$$

and k is the minimal number so that (*) is true. Set $H_j^i = A_{v_{i_j}^i}^{\pi_i}(H_F)$. Then H_j^i is equivalent to H_F^i as a representation of $U(\mathfrak{g}_C)$. Furthermore if $\sum H_j^i$ is not direct, then say $H_1^i \cap (\sum_{j=1}^k H_j^i) \neq (0)$. But H_1^i is irreducible, thus $H_1^i \subset \sum_{j=2}^k H_j^i$ contradicting the definition of k. 8.6.9 now implies the result.

8.6.11 *Definition* Let f be in $C_0^\infty(G)$. Then f is said to be left K-finite (resp. right K-finite) if the set $\{L_k f | k \text{ in } K\}$(resp. $\{R_k f | k \text{ in } K\}$) lies in a finite dimensional subspace of $C(G)$.

8.6.12 Let F be a finite subset of $\hat{K}$. Set $\chi_F = \sum_{\gamma \in F} d(\gamma)\bar{\chi}_\gamma$ where χ_γ is the character of γ.

8.6.13 If f is in $C_0^\infty(G)$ and if h is in $C^\infty(K)$ set $(h * f)(g) = \int_K h(k)f(k^{-1}g)\,dk$ and set $(f * h)(g) = \int_K f(gk)h(k)\,dk$.

8.6.14 *Lemma* f in $C_0^\infty(G)$ is left K-finite (resp. right K-finite) if and only if there is $F \subset K$, a finite subset so that $\chi_F * f = f$ (resp. $f * \chi_F = f$).

PROOF Let f be left K-finite. Let V_F be the space spanned by the $L_k f$, k in K. If u, w are in V_f set $\langle u, w\rangle = \int_G u(g)\overline{w(g)}dg$. Then $\langle\ ,\ \rangle$ is a K-invariant inner product on V_f. Thus $V_f = \sum_{\gamma \in F} n(\gamma)V_\gamma$ where $F \subset \hat{K}$ is a finite set. Thus in order to prove the lemma we need only show that if (π_γ, V_γ) is in γ then $d(\gamma)\pi_\gamma(\bar{\chi}_\gamma) = I$. Let $v, w \in V_\gamma$. Then

$$\langle d(\gamma)\pi(\bar{\chi}_\gamma)v, w\rangle = d(\gamma)\int_K \chi_\gamma(k)\langle\pi_\gamma(k)v, w\rangle dk$$

$$= d(\gamma)\sum_{j=1}^{d(\gamma)}\int_K \overline{\langle\pi_\gamma(k)v_i, v_i\rangle}\langle\pi_\gamma(k)v, w\rangle dk$$

(where $v_1, \ldots, v_{d(\gamma)}$ is an orthonormal basis of V_γ)

$$= \sum_{j=1}^{d(\gamma)}\langle v_i, v\rangle\langle w, v_i\rangle \text{ (by 2.9.3) } = \langle v, w\rangle.$$

The same argument works with right K-finite.

8.6.15 *Lemma* Let (π, H) be a K-finite representation of G. Then $E_\gamma v = \int_K d(\gamma)\overline{\chi_\gamma(k)}\pi(k)v\,dk$.

PROOF The argument is exactly the same as the proof of 8.6.14.

8.6.16 Recall that if (π, H) is a representation of G and if f is in $C_0^\infty(G)$ we define $\pi(f)v = \int_G f(g)\pi(g)v\,dg$.

8.6.17 Let $\mathscr{H}(G) = \mathscr{H}$ be the set of all left and right K-finite functions in

$C_0^\infty(G)$. Note that 8.6.14 implies that f is in $\mathscr{H}$ if and only if there is $F \subset \hat{K}$ so that $\bar{\chi}_F * f * \chi_F = f$.

8.6.18 *Lemma* Let f be in $\mathscr{H}$. Then there is $F \subset \hat{K}$ a finite subset so that if $E_F = \sum_{\gamma \in F} E_\gamma$ then

(1) $\pi(f) = E_F \circ \pi(f) \circ E_F$.

(2) $\operatorname{tr}\pi(f) = \sum \int_G \phi_\gamma^\pi(x) f(x)\, dx$.

Here $\operatorname{tr} \pi(f) = \sum_{i=1} \langle \pi(f)\varphi_i, \varphi_i \rangle$ for $\{\varphi_i\}$, an arbitrary orthonormal basis of H.

PROOF (1) Let v be in H. Then

$$\pi(f)v = \int_G f(g)\pi(g)v\, dg = \int_G (\bar{\chi}_F * f * \chi_F)(g)\pi(g)v\, dg,$$

here $F \subset \hat{K}$ is as in 8.6.17. Thus

$$\pi(f)v = \sum_{\gamma,\tau \in F} \int_{K \times G \times K} d(\gamma)d(\tau)\, \overline{\chi_\gamma(k_1)} f(k_1^{-1} g k_2)\chi_\tau(k_2)\pi(g)v\, dk_1\, dg\, dk_2$$

$$= \sum_{\gamma,\tau \in F} \int_{G \times K \times K} d(\gamma)d(\tau)\, \overline{\chi_\gamma(k_1)} f(g)\, \chi_\tau(k_2)\pi(k_1 g k_2^{-1})v\, dg\, dk_1 dk_2.$$

Now $\overline{\chi_\gamma(k_2)} = \chi_\gamma(k_2^{-1})$. Thus, we have

$$\pi(f)v = \sum_{\gamma \in F} \int_K d(\gamma)\overline{\chi_\gamma(k_1)}\pi(k_1)dk \circ \int_G f(g)\pi(g)\, dg$$
$$\circ \sum_{\gamma \in F} d(\gamma)\overline{\chi_\gamma(k_2)}\pi(k_2)v\, dk = E_F \circ \pi(f) \circ E_F.$$

We now prove (2). Let $\{\varphi_n\}$ be an orthonormal basis of H. Let $\{\psi_n\}$ be an orthonormal basis of H so that $\{\psi_1, \ldots, \psi_r\}$ is an orthonormal basis of $\sum_{\gamma \in F} H_\gamma = H_F$. Then

$$\sum_{i=1}^{\infty} \langle \pi(f)\psi_i, \psi_i \rangle = \sum_{i=1}^{\infty} \langle E_F\pi(f)E_F\psi_i, \psi_i \rangle$$
$$= \sum_{i=1}^{r} \langle \pi(f)E_F\psi_i, E_F\psi_i \rangle = \sum_{i=1}^{r} \langle \pi(f)E_F\psi_i, E_F\psi_i \rangle$$

since $E_F\psi_j = 0$, $j > r$. Thus

$$\sum_{i=1}^{\infty} \langle \pi(f)\psi_i, \psi_i \rangle = \sum_{i=1}^{r} \langle E_F\pi(f)E_F\psi_i, \psi_i \rangle$$
$$= \operatorname{tr} E_F \circ \pi(f) \circ E_F|_{H_F} = \sum_{\gamma \in F} \operatorname{tr} E_F \circ \pi(f) \circ E_F|_{H_\gamma}.$$

Now

$$E_\gamma \circ \pi(f) \circ E_\gamma = \int_G E_\gamma(f(g)\pi(g)\,dg)E_\gamma = \int_G f(g)E_\gamma\pi(g)E_\gamma\,dg.$$

Hence

$$\mathrm{tr}E_\gamma \circ \pi(f) \circ E_\gamma|_{H_\gamma} = \int_G f(g)\phi_\gamma^\pi(g)\,dg.$$

Now $\varphi_i = \sum_{j=1} \langle \varphi_i, \psi_j \rangle \psi_j$. on the other hand

$$\sum_{i,j,k \leq N} |\langle \varphi_i, \psi_j\rangle\langle \pi(f)\psi_j, \psi_k\rangle\langle \varphi_k, \varphi_i\rangle|$$

$$\leq (\sum_{j=1}^{N} |\langle \varphi_i, \varphi_j\rangle|^2)^{1/2} (\sum_{k=1}^{N} |\langle \psi_k, \varphi_i\rangle|^2)^{1/2} \sum_{j,k \leq N} |\langle \pi(f)\psi_i, \psi_j\rangle|$$

$$\leq \sum_{j,k \leq r} |\langle \pi(f)\psi_i, \psi_j\rangle|.$$

Here we have used Schwarz's lemma and the fact that $\langle \pi(f)\psi_i, \psi_j\rangle = 0$ if i or $j > r$. Thus the sum

$$(*) \sum_{i,j,k=1} \langle \varphi_i, \psi_j\rangle\langle \pi(f)\psi_j, \psi_k\rangle\overline{\langle \varphi_i, \psi_k\rangle}$$

is absolutely convergent. Since $\sum_{i=1}^{\infty} \langle \pi(f)\psi_i, \psi_i\rangle$ and $\sum_{i=1}^{\infty} \langle \pi(f)\varphi_i, \varphi_i\rangle$ are just rearrangements of (*), the lemma follows.

8.6.19 *Definition* Let (π, H) be a K-finite representation of G. Let Θ_π denote the linear functional on $\mathscr{H}$ given by $\Theta_\pi(f) = \mathrm{tr}\pi(f)$ for $f \in \mathscr{H}$. 8.6.18 says $\mathrm{tr}\pi(f)$ makes sense. Θ_π is called the character of the representation (π, H).

8.6.20 *Lemma* Let (π_1, H^1) and (π_2, H^2) be K-finite representations of G. Then $\Theta_{\pi_1} = \Theta_{\pi_2}$ if and only if $\phi_\gamma^{\pi_1} = \phi_\gamma^{\pi_2}$ for all γ in $\hat{K}$.

PROOF 8.6.18 implies that if $\phi_\gamma^{\pi_1} = \phi_\gamma^{\pi_2}$ for all γ in $\hat{K}$ then $\Theta_{\pi_1} = \Theta_{\pi_2}$. Suppose that $\Theta_{\pi_1} = \Theta_{\pi_2}$ and $\phi_\gamma^{\pi_1} \neq \phi_\gamma^{\pi_2}$ for some γ in $\hat{K}$. Then there is f in $C_0^\infty(G)$ so that $0 \neq \int_G (\phi_\gamma^{\pi_1}(x) - \phi_\gamma^{\pi_2}(x))f(x)\,dx$. Now

$$\int_K d(\gamma)\chi_\gamma(k)\,\phi_\gamma^{\pi_i}(k^{-1}x)dk$$

$$= \int_K d(\gamma)\chi_\gamma(k)\mathrm{tr}\,E_\gamma\pi_i(k)^{-1}\pi_i(x)E_\gamma dk$$

$$= \mathrm{tr}\,E_\gamma \int_K d(\gamma)\chi_\gamma(k)\pi_i(k)^{-1}\pi_i(x)E_\gamma dk$$

$$= \mathrm{tr}\,E_\gamma E_\gamma \pi_i(x)E_\gamma = \phi_\gamma^{\pi_i}(x).$$

Thus $\chi_\gamma * \phi_\gamma^{\pi_i} = \phi_\gamma^{\pi_i}$. Similarly $\phi_\gamma^{\pi_i} * \bar{\chi}_\gamma = \phi_\gamma^{\pi_i}$, $i = 1, 2$. Hence

$$0 \neq \int_G \chi_\gamma * \phi_\gamma^{\pi_1}(x) * \bar{\chi}_\gamma f(x)\,dx - \int_G \chi_\gamma * \phi_\gamma^{\pi_2}(x) * \bar{\chi}_\gamma f(x)\,dx.$$

On the other hand if ϕ is in $C^\infty(G)$,

$$\begin{aligned}\int_G d(\gamma)\chi_\gamma * \phi(x) f(x)\,dx &= \int_G \int_K d(\gamma)\chi_\gamma(k)\phi(k^{-1}x)dk\, f(x)\,dx \\ &= \int_K \left(\int_G d(\gamma)\bar{\chi}_\gamma(k)\phi(k^{-1}x) f(x) dx \right) dk \\ &= \int_K \left(\int_G d(\gamma)\bar{\chi}_\gamma(k)\phi(x) f(kx) dx \right) dk = \int_G \phi(x)\bar{\chi}_\gamma * f(x)\,dx.\end{aligned}$$

Thus if $\tilde{f} = \bar{\chi}_\gamma * f * \chi_\gamma$, then $0 \neq \int_G \phi_\gamma^{\pi_1}(x) f(x)\,dx - \int_G \phi_\gamma^{\pi_2}(x) f(x)\,dx$. But $\tilde{f}$ is in $\mathscr{H}$ and $\Theta_{\pi_i}(f) = \int_G \phi_\gamma^{\pi_i}(x) f(x)\,dx$ by 8.6.18(2). Since $\Theta_{\pi_1} = \Theta_{\pi_2}$ we have a contradiction.

8.6.21 *Theorem* Let (π_1, H^1) and (π_2, H^2) be K-finite unitary representations of G. $\Theta_{\pi_1} = \Theta_{\pi_2}$ if and only if (π_1, H^1) is unitarily equivalent with (π_2, H^2).

PROOF Suppose that (π_1, H^1) is unitarily equivalent with (π_2, H^2). Let $A: H^1 \to H^2$ be the unitary equivalence. Then $A: H_\gamma^1 \to H_\gamma^2$ for γ in $\hat{K}$ is an equivalence of K representations. Thus $E_\gamma \circ A = A \circ E_\gamma$. Hence

$$\begin{aligned}\phi_\gamma^{\pi_1}(x) = \operatorname{tr}(E_\gamma \circ \pi_1(x) \circ E_\gamma|_{H_\gamma^1}) &= \operatorname{tr} A^{-1}E_\gamma \circ \pi_2(x) \circ E_\gamma \circ A|_{H_\gamma^2} \\ &= \operatorname{tr} E_\gamma \circ \pi_2(x) \circ E_\gamma|_{H^2} = \phi_\gamma^{\pi^2}(x).\end{aligned}$$

Now 8.6.21 implies $\Theta_{\pi_1} = \Theta_{\pi_2}$.

Suppose that $\Theta_{\pi_1} = \Theta_{\pi_2}$. We show that π_1, π_2 are unitarily equivalent. We first need a lemma.

8.6.22 *Lemma* Let (π, H) be a unitary K-finite representation of G. Then $H = \sum_{j=1}^N H_j$ (orthogonal direct sum) of irreducible unitary representations of G, where N is finite or $N = \infty$.

PROOF As a representation of K, $H = \sum_{\gamma \in \hat{K}} H_\gamma$. Let

$$\hat{K}_1 = \{\gamma \text{ in } K | H_\gamma \neq 0\}.$$

Let $\hat{K}_1 = \{\gamma_j | j = 1, 2, \ldots\}$. If (π, H) is irreducible then there is nothing to prove. Suppose that (π, H) is reducible. There must be a first $r \geqslant 1$ so that there is a nonzero closed invariant subspace U of H so that $U \cap H_{\gamma_r} \neq H_{\gamma_r}$. Let $V \subset H_{\gamma_r}$ be so that $V = U \cap H_{\gamma_r}$ with U a closed invariant subspace of H, $\dim V > 0$ and $\dim V$ is minimal subject to this condition. Let $H_1 = \bigcap U$ where U is a closed invariant subspace of H such that $U \supset V$. We assert that H_1 is irreducible. Indeed if $H_1 \supset U$, a closed invariant subspace of H_1, then by the definition of V either $U \cap H_{\gamma_r} = (0)$ or $U \cap H_{\gamma_r} = V$. If $U \cap H_{\gamma_r} = V$ then, by definition of H_1, $U = H_1$ if $U \cap H_{\gamma_r} = (0)$, taking $H_1' = U^\perp \cap H_1$ we see that $U \neq (0)$ implies $H_1' \subset H_1$ and $H_1' \supset V$. This is a contradiction, thus $U = (0)$ or $U = H_1$. Proceeding as above with $H_1^\perp$ we find that after a finite number of stages

$$H = H_1 + \cdots + H_k \oplus U_k,$$

orthogonal direct sum such that $E_{\gamma_r} U_k = (0)$. Proceeding as above with U_k, we see that for each j there is an n_j so that $H = H_1 \oplus \cdots \oplus H_{n^j} \oplus W_j$, orthogonal direct sum, and H_i is irreducible for $i = 1, \ldots, n_j$, $E_{\gamma_i} W_j = (0)$, $i = 1, \ldots, j$. Since $W_j \supset W_{j+1}$ and for each k, $E_{\gamma_i} W_k = 0$, $i \leqslant k$, we see that $\bigcap_{j=1} W_j = (0)$ and hence $H = \sum_{i=1} H_i$, orthogonal direct sum of irreducible subrepresentations.

8.6.23 *Proof of Theorem 8.6.21 continued* According to 8.6.22, $H^i = \sum_{j=1}^{N_i} H_j^i$, $i = 1, 2$, orthogonal direct sum with H_j^i an irreducible unitary representation of G. Let π_j^i be the induced action on H_j^i, $i = 1, 2$, $j = 1, 2, \ldots$. Let $M_i = \{j \text{ integral } | 0 \leqslant j \leqslant N_i\}$. We construct a bijective mapping $\alpha\colon M_1 \to M_2$ so that H_j^1 is unitarily equivalent with $H_{\alpha(j)}^2$. This will clearly prove the theorem.

Let $\hat{K} = \{\gamma_j | j = 1, 2, \ldots\}$. Let $M_i(\gamma_j) = \{n \text{ in } M_i | H_n^i \cap H_{\gamma_j}^i \neq (0)\}$. Now $H_j^i = \sum_{\gamma \in \hat{K}} (H_j^i \cap H_\gamma^i)$. Thus $\bigcup_{j=1} M_i(\gamma_j) = M_i$, $i = 1, 2$. Set $M_{i,j} = \bigcup_{r=1}^{j} M_i(\gamma_r)$ and $M_{i,0} = \phi$. We will construct $\alpha\colon M_1 \to M_2$, a bijection satisfying:

(a) $\alpha(M_1(\gamma_j)) = M_2(\gamma_j), j \geqslant 1$,

(b) π_k^1 is unitarily equivalent with $\pi_{\alpha(k)}^2$.

We construct α inductively. Suppose that we have found $\alpha\colon M_{1,r-1} \to M_{2,r-1}$ satisfying (a), (b) for $j \leqslant r - 1$, k in $M_{1,r-1}$. We extend α to $M_{1,r}$. If $M_1(\gamma_r) = M_2(\gamma_r) = \phi$, there is nothing to do. Hence we may assume that at least one of $M_1(\gamma_r)$, $M_2(\gamma_r)$ is nonempty. Let $P_k^i\colon H^i \to H_k^i$ be the orthogonal projective and let $E_{i,\gamma}\colon H^i \to H_\gamma^i$ be the orthogonal projection. Set

$$\phi_{i,k}(x) = \operatorname{tr}(E_{i,\gamma_r} P_{i,k} \pi_i(x) E_{i,\gamma_r}|_{H\gamma_r{}^i}) = \phi_{\gamma_r}^{\pi_i{}^k}(x)$$

for k in M_i, x in G. Set $\phi_i = \phi_{\gamma_r}^{\pi_i}$, $i = 1, 2$. Clearly $\phi_i = \sum_{k \in M_i(\gamma_r)} \phi_{i,k}$, $i = 1, 2$. 8.6.20 implies that $\phi_1 = \phi_2$. We therefore see that

$$\sum_{k \in M_1(\gamma_r)} \phi_{1,k} = \sum_{k \in M_2(\gamma_r)} \phi_{2,k}.$$

Now suppose that k is in $M_1(\gamma_r) \cap M_{1,r-1}$. Since $\alpha(k)$ is already defined we see that $\phi_{1,k} = \phi_{2,\alpha(k)}$. Thus $\phi_{1,k}(1) \neq 0$ implies $\phi_{2,\alpha(k)}(1) \neq 0$. Thus $H^2_{\alpha(k)} \cap H^2_{\gamma_r} \neq (0)$. This implies $\alpha(k)$ is in $M_2(\gamma_r) \cap M_{2,r-1}$, and conversely, if l is in $M_2(\gamma_r) \cap M_{2,r-1}$. Since $\alpha\colon M_{1,r-1} \to M_{2,r-1}$ is a bijection there is exactly one k in $M_{1,r-1}$ so that $\alpha(k) = l$. Thus H^1_k is unitarily equivalent with H^1_l. Thus $\phi_{1,k} = \phi_{2,l}$. Thus as above $H^1_k \cap H^1_{\gamma_r} \neq (0)$. Thus k is in

$$M_{1,r-1} \cap M_1(\gamma_r).$$

We therefore see that α maps $M_1(\gamma_r) \cap M_{1,r-1}$ bijective, onto $M_2(\gamma_r) \cap M_{2,r-1}$. Let $M_i'(\gamma_r)$ be the complement of $M_i(\gamma_r) \cap M_{i,r-1}$ in $M_{i,r-1}$ for $i = 1, 2$. We have shown

$$\sum_{k \in M_1'(\gamma_r)} \phi_{1,k} = \sum_{k \in M_2'(\gamma_r)} \phi_{2,k}. \tag{1}$$

Let $\psi_1, \ldots, \psi_s$ be the distinct elements of $\phi_{i,k}$, k in $M_i'(\gamma_r)$, $i = 1, 2$. 8.6.10 implies that the ψ_j appear on both sides of (1) with the same coefficients. We may thus define a bijective map $\alpha\colon M_1'(\gamma_r) \to M_2'(\gamma_r)$ so that $\phi_{1,k} = \phi_{2,\alpha(k)} \neq 0$ for k in $M_1(\gamma_r)$. 8.6.10 now implies that H^1_k is infinitesimally equivalent with $H^2_{\alpha(k)}$. 8.6.7 now implies H^1_k is unitarily equivalent with $H^2_{\alpha(k)}$. We have therefore extended α to $M_{1,r}$ as a bijection $\alpha\colon M_{1,r} \to M_{2,r}$ satisfying (a), (b) above. The theorem is now completely proved.

8.7 Characters of Admissible Representations

8.7.1 *Definition* Let $(H, \langle\ ,\ \rangle)$ be a Hilbert space. Let T be a continuous linear operator on H. Then T is said to be of trace class if for each orthonormal basis $\{\varphi_n\}$ of H the sum $\sum_{n=1} \langle T\varphi_n, \varphi_n\rangle$ converges and is independent of the choice of $\{\varphi_n\}$. If T is of trace class define

$$\operatorname{tr} T = \textstyle\sum_{n=1} \langle T\varphi_n, \varphi_n\rangle.$$

8.7.2 *Lemma* If T is a continuous linear operator on H and if for a fixed orthonormal basis $\{\varphi_n\}$ of H, $\sum_{i,j} |\langle T\varphi_i, \varphi_j\rangle| < \infty$ then for any A, B continuous operators on H, ATB, TBA, and BAT are of trace class and $\operatorname{tr} ATB = \operatorname{tr} TBA = \operatorname{tr} BAT$.

PROOF Let $\|A\| = \sup\{\|Av\| \mid \|v\| = 1\}$. Set $a_{ij} = \langle A\varphi_i, \varphi_j\rangle$, $b_{ij} =$

$\langle B\varphi_i, \varphi_j\rangle$. Then

$$\sum_{i,j,n\leqslant N} |a_{in}b_{nj}\langle T\varphi_i, \varphi_j\rangle| \leqslant \sum_{i,j=1}^{N} ((\sum_{n=1}^{N} |a_{in}|^2)^{1/2} (\sum_{n=1}^{N} |b_{nj}|^2)^{1/2})|\langle T\varphi_i, \varphi_j\rangle|$$

by Schwarz's inequality. But $\sum_{n=1}^{N} |a_{in}|^2 \leqslant \|A^*\varphi_n\|^2 = \|A\varphi_n\|^2 = \|A\varphi_n\|^2 \leqslant \|A\|^2$. Thus we have

$$(1) \qquad \sum_{i,j,n\leqslant N} |a_{in}b_{nj}\langle T\varphi_i, \varphi_j\rangle| \leqslant \|A\|\ \|B\| \sum_{i,j\leqslant N} |\langle T\varphi_i, \varphi_j\rangle|.$$

Thus the sum

$$(2) \qquad \sum_{i,j,n=1}^{\infty} a_{in}b_{nj}\langle T\varphi_i, \varphi_j\rangle \text{ is absolutely convergent.}$$

But $\sum_{i=1}\langle ATB\varphi_i, \varphi_i\rangle$, $\sum_{i=1}\langle BAT\varphi_i, \varphi_i\rangle$, and $\sum_{i=1}\langle TBA\varphi_i, \varphi_i\rangle$ are just rearrangements of (2). They therefore have the same sum. Let now $U: H \to H$ be a unitary operator, then

$$\sum_{n=1} \langle ATBU\varphi_n, U\varphi_n\rangle = \sum_{n=1} \langle U^{-1}ATBU\varphi_n, \varphi_n\rangle$$
$$= \sum_{n=1} \langle UU^{-1}ATB\varphi_n, \varphi_n\rangle = \sum_{n=1} \langle ATB\varphi_n, \varphi_n\rangle.$$

Thus ATB, BAT, TBA are all trace class and have the same trace.

8.7.3 *Definition* Let G be a Lie group. Let (π, H) be a representation of G. Then (π, H) is said to be of trace class if for each f in $C_0^\infty(G)$, $\pi(f)$ is an operator of trace class. Set $\Theta_\pi(f) = \operatorname{tr} \pi(f)$. Then Θ_π is called the character of (π, H). (The Θ_π defined in 8.6 is just the restriction of this Θ_π to $\mathscr{H}$. We use the same symbol with the hope that there will be no confusion.)

8.7.4 *Theorem* Let G be a Lie group and let K be a compact subgroup of G. Let (π, H) be a representation of G so that

(1) as a representation of K, (π, H) is unitary,

(2) as a representation of K, $H = \sum_{\gamma\in\hat{K}} H_\gamma$ where H_γ is a multiple of an element of γ and $\dim H_\gamma \leqslant C(1 + \|\gamma\|^2)^k$ (see 5.6.5) for C, k real constants,

Then (π, H) is trace class. (Note that if $\dim H_\gamma \leqslant Cd(\gamma)^2$ for all γ in $\hat{K}$ ($d(\gamma) = \dim V_\gamma$, where V_γ is a representative of γ) then (2) is satisfied; (see the proof of 5.6.7.)

PROOF Let for f in $C_0^\infty(G)$, $\tilde{f}(x)(k_1, k_2) = f(k_1xk_2)$. Then $\tilde{f}(x)$ is in $C^\infty(K \times K)$. If γ, τ are in $\hat{K}$ set

$$\tilde{f}(x)^\wedge(\gamma, \tau)(k_1', k_2') = d(\gamma)d(\tau)\int_{K\times K} \overline{\chi_\gamma(k_1)\chi_\tau(k_2)}f(x)(k_1^{-1}k_1', k_2'k_2)\,dk_1\,dk_2.$$

Then arguing as in 5.7.8 and using the fact that f has compact support, we see that if we set $f_{\gamma,\tau}(x) = \tilde{f}(x)^\wedge(\gamma, \tau)(e, e)$ then

(1) for each $k > 0$, $l > 0$ there is a constant $C_{k,l}$ so that $|f_{\gamma,\tau}(x)| \leqslant C_{k,l}(1 + \|\gamma\|^2)^{-k}(1 + \|\tau\|^2)^{-l}$.

(1) implies that $\sum_{\gamma,\tau\in K} f_{\gamma,\tau}$ converges uniformly and absolutely to f (see 5.7.10). Thus $\pi(f) = \sum_{\gamma,\tau\in K} \pi(f_{\gamma,\tau})$.

Let $E_\gamma\colon H \to H_\gamma$ be the orthogonal projection. Then arguing as in 8.6.18 we find

$$\pi(f_{\gamma,\tau}) = E_\gamma \circ \pi(f) \circ E_\tau. \tag{2}$$

Using (1) we also have

(3) for each $k > 0$, $l > 0$ there is a constant $C'_{k,l}$ so that $\|\pi(f_{\gamma,\tau})\| \leqslant C'_{k,l}(1 + \|\gamma\|^2)^{-k}(1 + \|\tau\|^2)^{-l}$.

Let $v_1^\gamma, \ldots, v_{m(\gamma)}^\gamma$ be an orthonormal basis of H_γ for each γ in $\hat{K}$. Then for each γ, τ in $\hat{K}$,

$$\begin{aligned}
&\sum_{j=1}^{m(\gamma)} \sum_{i=1}^{m(\tau)} |\langle \pi(f_{\tau,\gamma})v_i^\gamma, v_j^\tau\rangle| \\
&\qquad \leqslant C'_{u,v}(1 + \|\gamma\|^2)^u(1 + \|\tau\|^2)^v m(\gamma)m(\tau) \\
&\qquad \leqslant CC'_{u,v}(1 + \|\gamma\|^2)^{u+k}(1 + \|\gamma\|^2)^{v+k} \text{ for each } u, v \text{ in } R, \text{ by (2).}
\end{aligned}$$

Take $u, v > \dim(K/2) + k$. Then 5.6.7 implies that

$$\sum_{\gamma,\tau\in\hat{K}} (1 + \|\gamma\|^2)^{-u-k}(1 + \|\tau\|^2)^{-v-k} < \infty.$$

This implies that the sum

$$\sum_{\gamma,\tau\in\hat{K}} \sum_{j=1}^{m(\gamma)} \sum_{i=1}^{m(\tau)} |\langle \pi(f_{\tau,\gamma})v_i^\gamma, v_i^\tau\rangle| < \infty. \tag{4}$$

But

$$\langle \pi(f_{\gamma,\tau})v_i^\gamma, v_j^\tau\rangle = \langle E_\tau(f)E_\gamma u_i^\gamma, u_i^\tau\rangle = \langle \pi(f)v_i^\gamma, v_i^\tau\rangle.$$

Let $\{\varphi_i\}$ be the orthonormal basis $\{v_i^\gamma | \gamma \text{ in } \hat{K}, i = 1, \ldots, m(\gamma)\}$ put in some order. (4) now implies

$$\sum_{i,j=1} |\langle \pi(f)\varphi_i, \varphi_j\rangle| < \infty. \tag{5}$$

The theorem now follows from 8.8.2.

8.7.5 *Corollary* Let (π, H) be K-finite representation of G (see 8.6.2) satisfying 8.7.4(2). Then (π, H) is trace class and the notions of character of 8.7.3 and 8.6.19 coincide.

8.7.6 *Corollary* Let G, K, A, N, M, $(\pi_{\xi,\nu}, H^{\xi,\nu})$ be as in 8.3.9. Then $(\pi_{\xi,\nu}, H^{\xi,\nu})$ is trace class.

PROOF $(\pi_{\xi,\nu}, H^{\xi,\nu})$ is K-finite. Furthermore $H_\gamma^{\xi,\nu} = H_\gamma^\xi$ and $\dim H_\gamma^\xi = m(\xi, \gamma)d(\gamma)$ (see 8.5.16). Clearly $m(\xi, \gamma) \leqslant d(\gamma)$. Thus $(\pi_{\xi,\nu}, H^{\xi,\nu})$ satisfies 8.7.4(1), (2).

8.8 The Character of a Principal Series Representation

8.8.1 We use the notation of 8.3 and 8.4.

8.8.2 *Theorem* Let ξ be in $\hat{M}$ and ν in $\mathfrak{a}_C^*$. If f is in $C_0^\infty(G)$ then

$$\Theta_{\pi_{\xi,\nu}}(f) = \int_{A\times M} F_f(ma)\operatorname{tr}\xi(m)e^{(-1)^{1/2}\nu(\log a)}dm\,da$$

where dm is normalized invariant measure on M, da is chosen independent of ξ, ν, f (but obviously dependent on dg), and F_f is as in 7.7.11.

PROOF We use the realization of $(\pi_{\xi,\nu}, H^{\xi,\nu})$ as $(\tilde{\pi}_{\xi,\rho+(-1)^{1/2}\nu}, \tilde{H}^{\xi,\rho+(-1)^{1/2}\nu})$ (see 8.4.2). Let f be in $C_0^\infty(G)$ and φ in $\tilde{H}^{\xi,\rho+(-1)^{1/2}\nu}$. Then

$$(\tilde{\pi}_{\xi,\rho+(-1)^{1/2}\nu}(f)\varphi)(k_1) = \int_G f(g)(\tilde{\pi}_{\xi,\rho+(-1)^{1/2}\nu}(g)\varphi)(k_1)dg.$$

Since φ is in $\tilde{H}^{\xi,\rho+(-1)^{1/2}\nu}$, $\xi(m)\varphi(gm) = \varphi(g)$ for all g in G, m in M. Thus

$$\begin{aligned}
&(\tilde{\pi}_{\xi,\rho+(-1)^{1/2}\nu}(f)\varphi)(k_1)\\
&\quad= \int_G \xi(m)f(g)(\tilde{\pi}_{\xi,\rho+(-1)^{1/2}\nu}(g)\varphi)(k_1 m)dg\\
&\quad= \int_G \xi(m)f(g)\varphi(g^{-1}k_1 m)dg = \int_G \xi(m)f(k_1 mg)\varphi(g^{-1})dg\\
&\quad= \int_G \xi(m)f(k_1 mg^{-1})\varphi(g)dg.
\end{aligned}$$

Here we have used the fact that G is unimodular (indeed, $\det(\mathrm{Ad}(g)) = 1$, see 2.5.7). 7.6.4 implies that

$$\int_G \xi(m)f(k_1mg^{-1})\varphi(g)dg$$

$$= \int_{K\times A\times N} \xi(m)f(k_1m(kan)^{-1})\varphi(kan)e^{2\rho(\log a)}dk\, da\, dn$$

$$= \int_{K\times A\times N} \xi(m)f(k_1mn^{-1}a^{-1}k^{-1})\varphi(k)e^{-(\rho+(-1)^{1/2}\nu)(\log a)}e^{2\rho(\log a)}\, dk\, da\, dn$$

$$= \int_{K\times A\times N} \xi(m)f(k_1mnak^{-1})\varphi(k)e^{(\rho+(-1)^{1/2}\nu)(\log a)}\, e^{-2\rho(\log a)}\, dk\, da\, dn.$$

(In the last equality we use the unimodularity of N and A to replace n^{-1} by n and a^{-1} by a.) We therefore have

$$\int_{K\times A\times N} \xi(m)f(k_1mnak^{-1})\varphi(k)e^{(-\rho(-1)^{1/2}\nu)(\log a)}\, dk\, da\, dn.$$

If $\mu\colon N \to N$ is given by $\mu(n) = a^{-1}na$, then $\mu^*dn = e^{-2\rho(\log a)}\, dn$. The last integral is thus

$$\int_{K\times A\times N} \xi(m)f(k_1ma(a^{-1}na)k^{-1})\varphi(k)e^{(-\rho+(-1)^{1/2}\nu)(\log a)}\, dk\, da\, dn$$

$$= \int_{K\times A\times N} \xi(m)f(k_1mank^{-1})\varphi(k)e^{(\rho+(-1)^{1/2}\nu)(\log a)}\, dk\, da\, dn.$$

Since the last integral is independent of M we may integrate M out and find

(1) $$(\tilde{\pi}_{\xi,\rho+(-1)^{1/2}\nu}(f)\varphi)(k_1)$$
$$= \int_{K\times M\times A\times N} \xi(m)f(k_1mank^{-1})\varphi(k)e^{(\rho+(-1)^{1/2}\nu)(\log a)}\, dk\, da\, dn.$$

Set

$$F_{f,\xi,\nu}(k_1, k_2) = \int_{M\times A\times N} \xi(m)e^{(\rho+(-1)^{1/2}\nu)(\log a)}f(k_1mank^{-1})\, dn\, da\, dn.$$

Then

(2) $$(\tilde{\pi}_{\xi,\rho+(-1)^{1/2}\nu}(f)\varphi)(k_1) = \int_K F_{f,\xi,\nu}(k_1, k)\varphi(k)\, dk.$$

We are thus left with the task of computing tr A (which is trace class by 8.7.6) where $A\colon H \to H$ and $(A\varphi)(k_1) = \int_K F_{f,\xi,\nu}(k_1, k)\varphi(k)\, dk$. We first note that

$$F_{f,\xi,\nu}(k_1m_1, k_2m_2) = \xi(m_1)^{-1}F_{f,\xi,\nu}(k_1, k_2)\xi(m_2).$$

Thus if φ is in $C(K; H_\xi)$ (the continuous maps from K to H_ξ) then

$\int_K F_{f,\xi,\nu}(k_1, k)\varphi(k)\,dk$ is in H^ξ. Thus A extends to a continuous map $\tilde{A}$, of $L^2(K) \otimes H_\xi$ to $H^\xi \subset L^2(K) \otimes H_\xi$, which is still of trace class. Furthermore $\operatorname{tr} \tilde{A} = \operatorname{tr} A$. Let γ be in $\hat{K}$. Let $v_1^\gamma, \ldots, v_{d(\gamma)}^\gamma$ be an orthonormal basis of V and let $w_1, \ldots, w_n$ be an orthonormal basis of H_ξ. Then the orthogonality relations (2.9.3) imply that if we set $\varphi_{ijl}^\gamma(k) = \sqrt{d(\gamma)}\langle \pi_\gamma(k)v_1^\gamma, v_i^\gamma\rangle$ then $\{\varphi_{ijl}^\gamma\}$ is an orthonormal basis of $L^2(K) \otimes H_\xi$. Thus

$$\begin{aligned}\operatorname{tr} \tilde{A} &= \sum_{\gamma\in\hat{K}} \sum_{l=1}^{n} \sum_{i,j=1}^{d(\gamma)} \langle \tilde{A}\varphi_{ijl}^\gamma, \varphi_{ijl}^r\rangle \\ &= \sum_{\gamma\in\hat{K}} \sum_{l=1}^{n} \sum_{i,j=1}^{d(\gamma)} \int_{K\times K} d(\gamma)\langle F_{f,\xi,\nu}(k_1, k_2) \\ &\quad \cdot\langle \pi(k_2)v_i^\gamma, v_j^\gamma\rangle w_l, \langle \pi(k_1)v_i^\gamma v_j^\gamma\rangle \omega_l > dk_1 dk_2 \\ &= \sum_{\gamma\in\hat{K}} \sum_{i,j=1}^{d(\gamma)} \int_{K\times K} d(\gamma) \operatorname{tr}(F_{f,\xi,\nu}(k_1, k_2)) \\ &\quad \cdot\langle \pi(k_2)v_i^\gamma, v_j^\gamma\rangle\langle \pi(k_1)v_i^\gamma, v_j^\gamma\rangle dk_1 dk_2.\end{aligned}$$

If φ is in $C^\infty(K \times K)$ we assert that

$$\begin{aligned}&\int_K \varphi(k, k)dk \\ &\quad = \sum_{\gamma\in K} \sum_{i,j=1}^{d(\gamma)} d(\gamma) \int_{K\times K} \varphi(k_1, k_2)\langle \pi(k_2)v_i^\gamma, v_j^\gamma\rangle\langle \pi(k_1)v_i^\gamma, v_j^\gamma\rangle dk_1 dk_2.\end{aligned}$$

Indeed,

$$\varphi(k_1, k_2) = \sum_{\gamma,\tau\in\hat{K}} \sum_{i,j=1}^{d(\gamma)} \sum_{r,s=1}^{d(\gamma)} a_{ij}b_{rs}\sqrt{d(\gamma)d(\tau)}\langle \pi(k_1)v_i^\gamma, v_j^\gamma\rangle\langle \pi(k_2)v_r^\tau, v_s^\tau\rangle$$

with convergence uniform and absolute. Thus

$$\begin{aligned}\sum_{\gamma\in\hat{K}} \sum_{i,j=1}^{d(\gamma)} d(\gamma) \int_{K\times K} \varphi(k_1, k_2)\langle \pi(k_2)v_i^\gamma, v_j^\gamma\rangle\langle \pi(k_1)v_i^\tau, v_j^\tau\rangle dk_1 dk_2 \\ = \sum_{\gamma\in\hat{K}} \sum_{i,j=1}^{d(\gamma)} a_{ij}^\gamma b_{ij}^\gamma.\end{aligned}$$

On the other hand

$$\begin{aligned}&\int_K \varphi(k, k)\,dk \\ &\quad = \sum_{\gamma,\tau\in K} \sum_{i,j=1}^{d(\gamma)} \sum_{r,s=1}^{d(\gamma)} a_{ij}b_{rs} \int_K \sqrt{d(\gamma)d(\tau)}\,\langle \pi(k)v_i^\gamma, v_j^\gamma\rangle\langle \pi(k)v_r^\tau, v_s^\tau\rangle dk \\ &\quad = \sum_{\gamma\in\hat{K}} \sum_{i,j=1}^{d(\gamma)} a_{ij}^\gamma b_{ij}^\gamma\end{aligned}$$

by 2.9.3. This proves our assertion. We therefore know

$$\text{(3)}\qquad \Theta_{\xi,\rho+(-1)^{1/2}\nu}(f) = \int_K \operatorname{tr}(F_{f,\xi,\nu}(k,k))\,dk$$
$$= \int_{K\times M\times A\times N} \operatorname{tr}\xi(m)e^{(\rho+(-1)^{1/2}\nu)(\log a)}$$
$$\cdot f(kmank^{-1})dk\,dm\,da\,dn$$
$$= \int_{M\times N} F_f(ma)\operatorname{tr}\xi(m)e^{(-1)^{1/2}\nu(\log a)}dm\,da$$

by the definition of F_f.

Now as representations of G, $(\tilde{\pi}_{\xi,\rho+(-1)^{1/2}\nu}, \tilde{H}^{\xi,\rho+(-1)^{1/2}\nu})$ and $(\pi_{\xi,\nu}, H^{\xi,\nu})$ are equivalent. Hence $\Theta_{\pi_{\xi,\nu+(-1)^{1/2}\nu}} = \Theta_{\pi_{\xi,\nu}}$. This proves the theorem.

8.8.3 *Corollary* Let ξ be in $\hat{M}$, ν in $\mathfrak{a}^*$. Let s be in $W(A)$. Then $(\pi_{\xi,\nu}, H^{\xi,\nu})$ and $(\pi_{\xi^s,s\nu}, H^{\xi^s,s\nu})$ are unitarily equivalent representations of G. ($\xi^s(m) = \xi(m^{*-1}mm^*)$ for some m^* in s, the class of ξ^s in $\hat{M}$ depends only on s.)

PROOF $\Theta_{\pi_{\xi,\nu}}(f) = \int_{A\times M} F_f(ma)\operatorname{tr}\xi(m)e^{(-1)^{1/2}\nu(\log a)}\,dmda$. Let s be in $W(A)$, then

$$\Theta_{\pi_{\xi^s,s\nu}}(f) = \int_{A\times M} F_f(ma)\operatorname{tr}\xi(m^{*-1}mm^*)e^{(-1)^{1/2}\nu(\log m^{*-1}am^*)}\,dm\,da$$

(here m^* is in s)

$$= \int_{A\times M} F_f(m^*mam^{*-1})\operatorname{tr}\xi(m)e^{(-1)^{1/2}\nu(\log a)}\,dn\,da$$

(since M^* acts as a compact group on M and A)

$$= \int_{A\times M} F_f(ma)\operatorname{tr}\xi(m)e^{(-1)^{1/2}\nu(\log a)}\,dm\,da,$$

by 7.7.11. Thus $\Theta_{\pi_{\xi,\nu}} = \Theta_{\pi_{\xi^s,s\nu}}$. The result now follows from 8.6.20.

8.9 The Weyl Group Revisited

8.9.1 We retain the notation of 8.3, 8.4, and 8.5. In particular we have G a semisimple Lie group with finite center. $G = KAN$, an Iwasawa decomposition of G. $\mathfrak{a}$ is the Lie algebra of A. M is the centralizer of A in K, M^* is the normalizer of A in K. $\mathfrak{g}_C$ is the complexification of $\mathfrak{g}$. $\mathfrak{m}$ is the Lie algebra of M. $\mathfrak{h}^-$ is maximal abelian in $\mathfrak{m}$. $\mathfrak{h}_0 = \mathfrak{h}^- + \mathfrak{a}$. $\mathfrak{h}$ is the complexification of $\mathfrak{h}_0$. Δ is the root system of $\mathfrak{g}_C$ relative to $\mathfrak{h}$. Δ^+ is a positive system so that if α is in Δ^+ and $\sigma\alpha \neq -\alpha$ (σ the conjugation of $\mathfrak{g}_C$ relative to $\mathfrak{g}$) then $\sigma\alpha$ is in Δ^+ (see 7.5.9). Let $\Sigma = \{\alpha \text{ in } \Delta^+ | \sigma\alpha \neq -\alpha\}$. We have seen in 7.5.11 that $\mathfrak{n} = (\sum_{\alpha\in\Sigma} \mathfrak{g}_\alpha) \cap \mathfrak{g}$.

8.9.2 Let Λ^+ be the positive restricted roots of $(\mathfrak{g}, \mathfrak{a})$ (see 7.3.5). Then $\Lambda^+ = \{\alpha|_\mathfrak{a} | \alpha \text{ in } \Sigma\}$. If λ is in Λ^+ then λ is called simple if λ cannot be written as a sum of two or more elements of Λ^+. Let $\pi = \{\alpha_1, \ldots, \alpha_l\}$ be the simple system of roots in Δ^+. We may assume that $\alpha_{l_1+1}, \ldots, \alpha_l$ are such that $\sigma\alpha_{l_1+j} = -\alpha_{l_1+j}$ and $\sigma\alpha_i \neq -\alpha_i$, $i = 1, \ldots, l_1$, $\sigma\alpha_i = \alpha_i$ for $l_0 < i \leqslant l_1$, $\sigma\alpha_i \neq \alpha_i$, $i \leqslant l_0$.

8.9.3 We have seen in the proof of 7.9.10 that $l_0 = 2p$ and after reordering, $\sigma\alpha_i = \alpha_{p+i} + \sum_{j+l_0+1}^{l} c_i^j \alpha_j$ with c_i^j nonnegative integers. Set $r = l_1 - l_0$, $q = p + r$

$$\lambda_i = \alpha_i|_\mathfrak{a},\ i = 1, \ldots, p, \qquad \lambda_{p+i} = \alpha_{l_0+i}|_\mathfrak{a},\ i = 1, \ldots, r.$$

If λ is in Λ^+ then $\lambda = \alpha|_\mathfrak{a}$, α in Σ. $\alpha = \sum n_i\alpha_i$ with $n_i \geqslant 0$, n_i an integer, $i = 1, \ldots, l$. Thus $\lambda = \sum_{i=1}^{l_1} n_i\alpha_i|_\mathfrak{a}$. On the other hand $\alpha_{p+i}|_\mathfrak{a} = \sigma\alpha_i|_\mathfrak{a} = \alpha_i|_\mathfrak{a}$. Hence we see that $\lambda = \sum_{i=1}^{q} m_i\lambda_i$ with $m_i \geqslant 0$, m_i an integer, $i = 1, \ldots, q$.

8.9.4 *Lemma* $\{\lambda_1, \ldots, \lambda_q\}$ is the set of simple roots of Λ^+. Furthermore $\{\lambda_1, \ldots, \lambda_q\}$ is a basis for $\mathfrak{a}^*$.

PROOF We first note that $\lambda_1, \ldots, \lambda_q$ form a basis of $\mathfrak{a}^*$. Indeed, $\{\alpha_1|_\mathfrak{a}, \ldots, \alpha_l|_\mathfrak{a}\}$ spans $\mathfrak{a}^*$. But $\{\alpha_1|_\mathfrak{a}, \ldots, \alpha_l|_\mathfrak{a}\} = \{\lambda_1, \ldots, \lambda_q\}$. Thus $\lambda_1, \ldots, \lambda_q$ span $\mathfrak{a}^*$. We must thus show that $\lambda_1, \ldots, \lambda_p$ are linearly independent. Suppose that $\sum_{i=1}^{q} c_i\lambda_i = 0$. Then $\sum_{i=1}^{q} c_i(\alpha_i + \sigma\alpha_i) = 0$. But then $\sum_{i=1}^{p} c_i(\alpha_i + \alpha_{i+p} + \sum_{j=l_0+1} c_i^j\alpha_j) = 0$. Since $\{\alpha_1, \ldots, \alpha_l\}$ is an independent set this clearly implies that $c_i = 0$, $i = 1, \ldots, q$.

If λ is not one of the λ_i, then $\lambda = \sum n_i\lambda_i$ $n_i \geqslant 0$, n_i an integer, and $\sum n_i \geqslant 2$. Thus λ cannot be simple. If $\lambda_i = \gamma + \mu$, γ, μ in Λ^+, then $\gamma = \sum n_j\lambda_j$, $\mu = \sum m_j\lambda_j$ with n_i, m_i nonnegative integers, $\sum n_j > 0$, $\sum m_j > 0$. Thus $\lambda_i = \sum (n_j + m_j)\lambda_j$ with $\sum (n_j + m_j) > 2$. But $\{\lambda_1, \ldots, \lambda_p\}$ is a linearly independent set. Thus λ_i is simple.

8.9.5 Let $M^*/M = W(A)$ and let $W(A)$ act on $\mathfrak{a}$ as in 7.5.6.

8.9.6 *Proposition* If λ is in Λ^+ define H_λ in $\mathfrak{a}$ by $B(H_\lambda, H) = \lambda(H)$ for all H in $\mathfrak{a}$. Define $s_\lambda(H) = H - (2\lambda(H)/\lambda(H_\lambda))H_\lambda$ for λ in Λ^+, H in $\mathfrak{a}$. Then

(1) s_λ is in $W(A)$ for λ in Λ^+.

(2) Set $s_i = s_{\lambda_i}$. Then $\{s_1, \ldots, s_q\}$ generates $W(A)$.

PROOF (1) is proved in a manner analogous to the beginning of the proof of 3.10.9. Let X be in $\mathfrak{g}_\lambda$ and let $Y = \theta X$ (θ the Cartan involution corresponding to $\mathfrak{g} = \mathfrak{k} \oplus \mathfrak{p}$). Then Y is in $\mathfrak{g}_{-\lambda}(\theta|_\mathfrak{a} = -I)$. Hence $[H, [X, Y]] = 0$ for all H in $\mathfrak{a}$. This implies (7.5.4) that $[X, Y]$ is in $\mathfrak{m} + \mathfrak{a}$. But $\theta[X, \theta X] = [\theta X, X] = -[X, \theta X]$. Thus $[X, Y]$ is in $(\mathfrak{m} + \mathfrak{a}) \cap \mathfrak{p} = \mathfrak{a}$. On the other hand $B([X, \theta X], H) = B(\theta X, [H, X]) = \lambda(H)B(\theta X, X)$. Thus we have

(a) If X is in $\mathfrak{g}_\lambda$, $[X, \theta X] = B(\theta X, X)H_\lambda$.

Since $B|_{\mathfrak{k}\times\mathfrak{k}}$ is negative definite, $B|_{\mathfrak{p}\times\mathfrak{p}}$ is positive definite and $\theta|_\mathfrak{k} = I$, $\theta|_\mathfrak{p} = -I$, we see that $\langle X, Y\rangle = -B(X, \theta Y)$ is positive definite. We may thus choose X in $\mathfrak{g}_\lambda$ so that $B(X, \theta X) = -1$. Set $Z = X + \theta X$. Then $\theta Z = Z$. Hence Z is in $\mathfrak{k}$. Thus $\exp tZ$ is in K. We compute $\mathrm{Ad}(\exp tZ)|_\mathfrak{a} = e^{t\,\mathrm{ad} Z}|_\mathfrak{a}$.

(i) If H is in $\mathfrak{a}$, $\lambda(H) = 0$, then $e^{t\,\mathrm{ad}Z}H = H + \sum_{k=1}^{\infty} (t^k/k!)(\mathrm{ad}Z)^{k-1}[Z, H]$. But $[Z, H] = [X + \theta X, H] = -\lambda(H)(X - \theta X) = 0$. Thus if $\lambda(H) = 0$, $e^{t\,\mathrm{ad}Z}H = H$.

(ii) We compute $e^{t\,\mathrm{ad}Z}H_\lambda$. We first note that $[Z, H_\lambda] = -\lambda(H)(X - \theta X)$. Set $W = X - \theta X$. Then $[Z, W] = -2[X, \theta X] = -2B(X, \theta X)H_\lambda = 2H_\lambda$. This gives

(iii) $(\mathrm{ad}Z)^{2k}H = (-1)^k 2^k \lambda(H_\lambda)^k H_\lambda$

(iv) $(\mathrm{ad}Z)^{2k+1}H = (-1)^{k+1}2^k \lambda(H_\lambda)^{k+1} W$.

Combining (iii) and (iv) we have

(v) $e^{t\,\mathrm{ad}Z}H = \cos(t((2\lambda(H_\lambda)^{1/2}))H_\lambda - \lambda(H_\lambda)\sin(t(2\lambda(H_\lambda))^{1/2})W$.

Hence if $t_0 = \pi/(2\lambda(H_\lambda))^{1/2}$ we see $e^{t_0\,\mathrm{ad}Z}H_\lambda = -H_\lambda$. Thus if $m^* = \exp(t_0 Z)$, $\mathrm{Ad}(m^*)\mathfrak{a} \subset \mathfrak{a}$, $\mathrm{Ad}(m^*)|_\mathfrak{a} = s_\lambda$. This proves (1).

To prove (2) we recall that $\mathfrak{a}' = \{H | H \text{ in } \mathfrak{a}, \lambda(H) \neq 0, \lambda \text{ in } \Lambda^+\}$.

(vi) If m^* is in M^*, if P is a connected component of $\mathfrak{a}'$, and if $\mathrm{Ad}(m^*)P = P$, then m^* is in M.

Let $s = \mathrm{Ad}(m^*)|_\mathfrak{a}$. Since $W(A)$ is a finite group there is k so that $s^k = I$. Let H_1 be in P, then $(1/k)(H_1 + sH_1 + \cdots + s^{k-1}H_1) = H$ is in P (P is convex) and $sH = H$. On the other hand $\mathrm{Ad}(m^*)\mathfrak{m} \subset \mathfrak{m}$. Thus $\mathrm{Ad}(m^*)$ fixes a regular element of $\mathfrak{m}$ (see 3.9). Hence there is m in M so that $\mathrm{Ad}(mm^*m^{-1})$ fixes a regular element of $\mathfrak{g}$ in $\mathfrak{h}_0$, hence a regular element of $\mathfrak{g}_C$ in $\mathfrak{h}_C$. The proof of 3.9.4 implies that $\mathrm{Ad}(mm^*m^{-1})|_{\mathfrak{h}_C} = I$. But $I = \mathrm{Ad}(mm^*m^{-1})|_\mathfrak{a} = \mathrm{Ad}(m^*)|_\mathfrak{a}$. Thus m^* is in M. (2) now follows from the same argument as the last part of the proof of 3.10.9.

8.9.7 *Lemma* Let λ be in Λ^+. If $k\lambda$ is in Λ^+ then $k = 1/2$, 1, or 2.

PROOF Let X be in $\mathfrak{g}_\lambda$ so that $B(X, \theta X) = -1$. Then 8.9.6(a) says that $[X, \theta X] = -H_\lambda$. Let $h = (2/\lambda(H_\lambda))H_\lambda$, $e = (2/\lambda(H_\lambda))X$, $f = -\theta X$. Then $[e, f] = h$, $[h, e] = 2e$, $[h, f] = -2f$. Thus $\{e, f, h\}$ spans a TDS (see 4.3.8). This implies that the eigenvalues of ad h are all integers. Now ad $h|_{\mathfrak{g}_{k\lambda}} = k\lambda(h)I = 2k\lambda(H_\lambda)/\lambda(H_\lambda) = 2k$. Thus $2k$ is a positive integer. Interchanging λ and $k\lambda$ we find that $2/k$ is a positive integer. Hence $k = n/2$, n a positive integer. Thus $4/n$ is a positive integer. This implies that n divides 4. Hence $n = 1, 2$, or 4. Thus $k = \frac{1}{2}$, 1, or 2.

8.9.8 Let $\Lambda_0^+ = \{\lambda \text{ in } \Lambda^+ | \frac{1}{2}\lambda \text{ is not in } \Lambda^+\}$. If s is in $W(A)$ let

$$\langle s \rangle = \{\lambda \text{ in } \Lambda_0^+ | -s\lambda \text{ is in } \Lambda^+\}.$$

8.9.9 If s is in $W(A)$ then 8.9.8(2) implies $s = s_{i_1} \ldots s_{i_r}$. Let $l(s)$ be the smallest such r.

8.9.10 If V is a finite set, denote by $[V]$ the cardinality of V.

8.9.11 *Lemma* Let $\Phi \subset \Lambda_0^+$. Then $\Phi = \langle s \rangle$ for some s in $W(A)$ if and only if

(1) whenever α is in Φ and $\alpha = \mu + \eta$, μ, η in Λ^+ then μ or η is in Φ.
(2) whenever α, β are in Φ and $\alpha + \beta$ is in Λ^+ then $\alpha + \beta$ is in Φ.

PROOF If $\Phi = \langle s \rangle$ and if α is in Φ, $\alpha = \mu + \eta$, μ, η in Λ^+ then $-s\alpha$ is in Λ^+. Thus $-s\mu - s\eta$ is in Λ^+. If $s\mu$ and $s\eta$ are in Λ^+ then $-s\mu - s\eta$ is not in Λ^+. Thus say $-s\mu$ is in Λ^+. Thus $\langle s \rangle$ satisfies (1). (2) is even easier.

To prove the sufficiency we prove the result by induction on $[\Phi]$.

(a) $[\Phi] = 1$. Then (1) implies that $\Phi = \{\lambda\}$ and λ is simple. We assert that $\Phi = \langle s_\lambda \rangle$. Suppose $s_\lambda = s_i$. We first note that $\lambda_i(H_{\lambda_j}) \leqslant 0$, $i \neq j$. Indeed, $s_i H_{\lambda_j} = H_{\lambda_j} - (2\lambda_i(H_{\lambda_j})/\lambda_i(H_{\lambda_i}))H_{\lambda_i}$. Thus since $s_i\Lambda = \Lambda$ we see that $\lambda_j - (2\lambda_i(H_{\lambda_j})/\lambda_i(H_{\lambda_j}))\lambda_i$ is in Λ. But 8.9.3 and 8.9.4 now imply that $\lambda_i(H_{\lambda_j}) \leqslant 0$. If λ is in Λ^+, $\lambda = \sum n_j\lambda_j$, $n_i \geqslant 0$, n_j an integer. $s_i\lambda = \sum_{j \neq i} n_j\lambda_j$ + a multiple of λ_i. Thus if λ is in Λ^+ and $\frac{1}{2}\lambda$ is not in Λ and $\lambda \neq \lambda_i$, $s_i\lambda$ is in Λ^+. Hence $\langle s_i \rangle = \langle \lambda_i \rangle$.

(b) Suppose that the result is true for $1 \leqslant [\Phi] < r$. If $[\Phi] = r$ and satisfies (1) and (2), then (1) implies that Φ contains a simple root, say, λ_i. Thus $s_i\Phi - \{-\lambda_i\} \subset \Lambda^+$. Set $\Phi' = s_i\Phi - \{-\lambda_i\}$. We assert that Φ' satisfies (1) and (2). Suppose γ is in Φ' and $\gamma = \mu + \eta$, μ, η in Λ^+. If say $\mu = \lambda_i$ then $s_i\gamma = -\lambda_i + s_i\eta$. Hence $-\lambda_i + s_i\eta$ is in $\Phi - \{\lambda_i\}$. Now λ_i is in Φ, thus (2) implies $s_i\eta$ is in Φ. Clearly $s_i\eta \neq \lambda_i$. Thus η is in Φ'. If $\mu, \eta \neq \lambda_i$ then $s_i\mu$ and

$s_i\eta$ are in Λ^+. Clearly $s_i\mu$, $s_i\eta \neq \lambda_i$. Thus $s_i\mu + s_i\eta$ is in Φ, hence (1) implies that say μ is in Φ'. Hence Φ' satisfies (1).

Suppose μ, η are in Φ' and $\mu + \eta$ is in Λ^+. Then $\mu + \eta \neq \lambda_i$ (λ_i is simple). Hence $s_i(\mu + \eta)$ is in Λ^+, $s_i\mu$, $s_i\eta$ are in $\Phi - \{\lambda_i\}$. Hence $s_i(\mu + \eta)$ is in $\Phi - \{\lambda_i\}$. Thus $\mu + \eta$ is in Φ'. Thus Φ' satisfies (1) and (2). Hence $\Phi' = \langle s' \rangle$ for some s' in $W(A)$. We leave it to the reader to check that $\langle s's_i \rangle = \Phi$.

8.9.12 *Corollary (to the proof of 8.9.11)* If s is in $W(A)$ and if λ_i is in $\langle s \rangle$ then $\langle ss_i \rangle = s_i\langle s \rangle - \{-\lambda_i\}$.

8.9.13 *Lemma* Let s be in $W(A)$. Then $s = s_{i_r} \ldots s_{i_1}$ and $\langle s \rangle = \{\lambda_{i_1}, s_{i_1}\lambda_{i_2}, \ldots, s_{i_1} \ldots s_{i_{r-1}}\lambda_{i_r}\}$ (a product of zero elements is 1).

PROOF By induction on $[\langle s \rangle]$, if $[\langle s \rangle] = 1$ the result is obvious. Now 8.9.12 implies that if λ_{i_1} is in $\langle s \rangle$, then $\langle ss_{i_1} \rangle = s_{i_1}\langle s \rangle - \{\lambda_{i_1}\}$. Thus $[\langle ss_{i_1} \rangle] < [\langle s \rangle]$. Applying the inductive hypothesis $ss_{i_1} = s_{i_r} \ldots s_{i_2}$ and $\langle ss_{i_1} \rangle = \{\lambda_{i_2}, s_{i_2}\lambda_{i_3}, \ldots, s_{i_2} \ldots s_{i_{r-1}}\lambda_{i_r}\}$. 8.9.12 completes the proof.

8.9.14 *Lemma* Let s in $W(A)$. Then $[\langle s \rangle] = l(s)$.

PROOF 8.9.13 implies that $[\langle s \rangle] \geqslant l(s)$. We show that if $s = s_{i_r} \ldots s_{i_1}$ then $[\langle s \rangle] \leqslant r$. We prove the result by induction or r. If $r = 1$ the result is clear. If true for r we prove for $r + 1$. Let $s' = ss_{i_1}$. Then the inductive hypothesis implies that $[\langle s' \rangle] \leqslant r$. Suppose that λ is in $\langle s \rangle$.

(a) If $\lambda \neq \lambda_{i_1}$ then $s_{i_1}\lambda$ is in Λ^+ and $s's_{i_1}\lambda = s\lambda$. But $-s\lambda$ is in Λ^+. Hence $s_{\lambda_{i_1}}\lambda$ is in $\langle s' \rangle$.

Hence by (a) $s_{i_1}(\langle s \rangle - \{\lambda_{i_1}\}) \subset \langle s' \rangle$. Thus $[\langle s \rangle] \leqslant [\langle s' \rangle] + 1 \leqslant r + 1$.

8.9.15 *Corollary* Let s be in $W(A)$. Suppose that $[\langle s \rangle] (= l(s)) = r$ and $s = s_{i_r} \ldots s_{i_1}$, then λ_{i_q} is not in $\langle s_{i_1} \ldots s_{i_{q-1}} \rangle$.

PROOF If λ_{i_q} is in $\langle s_{i_1} \ldots s_{i_{q-1}} \rangle$ then setting $s' = s_{i_1} \ldots s_{i_{q-1}}$ we see that $[\langle s's_{i_q} \rangle] \leqslant q - 2$. Thus $s' = s_{j_1} \ldots s_{j_t}s_{i_q}$ with $t \leqslant q - 2$. Now $s = (s_{i_r} \ldots s_{i_q})(s')^{-1} = s_{i_r} \ldots s_{i_{q-1}}s_{j_t} \ldots s_{j_1}$. But then $l(s) \leqslant r - q + t < r$. This is a contradiction.

8.9.16 *Lemma* Let s, s', s'' be in $W(A)$. Suppose that $s = s's''$ and $l(s) = l(s') + l(s'')$. Then $\langle s \rangle = \langle s'' \rangle \cup (s'')^{-1}\langle s' \rangle$.

PROOF This is an immediate consequence of 8.9.13.

8.10 The Intertwining Operators

8.10.1 We retain the notation of 8.9.

8.10.2 *Lemma* Let $\bar{\mathfrak{n}} = \theta(\mathfrak{n})$ (as usual). Let $\bar{\mathfrak{n}}_1 \subset \bar{\mathfrak{n}}$ be a subalgebra. Suppose that $\bar{\mathfrak{n}}_1 = \bar{\mathfrak{n}}_2 \oplus \bar{\mathfrak{n}}_3$ with $\bar{\mathfrak{n}}_2, \bar{\mathfrak{n}}_3$ subalgebras so that

$$\mathrm{ad}(\mathfrak{a})\bar{\mathfrak{n}}_i \subset \bar{\mathfrak{n}}_i,\ i = 1, 2.$$

Let $\bar{N}_i$ be the connected subgroup of $\bar{N}$ corresponding to $\bar{\mathfrak{n}}_i$. Then the map $\psi: \bar{N}_2 \times \bar{N}_3 \to \bar{N}_1$ given by $\psi(\bar{n}_2, \bar{n}_3) = \bar{n}_2\bar{n}_3$ is a surjective diffeomorphism.

PROOF We first note that $\psi_{*(e,e)}(X, Y) = X + Y$ for X in $\bar{\mathfrak{n}}_2$, Y in $\bar{\mathfrak{n}}_3$. There are thus neighborhoods U_2, U_3 of e in $\bar{N}_2$, $\bar{N}_3$, respectively, so that $\psi(U_2 \times U_3) = U_1$ is a neighborhood of e in $\bar{N}_1$ and $\psi: U_2 \times U_3 \to U_1$ is a diffeomorphism.

Let H be in $\mathfrak{a}$ so that $\lambda(H) > 0$ for all λ in Λ^+. Set $a_t = \exp tH$. Then $a_t\bar{N}_ia_t^{-1} \subset \bar{N}_i$ for $i = 1, 2, 3$.

(1) If $\bar{n}$ is in $\bar{N}$ then $\lim_{t\to\infty} a_t\bar{n}a_t^{-1} = e$. In fact. $\bar{n} = \exp X$ with X in $\bar{\mathfrak{n}}$ (see 7.4.3). $X = \sum_{\lambda\in\Lambda^+} X_{-\lambda}$ with $[H, X_{-\lambda}] = -\lambda(H)X_{-\lambda}$. Thus $a_t\bar{n}a^{-1} = \exp(\mathrm{Ad}(a_t)\sum X_{-\lambda}) = \exp(\sum e^{-\lambda(H)t}X_{-\lambda})$. Hence $\lim_{t\to\infty} a_t\bar{n}a_t^{-1} = e$.

(2) $\psi(a_t\bar{n}_2a_t^{-1}, a_t\bar{n}_3a_t^{-1}) = a_t\psi(\bar{n}_2, \bar{n}_3)a_t^{-1}$. This is obvious.

(3) ψ is surjective.

In fact, suppose $\bar{n}$ is in $\bar{N}_1$. Then if t is sufficiently large $a_t\bar{n}a_t^{-1}$ is in U_1 (by (1)). But then $a_t\bar{n}a_i^{-1} = \psi(\bar{n}_2, \bar{n}_3)$ for some $\bar{n}_i$ in $\bar{N}_i$, $i = 2, 3$. Hence $n = \psi(a_t^{-1}\bar{n}_2a_t, a_t^{-1}\bar{n}_3a_t)$ by (2).

(4) ψ is injective.

If $\psi(\bar{n}_2, \bar{n}_3) = \psi(\bar{n}_2', \bar{n}_3')$ let t be so large that $a_t\bar{n}_ia_t^{-1}$ and $a_t\bar{n}_i'a_t^{-1}$ are in U_i, $i = 1, 2$. Then $a_t\psi(\bar{n}_2, \bar{n}_3)a_i^{-1} = a_t\psi(\bar{n}_2', \bar{n}_3')a_t^{-1}$. Hence (2) and the injectivity of $\psi|_{U_2\times U_3}$ implies $a_t\bar{n}_ia_t^{-1} = a_t\bar{n}_i'a_t^{-1}$, $i = 2, 3$. Hence $\bar{n}_i = \bar{n}_i'$, $i = 2, 3$.

(5) ψ is everywhere regular.

Let $(\bar{n}_2', \bar{n}_3')$ be in $\bar{N}_2 \times \bar{N}_3$. Let t be so large that $(a_t\bar{n}_2'a_t^{-1}, a_t\bar{n}_3'a_t^{-1})$ is in $U_2 \times U_3$. Set $\xi(\bar{n}_2, \bar{n}_3) = (a_t\bar{n}_2a_t^{-1}, a_t\bar{n}_3a_t^{-1})$, and $\eta(\bar{n}_1) = a_t^{-1}\bar{n}_1a_t$ for $\bar{n}_i$ in $\bar{N}_i$, $i = 1, 2, 3$. Then $\psi = \eta \circ \psi \circ \xi$. Thus

$$\psi_{*(\bar{n}_2',\bar{n}_3')} = \eta_{*(a_t\bar{n}_2\bar{n}_3'a_t^{-1})}\psi_{*(a_t\bar{n}_2'a_t^{-1},a_t\bar{n}_3'a_t^{-1})}\xi_{*(\bar{n}_2',\bar{n}_3')}.$$

But $\psi_{*(x,y)}$ is of maximal rank for (x, y) in $U_2 \times U_3$. This proves (5).
(3), (4), and (5) prove the lemma.

8.10.3 If s is in $W(A)$ fix s^* in M^* such that $s^*M = s$.

8.10.4 If s is in $W(A)$ set $\bar{\mathfrak{n}}_s = \sum_{\lambda \in \langle s \rangle} \bar{\mathfrak{n}}_\lambda$ (the notation is as in 8.9.8 and $\bar{\mathfrak{n}}_\lambda = \mathfrak{g}_{-\lambda} + \mathfrak{g}_{-2\lambda}$). It is easily seen that $\bar{\mathfrak{n}}_s = \mathrm{Ad}(s^*)^{-1}\mathfrak{n} \cap \bar{\mathfrak{n}}$.

8.10.5 *Lemma* Let s, t, u be in $W(A)$. Suppose that $s = tu$ and $l(s) = l(t) + l(u)$ (see 8.9.9). Then $\bar{\mathfrak{n}}_s = \mathrm{Ad}(u^*)^{-1}\bar{\mathfrak{n}}_t \oplus \bar{\mathfrak{n}}_u$, a direct sum of subalgebras. Furthermore if $u = s_i$ for some i (see 8.9.6(2)), then $\bar{\mathfrak{n}}_u = \bar{\mathfrak{n}}_{\lambda_i}$ and $\bar{\mathfrak{n}}_{\lambda_i}$ normalizes $\mathrm{Ad}(s_i^*)^{-1}\bar{\mathfrak{n}}_t$ (i.e., $[\bar{\mathfrak{n}}_{\lambda_i}, \mathrm{Ad}(s_i^*)^{-1}\mathfrak{n}_t] \subset \mathrm{Ad}(s_i^*)^{-1}\bar{\mathfrak{n}}_t$).

PROOF To prove the first assertion we note that $\langle s \rangle = \langle u \rangle \cup u^{-1}\langle t \rangle$. Thus

$$\bar{\mathfrak{n}}_s = \sum_{\lambda \in \langle \mu \rangle} \bar{\mathfrak{n}}_\lambda \oplus \sum_{\lambda \in u^{-1}\langle t \rangle} \bar{\mathfrak{n}}_\lambda = \bar{\mathfrak{n}}_u \oplus \sum_{\lambda \in u^{-1}\langle t \rangle} \bar{\mathfrak{n}}_\lambda.$$

But $\mathrm{Ad}(u^*)^{-1}\mathfrak{g}_\lambda = \mathfrak{g}_{u^{-1}\lambda}$. Thus $\sum_{\lambda \in u^{-1}\langle t \rangle} \bar{\mathfrak{n}}_\lambda = \mathrm{Ad}(u^*)^{-1}\bar{\mathfrak{n}}_t$.

If $u = s_i$, for some i, then $\langle s_i \rangle = \langle \lambda_i \rangle$ and thus $\bar{\mathfrak{n}}_{s_i} = \bar{\mathfrak{n}}_{\lambda_i}$. Since $\bar{\mathfrak{n}}_s$ is a subalgebra of $\bar{\mathfrak{n}}$ we see that $[\bar{\mathfrak{n}}_{s_i}, \mathrm{Ad}(s_i^*)^{-1}\bar{\mathfrak{n}}_t] \subset \bar{\mathfrak{n}}_s$. But λ_i is simple, thus neither λ_i nor $2\lambda_i$ can be a sum of λ_i and an element of $s_i\langle t \rangle$. Thus

$$[\bar{\mathfrak{n}}_{s_i}, \mathrm{Ad}(s_i^*)^{-1}\bar{\mathfrak{n}}_t] \subset \mathrm{Ad}(s_i^*)^{-1}\mathfrak{n}_t.$$

8.10.6 Let for s in $W(A)$. $\bar{N}_s$ be the connected subgroup of $\bar{N}$ corresponding to $\bar{\mathfrak{n}}_s$.

8.10.7 *Lemma* If s, t, u are in $W(A)$ and $s = tu$ and $l(s) = l(t) + l(u)$, then the map

$$\psi: \bar{N}_t \times \bar{N}_u \to \bar{N}_s$$

given by $\psi(x, y) = u^{*-1}xu^*y$ is a surjective diffeomorphism. Furthermore if for each s in $W(A)$, $d\bar{n}_s$ is fixed invariant measure on $\bar{N}_s$ and if t, u, s are as above then if f is absolutely integrable on $\bar{N}_s$,

$$\int_{\bar{N}_s} f(\bar{n}_s)d\bar{n}_s = b(t, u) \int_{\bar{N}_t \times \bar{N}_u} f(u^{*-1}\bar{n}_t u^* \bar{n}_u)d\bar{n}_t\, d\bar{n}_s \tag{1}$$

with $b(t, u)$ a positive constant depending only on t, u.

PROOF The map $\bar{N}_t \to u^{*-1}\bar{N}_t u^*$ given by $x \mapsto u^{*-1}xu^*$ is a Lie isomorphism. $u^{*-1}\bar{N}_t u^*$ has Lie algebra $\mathrm{Ad}(u^*)^{-1}\mathfrak{n}_t$. Furthermore $\bar{\mathfrak{n}}_s = \mathrm{Ad}(u^*)^{-1}\bar{\mathfrak{n}}_t \oplus \bar{\mathfrak{n}}_u$. 8.10.2 now implies the first assertion. Now $\psi^* d\bar{n}_s = h(\bar{n}_t, \bar{n}_u) d\bar{n}_t d\bar{n}_s$ with h a positive C^∞ function. Since $\bar{N}_s$ is nilpotent it is unimodular. Thus $h(\bar{n}_t, \bar{n}_u) d\bar{n}_t d\bar{n}_u$ must be invariant under the action $x(\bar{n}_t, \bar{n}_u)y = (x\bar{n}_t, \bar{n}_u y)$, x, $\bar{n}_t$ in $\bar{N}_t$, y, $\bar{n}_u$ in $\bar{N}_u$. In fact, $\psi(x\bar{n}_t, \bar{n}_u y) = u^{*-1}xu^*\psi(\bar{n}_t, \bar{n}_u)y$. This implies that h is a constant which we denote by $b(t, u)$.

8.10.8 *Lemma* Let ξ be in $\hat{M}$, ν in $\mathfrak{a}_C^*$, s in W. Suppose that

(1) $$\int_{N/s^*Ns^{*-1}\cap N} |e^{-\nu(H(ns^*))}|\,dn$$

converges. Then

(2) $$(A(\xi, \nu, s)f)(g) = \int_{N/s^*Ns^{*-1}\cap N} f(gns^*)\,d\dot{n}$$

converges for all f in $\tilde{H}^{\xi,\nu}$, f continuous, and $A(\xi, \nu, s)f$ is continuous. Here $d\dot{n}$ is N-invariant measure on $N/s^*Ns^{*-1} \cap N$. (The integrals make sense for all f in $\tilde{H}^{\xi,\nu}$ since $f(gn) = f(g)$ for n in N, hence if n is in $s^*Ns^{*-1} \cap N, f(gns^*) = f(gs^*)$.) Furthermore

(3) $$A(\xi, \nu, s)f \text{ is in } \tilde{H}_F^{\xi^s, s(\nu-\rho)+\rho}$$

($\xi^s(m) = \xi(s^{*-1}ms^*)$ for m in M) for f in $\tilde{H}_F^{\xi,\nu}$.

(4) $$A(\xi, \nu, s)\tilde{\pi}_{\xi,\nu}(g)f = \tilde{\pi}_{\xi^s, s(\nu-\rho)+\rho}(g)A(\xi, \nu, s)f$$

for f in $\tilde{H}^{\xi,\nu}$, f continuous.

PROOF Assuming convergence of (2) we note that if f is in $\tilde{H}^{\xi,\nu}$, f continuous, then $(A(\xi, \nu, s)f)(gn) = (A(\xi, \nu, s)f)(g)$ for n in N since $d\dot{n}$ is N-invariant. Suppose that a is in A. Then

$$\begin{aligned}
(A(\xi, \nu, s)f)(ga) &= \int_{N/sNs^{-1}\cap N} f(gans^*)dn \\
&= \int_{N/sNs^{-1}\cap N} f(gana^{-1}as^*)dn \\
&= \int_{N/sNs^{-1}\cap N} f(gana^{-1}s^*s^{*-1}as^*)dn \\
&= e^{-\nu(H(s^{*-1}as^*))}\int_{N/sNs^{-1}\cap N} f(gana^{-1}s^*)dn \\
&= e^{-s\nu(H(a))}\int_{N/sNs^{-1}\cap N} f(gana^{-1}s^*)dn
\end{aligned}$$

Identifying $T(N/sNs^* \cap N)_{\bar{e}}$ with $\mathfrak{n}/\mathrm{Ad}(s^*)\mathfrak{n} \cap \mathfrak{n}$, we note that $n \mapsto ana^{-1}$ has differential $\mathrm{Ad}(a)^{\sim}: \mathfrak{n}/\mathrm{Ad}(s^*)\mathfrak{n} \cap \mathfrak{n} \to \mathfrak{n}/\mathrm{Ad}(s^*)\mathfrak{n} \cap \mathfrak{n}$, where $\mathrm{Ad}(a)^{\sim}$ is the induced map corresponding to $\mathrm{Ad}(a)$. Let $a = \exp H$. Then $\det \mathrm{Ad}(a)^{\sim} = e^{\mathrm{tr\,ad}^{\sim}(H)}$. But $\mathrm{tr\,ad}^{\sim}(H) = \sum_{\alpha\in P, s\alpha<0} \alpha = \rho - s\rho$. Thus

$$\int_{N/sNs^{-1}\cap N} f(gana^{-1}s^*)d\dot{n} = e^{\{s\rho-\rho\}(\log a)} \int_{N/sNs^{-1}\cap N} f(gns^*)d\dot{n}.$$

Hence

$$(A(\xi, v, s)f)(ga) = e^{-(s(v-\rho)+\rho)(\log a)}(A(\xi, v, s)f)(g).$$

If m is in M then

$$\int_{N/sNs^{-1}\cap N} f(gmns^*)d\dot{n}$$
$$= \int_{N/sNs^{-1}\cap N} f(g(mnm^{-1})s^*s^{*-1}ms^*)d\dot{n}$$
$$= \xi^s(m)^{-1} \int_{N/sNs^{-1}\cap N} f(gmnm^{-1}s^*)d\dot{n}.$$

Now M induces a compact group of diffeomorphism fixing e of $N/sNs^{-1} \cap N$ and therefore preserves the measure $d\dot{n}$. Thus we have

$$(A(\xi, v, s)f)(gman) = e^{-(s(v-\rho)+\rho)(\log a)}\, \xi^s(m)^{-1}(A(\xi, v, s)f)(g). \tag{2}$$

Now

$$\int_{N/sNs^{-1}\cap N} \|f(gns^*)\|d\dot{n}$$
$$= \int_{N/sNs^{-1}\cap N} \|f(k(gns^*)\exp H(gns^*)n(gns^*)\|\, d\dot{n}$$
$$= \int_{N/sNs^{-1}\cap N} |e^{-v(H(gns^*))}|\, \|f(k(gns^*)\|\, d\dot{n}$$
$$\leqslant \sup_{k\in K} \|f(k)\| \int_{N/sNs^{-1}\cap N} |e^{-v(H(gnm^*)}|\, d\dot{n}.$$

Now $g = kan_1$, thus $H(gns^*) = H(an_1ns^*)$. Thus

$$\int_{N/sNs^{-1}\cap N} \|f(gns^*)\|\, dn \leqslant \sup_{k\in K} \|f(k)\| \int_{N/sNs^{-1}\cap N} |e^{-v(H(ans^*)}|\, d\dot{n}$$
$$= \sup_{k\in K} \|f(k)\|\, |e^{(s(v-\rho)+\rho)(\log a)}| \int_{N/sNs^{-1}\cap N} |e^{-v(H(ns^*)}|\, d\dot{n}.$$

Thus if

$$\int_{N/sNs^{-1}\cap N} |e^{-\nu(H(ns^*))}|\, d\dot{n} < \infty$$

then $\int_{N/sNs^{-1}\cap N} f(gns^*)d\dot{n}$ converges uniformly and absolutely for g in compact sets. Thus $A(\xi, \nu, s)f$ is continuous. By the above if f is in $\tilde{H}_F^{\xi,\nu}$ and if γ is in $\hat{K}$ and $\chi_\gamma * f = f$ then $\chi_\gamma * A(\xi, \nu, s)f = A(\xi, \nu, s)f$, thus $A(\xi, \nu, s)f$ is in $\tilde{H}_F^{\xi^s, s(\nu-\rho)+\rho}$.

Let f be in $\tilde{H}^{\xi,\nu}$ and suppose that f is continuous. Then

$$(A(\xi, \nu, s)\tilde{\pi}_{\xi,\nu}(x)f)(g) = \int_{N/sNs^{-1}\cap N} f(x^{-1}gns^*)\, d\dot{n} = (A(\xi, \nu, s)f)(x^{-1}g)$$
$$= (\tilde{\pi}_{\xi^s, s(\nu-\rho)+\rho}(x)A(\xi, \nu, s)f)(g).$$

The lemma is now completely proved.

8.10.9 *Lemma* Let f be in $\tilde{H}^{\xi,\nu}$, then

$$\int_{N/sNs^{-1}\cap N} f(gns^*)d\dot{n} = \int_{\bar{N}_s} f(gs^*\bar{n})d\bar{n}_s.$$

in the sense that if one side converges then the other side converges and both sides are equal. Here $d\bar{n}_s$ has invariant measure on $\bar{N}_s$ normalized so that there are no constants in this equation.

PROOF Let $N_1 = N \cap s\bar{N}s^{-1}$, $N_2 = N \cap sNs^{-1}$, then N_1 and N_2 are closed subgroups of N. Let $\mathfrak{n}_1$ and $\mathfrak{n}_2$ be, respectively, the Lie algebras of N_1 and N_2. Then $\mathfrak{n}_1 \oplus \mathfrak{n}_2 = \mathfrak{n}$ and the hypotheses of 8.10.2 are satisfied with $\bar{N}$ replaced by N. 8.10.2 now implies that the natural map $N_1 \to N/N \cap sNs^{-1}$ is a surjective diffeomorphism. This implies that there is a C^∞ function $h: N_1 \to R$ so that if φ is integrable on $N/N \cap sNs^{-1} = N/N_2$ then $\int_{N/N_2} \varphi(\dot{n})d\dot{n} = \int_{N_1} \varphi(n_1N_2)h(n_1)dn_1$. But h must be invariant under the left action of N_1 on itself. Hence h is a constant. We normalize dn_1 so that $h \equiv 1$. Let φ be integrable on $N \cap s\bar{N}s^{-1}$. Then $\int_{N\cap s\bar{N}s^{-1}} \varphi(n)dn = \int_{\bar{N}\cap s^{-1}Ns} \varphi(s^*\bar{n}s^{*-1})d\bar{n}_s$ where $d\bar{n}_s$ is invariant measure on $\bar{N}_s$ normalized so that there are no constants in the formula. The result now follows.

8.10.10 We note that 8.10.9 implies that

$$(1) \qquad \int_{N/sNs^{-1}\cap N} e^{-\nu(H(ns^*))}dn = \int_{\bar{N}_s} e^{-\nu(H(\bar{n}))}d\bar{n}_s$$

in the sense that if either side of (1) converges then they both converge and are equal. Let $\mathfrak{a}_s^*$ be the set of all ν in $\mathfrak{a}_C^*$ such that either side of (1) converges absolutely.

8.10.11 *Theorem* If $s = tu$ with s, t, u in $W(A)$ and $l(s) = l(t) + l(u)$. Then $\mathfrak{a}_s^* \supset \mathfrak{a}_u^* \cap \tilde{u}\mathfrak{a}_t^*$ (here $\tilde{u}(\nu) = u(\nu - \rho) + \rho$ for ν in $\mathfrak{a}_c^*$). Furthermore if f is in $\tilde{H}^{\xi,\nu}$, γ in $\mathfrak{a}_s^*$, and if f is continuous then

$$A(\xi, \nu, s)f = b(t, u)A(\xi^u, u(\nu - \rho) + \rho, t)A(\xi, \nu, u)f \tag{1}$$

with $b(t, u)$ as in 8.10.7.

PROOF Suppose that ν is in $\mathfrak{a}_u^* \cap \tilde{u}\mathfrak{a}_t^*$ and f is in $\tilde{H}^{\xi,\nu}$, f continuous. Then

$$\begin{aligned}(A(\xi, \nu, s)f)(g) &= \int_{\bar{N}s} f(gs^*\bar{n}_s)d\bar{n}_s \\ &= b(t, u)\int_{\bar{N}_t \times \bar{N}_u} f(gs^*u^{*-1}\bar{n}_t u^*\bar{n}_u)d\bar{n}_t d\bar{n}_u \text{ by 8.10.7(1)} \\ &= b(t, u)\int_{\bar{N}_t \times \bar{N}_u} f(gt^*\bar{n}_t u^*\bar{n}_u)d\bar{n}_t d\bar{n}_u \\ &= b(t, u)\int_{\bar{N}_t} (A(\xi, \nu, u)f)(gt^*\bar{n}_t)d\bar{n}_t \\ &= b(t, u)(A(\xi^u, u(\nu - \rho) + \rho, t)A(\xi, \nu, u)f)(g)\end{aligned}$$

by 8.10.8(4). Now Fubini's theorem implies the result.

8.10.12 *Lemma* Let λ be in Λ_0^+. Suppose that 2λ is in Λ^+.

(1) $[\mathfrak{g}_\lambda, \mathfrak{g}_\lambda] = \mathfrak{g}_{2\lambda}$.

(2) If X is in $\mathfrak{g}_\lambda$, Y is in $\mathfrak{g}_{2\lambda}$, $X \neq 0$, $Y \neq 0$ then $X, Y, \theta X, \theta Y$ generate a Lie subalgebra of $\mathfrak{g}$, $\mathfrak{U}$, which is isomorphic with

$$su(2, 1) = \{W \text{ in } M_3(C) | WJ + J^t\,\overline{W} = 0, \text{ tr } W = 0\}$$

where

$$J = \begin{pmatrix} 1 & 0 & 0 \\ 0 & 1 & 0 \\ 0 & 0 & -1 \end{pmatrix}.$$

PROOF (1) Let Y be in $\mathfrak{g}_{2\lambda}$ and X in $\mathfrak{g}_\lambda$, $X \neq 0$. Then $[X, \theta X] = B(X, \theta X)H_\lambda$. Furthermore $B(X, \theta X) < 0$. Also $[\theta X, Y]$ is in $\mathfrak{g}_\lambda$. Hence

$[X, [\theta X, Y]] = [[X, \theta X], Y] + [\theta X, [X, Y]]$. Since $\mathfrak{g}_{3\lambda} = (0)$ and $[X, Y]$ is in $\mathfrak{g}_{3\lambda}$ we see that $[X, [\theta X, Y]] = B(X, \theta X)2\lambda(H_\lambda)Y$. Since $\lambda(H_\lambda) > 0$ this proves the result.

We now prove (2). It is convenient to normalize X, Y so that $B(X, \theta X) = -2/\lambda(H_\lambda)$, $B(Y, \theta Y) = -2/\lambda(H_\lambda)$. Let $H = [1/\lambda(H_\lambda)]H_\lambda$. Now $\theta[X, \theta X] = [\theta X, X] = -[X, \theta X]$, similarly $\theta[Y, \theta Y] = -[Y, \theta Y]$. Thus since $[X, \theta X]$, $[Y, \theta Y]$ are clearly in $\mathfrak{m} \oplus \mathfrak{a}$, $[X, \theta X]$ and $[Y, \theta Y]$ are in $\mathfrak{a}$. Let H_1 be in $\mathfrak{a}$,

$$B([X, \theta X], H_1) = B(\theta X, [H_1, X]) = \lambda(H_1)B(X, \theta X) = -2\lambda(H_1)/\lambda(H_\lambda).$$

We therefore see

(i) $[X, \theta X] = -2H$.

Similarly

(ii) $[Y, \theta Y] = -4H$.

Define Z by the formula $[\theta X, Y] = 2Z$. Now

$$\begin{aligned} B(Z, \theta Z) &= \frac{1}{4} B([\theta X, Y], [X, \theta Y]) = -\frac{1}{4} B([X, [\theta X, Y]], Y) \\ &= -\frac{1}{4} B([[X, \theta X], Y], \theta Y) \\ &= \frac{1}{2} B([H, Y], \theta Y) = B(Y, \theta Y) = -2/\lambda(H_\lambda). \end{aligned}$$

We therefore see that

(iii) $[Z, \theta Z] = -2H$.

On the other hand $[X, Z] = \frac{1}{2}[X, [\theta X, Y]] = \frac{1}{2}[[X, \theta X], Y] = -[H, Y] = -2Y$. Thus

(iv) $[X, Z] = -2Y$.

Let $2W = [\theta X, Z]$. Now W is in $\mathfrak{m} \oplus \mathfrak{a}$. If H_1 is in $\mathfrak{a}$,

$$B([\theta X, Z], H) = B(Z, [H, \theta X]) = -B(Z, \theta X) = -\frac{1}{2} B([\theta X, Y], \theta X] = 0.$$

Hence

(v) $\theta W = W$.

Furthermore

$$\begin{aligned} [W, X] &= \frac{1}{2}[[\theta X, Z], X] = \frac{1}{2}[[\theta X, X], Z] + \frac{1}{2}[\theta X, [Z, X]] \\ &= [H, Z] + [\theta X, Y] \quad \text{(by (i) and (iv))} = Z + 2Z = 3Z. \end{aligned}$$

Hence

(vi) $[W, X] = 3Z$.

Arguing similarly,

(vii) $[W, Z] = -3X$.
Since $[X, Z] = -2Y$ we see that
(viii) $[W, Y] = 0$.
Finally

$$[\theta Z, Y] = \frac{1}{3}[[W, \theta X], Y] = \frac{1}{3}[W, [\theta X, Y]] = \frac{2}{3}[W, Z] = -2X$$

by (vii). Hence
(ix) $[\theta Z, Y] = -2X$.

Using the fact that θ is an automorphism of $\mathfrak{g}$ leaving the subspace $\mathfrak{g}_1 = R\theta Y + R\theta X + R\theta Z + RW + RH + RX + RZ + RY$ invariant, we see that (i)–(ix) imply

(x) $\mathfrak{g}_1$ is a Lie subalgebra of $\mathfrak{g}$ and up to isomorphism $\mathfrak{g}_1$ is independent of $\mathfrak{g}$.

Consider now $\mathfrak{g}_0 = su(2, 1)$. Let

$$h = \begin{pmatrix} 0 & 0 & 0 \\ 0 & 0 & 1 \\ 0 & 1 & 0 \end{pmatrix}$$

$$x = \begin{pmatrix} 0 & -1 & 1 \\ 1 & 0 & 0 \\ 1 & 0 & 0 \end{pmatrix}$$

$$y = \begin{pmatrix} 0 & 0 & 0 \\ 0 & -i & i \\ 0 & -i & i \end{pmatrix}$$

Define $\theta V = -{}^t\overline{V}$. Then $\theta\mathfrak{g}_0 = \mathfrak{g}_0$ is an automorphism. Taking

$$z = \tfrac{1}{2}[\theta x, y],\ w = \tfrac{1}{2}[\theta x, z]$$

it is easily checked that

$$\mathfrak{g}_0 = R\theta y + R\theta x + R\theta z + Rw + Rh + Rz + Rx + Ry.$$

We leave it to the reader to check that (i)–(ix) are satisfied with the capital letters replaced by lower case letters.

This completes the proof of the lemma.

8.10.13 *Lemma* Let λ be in Λ_0^+. Let X be in $\mathfrak{g}_{-\lambda}$, Y be in $\mathfrak{g}_{-2\lambda}$ ($\mathfrak{g}_{-2\lambda}$

might be 0). Then if ν is in $\mathfrak{a}_C^*$

$$e^{\nu(H(\exp(X+Y))} = ((1 + (\lambda(H_\lambda)/2)\|X\|^2)^2 + 2\lambda(H_\lambda)\|Y\|^2)^{\nu(H_\lambda)/2\lambda(H_\lambda)},$$

here if V is in $\mathfrak{g}$, $\|V\|^2 = -B(V, \theta V)$.

PROOF Let $\mathfrak{g}_0 = su(2, 1)$.

$$\mathfrak{k}_0 = \left\{ \begin{bmatrix} Z & 0 \\ 0 & -\mathrm{tr} Z \end{bmatrix} \middle| Z \text{ in } M_2(C),\ {}^t\bar{Z} = -Z \right\}.$$

$\mathfrak{a}_0 = Rh$, $\mathfrak{n}_0 = Rx + Rz + Ry$. Here the notation is as in 8.10.12. We use the symbol θ for the Cartan involution of $\mathfrak{g}$ and $\mathfrak{g}_0$.

(a) $\mathfrak{g}_{2\lambda} \neq (0)$. Let X_0 be a multiple of θX so that $B(X_0, \theta X_0) = -2/\lambda(H_\lambda)$ (if $X = 0$, X_0 is arbitrary subject to the last condition). Let Y_0 be a multiple of θY so that $B(Y_0, Y_0) = -2/\lambda(H_\lambda)$. 8.10.12 implies that the map $X_0 \to x$, $Y_0 \to y$, $\theta X_0 \to \theta x$, $\theta Y_0 \to \theta y$ determines a θ-stable Lie algebra isomorphism of the Lie algebra generated by X_0, Y_0, θX_0, θY_0, with $su(2, 1)$.

(b) $\mathfrak{g}_{2\lambda} = (0)$. Let X_0 be as in (a). Then the map $X_0 \to x$, $\theta X_0 \to \theta x$ determines a θ-stable isomorphism of the Lie algebra spanned by X_0, Y_0, $H = 1/\lambda(H_\lambda)H_\lambda$ into $su(2, 1)$.

Let K_0 be the connected subgroup of $SU(2, 1)$ corresponding to $\mathfrak{k}_0$, A_0 the connected subgroup of $SU(2, 1)$ corresponding to $\mathfrak{a}_0$, and N_0 the connected subgroup of $SU(2, 1)$ corresponding to $\mathfrak{n}_0$. Then $SU(2, 1) = K_0 A_0 N_0$ is an Iwasawa decomposition of $SU(2, 1)$. Let $v_0 = (1, 0, 0)$, $v_1 = (0, 1/\sqrt{2}, 1/\sqrt{2})$, and $v_{-1} = (0, 1/\sqrt{2}, -1/\sqrt{2})$. Then $hv_i = iv_i$, $i = 0, 1, -1$. Let for $x = (x_1, x_2, x_3)$, $|x|^2 = |x_1|^2 + |x_2|^2 + |x_3|^2$. If g is in $SU(2, 1)$, $g = kan$, k in K_0, a in A_0, n in N_0, then $a = e^{th}$ with

$$e^t = |g \cdot v_1|.$$

We must therefore compute $|\exp t\theta x \exp s\theta y \cdot v_1|$. Now

$$\theta x = \begin{pmatrix} 0 & -1 & -1 \\ 1 & 0 & 0 \\ -1 & 0 & 0 \end{pmatrix}$$

$$\theta x^2 = \begin{pmatrix} 0 & 0 & 0 \\ 0 & -1 & -1 \\ 0 & 1 & 1 \end{pmatrix}$$

$$\theta x^3 = 0$$

and

$$\theta y = \begin{pmatrix} 0 & 0 & 0 \\ 0 & -i & -i \\ 0 & i & i \end{pmatrix}$$

$$(\theta y)^2 = 0.$$

Thus, $\theta x \cdot v_{-1} = \theta y \cdot v_{-1} = 0$, $\theta x \cdot v_1 = -(2/\sqrt{2})e_1$, $(\theta x)^2 v_1 = -2v_{-1}$, $\theta y \cdot v_1 = -2iv_{-1}$. Hence

$$\begin{aligned}(\exp t\theta x \exp s\theta y)v_1 &= (\exp t\theta x)(v_1 - 2iv_{-1}) \\ &= \exp t\theta x v_1 - 2isv_{-1} = v_1 - (2t/\sqrt{2})e_1 - t^2 v_{-1} - 2isv_{-1}.\end{aligned}$$

This gives

(c) $$|\exp t\theta x \exp s\theta y v_1| = ((1 + t^2)^2 + 4s^2)^{1/2}.$$

(c) implies that if ξ is in $\mathfrak{a}^*$ and $\xi(H) = 1$ $(H = \lambda(H_\lambda)^{-1}H)$ then $e^{\xi(H(\exp(t\theta X_0 + s\theta Y_0))} = ((1 + t^2)^2 + 4s^2)^{1/2}$. Note if we are in case (b) the formula is correct with $s = 0$.

Now $X = t\theta X_0$, $Y = s\theta Y_0$. $\|X\|^2 = -B(X, \theta X) = -t^2 B(X_0, \theta X_0) = 2t^2/\lambda(H_\lambda)$. Hence $t^2 = (\lambda(H_\lambda)/2)\|X\|^2$ and similarly $s^2 = (\lambda(H_\lambda)/2)\|Y\|^2$. Thus if $(H) = 1$ we see

(d) $$e^{\xi(H(\exp(X+Y)))} = ((1 + (\lambda(H_\lambda)/2)\|X\|^2)^2 + 2\lambda(H_\lambda)\|Y\|^2)^{1/2}.$$

(d) certainly implies the lemma.

8.10.14 Before proceeding we recall some facts about gamma and beta functions all of which can be found in Whittaker and Watson [1], pp. 235–264. The functions are defined by

(1) $$\Gamma(z) = \int_0^\infty t^{z-1} e^{-t}\, dt \text{ for } \operatorname{Re} z > 0.$$

(2) $$B(z, w) = \int_0^1 t^{z-1}(1 - t)^{w-1}\, dt \text{ for } \operatorname{Re} z > 0, \operatorname{Re} w > 0.$$

In (1) and (2) the integrals converge absolutely in the indicated domains. The equation relating the two functions is

(3) $$B(z, w) = \frac{\Gamma(z)\Gamma(w)}{\Gamma(z + w)} \text{ for } \operatorname{Re} z > 0, \operatorname{Re} w > 0.$$

Integrating the right-hand side of (1) by parts shows that Γ extends to a meromorphic function on C with simple poles at $-k$, k, a nonnegative integer. Hence B extends to a meromorphic function on $C \times C$.

Using the Euler product formula for $1/\Gamma$ we see

(4) $1/\Gamma$ is holomorphic on C. In particular Γ has no zeros on C

8.10.15 *Lemma* Let q, n, m be positive integers.

(1) $\int_0^\infty t^{q-1}(1 + t^2)^{-z}\, dt$ converges absolutely for Re $z > q/2$ and under this condition is equal to $\frac{1}{2}B(q/2, z - q/2)$.

(2) $$\int_0^\infty \int_0^\infty t^{m-1}s^{n-1}((1 + t^2)^2 + s^2)^{-z}\, dsdt = c(m, n; z)$$

converges absolutely for Re $z > (2n + m)/4$ and under this condition

$$c(m, n; z) = \frac{1}{4} B\left(\frac{m}{2}, 2z - n + 1 - \frac{m}{2}\right) B\left(\frac{n}{2}, z - \frac{n}{2}\right).$$

PROOF (1) Using the change of variable $t \to t^2$ we find that

$$\int_0^\infty t^{q-1}(1 + t^2)^{-z}\, dt = \frac{1}{2}\int_0^\infty t^{q/2-1}(1 + t)^{-z}\, dt.$$

Now letting for $0 < u < 1$, $t = u/(1 - u)$ we find

$$\int_0^\infty t^{q/2-1}(1 + t)^{-z}\, dt = \int_0^1 u^{(q/2)-1}(1 - u)^{z-(q/2)-1}\, du.$$

(1) now follows from 8.10.13(a).

To prove (2) we note that ds/s as a measure on $(0, \infty)$ is invariant under maps $s \mapsto \alpha s$ with α in R, $\alpha > 0$. Thus since

$$(1 + t^2)^2 + s^2 = (1 + t^2)^2(1 + (s/(1 + t^2))^2)$$

we see that

$$\int_0^\infty \int_0^\infty t^{m-1}s^{n-1}((1 + t^2)^2 + s^2)^{-z}\, dsdt$$
$$= \int_0^\infty \int_0^\infty t^{m-1}((1 + t^2)s)^{n-1}(1 + t^2)^{-2z}(1 + s^2)^{-z}\, ds\, dt$$
$$= \int_0^\infty t^{m-1}(1 + t^2)^{n-2z-1}\, dt \int_0^\infty s^{n-1}(1 + s^2)^{-z}\, ds.$$

Thus (1) implies $c(m, n; z)$ converges absolutely for z satisfying both Re$(2z - n + 1) > m/2$ and Re $z > n/2$, that is, for Re $z > (2n - 2 + m)/4$, Re $z > n/2$. (2) now follows from (1).

8.10.16 *Theorem* Let ν be in $\mathfrak{a}_C^*$. If for each λ in Λ_0^+, $\operatorname{Re}(\nu - \rho)(H_\lambda) > 0$ then

$$c(\nu) = \int_{\bar{N}} e^{-\nu(H(\bar{n}))}\, d\bar{n}$$

converges absolutely and $c(\nu) \neq 0$. Furthermore the map $\nu \to c(\nu)$ extends to a meromorphic function on $\mathfrak{a}_C^*$.

PROOF Let for ν in $\mathfrak{a}_C^*$, $1_\nu(g) = e^{-\nu(H(g))}$. Then 1_ν is an element of $\tilde{H}^{1,\nu}$. It clearly spans $\tilde{H}_1^{1,\nu}$ where 1 is the class of the trivial representation of K. Suppose that ν is in $\mathfrak{a}_s^*$. Then the formula for $A(1, \nu, s)$ and 8.10.8(3) imply that $A(1, \nu, s)1_\nu = c_s(\nu)1_{s(\nu-\rho)+\rho}$ where

(1) $$c_s(\nu) = \int_{\bar{N}s} e^{-\nu(H(\bar{n}_s))}\, d\bar{n}_s.$$

Now 8.10.11 implies that if s is in $W(A)$ and $s = tu$ with $l(s) = l(u) + l(t)$ and if ν is in $\mathfrak{a}_t^*$, $u(\nu - \rho) + \rho$ is in $\mathfrak{a}_u^*$ then ν is in $\mathfrak{a}_s^*$ and

(2) $$c_s(\nu) = b(t, u)c_t(u(\nu - \rho) + \rho)c_u(\nu).$$

Let s_0 in $W(A)$ be the unique element such that $s_0\Lambda^+ = -\Lambda^+$. Then $c(\nu) = c_{s_0}(\nu)$. Let $s_0 = s_{i_r} \ldots s_{i_1}$ in the notation of 8.9.13 with $l(s_0) = r = [\Lambda^+)$. We need the following.

(3) If ν satisfies $\operatorname{Re}(\nu - \rho)(H_\lambda) > 0$ and if $u = s_{i_k} \ldots s_{i_1}$ then

$$\operatorname{Re}(u(\nu - \rho) + \rho - \frac{1}{2}(m_{\lambda_{i_{k+1}}} + 2m_{2\lambda_{i_{k+1}}})\lambda_{i_{k+1}})(H_{\lambda_{i_{k+1}}}) > 0$$

where

$$m_{\lambda_{i_{k+1}}} = \dim \mathfrak{g}_{\lambda_{i_{k+1}}}, m_{2\lambda_{i_{k+1}}} = \dim \mathfrak{g}_{2\lambda_{i_{k+1}}}.$$

In fact $\operatorname{Re}(u(\nu - \rho))(H_{\lambda_{i_{k+1}}}) = \operatorname{Re}(\nu - \rho)(u^{-1}H_{\lambda_{i_{k+1}}})$. 8.9.15 implies that $u^{-1}H_{\lambda_{i_{k+1}}} = H_\mu$ with μ in Λ_0^+. Thus

$$\operatorname{Re}(u(\nu - \rho) + \rho - \frac{1}{2}(m_{\lambda_{i_{k+1}}} + 2m_{2\lambda_{i_{k+1}}})\lambda_{i_{k+1}})(H_{\lambda_{i_{k+1}}})$$

$$> \rho(H_{\lambda_{i_{k+1}}}) - 12(m_{\lambda_{i_{k+1}}} + 2m_{\lambda_{i_{k+1}}})\lambda_{i_{k+1}}(H_{\lambda_{i_{k+1}}}).$$

Now $2\rho = \sum_{\lambda\in\Lambda^+} m_\lambda \lambda$ with $m_\lambda = \dim \mathfrak{g}_\lambda$. Since

$$s_{\lambda_{i_{k+1}}}\Lambda^+ = \{\Lambda^+ - \{\lambda_{i_{k+1}}, 2\lambda_{i_{k+1}}\}\} \cup \{-\lambda_{i_{k+1}}, -2\lambda_{i_{k+1}}\}$$

we see that

$$s_{\lambda_{i_{k+1}}}\rho = \rho - (m_{\lambda_{i_{k+1}}} + 2m_{\lambda_{i_{k+1}}})\lambda_{i_{k+1}}.$$

Hence

$$\rho(H_{\lambda_{i_{k+1}}}) = \frac{1}{2}(m_{\lambda_{i_{k+1}}} + 2m_{2\lambda_{i_{k+1}}})\lambda_{i_{k+1}}(H_{\lambda_{i_{k+1}}}).$$

This certainly implies (2).

Using (2) and 8.10.11, 8.10.16 will follow if we can show the following.

(4) Let λ_i be simple. Let ν be in $\mathfrak{a}_C^*$ such that

$$\operatorname{Re} \nu(H_{\lambda_i}) > \tfrac{1}{2}(m_{\lambda_i} + 2m_{2\lambda_i})\lambda_i(H_{\lambda_i}),$$

then $c_{s_i}(\nu)$ is absolutely convergent, $c_{s_i}(\nu) \neq 0$ and continues to a meromorphic function on $\mathfrak{a}_C^*$.

To prove (3) we must therefore compute

(5) $$\int_{\bar{N}_{s_i}} e^{-\nu(H(\bar{n}))}\, d\bar{n}.$$

Now $\bar{\mathfrak{n}}_{s_i} = \mathfrak{g}_{-\lambda_i} + \mathfrak{g}_{-2\lambda_i}$. The map $\bar{\mathfrak{n}}_{s_i} \xrightarrow{\psi} \bar{N}_{s_i}$ given by $\psi(X) = \exp X$ is a diffeomorphism (see 7.4.3). Furthermore we may normalize $d\bar{n}$ on $\bar{N}_{s_i}$ so that $\psi^* d\bar{n} = dX(\bar{n} = \psi(X))$. Thus (5) becomes

(6) $$\int_{\mathfrak{g}_{-\lambda_i} \times \mathfrak{g}_{-2\lambda_i}} e^{-\nu(H(\exp(X+Y)))}\, dX\, dY.$$

Now 8.10.12 implies that (6) is

(7) $$\int_{\mathfrak{g}_{-\lambda_i} \times \mathfrak{g}_{-2\lambda_i}} (1 + (\lambda_i(H_{\lambda_i})/2)\|X\|^2)^2 + 2\lambda_i(H_{\lambda_i})\|Y\|^2)^{-\nu(H_{\lambda_i})/2\lambda_i(H\nu_i)} \cdot dX dY.$$

Let $S_1 = \{X \text{ in } \mathfrak{g}_{-\lambda_i} | (\lambda_i(H_{\lambda_i})/2)\|X\|^2 = 1\}$, $S_2 = \{Y \text{ in } \mathfrak{g}_{-2\lambda_i} | 2\lambda_i(H_{\lambda_i})\|Y\|^2 = 1\}$. Then $\dim S_1 = m_{\lambda_i} - 1$, $\dim S_2 = m_{2\lambda_i} - 1$. The usual formula for integration in polar coordinates (cf. Loomus-Sternberg ( )) shows that (7) is just

(8) $$c_1 c_2 \int_0^\infty \int_0^\infty t^{m_{\lambda_i}-1} s^{m_{2\lambda_i}-1}((1 + t^2)^2 + s^2)^{-\nu(H_{\lambda_i})/2\lambda_i(H_{\lambda_i})}\, dt ds,$$

with c_i the Euclidean volume of S_i.

8.10.15 implies that (8) is absolutely convergent if

$$\operatorname{Re} \nu(H_{\lambda_i})/2\lambda_i(H_{\lambda_i}) > (m_{\lambda_i} + 2m_{2\lambda_i})/4,$$

that is, if $\operatorname{Re}(H_{\lambda_i}) > \frac{1}{2}(m_{\lambda_i} + 2m_{2\lambda_i})\lambda_i(H_{\lambda_i})$, and that (8) is equal to

$$\frac{c_1 c_2}{4} B\left(\frac{m_{\lambda_i}}{2}, \frac{\nu(H_{\lambda_i})}{\lambda(H_{\lambda_i})} - m_{2\lambda_i} - \frac{m_{\lambda_i}}{2} + 1\right) B\left(\frac{m_{2\lambda_i}}{2}, \frac{\nu(H_{\lambda_i})}{2\lambda_i(H_{\lambda_i})} - \frac{m_{2\lambda_i}}{2}\right).$$

Now 8.10.14(3) and (4) imply that (8) is absolutely convergent and nonzero for Re $\nu(H_{\lambda_i}) > m_{2\lambda_i} + m_{\lambda_i}/2$ and that c_{s_i} extends to a meromorphic function on $\mathfrak{a}_C^*$.

8.10.17 *Corollary* Let ξ be in $\hat{M}$ and let ν be in $\mathfrak{a}_C^*$ such that $\text{Re}(\nu - \rho)(H_\lambda) > 0$ for all λ in Λ^+. Then $A(\xi, \nu, s)$ is defined for s in $W(A)$ and has the properties of 8.10.8.

8.11 The Analytic Continuation of the Intertwining Operators

8.11.1 In 8.10 we showed that if ξ is in $\hat{M}$ and ν is in $\mathfrak{a}^*$ such that $\text{Re}(\nu - \rho)(H_\lambda) > 0$ for all ν in Λ^+, then we could define

$$A(\xi, \nu, s)\colon \tilde{H}_F^{\xi,\nu} \to \tilde{H}_F^{\xi s, s(\nu-\rho)+\rho}$$

such that if X is in $U(\mathfrak{g})$,

$$A(\xi, \nu, s)\tilde{\pi}_{\xi,\nu}(X) = \tilde{\pi}_{\xi s, s(\nu-\rho)+\rho}(X)A(\xi, \nu, s).$$

Furthermore it is clear from the definition of $A(\xi, \nu, s)$ and the proof of 8.10.8 that if for each γ in K we identify $\tilde{H}_\gamma^{\xi,\nu}$ with H_γ^ξ than

$$A(\xi, \nu, s)\colon H_\gamma^\xi \to H_\gamma^{\xi s}$$

is a K-intertwining operator. Setting $T_\gamma(\xi, \nu, s) = A(\xi, \nu, s)|_{H_\gamma^\xi}$ then $T_\gamma(\xi, \nu, s)$ is in $\text{Hom}_K(H_\gamma^\xi, H_\gamma^{\xi s})$. The proof of 8.10.8 implies that the map

$$\nu \to T_\gamma(\xi, \nu, s)$$

of $\{\nu \text{ in } \mathfrak{a}_C^* | \text{Re}(\nu - \rho)(H_\lambda) > 0 \text{ for all } \lambda \text{ in } \Lambda^+\}$ is holomorphic.

8.11.2 In this section we show that the maps $\nu \mapsto T_\gamma(\xi, \nu, s)$ extend to meromorphic functions on $\mathfrak{a}_C^*$. To prove this result we need more fine structure of semisimple Lie groups.

8.11.3 *Theorem* Let for X, Y in $\langle X, Y\rangle = -B(X, \theta Y)$. Let λ be in Λ^+.

Let $S_\lambda^1 = \{X \text{ in } \mathfrak{g}_{-\lambda} | \langle X, X\rangle = 1\}$, $S_\lambda^2 = \{X \text{ in } \mathfrak{g}_{-2\lambda} | \langle X, X\rangle = 1\}$. Then

(1) If $\dim \mathfrak{g}_{2\lambda} \leqslant 1$, and $\dim \mathfrak{g}_\lambda > 1$ then $\text{Ad}(M)|_{\mathfrak{g}_{-\lambda}}$ acts transitively on S_λ^1.

(2) If $\dim \mathfrak{g}_{2\lambda} > 1$ then $\text{Ad}(M)|_{\mathfrak{g}_{-\lambda}\oplus\mathfrak{g}_{-2\rho}}$ acts transitively on $S_\lambda^1 \oplus S_\lambda^2 = \{X + Y | X \text{ in } S_\lambda^1, Y \text{ in } S_\lambda^2\}$.

PROOF We first make an observation. Let U be a compact Lie subgroup of $O(n)$ (the orthogonal group of R^n). Let $\mathfrak{U}$ be the Lie algebra of U.

(1) If for some v in R^n, $\langle v, v\rangle = 1$, $\mathfrak{U} \cdot v = \{w \text{ in } R^n | \langle v, w\rangle = 0\}$ and if $n > 1$ then $U \cdot v = S^{n-1}$.

In fact, let $\psi: U \to S^{n-1}$ be defined by $\psi(u) = u \cdot v$. Then $\psi_{*u}(X) = d/dt\, \psi(u \exp tX)|_{t=0} = uXv$. Thus $\psi_{*u}: \mathfrak{U} \to \{z \text{ in } R^n | \langle z, u \cdot v\rangle = 0\} = T(S^{n-1})_{u\cdot v}$ is a surjection. If $n > 1$ then the implicit function theorem implies $\psi(U)$ is open in S^{n-1}. Since $\psi(U)$ is clearly closed in S^{n-1} and if $n > 1$, S^{n-1} is connected, $\psi(U) = S^{n-1}$. This proves (1).

Now let X be in $\mathfrak{g}_{-\lambda}$, $\langle X, X\rangle = 2/\lambda(H_\lambda)$. Then

(2) $$[X, \theta X] = h = (2/\lambda(H_\lambda))H_\lambda,\ [h, X] = -2X,\ [h, \theta X] = 2\theta X.$$

(2) follows from the proof of 8.10.12.

Let Z be in $\mathfrak{g}_{-2\lambda}$. Then $[\theta X, Z]$ is in $\mathfrak{g}_{-\lambda}$. Also

$$[X, [\theta X, Z]] = [[X, \theta X], Z] = [h, Z] = -4Z.$$

$$\begin{aligned}[X, [\theta X, [\theta X, Z]] &= [[X, \theta X], [\theta X, Z]] + [\theta X, [X, [\theta X, Z]]\\ &= -2[\theta X, Z] - 4[\theta X, Z] = -6[\theta X, Z].\end{aligned}$$

Similarly

$$\operatorname{ad}X\cdot(\operatorname{ad}\theta X)^3 Z = \operatorname{ad} h(\operatorname{ad}\theta X)^2 Z + \operatorname{ad}\theta X \operatorname{ad} X(\operatorname{ad}\theta X)^2 Z = -6(\operatorname{ad}\theta X)^2 Z.$$

$$\begin{aligned}\operatorname{ad} X(\operatorname{ad}\theta X)^4 Z &= \operatorname{ad} h(\operatorname{ad}\theta X)^3 Z + \operatorname{ad}\theta X \operatorname{ad} X(\operatorname{ad}\theta X)^3 Z\\ &= 2(\operatorname{ad}\theta X)^3 Z - 6(\operatorname{ad}\theta X)^3 Z = -4(\operatorname{ad}\theta X)^3 Z.\end{aligned}$$

Finally $(\operatorname{ad}\theta X)^4 Z$ is in $\mathfrak{g}_{2\lambda}$. Hence $(\operatorname{ad}\theta X)^5 Z = 0$. We have the following.

(3) If Z is in $\mathfrak{g}_{-2\lambda}$ then $(\operatorname{ad}\theta X)^5 Z = 0$ and $\operatorname{ad}X\cdot(\operatorname{ad}\theta X)^k Z = C_k(\operatorname{ad}\theta X)^{k-1} Z$ for $k = 1, 2, 3, 4$ with $C_1 = -4$, $C_2 = -6$, $C_3 = -6$, $C_4 = -4$.

Let

$$V_2 = \mathfrak{g}_{-2\lambda} + (\operatorname{ad}\theta X)\mathfrak{g}_{-2\lambda} + (\operatorname{ad}\theta X)^2\mathfrak{g}_{-2\lambda} + (\operatorname{ad}\theta X)^3\mathfrak{g}_{-2\lambda} + (\operatorname{ad}\theta X)^4\mathfrak{g}_{-2\lambda}.$$

Then $\operatorname{ad}X \cdot V_2 \subset V_2$, $\operatorname{ad}\theta X \cdot V_2 \subset V_2$. Let $\mathfrak{g}^2_{-\lambda} = V_2 \cap \mathfrak{g}_{-\lambda} = (\operatorname{ad}\theta X)\mathfrak{g}_{-2\lambda}$.

(4) Let $\mathfrak{g}^1_{-\lambda} = \{Z \text{ in } \mathfrak{g}_{-\lambda} | \langle Z, \mathfrak{g}^2_{-\lambda}\rangle = 0\}$. Then $\mathfrak{g}^1_{-\lambda} = \{Z \text{ in } \mathfrak{g}^1_{-\lambda} | [X, Z] = 0\}$.

In fact if $[X, Z] = 0$ and W is in $\mathfrak{g}_{-\lambda}$ then $\langle Z, W\rangle = -B(Z, \theta W) = -B(Z, \theta[\theta X, Y])$ for some Y in $\mathfrak{g}_{-2\lambda}$. Thus $\langle Z, W\rangle = -B(Z, [X, \theta Y]) = B([X, Z], \theta Y) = 0$. Now (3) implies ad $X: \mathfrak{g}_{-\lambda} \to \mathfrak{g}_{-2\lambda}$ is surjective and that ad $X: \mathfrak{g}^1_{-\lambda} \to \mathfrak{g}_{-2\lambda}$ is bijective. Hence $\dim(\ker \operatorname{ad}X|_{\mathfrak{g}_{-\lambda}}) = \dim \mathfrak{g}^1_{-\lambda}$. This proves (4).

If Z is in $\mathfrak{g}_{-\lambda}$ then $[X, [\theta X, Z]] = [[X, \theta X], Z] = -2Z$.

$$[X, [\theta X, [\theta X, Z]]] = [[X, \theta X], [\theta X, Z]] + [\theta X, [X, [\theta X, Z]] = -2[\theta X, Z].$$

$(\operatorname{ad}\theta X)^3 Z$ is in $\mathfrak{g}_{2\lambda}$. Since $(\operatorname{ad}\theta X)^4\mathfrak{g}_{-2\lambda} = \mathfrak{g}_{2\lambda}$ by (3) and since $\operatorname{ad}X(\operatorname{ad}\theta X)^3 Z =$

$\mathrm{Ad}h(\mathrm{ad}\theta X)^2Z + \mathrm{Ad}\theta X\ \mathrm{ad}X(\mathrm{ad}\theta X)^2Z$ we see that if $\mathrm{ad}(\theta X)^3Z \neq 0$ then $\mathfrak{g}^1_{-\lambda} \cap \mathfrak{g}^2_{-\lambda} \neq (0)$ which is absurd. Thus $(\mathrm{ad}\theta X)^3Z = 0$. We have the following.

(5) Let Z be in $\mathfrak{g}^1_{-\lambda}$. Then

$$(\mathrm{ad}\,\theta X)^3Z = 0 \text{ and } \mathrm{ad}\,X\,\mathrm{ad}\,\theta XZ = -2Z,\ \mathrm{ad}\,X(\mathrm{ad}\,\theta X)^2Z = -2\,\mathrm{ad}\,\theta X\cdot Z.$$

Set $V_1 = \mathfrak{g}^1_{-\lambda} + \mathrm{ad}\theta X\mathfrak{g}^1_{-\lambda} + (\mathrm{ad}\theta X)^2\mathfrak{g}^1_{-\lambda}$. Then as above $\mathrm{ad}X \cdot V_1 \subset V_1$, $\mathrm{ad}\theta X \cdot V_1 \subset V_1$. Also the arguments above imply $V_1 \cap V_2 = (0)$.

Let $V^\lambda = \mathfrak{g}_{-2\lambda} + \mathfrak{g}_{-\lambda} + \mathfrak{m} + RH_\lambda + \mathfrak{g}_\lambda + \mathfrak{g}_{2\lambda}$. Then $\mathrm{ad}X \cdot V^\lambda \subset V^\lambda$, $\mathrm{ad}\theta X \cdot V^\lambda \subset V^\lambda$. Also $\langle\ ,\ \rangle$ is positive definite on V^λ.

(6) Let $V_0 = \{Z \text{ in } V^\lambda | \langle Z, V_1 + V_2\rangle = 0\}$. Then $V_0 \subset \mathfrak{m}$ and $\mathrm{ad}X \cdot V_0 = \mathrm{ad}\theta X \cdot V_0 = 0$.

In fact, $V_1 + V_2 \supset \mathfrak{g}_{-2\lambda} + \mathfrak{g}_{-\lambda} + \mathfrak{g}_\lambda + \mathfrak{g}_{2\lambda}$ by their definitions and (3), (5). Thus $V_0 \subset \mathfrak{m} + Rh$. Now h is in V_1 and $\langle \mathfrak{m}, h\rangle = 0$. Thus $V_0 \subset \mathfrak{m}$. Now $\theta(V_1) \subset V_1$, $\theta V_2 \subset V_2$. Thus $V_0 = \{Z \text{ in } V^\lambda | B(Z, V_1 + V_2) = 0\}$. But then $\mathrm{ad}X \cdot V_0 \subset V_0$, $\mathrm{ad}X \cdot V_1 \subset V_1$. Now $\mathrm{ad}X \cdot \mathfrak{m} \subset \mathfrak{g}_{-\lambda}$, $\mathrm{ad}\theta X \cdot \mathfrak{m} \subset \mathfrak{g}_\lambda$. Hence $\mathrm{ad}X \cdot V_0 = \mathrm{ad}\theta X \cdot V_0 = (0)$.

(7) $V^\lambda = V_0 \oplus V_1 \oplus V_2$ and $\{\theta X, h, X\} \subset V_1$. $\mathfrak{m} = \mathfrak{m}_0 + \mathfrak{m}_1 + \mathfrak{m}_2$, $\mathfrak{m}_i = \mathfrak{m} \cap V_i$.

With these preliminaries (actually only really needed to prove the second assertion of the theorem) we begin the proof of the theorem.

Case (1) $\dim \mathfrak{g}_{2\lambda} = (0)$. Then $\mathfrak{g}_{-\lambda} = \mathfrak{g}^1_{-\lambda}$. Furthermore (5) implies $\mathrm{ad}X \cdot (\mathfrak{m}_1 + Rh) = \mathfrak{g}_{-\lambda}$. $[Rh, X] \subset RX$ and $[\mathfrak{m}_1, X] \subset \{Z \text{ in } \mathfrak{g}_{-\lambda} | \langle Z, X\rangle = 0\}$. Thus $[\mathfrak{m}_1, X] = \{Z \text{ in } \mathfrak{g}_{-\lambda} | \langle Z, X\rangle = 0\}$. (1) now implies the result in this case.

Case (2) $\dim \mathfrak{g}_{2\lambda} = 1$. Then $\dim \mathfrak{g}_{-\lambda} = \dim \mathfrak{g}^1_{-\lambda} + \dim \mathfrak{g}^2_{-\lambda}$. $\dim \mathfrak{g}^2_{-\lambda} = \dim \mathfrak{g}_{2\lambda} = 1$. $\dim \mathfrak{g}^1_{-\lambda} \geqslant 1$ since X is in $\mathfrak{g}^1_{-\lambda}$. Hence $\dim \mathfrak{g}_{-\lambda} \geqslant 2$. Now using (3) and (5) $\mathrm{ad}X \cdot \mathfrak{m}_1 = \{Z \text{ in } \mathfrak{g}^1_{-\lambda} | \langle Z, X\rangle = 0\}$ $\mathrm{ad}X \cdot \mathfrak{m}_2 = \mathfrak{g}^2_{-\lambda}$. Thus $[\mathfrak{m}, X] = \{Z \text{ in } \mathfrak{g}_{-\lambda} | \langle Z, X\rangle = 0\}$. Hence the result in this case again follows from (1).

Case (3) $\dim \mathfrak{g}_{2\lambda} > 1$. Then as above $\dim \mathfrak{g}_{-\lambda} > 1 + \dim \mathfrak{g}_{2\lambda} > 1$. Also arguing as in Case (2) we see that $[\mathfrak{m}_1 + \mathfrak{m}_2, X] = \{Z \text{ in } \mathfrak{g}_{-\lambda} | \langle Z, X\rangle = 0\}$. This implies that $\mathrm{Ad}(M)$ acts transitively on S^1_λ by (1). Let $M_0 = \{m \text{ in } M | \mathrm{ad}m \cdot X = X\}$. Then clearly the Lie algebra of M_0 is $\mathfrak{m}_0$. We must show that $\mathrm{Ad}(M_0)$ acts transitively on S^2_λ.

If Z is in V^λ let $Z = Z_0 + Z_1 + Z_2$, Z_i in V_i $i = 0, 1, 2$.

(8) Let u be in $\mathfrak{m}_2$. Then $[\theta X, (\mathrm{ad}X)^2u] = -4\mathrm{ad}Xu$, $[\theta X, \mathrm{ad}Xu] = -6u$.

In fact, $u = (\mathrm{ad}\theta X)^2v$ with v in $\mathfrak{g}_{-2\lambda}$. Thus (3) implies that $\mathrm{ad}X \cdot u =$

$-6\operatorname{ad}X \cdot v$, $(\operatorname{ad}X)^2u = 24v$. Hence $[\theta X, (\operatorname{ad}X)^2u] = 24[\theta X, v] = -4\operatorname{ad}Xu$. Also $[\theta X, \operatorname{ad}X \cdot u] = -6(\operatorname{ad}\theta X)^2 v = -6u$. This proves (8).

(9) If u, v are in $\mathfrak{m}_2$ then $[u, (\operatorname{ad}X)^2v] = 2/3[\operatorname{ad}Xv, \operatorname{ad}Xu]$.

Indeed,

$$\begin{aligned}[u, (\operatorname{ad}X)^2v] &= -1/6[[\theta X, \operatorname{ad}Xu], (\operatorname{ad}X)^2v]]\\ &= -1/6[[\theta X, (\operatorname{ad}X)^2v], \operatorname{ad}Xu] \text{ (since } [\mathfrak{g}_{-\lambda}, \mathfrak{g}_{-2\lambda}] = 0)\\ &= 4/6[\operatorname{ad}Xv, \operatorname{ad}Xu].\end{aligned}$$

(10) If u, v are in $\mathfrak{m}_2$ then

$$[v, (\operatorname{ad}X)^2u] = [(\operatorname{ad}X)^2v, u] = 2/3[\operatorname{ad}Xu, \operatorname{ad}Xv].$$

In fact,

$$\begin{aligned}[v, (\operatorname{ad}X)^2u] &= 2/3[\operatorname{ad}Xu, \operatorname{ad}Xv]\\ &= -2/3[\operatorname{ad}Xv, \operatorname{ad}Xu] = -[u, (\operatorname{ad}X)^2v].\end{aligned}$$

(11) If u, v are in $\mathfrak{m}_2$ then $(\operatorname{ad}X)^2[u, v] = 2/3[\operatorname{ad}Xu, \operatorname{ad}Xv]$.

Indeed,

$$\begin{aligned}(\operatorname{ad}X)^2[u, v] &= \operatorname{ad}X\cdot\{[\operatorname{ad}Xu, v] + [u, \operatorname{ad}Xv]\}\\ &= [(\operatorname{ad}X)^2u, v] + 2[\operatorname{ad}Xu, \operatorname{ad}Xv] + [u, (\operatorname{ad}X)^2v]\\ &= 2/3[\operatorname{ad}Xu, \operatorname{ad}Xv] \text{ by (10).}\end{aligned}$$

(12) Let u, v be in $\mathfrak{m}_2$, then $[u, v]_1 = 0$ (recall that if X is in V^λ, $X = X_1 + X_2 + X_2$, X_i in V_i).

In fact, $(\operatorname{ad}X)^2[u, v] = 2/3[\operatorname{ad}Xu, \operatorname{ad}Xv]$ by (11). Thus

$$\begin{aligned}\operatorname{ad}\theta X(\operatorname{ad}X)^2[u, v] &= 2/3[\operatorname{ad}\theta X\operatorname{ad}Xu, \operatorname{ad}Xv] + 2/3[\operatorname{ad}Xu, \operatorname{ad}\theta X\operatorname{ad}Xv]\\ &= -4[u, \operatorname{ad}Xv] - 4[\operatorname{ad}Xu, v] = -4\operatorname{ad}X[u, v].\end{aligned}$$

Now $[u, v] = [u, v]_0 + [u, v]_1 + [u, v]_2$. By the definition of the V_i we see that. $(\operatorname{ad}X)^2[u, v] = (\operatorname{ad}X)^2[u, v]_2$. $(\operatorname{ad}\theta X)(\operatorname{ad}X)^2[u, v]_2 = -4\operatorname{ad}X \cdot [u, v]_2$ by (11) Hence $\operatorname{ad}X\cdot[u, v] = \operatorname{ad}X\cdot[u, v]_2$ by the above. This implies that $\operatorname{ad}X\cdot[u, v]_1 = 0$. But then $[u, v]_1 = 0$ by (5).

(13) Let u, v be in $\mathfrak{m}_2$. Then $[[u, v]_0, (\operatorname{ad}X)^2v] = -2[[u, v]_2, (\operatorname{ad}X)^2v]$.

In fact,

$$\begin{aligned}&[[u, v]_2, (\operatorname{ad}X)^2v]\\ &\quad= [(\operatorname{ad}X)^2[u, v]_2, v]\\ &\quad= [(\operatorname{ad}X)^2[u, v], v]] \text{ (here we have used (10 and } (\operatorname{ad}X)^2(\mathfrak{m}_0 + \mathfrak{m}_1) = 0)\\ &\quad= -[[u, (\operatorname{ad}X)^2v], v] \text{ (by (11), (10))}\\ &\quad= -[[u, v], (\operatorname{ad}X)^2v] - [u, [(\operatorname{ad}X)^2v, v]].\end{aligned}$$

Now $[(\mathrm{ad}X)^2v, v] = 0$ by (9). Thus

$$\begin{aligned}[[u, v]_2, (\mathrm{ad}X)^2v] &= -[[u, v], (\mathrm{ad}X)^2v] \\ &= -[[u, v]_0, (\mathrm{ad}X)^2v] - [[u, v]_2, (\mathrm{ad}X)^2v]\end{aligned}$$

by (12). This proves (13).

Now let z, w be in $\mathfrak{g}_{-2\lambda}$, $\langle z, w\rangle = 0$ and $\langle z, z\rangle = \langle w, w\rangle = 1/\lambda(H_\lambda)$. Let $u = (\mathrm{ad}\theta X)^2z$, $v = (\mathrm{ad}\theta X)^2w$. We first note that $[\theta z, w]$ is in $\mathfrak{m}$ and

$$\begin{aligned}[[\theta z, w], z] &= [[\theta z, z], w] \\ &= -4w \text{ (in fact, } [\theta z, z] \text{ is in } \mathfrak{a} \text{ and } B([\theta z, z], H) = -B(\theta z, [H, z]) \\ &= -2\lambda(H)B(\theta z, z) = 2\lambda(H)/\lambda(H_\lambda) = B(h, H)).\end{aligned}$$

Now

$$\begin{aligned}\mathrm{ad}\,Xu &= (\mathrm{ad}\,X)(\mathrm{ad}\,\theta X)^2z = \mathrm{ad}\,h\,\mathrm{ad}\,\theta XZ + \mathrm{ad}\,\theta X\,\mathrm{ad}\,X\,\mathrm{ad}\,\theta Xz \\ &= -6\mathrm{ad}\theta X\cdot Z.\end{aligned}$$

Hence $(\mathrm{ad}\,X)^2u = 24z$. Similarly $(\mathrm{ad}\,X)^2v = 24w$. Thus

$$[\theta z, w] = (1/288)[(\mathrm{ad}\,\theta X)^2u, (\mathrm{ad}\,X)^2v].$$

On the other hand

$$\begin{aligned}&(\mathrm{ad}\,X)^2[(\mathrm{ad}\,\theta X)^2u, (\mathrm{ad}\,X)^2v] \\ &= [(\mathrm{ad}\,X)^2(\mathrm{ad}\,\theta X)^2u, (\mathrm{ad}\,X)^2v] = [\mathrm{ad}\,X\,\mathrm{ad}\,X(\mathrm{ad}\,\theta X)^2u, (\mathrm{ad}\,X)^2v] \\ &= -4[\mathrm{ad}\,X\,\mathrm{ad}\,\theta Xu, (\mathrm{ad}\,X)^2v] = 24[u, (\mathrm{ad}\,X)^2v] = 24(\mathrm{ad}\,X)^2[u, v].\end{aligned}$$

Hence $[\theta z, w]_2 = (1/12)[u, v]_2$.

On the other hand

$$\begin{aligned}&\mathrm{ad}\,X\cdot[(\mathrm{ad}\,\theta X)^2u, (\mathrm{ad}\,X)^2v] \\ &= -4[\mathrm{ad}\,\theta Xu, (\mathrm{ad}\,X)^2v] = -4\mathrm{ad}\,\theta X[u, (\mathrm{ad}\,X)^2v] - 16[u, \mathrm{ad}\,Xv] \\ &= -4(\mathrm{ad}\,\theta X)(\mathrm{ad}\,X)^2[u, v] - 16[u, \mathrm{ad}\,Xv]\end{aligned}$$

by (11). Thus we see that $\mathrm{ad}X \cdot [(\mathrm{ad}\theta X)^2u, (\mathrm{ad}X)^2v]_1 = -16[u, \mathrm{ad}Xv]_1$. But $\theta[\theta z, w] = [\theta z, w]$ since $[\theta z, w]$ is in $\mathfrak{m}$. We therefore see that

$$\begin{aligned}\mathrm{ad}\,X\cdot[(\mathrm{ad}\theta X)^2u, (\mathrm{ad}\,X)^2v]_1 &= \mathrm{ad}\,X\cdot[(\mathrm{ad}\,X)^2u, (\mathrm{ad}\,\theta X)^2v]_1 \\ &= -\mathrm{ad}\,X\cdot[(\mathrm{ad}\,\theta X)^2v, (\mathrm{ad}\,X)^2u]_1 = 16[v, \mathrm{ad}\,Xu]_1\end{aligned}$$

by the above. Hence

$$\begin{aligned}\mathrm{ad}\,X\cdot[(\mathrm{ad}\,\theta X)^2u, (\mathrm{ad}\,X)^2v]_1 &= -8([u, \mathrm{ad}\,Xv]_1 + [\mathrm{ad}\,Xu, v]_1) \\ &= -8\,\mathrm{ad}\,X\cdot[u, v]_1.\end{aligned}$$

Now (12) implies that $[u, v]_1 = 0$. Hence $\operatorname{ad}X \cdot [\theta z, w]_1 = 0$. Hence $[\theta z, w]_1 = 0$.

Now

$$[[\theta z, w], w] = [[\theta z, w]_0 + [\theta z, w]_2, w] = [[\theta z, w]_0 + (1/12)[u, v]_2, w].$$

Also, $w = 1/24(\operatorname{ad}X)^2 v$. Thus

$$\begin{aligned}[[\theta z, w], w] &= [[\theta z, w]_0, w] + 1/288[[u, v]_2, (\operatorname{ad}X)^2 v] \\ &= [[\theta z, w]_0, w] - 1/144[[u, v]_0, (\operatorname{ad}X)^2 v] \\ &= [[\theta z, w]_0, w]] - 1/6[[u, v]_0, w]].\end{aligned}$$

We have thus proved the following.

(14) If z, w are in $\mathfrak{g}_{-2\lambda}$, $\langle z, w\rangle = 0$, $z, w \neq 0$ then there is u in $\mathfrak{m}_0$ such that $[u, z] = w$.

(14) implies that if z is in $\mathfrak{g}_{-2\lambda}$, $\langle z, z\rangle = 1$, then

$$[\mathfrak{m}_0, z] = \{w \text{ in } \mathfrak{g}_{-2\lambda} | \langle w, z\rangle = 0\}.$$

This combined with the remarks at the beginning of the proof of Case (3) proves the theorem.

8.11.4 Let γ be in $\hat{K}$. Let ξ be in $\hat{M}$ and let A be in $\operatorname{Hom}_M(V_\gamma, H_\xi)$. Let for v in $\mathfrak{a}_C^*$, v in V_γ, $L(A, v, \nu; kan) = e^{-\nu(\log a)} A(\pi_\gamma(k)^{-1} v)$ for k in K, a in A, n in N. Then an easy computation shows that $L(A, v, \nu)$ is in $\tilde{H}^{\xi,\nu}$. Furthermore the map $V_\gamma \otimes \operatorname{Hom}_M(V_\gamma, H_\xi) \to \tilde{H}^{\xi,\nu}$ given by $v \otimes A \mapsto L(A, v, \nu)$ is a bijective K-intertwining operator.

8.11.5 *Lemma* If ν is in $\mathfrak{a}_C^*$, $\operatorname{Re}(\nu - \rho)(H_\lambda) > 0$ for all λ in Λ^+ then

(1) $$T_\gamma(\xi, \nu, s)L(A, v, \nu) = L(A \circ B_\gamma(\nu, s)\pi_\gamma(s^*)^{-1}, v, s(\nu - \rho) + \rho)$$

where

(2) $$B_\gamma(\nu, s) = \int_{\bar{N}_s} \pi_\gamma(k(\bar{n}_s))^{-1} e^{-\nu(H(\bar{n}_s))}\, d\bar{n}_s.$$

PROOF We have already seen that if f is in $\tilde{H}_\gamma^{\xi,\nu}$ then $T_\gamma(\xi, \nu, s)f$ is in $\tilde{H}_\gamma^{\xi,\nu,s(\nu-\rho)+\rho}$. To prove the lemma we therefore need only show that

$$(T_\gamma(\xi, \nu, s)L(A, v, \nu))(k) = A(B_\gamma(\nu, s)\pi_\gamma(s^*)^{-1}\pi_\gamma(k)^{-1}v).$$

Now

$$(*)\qquad (T_\gamma(\xi, v, s)L(A, v, v))(k) = \int_{\bar{N}_s} L(A, v, v)(ks^*\bar{n}_s)d\bar{n}_s$$
$$= \int_{\bar{N}_s} e^{-v(H(s^*\bar{n}_s))}A(\pi_\gamma(k(ks^*\bar{n}_s))^{-1})v)d\bar{n}_s.$$

But $H(s^*\bar{n}_s) = H(\bar{n}_s)$ and $k(ks^*\bar{n}_s) = ks^*k(\bar{n}_s)$. Thus (*) is equal to

$$\int_{\bar{N}_s} e^{-v(H(\bar{n}_s))}A(\pi_\gamma(k(\bar{n}_s))^{-1}\pi_\gamma(s^*)^{-1}\pi_\gamma(k)^{-1}v)d\bar{n}_s$$
$$= A\left(\left(\int_{\bar{N}_s} e^{-v(H(\bar{n}_s))}\pi_\gamma(k(\bar{n}_s))^{-1}\, d\bar{n}_s\pi_\gamma(s^*)^{-1}\pi_\gamma(k)^{-1}v\right)\right.$$
$$= A(B_\gamma(v, s)\pi_\gamma(s^*)^{-1}\pi_\gamma(k)^{-1}v).$$

This proves the lemma.

8.11.6 *Lemma* Let s, t, u be in $W(A)$. Suppose that $s = tu$ and $l(s) = l(t) + l(u)$. Then if $\mathrm{Re}(v - \rho)(H_\lambda) > 0$ for all λ in Λ^+,

$$(1)\qquad B_\gamma(v, s)\,\pi_\gamma(s^*)^{-1} = b(t, u)B_\gamma(u(v - \rho) + \rho, t)\pi_\gamma(t^*)^{-1}B_\gamma(v, u)\pi_\gamma(u^*)^{-1}$$

where $b(t, u)$ is as in 8.10.11.

PROOF This lemma is an immediate consequence of 8.10.11 and 8.11.5.

8.11.7 8.11.6 combined with 8.11.5 implies that if we wish to show that $v \mapsto T_\gamma(\xi, v, s)$ has a mermorphic continuation to $\mathfrak{a}_C^*$ we need only show that if $s = s_i$ then $v \to B(v, s_i)$ has a meromorphic continuation to $\mathfrak{a}_C^*$.

8.11.8 *Lemma* Let $U(2)$ be the group of all 2×2 unitary matrices. Let for each $l \geqslant 0$, l an integer $\mathscr{P}^l$ be the space of all complex polynomials in z_1, z_2 (the coordinates of C^2) homogeneous of degree l. Let for g in $U(2)$, $(\pi^l(g)f)(z) = f(g^{-1}z)$ for z in C^2, f in $\mathscr{P}^l$. Let for each k an integer, f in $\mathscr{P}^l$, $\pi^{k,l}(g)f = \det(g)^k\pi^l(g)f$. If $K = U(2)$, $\hat{K} = \{[\pi^{k,l}]\,|k, l \text{ in } z, l \geqslant 0\}$ (here $[\pi^{k,l}]$ is the equivalence class of $(\pi^{k,l}, \mathscr{P}^l)$).

PROOF Let $T \subset U(2)$ be the subgroup of all diagonal matrices

$$t = \begin{bmatrix} t_1 & 0 \\ 0 & t_2 \end{bmatrix}.$$

Let $f_{j,l}(z) = z_1^{l-j}z_2^j$. Then $\pi^{k,l}(t)f_{j,l} = t_1^{k+j-l}t_2^{k-j}f_{j,l}$.

We now prove that $\pi^{k,l}$ is irreducible. Indeed we prove that it is an irreducible representation of $SU(2) = \{g \text{ in } U(2) \mid \det(g) = 1\}$. In fact, $SU(2)$ is the group of all

$$g = \begin{bmatrix} a & b \\ -\bar{b} & \bar{a} \end{bmatrix}$$

with a, b in C, $|a|^2 + |b|^2 = 1$.

Suppose now $f = \sum a_j f_{j,l}$ is in $\mathscr{P}^l$, $f \neq 0$. Let V_f be the smallest invariant subspace of $\mathscr{P}^l$ containing f. We show that $V_f = \mathscr{P}^l$ and hence prove the irreducibility. Now

$$(\pi(g)^{-1}f_{j,l})(z_1, z_2) = (az_1 + bz_2)^{l-j}(-\bar{b}z_1 + \bar{a}z_2)^j$$

for

$$g = \begin{bmatrix} a & b \\ -\bar{b} & \bar{a} \end{bmatrix}$$

as above. Thus if

$$g = \begin{bmatrix} a & 0 \\ 0 & \bar{a} \end{bmatrix}$$

then

$$\pi(g)^{-1}f = \sum a^{l-j}\bar{a}^j f_{j,l} = \sum a^{l-2j} f_{j,l}.$$

Now $a = e^{i\theta}$. Thus if $a_j \neq 0$ then

$$\frac{1}{2\pi}\int_0^{2\pi} e^{(2j-l)i\theta}\, \pi \begin{bmatrix} e^{i\theta} & 0 \\ 0 & e^{-i\theta} \end{bmatrix}^{-1} f = a_j f_{j,l}.$$

Thus $f_{j,l}$ is in V_f. Now

$$\begin{aligned} \pi(g)^{-1}f_{j,l} &= (az_1 + bz_2)^{l-j}(-\bar{b}z_1 + \bar{a}z_2)^j \\ &= \sum_{r=0}^{l-j}\sum_{s=0}^{j}\binom{l-j}{r}\binom{j}{s} a^{l-j-r}b^r(-\bar{b})^{j-s}\bar{a}^s f_{r+s,l} \\ &= \sum_{u=0}^{l}\left(\sum_{r+s=u}\binom{l-j}{r}\binom{j}{s} a^{l-j-r}\bar{a}^s b^r(-\bar{b})^{j-s}\right) f_{u,l}. \end{aligned}$$

Setting $a = 2^{-1/2}e^{i\theta}$, $b = 2^{-1/2}e^{i\theta}$ we see that $\sum_{u=1}^{l}(\sum_{r+s=u}\binom{l-j}{r}\binom{j}{s})f_{u,l}$ is in V_f. Hence by the above $f_{u,l}$ is in V_f for all $u = 0, 1, \ldots, l$. Hence $V_f = \mathscr{P}^l$ as asserted.

Let $\mathfrak{h}$ be the Lie algebra of T. Then

$$\mathfrak{h} = \left\{ \begin{bmatrix} it_1 & 0 \\ 0 & it_2 \end{bmatrix} \middle| t_1, t_2 \text{ in } R \right\}.$$

The unit lattice of T is

$$\left\{ \begin{bmatrix} i2\pi k_1 & 0 \\ 0 & i2\pi k_2 \end{bmatrix} \middle| k_1, k_2 \text{ in } Z \right\}.$$

$\mathfrak{h}_1$ in the notation of 4.6.12 is

$$\left\{ \begin{bmatrix} it & 0 \\ 0 & -it \end{bmatrix} \middle| t \text{ in } R \right\}.$$

Γ_1, the unit lattice of $\mathfrak{h}_1$, is

$$\left\{ \begin{bmatrix} i2\pi k & 0 \\ 0 & -i2\pi k \end{bmatrix} \middle| k \text{ in } Z \right\}.$$

We choose as a Weyl chamber of $\mathfrak{h}_1$,

$$\left\{ \begin{bmatrix} it & 0 \\ 0 & -it \end{bmatrix} \middle| t > 0 \right\}.$$

Then Λ is dominant integral if

$$\Lambda\left(\begin{bmatrix} it & 0 \\ 0 & -it \end{bmatrix} \right] = ilt$$

with $l \geqslant 0$, an integer. The weights of $\pi^{k,l}$ are all of the form

$$\lambda\left\{ \begin{bmatrix} it_1 & 0 \\ 0 & it_2 \end{bmatrix} \right] = i\{(k + j)t_1 + (k + l - j)t_2\}.$$

The highest weight of $\pi^{k,l}$ is a weight of $\mathfrak{h}_1$ and is therefore

$$\Lambda_{k,l}\left(\begin{bmatrix} it & 0 \\ 0 & -it \end{bmatrix} \right] = ilt.$$

As a weight of $\mathfrak{h}$ it is given by

$$\Lambda_{k,l}\left(\begin{bmatrix} it_1 & 0 \\ 0 & it_2 \end{bmatrix} \right) = i\{(k + l)t_1 + kt_2\}.$$

Suppose Λ is an integral weight on $\mathfrak{h}$ so that $\Lambda|_{\mathfrak{h}_1}$ is dominant integral. Then

$$\Lambda\left(\begin{bmatrix} it_1 & 0 \\ 0 & it_2 \end{bmatrix} \right) = i(k_1 t_1 + k_2 t_2),$$

k_1, k_2 integers and $k_1 - k_2 \geqslant 0$. But then setting $k = k_2$, $l = k_1 - k_2$ we find $\Lambda = \Lambda_{k,l}$. 4.6.12 now implies the lemma.

8.11.9 *Theorem* Suppose that G has a finite dimensional faithful representation. Let γ be in $\hat{K}$, then $\nu \to B_\gamma(\nu, s_i)$ extends to a meromorphic function on $\mathfrak{a}_C^*$. Furthermore if $\nu_0(H_{\lambda_i})$ is not an integer or a half integer then there is an open neighborhood of ν_0 on which $\nu \to B_\gamma(\nu, s_i)$ is holomorphic.

PROOF Recall that $B_\gamma(\nu, s_i) = \int_{\bar{N}_{s_i}} e^{-\nu(H(\bar{n}_{s_i}))}\pi_\gamma(k(\bar{n}_{s_i}))^{-1} d\bar{n}_{s_i}$. The Lie algebra of $\bar{N}_{s_i}$ is just $\mathfrak{g}_{-\lambda_i} + \mathfrak{g}_{-2\lambda_i}$.

(1) $\mathfrak{g}_{-2\lambda_i} = (0)$. Then let X be in $\mathfrak{g}_{-\lambda_i}$ such that $B(X, \theta X) = -1$. Then $\bar{N}_{s_i} = \exp(R(\mathrm{Ad}M)X)$ by 8.11.3. Furthermore $k(m\bar{n}m^{-1}) = mk(\bar{n})m^{-1}$. Hence using the formula for integration in polar coordinates we find

$$B_\gamma(\nu, s_i) = c\int_M \pi_\gamma(m)\int_0^\infty t^{m_{\lambda_i}-1}e^{-\nu(H(\exp tX))}\,\pi_\gamma(k(\exp tX))^{-1}dt\,\pi_\gamma(m)^{-1}dm$$

where c is the volume of the unit sphere in $\bar{\mathfrak{n}}_{s_i}$, $m_{\lambda_i} = \dim \mathfrak{g}_{-\lambda_i}$, and $\alpha = 0$, if $\dim \mathfrak{g}_{-\lambda_i} > 1$, $\alpha = -\infty$ if $\dim \mathfrak{g}_{-\lambda_i} = 1$.

(2) If $\mathfrak{g}_{-2\lambda_i} \neq (0)$ then let X be in $\mathfrak{g}_{-\lambda_i}$, Y in $\mathfrak{g}_{-2\lambda_i}$ so that $B(X, \theta X) = B(Y, \theta Y) = -1$. Then as above

$$B_\gamma(\nu, s_i) = c\int_M \pi_\gamma(m)\int_0^\infty\int_\alpha^\infty t^{m_{\lambda_i}-1}s^{m_{2\lambda_i}-1}e^{-\nu(H(\exp(tX+sY))}$$
$$\cdot\pi_\gamma(k(\exp(tX + sY))^{-1}dxdt\,\pi_\gamma(m)^{-1}dm$$

with $\alpha = 0$ if $\dim \mathfrak{g}_{-2\lambda_i} > 1$, $\alpha = -\infty$ if $\dim \mathfrak{g}_{-2\lambda_i} = 1$.

Set in case (1),

$$c_\gamma(\nu, s_i) = \int_\alpha^\infty t^{m_{\lambda_i}-1}e^{-\nu(H(\exp tX))}\pi_\gamma(k(\exp tX))^{-1}dt$$

and in case (2),

$$c_\gamma(\nu, s_i) = \int_0^\infty\int_\alpha^\infty t^{m_{\lambda_i}-1}s^{m_{2\lambda_i}-1}e^{-\nu(H(\exp(tX+sY))}\pi_\gamma(k(\exp tX + sY))^{-1}dt.$$

The theorem will be proved if we can show that the c_γ's have the properties asserted for the B_γ's.

Now in case (1), X, θX generate a TDS in $\mathfrak{g}$ and in case (2), X, θX, Y, θY generate a Lie algebra isomorphic with the Lie algebra of $SU(2, 1)$ (see 8.10.12). Thus in either case (1) or case (2) the vectors in question generate a subalgebra isomorphic with a Lie subalgebra $\mathfrak{g}_i$ of $SU(2, 1)$. Let G_i be the connected subgroup of G corresponding to $\mathfrak{g}_i$. Then G_i is covered by a the corresponding subgroup of $SU(2, 1)$ (since we are assuming that G has a faithful representation). Let $K_i = K \cap G_i$. Then K_i is isomorphic to a closed subgroup of $U(2)$ (the K for $SU(2, 1)$ is $U(2)$). Thus as a representation of K_i, (π_γ, V_γ) splits into a sum of representations which are subrepresentations of the $\pi^{k,l}$ of 8.11.8. We have thus reduced the computation of $c_\gamma(v, s_i)$ to $SU(2, 1)$ and π_γ to $\pi^{k,l}$. We assume that in case (1) or (2), $(2/\lambda(H_{\lambda_i}))X = x$, $(2/\lambda(H_{\lambda_i}))Y = y$ (as in the proof of 8.10.12). We compute $k(\exp(tx + sy))$. A simple computation using the results of the proof of 8.10.12 gives

$$\bar{n}(t, s) = \exp(tx + sy) = \begin{pmatrix} 1 & -t & -t \\ t & 1 - is - t^2/2 & -is - t^2/2 \\ -t & is + t^2/2 & 1 + is + t^2/2 \end{pmatrix}$$

Now

$$\bar{n}(t, s) = k(\bar{n}(t, s))\exp(H(\bar{n}(t, s))n(\bar{n}(t, s)).$$

Thus

$$\bar{n}(t, s)(e_2 + e_3) = ((1 + t^2)^2 + 4s^2)^{1/2}k(\bar{n}(t, s))(e_2 + e_3).$$

Here we have used 8.10.13 (or at least its proof). Hence

$$k(\bar{n}(t, s))(e_2 + e_3) = ((1 + t^2)^2 + 4s^2)^{-1/2} \cdot(-2t, 1 - 2is - t^2, 1 + 2is + t^2).$$

Now

$$k(\bar{n}(t, s)) = \begin{bmatrix} A(t, s) & 0 \\ 0 & d((t, s)) \end{bmatrix}$$

with $d(t, s) = \det(A(t, s))^{-1}$ and $A(t, s)$ in $U(2)$. Thus

$$A(t, s)(0, 1) = ((1 + t^2) + 4s^2)^{-1/2}(-2t, 1 - 2is - t^2)$$

and

$$d(t, s) = (1 + 2is + t^2)/((1 + t^2)^2 + 4s^2)^{1/2}.$$

From this it is not hard to see that

$$k(\bar{n}(t,s)) = \begin{pmatrix} \dfrac{(1 + 2is - t^2)(1 - 2is + t^2)}{(1 + t^2)^2 + 4s^2} & \dfrac{-2t}{((1 + t^2)^2 + 4s^2)^{1/2}} & 0 \\ \dfrac{2t(1 - 2is + t^2)}{(1 + t^2)^2 + 4s^2} & \dfrac{1 - 2is - t^2}{((1 + t^2)^2 + 4s^2)^{1/2}} & 0 \\ 0 & 0 & \dfrac{1 + 2is + t^2}{((1 + t^2)^2 + 4s^2)^{1/2}} \end{pmatrix}$$

(1) If $s = 0$ then setting $\bar{n}(t) = \bar{n}(t, 0)$ we see that

$$k(\bar{n}(t)) = \begin{pmatrix} \dfrac{1-t^2}{1+t^2} & \dfrac{-2t}{1+t^2} & 0 \\ \dfrac{2t}{1+t^2} & \dfrac{1-t^2}{1+t^2} & 0 \\ 0 & 0 & 1 \end{pmatrix}.$$

Thus $\pi^{k,l}(k(\bar{n}(t))$ is diagonalizable with diagonal entries of the form

$$\left(\frac{1 - t^2 - 2it}{1 + t^2}\right)^s \quad \text{or} \quad \left(\frac{1 - t^2 + 2it}{1 + t^2}\right)^s$$

with s an integer and $s \geqslant 0$. From this we see that $c_\gamma(v, s_i)$ is diagonalizable with diagonal entries

$$\int_\alpha^\infty t^{m_{\lambda_i}-1}(1 + t^2)^{-v(H_{\lambda_i}))2\lambda_i(H_{\lambda_i})}(1 - t^2 \pm 2it)^s(1 + t^2)^{-s}\,dt.$$

Thus $c_\gamma(v, s_i)$ has diagonal entries which are sums of terms of the form

$$\int_0^\infty t^{p-1}(1 + t^2)^{-v(H_{\lambda_i}))2\lambda_i(H_{\lambda_i})-s}\,dt$$

with p, s integers, $p \geqslant m_{\lambda_i}$. The result in this case now follows from 8.10.15(1) and 8.10.14.

(2) In this case we compute the matrix entries of $\pi^{k,l}(k(\bar{n}(t, s))$. Let $f_{j,l}$ be as in the proof of 8.11.8. We leave it to the reader to check that if

$$\pi^{k,l}(k(\bar{n}(t, s))^{-1}f_{j,l} = \sum \pi^{k,l}_{u,j}(t, s)f_{u,l}$$

then

$$\pi^{k,l}_{u,j}(t, s) = \left(\frac{i + 2is + t^2}{((1 + t^2)^2 + 4s^2)^{1/2}}\right)^k \sum_{r+s=u} \binom{l-j}{r}\binom{j}{s}(-1)^r s^{r+j-s}$$
$$\cdot((1 + t^2)^2 + 4s^2)^{(u/2)-l}(1 - 2is + t^2)^{l-u}t^{r+j-s}$$
$$\cdot(1 - 2is - t^2)^s(1 + 2is - t^2)^{l-j-r}.$$

This implies that one-matrix entries of $c_\gamma(v, s_i)$ are linear combinations of integrals of the form

$$\int_0^\infty \int_0^\infty t^{n-1}s^{m-1}((1 + t^2)^2 + 4s^2)^{-(v(H_{\lambda_i})/2\lambda_i(H_{\lambda_i}))+r}\,dt\,ds$$

with n, m positive integers and r an integer or a half integer. Thus the result in case (2) now follows from 8.10.15 and 8.10.14. Q.E.D.

8.11.10 *Corollary* The map $\nu \to T_\gamma(\xi, \nu, s)$, $\mathrm{Re}(\nu - \rho)(H_\lambda) > 0$ for λ in Λ^+ extends to a meromorphic operator valued function on $\mathfrak{a}_C^*$. Furthermore if $\nu_0(H)$ is not an integer or a half integer for each λ in Λ^+ then $\nu \mapsto T_\gamma(\xi, \nu, s)$ is holomorphic in a neighborhood of ν_0.

8.12 The Asymptotics of the Principal Series for Semisimple Lie Groups of Split Rank 1

8.12.1 Let G be a semisimple Lie group. Let $G = KAN$ be an Iwasawa decomposition of G. G is said to be of split rank 1 if $\dim A = 1$.

8.12.2 Let us retain the notation of the previous sections. Let ν be in $\mathfrak{a}_C^*$. Let γ, τ be in $\hat{K}$ and let $A: V_\tau \to V_\gamma$ be a linear map. Define

$$(1) \qquad E_{\gamma,\tau}(A: \nu: x) = \int_K e^{-(\nu+\rho)(H(xk))}\, \pi_\gamma(h(xk)) A \pi_\tau(k)^{-1} dk.$$

8.12.3 *Lemma* If $A: V_\tau \to V_\gamma$ let $A^\circ = \int_M \pi_\gamma(m) A \pi_\tau(m)^{-1} dm$ (here $\int_M dm = 1$). Then $E_{\gamma,\tau}(A: \nu) = E_{\gamma,\tau}(A^\circ: \nu)$.

PROOF

$$\begin{aligned} E_{\gamma,\tau}(A: \nu: x) &= \int_K e^{-(\nu+\rho)(H(xk))} \pi_\gamma(k(xk)) A \pi_\tau(k)^{-1} dk \\ &= \int_K e^{-(\nu+\rho)(H(xkm))} \pi_\gamma(k(xkm)) A \pi_\tau(km)^{-1} dk \text{ (for } m \text{ in } M) \\ &= \int_K e^{-(\nu+\rho)(H(xk))}\, \pi_\gamma(k(xk)) \pi_\gamma(m) A \pi_\tau(m)^{-1} \pi_\tau(k)^{-1} dk. \end{aligned}$$

Since this equality is true for all m in M we may integrate out the m and find

$$\begin{aligned} E_{\gamma,\tau}(A: \nu: x) &= \int_K e^{-(\nu+\rho)(H(xk))} \pi_\gamma(k(xk)) \int_M \pi_\gamma(m) A\, \pi_\tau(m)^{-1} dm \pi_\tau(k)^{-1} dk \\ &= E_{\gamma,\tau}(A^\circ: \nu: x). \end{aligned}$$

8.12.4 The $E_{\gamma,\tau}(A: \nu)$ are called Eisenstein integrals or generalized spherical functions. The importance of these integrals for us is that they are essentially the matrix entries of the principal series. In fact, let ξ be in $\hat{M}$. Let ν be in $\mathfrak{a}_C^*$. Let γ, τ be in $\hat{K}$ and let A be in $\mathrm{Hom}_M(V_\gamma, H_\xi)$, B be in $\mathrm{Hom}_M(V_\tau, H_\xi)$. Let v be in V_γ, w be in V_τ. Let

$$f(kan) = e^{-(\nu+\rho)(\log a)}A(\pi_\gamma(k)^{-1}v)$$

$$h(kan) = e^{-(\nu+\rho)(\log a)}B(\pi_\tau(k)^{-1}w)$$

for k in K, a in A, and n in N. Then f, h are in, respectively, $H_\gamma^{\xi,\rho+\nu}$ and $H_\tau^{\xi,\rho+\bar{\nu}}$. We compute

$$\begin{aligned}\langle \tilde{\pi}_{\xi,\rho+\bar{\nu}}(g^{-1})f, h\rangle &= \int_K \langle (\tilde{\pi}_{\xi,\rho+\bar{\nu}}(g^{-1})f)(k), h(k)\rangle\, dk \\ &= \int_K \langle f(gk), h(k)\rangle\, dk \\ &= \int_K e^{-(\rho+\bar{\nu})(H(gk))}\langle A(\pi_\gamma(k(gk))^{-1}v), B(\pi_\tau(k)^{-1}w\rangle dk.\end{aligned}$$

Let $A^*: H_\xi \to V_\gamma$ be defined by $\langle Av, u\rangle = \langle v, A^*u\rangle$ for v in V_γ, u in H_ξ. Then $A^*B: V_\tau \to V_\gamma$ and clearly $(A^*B)^\circ = A^*B$. We therefore see

$$\begin{aligned}\langle \tilde{\pi}_{\xi,\rho+\bar{\nu}}(g)^{-1}f, h\rangle &= \int_K e^{-(\rho+\bar{\nu})(H(gk)}\langle v, \pi_\gamma(k(gk))A^*B\pi_\tau(k)^{-1}\rangle dk \\ &= \langle v, E_{\gamma,\tau}(A^*B: \nu: g)w\rangle.\end{aligned}$$

8.12.5 We now assume that G is of split rank 1. Then $\mathfrak{a}_C^* = C\rho$. Also $W(A) = \{1, s_0\}$. Let s^* in M^* be chosen once and for all so that $s^*M = s_0$. Let γ be in $\hat{K}$, z in C. Define $B_\gamma(z) = B_\gamma((1 + z)\rho, s_0)$ (see 8.11.5). If τ is in $\hat{K}$ set for $A: V_\tau \to V_\gamma$ linear,

$$E_{\gamma,\tau}(A: z: x) = E_{\gamma,\tau}(A: z\rho: x)$$

for x in G.

8.12.6 *Theorem* Let γ, τ be in $\hat{K}$. Let H in $\mathfrak{a}$ be so that $\rho(H) > 0$. Set $a_t = \exp tH$. Let A be in $L(V_\tau, V_\gamma)$. If ν is in R, $|\nu| \geqslant \varepsilon > 0$ then

$$\begin{aligned}&\|e^{\rho(\log a_t)}E_{\gamma,\tau}(A: i\nu: a_t) - e^{-i\nu\rho(\log a_t)}A^\circ B_\tau(-i\nu) \\ &\qquad - e^{i\nu\rho(\log a_t)}\pi_\gamma(s^*)B_\gamma(i\nu)^*A^\circ\pi_\tau(s^*)^{-1}\| \leqslant C_\varepsilon e^{-\delta t}\end{aligned}$$

for some $\delta > 0$ independent of ν and ε, C_ε depending only on ε.

PROOF We have seen that $E(A: z: a_t) = E(A^\circ: z: a_t)$. We may thus assume that $A = A^\circ$. As an M representation $V_\gamma = \sum_{\xi \in M} n_\gamma(\xi) H_\xi$. Let $P_\xi: V_\gamma \to n_\gamma(\xi)H_\xi$ be the corresponding projection. We may thus assume that $A = P_\xi A$ (in fact, $A = \sum_{\xi \in M} P_\xi A$).

Let now Ω be the left invariant differential operator on G defined as follows: Let $X_1, \ldots, X_n$ be a basis of $\mathfrak{g}$. Let $Y_1, \ldots, Y_n$ be defined by $B(X_i, Y_j) = \delta_{ij}$, where B is the Killing form of $\mathfrak{g}$. Set $\Omega = \sum X_i Y_i$. Clearly Ω is independent of the choice of basis and hence $\Omega = \sum \mathrm{Ad}(g)X_i \mathrm{Ad}(g) Y_i = \mathrm{Ad}(g)\Omega$.

We now compute $\Omega E(A: z: x)$. We first set

$$f_k(x) = e^{-(1+z)\rho(H(xk))}\pi_\gamma(k(xk))A\pi_\tau(k)^{-1}.$$

Then $E(A: z: x) = \int_K f_k(x)dk, \Omega E(A: z: x) = \int_K (\Omega f_k)(x)dk$. On the other hand $f_k(x) = f_e(xk)\pi_\tau(k)^{-1}$. Since $(\Omega f_e)(xk) = ((\mathrm{Ad}(k)\Omega) f_k(x))\pi_\tau(k)$, we see that $\Omega f_k(x) = (\Omega f_e)(xk)\pi_\tau(k)^{-1}$. We therefore compute Ωf_e. Set $f = f_e$.

Let $\mathfrak{n} = \mathfrak{n}_\lambda + \mathfrak{n}_{2\lambda}$ be the root space decomposition of $\mathfrak{n}$. (Note that $\mathfrak{n}_{2\lambda}$ might be zero.) Let $p = \dim \mathfrak{n}_\lambda$, $q = \dim \mathfrak{n}_{2\lambda}$. Let $X_1, \ldots, X_p, X_{p+1}, \ldots, X_r$ $(r = p + q)$ be a basis of $\mathfrak{n}$ so that $X_1, \ldots, X_p$ are in $\mathfrak{n}_\lambda$, $X_{p+1}, \ldots, X_r$ are in $\mathfrak{n}_{2\lambda}$, and $-B(X_i, \theta X_j) = \delta_{ij}$. Let H_λ in $\mathfrak{a}$ be defined by $B(h, H_\lambda) = \lambda(h)$ for h in $\mathfrak{a}$. Then $[X_i, \theta X_i] = -H_\lambda$, $i = 1, \ldots, p$, $[X_i, \theta X_i] = -2H_\lambda$, $i = p + 1, \ldots, r$.

Now $\mathfrak{g} = \bar{\mathfrak{n}} \oplus \mathfrak{m} \oplus \mathfrak{a} \oplus \mathfrak{n}$. Let H_0 in $\mathfrak{a}$ be such that $B(H_0, H_0) = 1$. Let $U_1, \ldots, U_r$ be a basis of $\mathfrak{m}$ so that $-B(U_i, U_j) = \delta_{ij}$. Then

$$\Omega = -\sum X_i \theta X_i - \sum \theta X_i X_i - \sum U_i^2 + H_i^2.$$

Set $\Omega_0 = \sum U_i^2$. We note that if m is in M then $\mathrm{Ad}(m)\Omega_0 = \Omega_0$. Also by definition of f, $f(xn) = f(x)$ for n in N. Since $X_i\theta X_i = -H_\lambda + \theta X_i X_i$ for $1 \leqslant i \leqslant p$, $X_i \theta X_i = -2H_\lambda + \theta X_i X_i$ for $p + 1 \leqslant i \leqslant r$, we see that

$$\Omega f = (p + 2q)H_\lambda f - \Omega_0 f + H_0^2 f.$$

On the other hand

$$\begin{aligned}(\Omega_0 f)(x) &= \sum_{i=1}^m \frac{d^2}{dt^2} f(x \exp tU_i)\bigg|_{t=0} \\ &= \sum_{i=1}^m e^{-(1+z)\rho(H(x))}\, \pi_\gamma(k(x))\pi_\gamma(U_i)^2 A \\ &= e^{-(1+z)\rho(H(x))}\pi_\gamma(k(x))\pi_\gamma(\Omega_0)A.\end{aligned}$$

But $\pi(\Omega_0)$ acts by a scalar on each irreducible M subrepresentation of V_γ. Hence $\pi(\Omega_0)P_\xi = \lambda_\xi P_\xi$ with λ_ξ a scalar depending only on ξ.

Also

$$H_\lambda f(x) = \frac{d}{dt} f(x \exp tH_\lambda)$$

$$= \frac{d}{dt} e^{-(1+z)\rho(H(x))} e^{-(1+z)\rho(H_\lambda)} \pi_\gamma(k(x))A$$

$$= -(1 + z)\rho(H_\lambda) f(x).$$

Similarly, $H^2 f(x) = ((1 + z)\rho(H_0))^2 f(x)$. We have the formula

$$\Omega f = -(p + 2q)(1 + z)\rho(H_\lambda) f + ((1 + z)\rho(H_0))^2 f - \lambda_\xi f.$$

Hence

(1) $\Omega E(A: z: x)$

$$= \{((1 + z)\rho(H_0))^2 - (p + 2q)(1 + z)\rho(H_\lambda) - \lambda_\xi\} E(A: z: x).$$

Suppose now that $\varphi: G \to L(V_\tau, V_\gamma)$ is C^∞ and suppose that $\varphi(k_1 x k_2) = \pi_\gamma(k_1)\varphi(x)\pi_\gamma(k_2)$ (in particular $E(A: z: x) = \varphi(x)$ satisfies this condition). Let a be in A. We compute $(\Omega\varphi)(a)$. To do this it is convenient to introduce some more notation. Let $Z_i = 2^{-1/2}(X_i + \theta X_i)$, $Y_i = 2^{-1/2}(X_i - \theta X_i)$. Then $B(Z_i, Z_j) = -\delta_{ij}$, $B(Y_i, Y_j) = \delta_{ij}$. In this notation

$$\Omega = -\sum_{i=1}^{r} Z_i^2 - \Omega_0 + H_0^2 + \sum Y_i^2.$$

Now if a is in A then

$$\mathrm{Ad}(a)^{-1} Z_i = 2^{-1/2}(\mathrm{Ad}(a)^{-1} X_i + \mathrm{Ad}(a)^{-1} \theta X_i)$$

$$= \begin{cases} 2^{-1/2}(e^{-\lambda(\log a)} X_i + e^{\lambda(\log a)}\, \theta X_i), & 1 \leqslant i \leqslant p \\ 2^{-1/2}(e^{-2\lambda(\log a)}\, X_i + e^{2\lambda(\log a)}\, \theta X_i), & p + 1 \leqslant i \leqslant r. \end{cases}$$

Now $X_i = 2^{-1/2}(Z_i + Y_i)$. Hence

$$\mathrm{Ad}(a)^{-1} Z_i = \begin{cases} \cosh(\lambda(\log a)) Z_i - \sinh(\lambda(\log a)) Y_i, & 1 \leqslant i \leqslant p \\ \cosh(2\lambda(\log a)) Z_i - \sinh(2\lambda(\log a)) Y_i, & p + 1 \leqslant i \leqslant r. \end{cases}$$

Here $\cosh t = \frac{1}{2}(e^t + e^{-t})$, $\sinh t = \frac{1}{2}(e^t - e^{-t})$.

Now $(d^2/dt^2)\varphi(\exp tZ_i a)|_{t=0} = \pi_\gamma(Z_i^2)\varphi(a)$. On the other hand

$$\left.\frac{d^2}{dt^2} \varphi(\exp tZ_i a)\right|_{t=0} = \left.\frac{d^2}{dt^2} \varphi(a \exp t\,\mathrm{Ad}(a)^{-1} Z_i))\right|_{t=0}$$

$$= ((\mathrm{Ad}(a)^{-1} Z_i)^2 \varphi)(a)$$

$$= (\cosh \lambda_i(\log a) Z_i - \sinh \lambda_i(\log a) Y_i)^2 \varphi(a)$$

with $\lambda_i = \lambda$, $1 \leqslant i \leqslant p$, $\lambda_i = 2\lambda$, $p < i \leqslant r$. This implies that

(2) $$\pi_\gamma(Z_i)^2\varphi(a) = (\cosh \lambda_i(\log a))^2(Z_i^2\varphi)(a)$$
$$- \cosh \lambda_i(\log a)\sinh \lambda_i(\log a)((Z_iY_i + Y_iZ_i)\varphi)(a)$$
$$+ (\sinh \lambda_i(\log a))^2(Y_i^2\varphi)(a).$$

Now $X_i = 2^{-1/2}(Z_i + Y_i)$, $\theta X_i = 2^{-1/2}(Z_i - Y_i)$, hence $H_{\lambda_i} = [\theta X_i, X_i] = [Y_i, Z_i]$. Thus $Y_iZ_i = H_{\lambda_i} + Z_iY_i$. Also

$$Y_i = \coth \lambda_i(\log a)Z_i - \sinh \lambda_i(\log a)^{-1}\mathrm{Ad}(a)^{-1}Z_i$$

($\coth t = \cosh t/\sinh t$). These observations combined with (2) give

(3) $(Y_i^2\varphi)$ (log a)
$$= (\sinh \lambda_i(\log a))^{-2}\pi_\gamma(Z_i)^2\varphi(a)$$
$$+ (\coth \lambda_i(\log a))^2\varphi(a)\pi_\tau(Z_i)^2 + \coth \lambda_i(\log a)(H_{\lambda_i}\varphi)(\log a)$$
$$- 2 \sinh \lambda_i(\log a)^{-1} \coth \lambda_i(\log a)\pi_\gamma(Z_i)\varphi(a)\pi_\tau(Z_i).$$

For convenience we take $H_0 = H$, $a_t = \exp tH_0$. $B(H_\lambda, H_0) = \lambda(H_0)$. Thus $H_\lambda = \lambda(H_0)H_0$. Now

$$(\Omega\varphi)(a_t) = \sum_{i=1}^{r} (Y_i^2\varphi)(a_t) + (H^2\varphi)(a_t) - (\Omega_0\varphi)(a_t) - \sum_{i=1}^{r} (Z_i^2\varphi)(a_t)$$
$$= \frac{d^2}{dt^2}\varphi(a_t) + (p \coth \lambda(\log a_t)\lambda(H_0) + 2q \coth 2\lambda(\log a_t)\lambda(H_0))\frac{d\varphi(a_t)}{dt}$$
$$+ (\sinh \lambda(\log a_t))^{-2} \sum_{i=1}^{p} \pi_\gamma(Z_i)^2\varphi(a_t)$$
$$+ (\sinh 2\lambda(\log a_t))^{-2} \sum_{i=p+1}^{r} \pi_\gamma(Z_i)^2\varphi(a)$$
$$+ (\coth \lambda(\log a))^2 \sum_{i=1}^{p} \varphi(a)\pi_\tau(Z_i)^2$$
$$+ (\coth 2\lambda(\log a))^2 \sum_{i=p+1}^{r} \varphi(a)\pi_\tau(Z_i)^2$$
$$- 2 \sinh \lambda(\log a_t)^{-1} \coth \lambda(\log a_t) \sum_{i=1}^{p} \pi_\gamma(Z_i)\varphi(a_t)\pi_\tau(Z_i)$$
$$- 2 \sinh 2\lambda(\log a_t)^{-1} \coth 2\lambda(\log a_t) \sum_{i=p+1}^{r} \pi_\gamma(Z_i)\varphi(a_t)\pi_\tau(Z_i)$$
$$- \varphi(a_t)\pi_\tau(\Omega_0) - \sum_{i=1}^{r} \varphi(a_t)\pi_\tau(Z_i)^2.$$

Let for v in $L(V_\tau, V_\gamma)$,

$$\begin{aligned} Q_1(t)v = {} & \sinh \lambda(\log a_t)^{-2} \sum_{i=1}^{p} \pi_\gamma(Z_i)^2 v + \sinh 2\lambda(\log a_t)^{-2} \sum_{i=p+1}^{r} \pi_\gamma(Z_i)^2 v \\ & + (\coth \lambda(\log a_t)^2 - 1) \sum_{i=1}^{p} v\pi_\tau(Z_i)^2 \\ & + (\coth 2\lambda(\log a_t)^2 - 1) \sum_{i=p+1}^{r} v\pi_\tau(Z_i)^2 \\ & - 2 \sinh \lambda(\log a_t)^{-1} \coth \lambda(\log a_t) \sum_{i=1}^{p} \pi_\gamma(Z_i) v \pi_\tau(Z_i) \\ & - 2 \sinh 2\lambda(\log a_t)^{-1} \coth 2\lambda(\log a_t) \sum_{i=p+1}^{r} \pi_\gamma(Z_i) v\, \pi_\tau(Z_i). \end{aligned}$$

Then

$$\begin{aligned} (4)\quad (\Omega\varphi)(a_t) = {} & \frac{d^2\varphi(a_t)}{dt^2} + (p \coth \lambda(\log a_t)\lambda(H_0) + 2q \coth 2\lambda(\log a_t)\lambda(H_0)) \\ & \cdot \frac{d\varphi}{dt}(a_t) + Q_1(t)\varphi(a_t) - \varphi(a_t)\pi_\tau(\Omega_0). \end{aligned}$$

We now return to the case $\varphi(x) = E(A : z : a_t)$ with $A = A^\circ = P_\xi A^\circ$. Clearly $A^\circ P_\xi = P_\xi A^\circ$. Hence $E(A : z : a_t)\pi_\tau(\Omega_0) = \lambda_\xi E(A : z : a_t)$. Combining (4) with (1) we find that if $f(t) = E(A : z : a_t)$ then for $t > 0$

$$\begin{aligned} (5)\quad & f''(t) + \lambda(H_0)(p \coth \lambda(\log a_t) + 2q \coth 2\lambda(\log a_t))f'(t) + Q_1(t)f(t) \\ & \quad = (((1 + z)\rho(H_0))^2 - (p + 2q)(1 + z)\rho(H_\lambda))f(t). \end{aligned}$$

Now $\log a_t = tH_0$. Thus $\coth \lambda(\log a_t) = \coth \lambda(H_0)t$. Let $\Delta(t) = (\sinh \lambda(H_0)t)^p(\sinh 2\lambda(H_0)t)^q$. Set $h(t) = \Delta^{1/2}(t)f(t)$ for $t > 0$. A straightforward computation gives, for $t > 0$,

$$\begin{aligned} (6)\qquad & h''(t) + \frac{\lambda(H_0)^2}{2}(p(\sinh \lambda(H_0)t)^{-2} + 2q(\sinh 2\lambda(H_0)t)^{-2})h(t) \\ & \quad - \frac{\lambda(H_0)^2}{4}(p \coth \lambda(H_0)t + 2q \coth 2\lambda(H_0)t)^2 h(t) + Q_1(t)h(t) \\ & = (((1 + z)\rho(H_0))^2 - (2p + q)(1 + z)\rho(H_\lambda))h(t). \end{aligned}$$

Set now

$$Q_2(t) = \frac{(\lambda(H_0)^2}{4}(p + 2q)^2 - \frac{\lambda(H_0)^2}{4}(p \coth \lambda(H_0)t + 2q \coth 2\lambda(H_0)t)^2)I.$$

Using

$$(1 + z)^2\rho(H_0)^2 - (p + 2q)(1 + z)\rho(H_0) + \frac{\lambda(H_0)^2}{4}(p + 2q)^2 = z^2\rho(H_0)^2,$$

we find that if $Q(t) = -Q_1(t) - Q_2(t)$ for $t > 0$, then

(7) $$-h''(t) + Q(t)h(t) = -z^2\rho(H_0)^2h(t).$$

Using the fact that $(\coth t)^2 - 1 = (\sinh t)^{-2}$ and $\coth 2t - 1 = 2e^{-2t}(1 - e^{-2t})^{-1}$, that $(\sinh t)^{-1} = 2(e^t - e^{-t})^{-1} = 2e^{-t}(1 - e^{-2t})$ and $\coth t = (1 + e^{-2t})/(1 - e^{-2t})$, we see that if $t > a > 0$, then $\|Q(t)\| \leqslant C_1e^{-\lambda(H_0)t}$ with C_1 a positive constant depending only on a. Thus if $s > a$, then

(8) $$\int_s^\infty \|Q(t)\|(1 + t)dt \leqslant C_2e^{-\lambda(H_0)s} + C_3se^{-\lambda(H_0)s}.$$

Let $\varepsilon_1 > 0$ be such that $\lambda(H_0) > \varepsilon_1$. Then there is a constant C_4 so that $s < C_4e^{\varepsilon_1 s}$. Hence

(9) $$\int_s^\infty \|Q(t)\|(1 + t)dt < C_4e^{-(\lambda(H_0)-\varepsilon_1)s}.$$

Let $\delta = \lambda(H_0) - \varepsilon_1$. A.8.2.3 now implies that if $|v| > \varepsilon$, in R there is a constant depending only on ε and A_1, A_2 in $L(V_\tau, V_\gamma)$ such that

(10) $$\|\Delta^{1/2}(t)E(A: iv: a_t) - e^{iv\rho(H_0)t}A_1 - e^{-iv(H_0)t}A_2\| < C_\varepsilon e^{-\delta t}.$$

$\lim_{t\to\infty} |\Delta^{1/2}(t) - e^{\rho(H_0)t}| = 0$. Hence we have

(11) $$|e^{\rho(H_0)t}E(A: iv: a_t) - e^{iv\rho(H_0)t}A_1 - e^{-iv\rho(H_0)t}A_2\| < C_\varepsilon e^{-\delta t},$$

for $t > a$ some positive constant and $|v| \geqslant \varepsilon > 0$, C_ε depending only on ε.

We are left with the computation of A_1 and A_2. We also note from the proof of A.8.2.3 that $A_1 = A_1(v)$ and $A_2 = A_2(v)$ depend continuously on v for v in R, $v \neq 0$. Furthermore in the notation of A.8.2.18 we see that $A_\mu(v, w)$ is defined for μ in H_ε^+ and depends continuously on μ. Thus there is a neighborhood U of v in H_ε^+ such that if μ is in U, A_μ: $L(V_\tau, V_\gamma) \times L(V_\tau, V_\gamma) \to V_\mu$ is a linear isomorphism. Applying A.8.2.12 and A.8.2.16 we see that A_2 extends to a continuous function on U and that if μ is in U, $\operatorname{Im} \mu > 0$ then

$$\lim_{\substack{t\to\infty \\ t>0}} |e^{\rho(H_0)t}E(A: i\mu: a_t) - e^{-i\mu(H_0)t}A_2(\mu)\| = 0.$$

We therefore compute for $\operatorname{Re} z > 0$,

$$e^{(1+iz)\rho(H_0)t}E(A: iz: a_t) = e^{(1+iz)\rho(H_0)t}\int_K e^{-(1+iz)\rho(H(a_tk))}\pi_\gamma(k(a_tk))A\pi_\tau(k)^{-1}dk.$$

Now letting f_k be as in the beginning of the proof we see that since $A = A^{\circ}$, $f_{km} = f_k$ for m in M. Now applying 7.6.8 we see that

(12) $e^{(1+iz)\rho(H_0)t}E(A: iz: a_t)$

$$= e^{(1+iz)\rho(H_0)t}\int_{\bar{N}} e^{-(1+iz)\rho(H(a_t k(\bar{n})))}\pi_\gamma(k(a_t k(\bar{n}))A\pi_\tau(k(\bar{n}))^{-1}e^{-2\rho(H(k))}d\bar{n}.$$

Here $x = k(x)\exp(H(x))n(x)$ with $k(x)$ in K, $H(x)$ in $\mathfrak{a}$, $n(x)$ in N. Now $\bar{n} = k(\bar{n})\exp H(\bar{n})n(\bar{n})$. Thus $k(\bar{n})$ is in $\bar{n}\exp(-H(\bar{n}))N$. Hence

(i) $H(a_t k(\bar{n})) = H(a_t\bar{n}) - H(\bar{n})$

(ii) $k(a_t k(\bar{n})) = k(a_t\bar{n}) = k(a_t\bar{n}a^{-1})$.

Also

(iii) $H(a_t\bar{n}) = H(a_t\bar{n}a_0^{-1}) + H(a_t)$.

Combining (12) with (i), (ii), and (iii) we find

(13) $e^{(1+iz)\rho(H_0)t}E(A: iz: a_t)$

$$= \int_{\bar{N}} e^{-(1+iz)\rho(H(a_t\bar{n}a_t^{-1}))}e^{-(1-iz)\rho(H(\bar{n}))}\pi_\gamma(k(a_t\bar{n}a_t^{-1}))A\pi_\tau(k(\bar{n}))^{-1}d\bar{n}.$$

Now

(*) $$\|e^{-(1+iz)\rho(H(a_t\bar{n}a_t^{-1})}e^{-(1-iz)\rho(H(\bar{n}))}\pi_\gamma(k(a_t\bar{n}a_t^{-1})A\pi_\tau(k(\bar{n}))^{-1}\| \leqslant \|A\|e^{-(1+\mathrm{Im}z)\rho(H(\bar{n}))}e^{-(1-\mathrm{Im}z)\rho(H(a\bar{n}_t a^{-1})}.$$

Thus if we take z such that $0 < \mathrm{Im}\, z < 1$,

(*) $$\leqslant \|A\|\, e^{-(1+\mathrm{Im}z)\rho(H(\bar{n}))}.$$

(Here we use 8.13.7). On the other hand 8.10.16 implies $\int_{\bar{N}} e^{-(1+\mathrm{Im}z)\rho(H(\bar{n}))}d\bar{n} < \infty$ for $\mathrm{Im}z > 0$. We may thus apply the Lebesque dominated convergence theorem to the right-hand side of (13) to find

(14) If $0 < \mathrm{Im}z < 1$, then

$$\lim_{\substack{t\to\infty \\ t>0}} e^{(1+iz)\rho(H_0)t}E(A: iz: a_t)$$

$$= \int_{\bar{N}} e^{-(1-iz)\rho(H(\bar{n}))}A\pi_\tau(k(\bar{n}))^{-1}d\bar{n} = AB_\tau(-iz)$$

by 8.11.5. Hence $A_2(z) = AB_\tau(-iz)$ for z in U.

We will complete the proof by computing A_1. To do this we return to the formula

$$E(A: iz: a_t) = \int_K e^{-(1+iz)\rho(H(a_t k))}\pi_\gamma(k(a_t k))A\pi_\tau(k)^{-1}dk.$$

We make the change of variables $k = s^{*-1}k$ and find that

$$E(A: iz: a_t) = \int_K e^{-(1+iz)\rho(H(a_t s^* k))}\pi_\gamma(k(a_t s^* k))A\pi_\tau(k)^{-1}\pi_\tau(s^*)^{-1}dk.$$

Now

$$a_t s^* k = s^* s^{*-1} a_t s^* k = s^* a_{-t} k \ (\mathrm{Ad}(s^*)|_{\mathfrak{a}} = -I).$$

Thus

$$\begin{aligned} E(A: iz: a_t) &= \int_K e^{-(1+iz)\rho(H(a_{-t}k))} \\ &\quad \cdot \pi_\gamma(s^*)\pi_\gamma(k(a_{-t}k))A\pi_\tau(k)^{-1}\pi_\tau(s^*)^{-1}dk \\ &= \int_{\bar{N}} e^{-(1+iz)\rho(H(a_{-t}k(\bar{n})))}\pi_\gamma(s^*) \\ &\quad \cdot \pi_\gamma(k(a_{-t}k(\bar{n}))A\pi_\tau(k(\bar{n}))^{-1}\pi_\tau(s^*)^{-1}e^{-2\rho(H(\bar{n}))}d\bar{n} \\ &= e^{(1+iz)\rho(H_0)t}\int_{\bar{N}} e^{-(1+iz)\rho(H(a_t^{-1}\bar{n}a_t))}e^{-(1-iz)\rho(H(\bar{n}))} \\ &\quad \cdot \pi_\gamma(s^*)\pi_\gamma(k(a_t^{-1}\bar{n}a_t)A\pi_\tau(k(\bar{n}))^{-1} \ \pi_\tau(s^*)^{-1}d\bar{n}, \end{aligned}$$

here we have used (i), (ii) and (iii) again.

We now make the change of variables $a_t^{-1}\bar{n}a_t \to \bar{n}$. We therefore find that

$$\begin{aligned} E(A: iz: a) &= e^{-(1-iz)\rho(H_0)t}\int_{\bar{N}} e^{-(1+iz)\rho(H(\bar{n}))}e^{-(1-iz)\rho(a_t\bar{n}a_t^{-1})} \\ &\quad \cdot \pi_\gamma(s^*)\pi_\gamma(k(\bar{n}))A\pi_\tau(k(a_t\bar{n}a_t^{-1})^{-1}\pi_\tau(s^*)^{-1}d\bar{n}. \end{aligned}$$

Hence

$$\begin{aligned} &e^{(1-iz)\rho(H_0)t}E_{\gamma,\tau}(A: iz: a_t) \\ &\quad = (\pi_\tau(s^*)e^{(1-iz)\rho(H_0)t}E_{\tau,\gamma}(A^*: i\bar{z}: a_t)\pi_\gamma(s^*)^{-1})^*. \end{aligned}$$

This implies that $A_1(v)$ extends to a neighborhood of v in $\bar{H}_\varepsilon^+$ and $A_1(z) = \pi_\gamma(s^*)B_\gamma(i\bar{z})^*A_\tau(s^*)^{-1}$, z sufficiently near in $\bar{H}_\varepsilon^+$. This implies that $A_1(v) = \pi_\gamma(s^*)B_\gamma(iv)^*A\pi_\tau(s^*)^{-1}$ for v in R, $v \neq 0$. Q.E.D.

8.13 The Composition Series of the Principal Series

8.13.1 Let G be a Lie group. Let (π, H) be a representation of G. Then a composition series for π is a sequence $H = H_0 \supset H_1 \supset \cdots \supset H_k \supset \{0\}$ such that H_i is a closed invariant subspace of H_{i-1} and the representation on H_i/H_{i+1} induced by π is irreducible ($H_{k+1} = \{0\}$).

8.13.2 Let G be a connected semisimple Lie group with a faithful finite dimensional representation. Let K, A, N, M, $\mathfrak{a}$, ρ be as in 8.9.

8.13.3 *Theorem* Let ξ be in $\hat{M}$, $\nu \in \mathfrak{a}_C^*$, then $(\tilde{\pi}_{\xi,\nu}, \tilde{H}^{\xi,\gamma})$ has a composition series.

8.13.4 Before we prove 8.13.3 we establish some preliminary results.

8.13.5 Let (π, H) be a representation of G. Let f be in H. Let H_f be the smallest closed invariant subspace of H containing f. f is said to be a cyclic vector for π if $H_f = H$.

8.13.6 *Proposition* If ν is in $\mathfrak{a}_C^*$ and $\text{Re}(\nu - 2\rho)(H_\lambda) \geqslant 0$ for all λ in Λ^+, then 1_ν is a cyclic vector for $\tilde{H}^{1,\nu}$ (see 8.5.8 for notation).

PROOF Suppose f is a continuous function on K/M so that $\int_{KtM}(\tilde{\pi}_{1\,i\nu}(g)1_\nu)(x)f(x)dx = 0$ for all g in G. We show that $f(eK) = 0$. This will prove the lemma since if 1_ν is not cyclic than the orthogonal complement to the cyclic space for 1_ν would contain nonzero K-finite functions (which are continuous) and it would be K-invariant. Let H in $\mathfrak{a}$ be so that $\lambda(H) > 0$ for all λ in Λ^+. Let $a_t = \exp tH$. We compute

$$(1) \qquad \int_{K/M} (\tilde{\pi}_{1,\nu}(a_t)1_\nu)(x)f(x)dx = \int_{K/M} e^{-\nu(H(a_t{}^{-1}k))}f(kM)d(kM).$$

Now transforming the integral to one over $\bar{N}$ we have (see 7.6.8)

$$(2) \qquad \int_{K/M} e^{-\nu(H(a_t{}^{-1}k))}f(kM)d(kM) = \int_{\bar{N}} e^{-2\rho(H(\bar{n}))}e^{-\nu(H(a_t{}^{-1}k(\bar{n}))}f(k(\bar{n})M)d\bar{n}.$$

Now $k(\bar{n}) \exp H(\bar{n})n(\bar{n}) = \bar{n}$. Hence $k(n) = \bar{n} \exp(-H(\bar{n}))n$ with n in N. Thus $H(a_t^{-1}k(\bar{n})) = H(a_t^{-1}\bar{n} \exp(-H(\bar{n}))) = H(a_t^{-1}\bar{n}) - H(\bar{n})$. Thus

$$(3) \qquad \int_{K/M} e^{-\nu(H(a_t{}^{-1}k))}f(kM)\, d(kM)$$
$$= \int_{\bar{N}} e^{(\nu-2\rho)(H(\bar{n}))}e^{-\nu(H(a_t{}^{-1}k(\bar{n}))}f(k(\bar{n})M)d\bar{n}$$
$$= e^{t\nu(H)} \int_{\bar{N}} e^{(\nu-2\rho)(H(\bar{n}))}e^{-\nu(H(a_t{}^{-1}\bar{n}a_t)}f(k(\bar{n})M)d\bar{n}.$$

Suppose that $\int_{K/M} (\tilde{\pi}_{1,\nu}(g)1_\nu)(x)f(x)dx = 0$ for all g. Then

$$\int_{\bar{N}} e^{(\nu-2\rho)(H(\bar{n}))}e^{-\nu(H(a_t^{-1}\bar{n}a_t))}f(k(\bar{n})M)d\bar{n} = 0 \tag{4}$$

for all t.

Transforming the integral by the transformation $a_t^{-1}\bar{n}a_t \to \bar{n}$ we see that

$$\int_{\bar{N}} e^{(\nu-2\rho)(H(a_t\bar{n}a_t^{-1}))}e^{-\nu(H(\bar{n}))}f(k(a_t\bar{n}a_t^{-1})M)d\bar{n} = 0 \tag{5}$$

for all t.

Before proceeding we need the following result.

8.13.7 *Lemma* Let ${}^+\Lambda = \{\mu \text{ in } \mathfrak{a}^* | \mu(H_\lambda) \geqslant 0 \text{ for } \lambda \text{ in } \Lambda^+\}$. Let h be in $\mathfrak{a}$ so that $\lambda(h) > 0$ for all λ in Λ^+. If $\bar{n}$ is in $\bar{N}$ and μ is in ${}^+\Lambda$ then

(1) $\mu(H(\bar{n})) \geqslant 0$

(2) $\mu(H(\bar{n}) - H(a\bar{n}a^{-1})) \geqslant 0$, $a = \exp h$.

PROOF We return to the notation of 8.5.8. Let (π, V) be an irreducible representation of $\tilde{G}$ with highest weight Λ (we use the order of 7.5.9). Let v in V correspond to the weight space Λ and let $(\ ,\ )$ be a $\tilde{K}$-invariant inner product on V. Then $(\pi(g)v, \pi(g)v)$ is well defined for g in G (see the proof of 8.5.8). Now

$$(\pi(kan)v, \pi(kan)v) = (\pi(k)\pi(a)\pi(n)v, \pi(k)\pi(a)\pi(n)v) = e^{2\Lambda(\log a)}(v, v)$$

for k in K, a in A, n in N. Thus $(\pi(g)v, \pi(g)v) = e^{2\Lambda(H(g))}(v, v)$ for g in G. Let X be in $\bar{\mathfrak{n}}$, and let $a = \exp h$ with h as in the statement of the lemma.

X is in $\sum_{\alpha\in\Sigma} \mathfrak{g}_\alpha$. Thus if $k > 0$, $\pi(X)^k \cdot v$ is in a sum of weight spaces of V corresponding to the weights $\Lambda - Q$ where Q is a sum of k positive roots. Thus

$$\pi(\exp X)v = v + \sum_{k=1} \frac{1}{k!}\pi(X)^k v = v + \sum v_{\Lambda-Q}$$

with $v_{\Lambda-Q}$ in the $\Lambda - Q$ weight space for π. Thus

$$\pi(a \exp Xa^{-1})v = e^{-\Lambda(\log a)}\pi(a \exp X)v = v + \sum e^{-Q(\log a)}v_{\Lambda-Q}.$$

Now $\pi(\tilde{G}) \subset G_C$ and $\pi(\tilde{K}) \subset G_u$ (see 8.5.8 for notation). $(\ ,\ )$ may thus be taken to be a G_u-invariant inner product. But $\exp(\sqrt{-1}\pi(\mathfrak{a})) \subset G_u$. Thus the $v_{\Lambda-Q}$ are mutually orthogonal (Schur orthogonality). Thus we see that

$$e^{2\Lambda(H(a \exp Xa^{-1}))}(v, v) = (v, v) + \sum e^{-2Q(\log a)}(v_{\Lambda-Q}, v_{\Lambda-Q}),$$

$$e^{2\Lambda(H(\exp X))}(v, v) = (v, v) + \sum (v_{\Lambda-Q}, v_{\Lambda-Q}).$$

Since $Q(h) \geqslant 0$ for all Q, we see that $e^{2\Lambda(H(a \exp Xa^{-1}))} \leqslant e^{2\Lambda(H(\exp X))}$. Since $\exp\colon \bar{\mathfrak{n}} \to \bar{N}$ is surjective we see that $\Lambda(H(\bar{n}) - H(a\bar{n}a^{-1})) \geqslant 0$ for all $\bar{n}$ in $\bar{N}$ and a as above. Also it is clear that $(v, v) + \sum (v_{\Lambda-Q}, v_{\Lambda-Q}) \geqslant (v, v)$, thus $\Lambda(H(\bar{n})) \geqslant 0$.

If λ is in Λ^+ then $\lambda = \alpha|_{\mathfrak{a}}$ for some α in Σ. Let $\Lambda_1, \ldots, \Lambda_l$ be the basic highest weights, i.e., $2\Lambda_i(H_{\alpha_j})/\alpha_j(H_{\alpha_j}) = \delta_{ij}$ where $\{\alpha_1, \ldots, \alpha_l\}$ are the simple roots of Δ^+. Let $\alpha_1, \ldots, \alpha_{l_0}$ be so that $\sigma\alpha_i \neq -\alpha_i$. Set $\lambda_i = \frac{1}{2}(\alpha_i + \sigma\alpha_i)$, $i = 1, \ldots, l_0$. Then every λ in Λ^+ is a nonnegative integral combination of the λ_i. Also, $\sigma\alpha_i = \sum_{j=1}^{l} c_j^i\alpha_j$ with c_i^j a nonnegative integer. $\sigma^2\alpha_i = \sum_j c_j^k c_i^j \alpha_k = \alpha_i$, $i = 1, \ldots, l_0$. Thus since the c_i^j are nonnegative integers, we see that at most one of the c_i^j is nonzero for $1 \leqslant j \leqslant l_0$. Thus $\sigma\alpha_i = \alpha_{i'} + \sum_{j=r_0+1}^{l} m_i^j\alpha_j$. Let $\alpha_1, \ldots, \alpha_{r_i}$ be so that $\sigma\alpha_i = \alpha_i$, $i = 1, \ldots, r$ and $\alpha_{r_1+1}, \ldots, \alpha_{r_1+2r_2}$ $(r_1 + 2r_2 = l_0)$ be so that $\sigma\alpha_i \neq \alpha_i$. We may assume that $(r + s)' = r + r_1 + s$. Let $\delta_i = \frac{1}{2}(\Lambda_i + \sigma\Lambda_i)$ for

$$i = 1, \ldots, r_1 + r_2(r_1 + r_2 = \dim \mathfrak{a}).$$

Now $\delta_i(\lambda_j) = \Lambda_i(\lambda_j) = \frac{1}{2}\alpha_j(H_{\alpha_j})\delta_{ij}$ for $i, j = 1, \ldots, r_1 + r_2$. Thus if λ is in ${}^+\Lambda$, then $\lambda = \sum c_i\delta_i$ with $c_i \geqslant 0$. Now $\delta_i = \Lambda_i|_{\mathfrak{a}}$. We have seen that

$$2\Lambda_i(H(\bar{n}) - H(a\bar{n}a^{-1})) \geqslant 0$$

and $2\Lambda_i(H(\bar{n})) \geqslant 0$ for all $\bar{n}$ in $\bar{N}$ and a as above. This proves the lemma.

8.13.8 We now conclude the proof of 8.13.6. 8.13.6(5) can be rewritten

$$(6)\qquad \int_{\bar{N}} e^{-(\nu-2\rho)(H(\bar{n}))-H(a_t\bar{n}a_t^{-1}))}e^{-2\rho(H(\bar{n}))}f(k(a_t\bar{n}a_t^{-1})M)d\bar{n} = 0$$

for all t in R.

Now applying 8.13.7 we see that $|e^{-(\nu-2\rho)(H(\bar{n})-H(a_t\bar{n}a_t^{-1}))}| \leqslant 1$ for all $t > 0$ and $\bar{n}$ in $\bar{N}$. The Lebesque dominated convergence theorem implies that

$$(7)\qquad 0 = \lim_{\substack{t\to\infty\\ t>0}} \int_{\bar{N}} e^{-(\nu-2\rho)(H(\bar{n})-H(a_t\bar{n}a_t^{-1}))}e^{-2\rho(H(\bar{n}))}f(k(a_t\bar{n}a_t^{-1})M)d\bar{n}$$

$$= \int_{\bar{N}} \lim_{\substack{t\to\infty\\ t>0}} e^{-(\nu-2\rho)(H(\bar{n})-H(a_t\bar{n}a_t^{-1}))}e^{-2\rho(H(\bar{n}))}f(k(a_t\bar{n}a_t^{-1})M)d\bar{n}.$$

Now $\lim_{t+\infty, t>0} a_t\bar{n}a_t^{-1} = e$. We therefore find

(8) $$0 = \left(\int_{\bar{N}} e^{-\nu(H(\bar{n}))} d\bar{n}\right) f(eM) = c(\nu) f(eM).$$

8.10.16 now implies $f(eM) = 0$. The result now follows.

8.13.9 *Lemma* Let ν be in $\mathfrak{a}^*$ and let ξ be in $\hat{M}$. Suppose that 1_ν is a cyclic vector for $\tilde{H}^{1,\nu}$. Let λ be in $\mathfrak{a}_C^*$ such that $V_{\xi,\lambda} \neq (0)$ (see 8.5.6). If f is in $V_{\xi,\lambda}^N$ (see 8.5 for notation), $f \neq 0$, then $1_\nu \cdot f$ is a cyclic vector for $\tilde{H}^{\xi,\lambda+\nu}$.

PROOF We note that $\bar{N}AK = G$ by 7.4.3. .Also $\tilde{\pi}_{\xi,\lambda}(a)f = e^{\lambda(\log a)}f$. Thus since $\tilde{\pi}_{1,\nu}(G)1_\nu = \tilde{\pi}_{1,\nu}(\bar{N}A)1_\nu$. We see that the cyclic space for $1_\nu f$ in $\tilde{H}^{\xi,\lambda+\nu}$ contains $(\tilde{\pi}_{1,\nu}(G)1_\nu)f$. Differentiating we see that the cyclic space contains $(\tilde{\pi}_{1,\nu}(U(\mathfrak{g}))1_\nu)f$.

Now $\tilde{\pi}_{1,\nu}(U(\mathfrak{g})1_\nu)|_K$ is uniformly dense in $C(K/M)$ since 1_ν is cyclic. We therefore see that if V is the cyclic space for $1_\nu f$, $V|_K$ contains $C(K/M)\cdot f|_K$. But then $V|_K$ contains $C(K/M) \cdot \pi_\xi(K)f|_K$. But then

$$V|_K \supset C(K;\xi) = \{h\colon K \to H_\xi | h(km) = \xi(m)^{-1}h(k) \text{ for } k \text{ in } K,\ m \text{ in } M\}.$$

This implies that V is dense in $\tilde{H}^{\xi,\lambda+\nu}$. The result is now proved.

8.13.10 We now prove 8.13.3. Let ξ be in $\hat{M}$, ν in $\mathfrak{a}_C^*$. There is μ in $\mathfrak{a}^*$ so that $V_{\xi,\mu} \neq (0)$ (see 8.5). Hence $V_{\xi,\mu-2k\rho} \neq (0)$ for all $k \geq 0$, k an integer. Now $1_{\nu-\mu+2k\rho} \cdot V_{\xi,\mu-2k\rho} \subset \tilde{H}^{\xi,\nu}$. Also $1_{2\rho-\bar{\nu}-\mu+2k\rho} \cdot V_{\xi,\mu-2k\rho} \subset \tilde{H}^{\xi,2\rho-\bar{\nu}}$. Let $k > 0$ an integer be so large that

$$\operatorname{Re}(\nu - \mu + 2k\rho)(H_\lambda) > 0$$

$$\operatorname{Re}(\bar{\nu} - \mu + 2k\rho)(H) > 0$$

for all λ in Λ^+. If f is in $V_{k,\mu-2(k+1)\rho}^N$ then 8.13.9 implies that $1_{\nu-\mu+2(k+1)\rho}f$ is a cyclic vector for $\tilde{H}^{\xi,\nu}$ and $1_{2\rho-\bar{\nu}-\mu+2(k+1)\rho}f$ is a cyclic vector for $\tilde{H}^{\xi,2\rho-\bar{\nu}}$.

Let V_1 be the smallest closed K-invariant subspace of $\tilde{H}^{\xi,\nu}$ containing $1_{\nu-\mu+2(k+1)\rho}f$ and let V_2 be the smallest closed K-invariant subspace of $\tilde{H}^{\xi,2\rho-\bar{\nu}}$ containing $1_{2\rho-\bar{\nu}-\mu+2(k+1)}f$. We note that $\dim V_1 = \dim V_2 < \infty$ and as K representations V_1 and V_2 are equivalent.

As a representation of K, $V_i = \sum_{\gamma\in\hat{K}} m(\gamma)V_\gamma$. Let

$$\tilde{V}_i = \sum_{\substack{\gamma\in k \\ m(\gamma)\neq 0}} m_\xi(\gamma)v \;(m_\xi(\gamma) = \dim \operatorname{Hom}_M(V_\gamma, H_\xi)).$$

(1) If U is a nonzero closed invariant subspace of $\tilde{H}^{\xi,\nu}$ then $U \cap \tilde{V}_1 \neq (0)$. In fact if $U \cap \tilde{V}_1 = (0)$ then as a K representation U contains no V_γ such

that $m(\gamma) \neq 0$. Thus U is orthogonal to $\tilde{V}_1$. This implies that $U^\perp$ in $\tilde{H}^{\xi,2\rho-\bar{\nu}}$ contains $\tilde{V}_2$. But $\tilde{V}_2$ contains a cyclic vector for $\tilde{H}^{\xi,2\rho-\bar{\nu}}$. This implies $U^\perp = \tilde{H}^{\xi,2\rho-\bar{\nu}}$. Hence $U = (0)$, a contradiction.

By the above $\tilde{H}_F^{\xi,\nu}$ and $\tilde{H}_F^{\xi,2\rho-\bar{\nu}}$ are finitely generated as representations of $U(\mathfrak{g}_C)$. Hence (see Jacobson [2], p. 166) $\tilde{H}_F^{\xi,\nu}$ and $\tilde{H}_F^{\xi,2\rho-\bar{\nu}}$ satisfy the ascending chain condition as representations of $U(\mathfrak{g}_C)$. But the duality between $\tilde{H}_F^{\xi,\nu}$ and $\tilde{H}_F^{\xi,2\rho-\bar{\nu}}$ as representations of $U(\mathfrak{g}_C)$ implies that $\tilde{H}_F^{\xi,\nu}$ satisfies the descending chain condition. This implies that $\tilde{H}^{\xi,\nu}$ has a composition series as a representation of $U(\mathfrak{g}_C)$. Closing the composition series constituents of $\tilde{H}_F^{\xi,\nu}$ gives a composition series for $\tilde{H}^{\xi,\nu}$.

8.13.11 *Corollary* (*to the proof of 8.13.3*) Let ξ be in $\hat{M}$ and let ν be in $\mathfrak{a}^*$. Then there is a finite subset $F_{\nu,\xi}$ of $\hat{K}_\xi = \{\gamma \text{ in } \hat{K} | m_\xi(\gamma) \neq 0\}$ and an open neighborhood U of ν in $\mathfrak{a}^*$ so that if ω is in $U + \sqrt{-1}\mathfrak{a}^*$ and if V is a closed invariant subspace of $\tilde{H}^{\xi,\omega}$ such that $V \supset \sum_{\gamma \in F_{\xi,\nu}} \tilde{H}_\gamma^{\xi,\omega}$ then $V = \tilde{H}^{\xi,\omega}$.

If W is a closed invariant subspace of $\tilde{H}^{\xi,\omega}$ (ω in $U + \sqrt{-1}\mathfrak{a}^*$) and $W \cap \sum_{\gamma \in F_{\xi,\nu}} \tilde{H}_\gamma^{\xi,\omega} = (0)$ then $W = (0)$.

PROOF Let $\mu \in \mathfrak{a}^*$ be such that $V_{\xi,\mu} \neq 0$. Let k be so large that $(\nu - \mu + 2k\rho)(H_\lambda) > 0$ for all λ in Λ^+. Let

$$U = \{\omega \in \mathfrak{a}^* | (\omega - \mu + 2k\rho)(H_\lambda) > 0 \text{ for all } \lambda \text{ in } \Lambda^+\}.$$

Let $F_{\nu,\xi} = \{\gamma \in \hat{K} | \mathrm{Hom}_K(V_{\xi,\mu-2(k+1)\rho}, V_\gamma) \neq 0\}$. The result now follows from the proof of 8.13.3.

8.13.12 *Theorem* Let ξ be in $\hat{M}$. Then the set of all ν in $\mathfrak{a}_C^*$ such that $\tilde{H}^{\xi,\nu}$ is reducible is contained in a countable union of nontrivial algebraic complex hypersurfaces of $\mathfrak{a}_C^*$. In particular, if $\mathfrak{a}_\xi^* = \{\nu \text{ in } \mathfrak{a}^* | (\pi_{\xi,\nu}, H^{\xi,\nu}) \text{ is not irreducible}\}$ then $\mathfrak{a}^*$ has measure zero in $\mathfrak{a}^*$ relative to Lebesque measure.

PROOF Let ν be in $\mathfrak{a}^*$. Let U and $F_{\xi,\nu}$ be as in 8.13.11. Let for γ in $\hat{K}$, $E_\gamma\colon \tilde{H}^{\xi,\omega} \to \tilde{H}_\gamma^{\xi,\omega}$ be the K-invariant orthogonal projection. Let ω be in $\mathfrak{a}_C^*$ and let γ_1, γ_2 be in $F_{\xi,\nu}$. Let $\mathscr{H}_{\gamma_1,\gamma_2}(\omega)$ be the space of linear maps of $\tilde{H}_{\gamma_1}^{\xi,\omega}$ to $\tilde{H}_{\gamma_2}^{\xi,\omega}$ spanned by the set $\{E_{\gamma_2}\tilde{\pi}_{\xi,\omega}(g)|_{\tilde{H}_{\gamma_1}^{\xi,\omega},F} g \text{ in } G\}$. Clearly, 8.13.13 implies that if ω is in $U + \sqrt{-1}\mathfrak{a}^*$, $\tilde{H}^{\xi,\omega}$ is irreducible if and only if $\mathscr{H}_{\gamma_1,\gamma_2}(\omega) = L(\tilde{H}_{\gamma_1}^{\xi,\omega}, \tilde{H}_{\gamma_2}^{\xi,\omega})$ for all γ_1, γ_2 in $F_{\xi,\nu}$. The result will be proved if we can show that there is δ in $\mathfrak{a}_C^*$ so that $\mathscr{H}_{\gamma_1,\gamma_2}(\delta) = L(\tilde{H}_{\gamma_1}^{\xi,\delta}, \tilde{H}_{\gamma_2}^{\xi,\delta})$ for all $\gamma_1, \gamma_2 \in F_{\xi,\nu}$. But 8.5.15 implies that if μ in $\mathfrak{a}^*$ is such that $V_{\xi,\mu} \neq (0)$ and if k is sufficiently large than $V_{\xi,\mu-2k\rho}$ contains each γ in $F_{\xi,\nu}$ with multiplicity $m(\xi, \gamma)$.

Since $V_{\xi,\mu-2k\rho}$ is irreducible we see that if $\delta = \mu - 2k\rho$ then δ satisfies the required conditions.

8.14 The Normalization of the Intertwining Operators

8.14.1 We retain the notation of the previous sections and assume that G is a connected semisimple Lie group with a faithful finite dimensional representation.

8.14.2 Let ξ be in $\hat{M}$ and s in W. Let ν be in $\mathfrak{a}^*$. Then

$$A(\xi, \nu, s)\colon \tilde{H}_F^{\xi s,\nu} \to \tilde{H}_F^{\xi s, s(\nu-\rho)+\rho}.$$

Hence $A(\xi^s, s(\nu - \rho) + \rho, s^{-1})A(\xi, \nu, s)\colon \tilde{H}_F^{\xi,\nu} \to \tilde{H}_F^{\xi,\nu}$ is an intertwining operator. Since 8.13.13 implies that almost all $\tilde{H}^{\xi,\nu}$ are infinitesimally irreducible we see that $A(\xi^s, s(\nu - \rho) + \rho, s^{-1})A(\xi, \nu, s)$ is a scalar for almost all ν. Since the work of 8.12 implies that for each γ in $\hat{K}$, $\nu \to A(\xi, \nu, s)|_{\tilde{H}_\gamma{}^{\xi,\nu}}$ is a meromorphic function from $\mathfrak{a}_C^*$ to $L(\tilde{H}_\gamma^{\xi,\nu}, H^{\xi^s, s(\nu-\rho)+\rho})$, we see that $A(\xi^s, s(\nu - \rho) + \rho, s^{-1})A(\xi, \nu, s) = C(\xi, \nu, s)I$ for all $\nu \in \mathfrak{a}_C^*$ with $C(\xi, \nu, s)$ a meromorphic function of ν.

8.14.3 Now 8.12.6 and 8.10.12 combined imply that $A(\xi, \nu, s)$ is not identically zero for each ξ, s. Thus $C(\xi, \nu, s)$ is not identically zero.

8.14.4 Let s, t, u be in $W(A)$, $s = tu$, $l(s) = l(t) + l(u)$. Then

$$A(\xi, \nu, s) = b(t, u)A(\xi^u, u(\nu - \rho) + \rho, t)A(\xi, \nu, u)$$

according to 8.10.12. Also $s^{-1} = u^{-1}t^{-1}$. Clearly $l(s^{-1}) = l(u^{-1}) + l(t^{-1})$. Hence

$$\begin{aligned} &A(\xi^s, s(\nu - \rho) + \rho, s^{-1}) \\ &\quad = b(u^{-1}, t^{-1})A(\xi^u, u(\nu - \rho) + \rho, u^{-1})A(\xi^s, s(\nu - \rho) + \rho, t^{-1}). \end{aligned}$$

Thus

$$C(\xi, \nu, s) = b(t, u)b(u^{-1}, t^{-1})C(\xi^u, u(\nu - \rho) + \rho, t)C(\xi, \nu, u).$$

8.14.5 8.14.4 implies that there is a product formula for the $C(\xi, \nu, s)$ analogous to that for the $A(\xi, \nu, s)$ over the minimal product of simple reflections that give s.

8.14.6 In order to simplify our computation of $C(\xi, \nu, s)$ we make the assumption that if ξ is in $\hat{M}$ there is γ in $\hat{K}$ so that $\dim \operatorname{Hom}_M(V_\gamma, H_\xi) = 1$. This is not an inordinately restrictive assumption. Indeed, it seems quite likely that this is true for semisimple linear Lie groups. We show that this is true at least for semisimple Lie groups with one conjugacy class of Cartan subalgebra.

8.14.7 *Lemma* If G has one conjugacy class of Cartan subalgebra and if ξ is in $\hat{M}$ then there exists γ in $\hat{K}$ so that $\dim \operatorname{Hom}_M(V_\gamma, H_\xi) = 1$.

PROOF Let $T \subset M$ be a maximal torus of M. 7.8.5 implies that T is a maximal torus of K. Let $W(T)$ be the Weyl group of (K, T). Let Δ_K be the root system of K relative to T and let Δ_M be the root system of M relative to T. Then $\Delta_K \supset \Delta_M$. Let Δ_K^+ be a system of positive roots for Δ_K. Then $\Delta_K^+ \cap \Delta_M$. Δ_M^+ is a system of positive roots for Δ_M. Let Λ be the highest weight of (ξ, H_ξ) relative to Δ_M^+. There is s in $W(T)$ so that $s \cdot \Lambda$ is a dominant integral form for (K, T, Δ^+). Let (π, H) be the irreducible representation of K with highest weight $s \cdot \Lambda$. Then Λ is the highest weight of (π, H) relative to the system of positive roots $s \cdot \Delta_K^+$ for K. Thus the weight space H_Λ of (π, H) relative to T is one-dimensional.

Now $\mathfrak{m}_C = \mathfrak{h}^- + \sum_{\alpha \in \Delta_M} (\mathfrak{g}_C)_\alpha$. If α is in Δ_M then one of $\Lambda + \alpha$ and $\Lambda - \alpha$ is not a weight of (π, H) relative to T. If α is in Δ_M^+ and $\Lambda + \alpha$ is a weight, then since $\Lambda(H_\alpha) \geqslant 0$ the proof of 4.5.3 implies that $\Lambda - \alpha$ is a weight. Thus $\Lambda + \alpha$ is not a weight for each α in Δ_M. This implies that the M-cyclic space of the Λ weight space of (π, H) is the irreducible representation of M with highest weight Λ. This clearly implies that, if $\gamma = [(\pi, H)]$, $\dim_M(V_\gamma, H_\xi) = 1$.

8.14.8 Returning to our assumption in 8.14.7 we note that if we choose γ_ξ in K so that $\dim \operatorname{Hom}_M(V_{\gamma_\xi}, H_\xi) = 1$ then we may choose γ_{ξ^s} to equal γ_ξ. In fact if s^* is a representative of s and (π_γ, V_γ) is a representative of γ in $\hat{K}$ and if A is in $\operatorname{Hom}_M(V_\gamma, H_\xi)$ then $A \cdot \pi_\gamma(s^*)$ is in $\operatorname{Hom}_M(V_\gamma, H_{\xi^s})$.

8.14.9 Fix for each ξ in $\hat{M}$, γ_ξ in $\hat{K}$ so that $\gamma_{\xi^s} = \gamma_\xi$ for s in $W(A)$ and such that $\dim \operatorname{Hom}_M(V_{\gamma_\xi}, H_\xi) = 1$.

8.14.10 Let ξ be in $\hat{M}$, ν in $\mathfrak{a}_C^*$, then

$$A(\xi, \nu, s)\colon \tilde{H}_{\gamma_\xi}^{\xi,\nu} \to \tilde{H}_{\gamma_\xi}^{\xi^s, s(\nu-\rho)+\rho} = \tilde{H}_{\gamma_{\xi^s}}^{\xi^s, s(\nu-\rho)+\rho}.$$

Since $A(\xi, \nu, s)$ commutes with the action of K and since the assumption on γ_ξ implies that $\tilde{H}_{\gamma_\xi}^{\xi,\nu}$ and $\tilde{H}_{\gamma_\xi}^{\xi^s, s(\nu-\rho)+\rho}$ are irreducible as representations of K we see that $A(\xi, \nu, s)|_{H_{\gamma_\xi}^{\xi,\nu}} = d(\xi, \nu, s)I$, with $d(\xi, \nu, s)$ a meromorphic function of ν.

8.14.11 It is now obvious that $C(\xi, \nu, s) = d(\xi^s, s(\nu - \rho) + \rho, s^{-1})d(\xi, \nu, s)$. Thus if we see $\mathscr{A}(\xi, \nu, s) = d(\xi, \nu, s)^{-1}A(\xi, \nu, s)$ then $\nu \mapsto \mathscr{A}(\xi, \nu, s)|_{\tilde{H}_\gamma^{\xi,\nu}}$ is meromorphic in ν and $\mathscr{A}(\xi^s, s(\nu - \rho) + \rho, s^{-1})\mathscr{A}(\xi, \nu, s) = I$.

8.14.12 *Theorem* Let G satisfy the condition of 8.14.6. Let for ξ in $\hat{M}$, ν in $\mathfrak{a}_C^*$, s in $W(A)$, $\mathscr{A}(\xi, \nu, s)$ be as in 8.14.11. If $\operatorname{Re}\nu = \rho$ (i.e., $\tilde{H}^{\xi,\nu}$ is unitary) then $\mathscr{A}(\xi, \nu, s)$ extends to a unitary operator from $\tilde{H}^{\xi,\nu}$ to $\tilde{H}^{\xi^s, s(\nu-\rho)+\rho}$.

PROOF We have seen that for all but a set of measure zero of $\mathfrak{a}^*$ the representation $(\tilde{\pi}_{\xi,\rho+i\nu}, \tilde{H}^{\xi,\rho+i\nu})$ is irreducible. Thus $A(\xi, \rho + i\nu, s)$ is a scalar multiple of a unitary operator for ν off of a set of measure zero. But then the definition of $\mathscr{A}(\xi, \rho + i\nu, s)$ implies that $\mathscr{A}(\xi, \rho + i\nu, s)$ is unitary for ν in $\mathfrak{a}^*$.

8.14.13 It can be shown that the matrix entries of the $\mathscr{A}(\xi, \nu, s)|_{H_\gamma^{\xi,\nu}}$ for γ in $\hat{K}$ are actually rational functions of ν. We will develop the proof of this result in exercise 8.16.7.

8.14.14 We note that the $\mathscr{A}(\xi, \nu, s)$ for $\operatorname{Re}\nu = \rho$ give the unitary intertwining operators predicted by our computation of characters in 8.8.

8.15 The Plancherel Measure

8.15.1 Let G be a Lie group. Let $\hat{G}$ be the set of all equivalence classes of trace class, irreducible, unitary representations of G. Let for each ω in $\hat{G}$, Θ_ω be the character of a representative of ω. The Plancherel problem is to find, explicitly, a measure μ on $\hat{G}$ (the Plancherel measure) so that

$$f(e) = \int_{\hat{G}} \Theta_\omega(f)d\mu(\omega)$$

for all f in $C_0^\infty(G)$.

8.15.2 *Example* Let G be a compact Lie group. If γ is in $\hat{G}$ then if f is in $C^\infty(G)$, $\Theta_\gamma(f) = \int_G \chi_\gamma(g)f(g)dg$ where χ_γ is the character of γ in the usual sense. Then 5.7.10 implies that

$$f(e) = \sum_{\gamma \in G} d(\gamma)\Theta_\gamma(f).$$

Thus the Plancherel measure on $\hat{G}$ assigns to each γ in $\hat{G}$, $\mu(\gamma) = d(\gamma)$.

8.15.3 We now give a solution of the Plancherel problem for G a connected semisimple Lie group with one conjugacy class of Cartan subalgebra.

8.15.4 *Theorem* Let G be a connected semisimple Lie group with one conjugacy class of Cartan subalgebra. Then there is a nonzero (but not necessarily positive) normalization of $d\nu$, Euclidean measure on $\mathfrak{a}^*$ so that if f is in $C_0^\infty(G)$ then

$$f(e) = \sum_{\xi \in \hat{M}} \int_{\mathfrak{a}^*} \Theta_{\pi_{\xi,\nu}}(f)m(\xi, \nu)\, d\nu$$

with $m(\xi, \nu) = d(\xi)\prod_{\alpha \in \Sigma} (\Lambda_\xi + \rho_M + (-1)^{1/2}\nu)(H_\alpha)$ where the notation is as in 7.12.

PROOF In the notation of 7.12.4, $\Theta_{\pi_{\xi,\nu}} = \Theta_{\xi,\nu}$ according to 8.9.2. The result is thus a restatement 7.12.9.

8.15.5 We now assume that G has split rank 1 ($\dim \mathfrak{a} = 1$) and has one conjugacy class of Cartan subalgebra. This implies that G is either $SO(2n + 1, 1)$ or G is the twofold (simply connected) covering of $SO(2n + 1, 1)$, $\mathrm{Spin}(2n + 1, 1)$ (see 7.9.6(3)). We will show in the remainder of this section that if G is as we have assumed then $|m(\xi, \nu)| = |C(\xi, \rho + i\nu, s)|^{-1}$ (up to a constant) where s is the nontrivial element of the Weyl group of A and $C(\xi, \nu, s)$ is as in 8.13.

8.15.6 Setting $\nu = t\rho$ with t in R we set $P_\xi(t) = m(\xi, t\rho)c$ where c is the constant such that 8.15.4 says

(1) $$f(e) = \sum_{\xi\in\hat{M}} \int_{-\infty}^{\infty} \Theta_{\pi_{\xi,t\rho}}(f)P_\xi(t)dt.$$

Note that $P_\xi(t) \geqslant 0$.

The formula for $P_\xi(t)$ in 8.15.4 implies that $P_\xi(t) > 0$ if $t \neq 0$.

8.15.7 Let for ξ in $\hat{M}$, (π_ξ, H^ξ) denote the restriction of $(\pi_{\xi,\nu}, H^{\xi,\nu})$ to K. That is, $\pi_\xi = \pi_{\xi,\nu}|_K$ and H^ξ is the underlying Hilbert space of $H^{\xi,\nu}$ which we have seen in 8.3 to be independent of ν. Let $\pi_{\xi,t} = \pi_{\xi,t\rho}$ and $H^{\xi,t} = H^{\xi,t\rho}$ for t in C.

8.15.8 Suppose that H^ξ_γ and $H^\xi_\tau \neq 0$. Let $t \mapsto B(t)$ be a C^∞ mapping, with compact support, of R into $L(H^\xi_\tau, H^\xi_\gamma)$. We look at $B(t)$ as a linear operator on H^ξ by the formula $B(t)\varphi = E_\tau B(t)(E_\gamma\varphi)$. Set

$$f(g) = \int_{-\infty}^{\infty} \operatorname{tr}(B(t)\pi_{\xi,t}(g)^{-1})P_\xi(t)dt.$$

We note that $\operatorname{tr} B(t)\,\pi_{\xi,t}(g)^{-1} = \operatorname{tr} E_\tau B(t)E_\gamma \pi_{\xi,t}(g)^{-1}$ which exists by 8.8.2 and equals $\operatorname{tr}(E_\tau B(t)E_\gamma)(E_\gamma \pi_{\xi,t}(g)^{-1}E_\tau)$. Thus for fixed g, $t \to \operatorname{tr}(B(t)\pi_{\xi,t}(g)^{-1})$ is continuous and compactly supported. f is therefore a continuous function on G.

Let h be in $C_0^\infty(G)$. We compute

$$\begin{aligned}\int_G f(x)\overline{h(x)}dx &= \int_G \int_{-\infty}^{\infty} \operatorname{tr}(B(t)\pi_{\xi,t}(g)^{-1})P_\xi(t)dt\,\overline{h(g)}dg \\ &= \int_{-\infty}^{\infty} \int_G \operatorname{tr}(B(t)(\pi_{\xi,t}(g)^{-1}\overline{h(g)}))dg\,P_\xi(t)dt \\ &= \int_{-\infty}^{\infty} \operatorname{tr}(B(t)(\pi_{\xi,t}(h))^*)P_\xi(t)dt.\end{aligned}$$

Now $\operatorname{tr} B(t)\pi_{\xi,t}(h)^* \leqslant C\sqrt{\operatorname{tr} \pi_{\xi,t}(h)\pi_{\xi,t}(h)^*}$ since B is compactly supported. Now

$$\sum_{\xi\in\hat{M}} \int_{-\infty}^{\infty} \operatorname{tr}(\pi_{\xi,t}(h)\pi_{\xi,t}(h)^*)P_\xi(t)dt = \int_G |h(x)|^2 dx$$

(This follows since if we define $h^*(x) = \overline{h(x^{-1})}$ and $h*h^*(g) = \int_G h(x)h^*(x^{-1}g)dx$ then $\pi_{\xi,t}(h*h^*) = \pi_{\xi,t}(h)\pi_{\xi,t}(h)^*$; the rest now follows from the Plancherel theorem.) We therefore see that

$$\left|\int_G f(x)\overline{h(x)}dx\right| \leqslant C\|h\| = C\left(\int_G |h(x)|^2 dx\right)^{1/2}.$$

This implies that f agrees with a function in $L^2(G)$ almost everywhere on G. Arguing as above for h we now find that $\int_G |f(x)|^2 dx = \int_{-\infty}^{\infty} \operatorname{tr}(B(t)B(t)^*)P_\xi(t)dt$. We have proved the following lemma.

8.15.9 *Lemma* Let ξ be in $\hat{M}$ and γ, τ in $\hat{K}$. Let $t \mapsto B(t)$ be an element of $C_0^\infty(R; L(H_\gamma^\xi, H_\tau^\xi))$. Let $f(g) = \int_{-\infty}^{\infty} \operatorname{tr}(B(t)\pi_{\xi,t}(g)^{-1})dt$. Then f is in $L^2(G)$ and

$$\int_G |f(x)|^2 dx = \int_{-\infty}^{\infty} \operatorname{tr}(B(t)B(t)^*)P_\xi(t)dt.$$

8.15.10 *Lemma* Let φ be in $C_0^\infty(R - (-\varepsilon, \varepsilon))$ for some $\varepsilon > 0$. Let ξ be in $\hat{M}$ and γ in $\hat{K}$. Let $B_\gamma^\xi(\varphi)(g) = \int_{-\infty}^{\infty} \varphi(t)E_\gamma \pi_{\xi,t}(g)^{-1}E_\gamma dt$. Then

$$\int_G \operatorname{tr}(B_\gamma^\xi(g)(B_\gamma^\xi(g))^*)dg = \dim(H_\gamma^\xi)^2 \int_{-\infty}^{\infty} |\varphi(t)|^2 P_\xi(t)^{-1} dt.$$

PROOF Let $v_1, \ldots, v_n$ be an orthonormal basis of H_γ^ξ. Let $E_{ij}v_k = \delta_{jk}v_i$. Set $B_{ij}(t) = \varphi(t)P_\xi(t)^{-1}E_{ij}$. Then B_{ij} satisfies the hypothesis of 8.15.9. Thus there is a function f_{ij} in $L^2(G)$ so that

$$f_{ij}(g) = \int_{-\infty}^{\infty} \operatorname{tr}(B_{ij}(t)\pi_{\xi,t}(g)^{-1})dt$$

and

$$\int_G |f_{ij}(g)|^2 dg = \int_{-\infty}^{\infty} |\varphi(t)|^2 P_\xi(t)^{-1} dt.$$

Now

$$\begin{aligned} B_v(\varphi)(g) &= \int_{-\infty}^{\infty} \varphi(t)E_\gamma \pi_{\xi,t}(g)^{-1}E_\gamma dt \\ &= \int_{-\infty}^{\infty} \varphi(t)E_\gamma \pi_{\xi,t}(g)^{-1}E_\gamma P_\xi(t)^{-1}P_\xi(t)dt \text{ (since 0 is not in supp } \varphi) \\ &= \sum_{i,j=1} \left(\int_{-\infty}^{\infty} \operatorname{tr}(B_{ij}(t)\pi_{\xi,t}(g)^{-1})dt \right) E_{ij} = \sum_{i,j=1}^{n} f_{ij}(g)E_{ij}. \end{aligned}$$

Thus

$$\int_G \operatorname{tr}(B_\gamma^\xi(\varphi)B_\gamma^\xi(\varphi)^*(g))dg$$

$$= \sum_{i,j=1}^{n} \int_G |f_{ij}(g)|^2 dg = n^2 \int_{-\infty}^{\infty} |\varphi(t)|^2 P_\xi(t)^{-1} dt.$$

8.15.11 *Lemma* There is a normalization of dg so that if f is in $C_0(G)$

then

$$\int_G f(g)dg = \int_0^\infty (\sinh t)^q \int_{K\times K} f(k_1 \exp tHk_2)\, dk_1\, dk_2\, dt$$

($q = \dim N$, $\lambda(H) = 1$, λ the positive root of $\mathfrak{a}$).

PROOF We will develop a proof of the most general form of this lemma in exercises 8.16.4, 5, 6.

8.15.12 *Theorem* Let G be split rank 1 with one conjugacy class of Cartan subalgebra (i.e., $G = SO(2n + 1, 1)$ or $\mathrm{Spin}(2n + 1, 1)$). Then there is a positive constant c so that $|d(\xi, (1 + it)\rho, s)|^2 = cp(t)^{-1}$. Here s is the nontrivial element of $W(A)$ and $d(\xi, \nu, s)$ is as in 8.14.10.

PROOF Let φ be in $C_0(R - (-\varepsilon, \varepsilon))$ for some $\varepsilon > 0$. Let $B_\gamma^\xi(\varphi)$ be as in 8.15.10. Then

$$\int_G \mathrm{tr}(B_\gamma^\xi(\varphi)(g)B_\gamma^\xi(\varphi)(g)^*)dg = \int_{-\infty}^{\infty} (\sinh t)^q \int_{K\times K} \mathrm{tr}(B_\gamma^\xi(\varphi)(k_1 \exp tHk_2) \cdot B_\gamma^\xi(\varphi)(k_1 \exp tHk_2)^*)dk_1 dk_2 dt$$

($q = \dim N = 2\rho(H_\lambda)/\lambda(H_\lambda)$ where λ is the positive root of $\mathfrak{a}$). Now $B_\gamma^\xi(\varphi)(k_1 g k_2) = \pi_{\xi,t}(k_2)^{-1} B_\gamma^\xi(\varphi)(g)\pi_{\xi,t}(k_1)^{-1}$. Hence

$$(1)\qquad \int_G \mathrm{tr}(B_\gamma^\xi(\varphi)(g)(B_\gamma^\xi(\varphi)(g))^*)dg = \int_0^\infty (\sinh t)^q\, \mathrm{tr}(B_\gamma^\xi(\varphi)(\exp tH))(B_\gamma^\xi(\varphi)(\exp tH))^*)dt.$$

Now Let A, B be in $\mathrm{Hom}_M(V_\gamma, H_\xi)$ and let v, u be in V_γ. Let $f_{A,v}(k) = A(\pi_\gamma(k)^{-1}v)$, $f_{B,u}(k) = B(\pi_\gamma(k)^{-1}u)$. Then

$$\begin{aligned}
&\langle B_\gamma^\xi(\varphi)(g)f_{A,v}, f_{B,u}\rangle \\
&= \int_{-\infty}^{\infty} \varphi(t)\, \langle \tilde{\pi}_{\xi,(1+it)\rho}(g)^{-1} f_{A,v}, f_{B,u}\rangle\, dt \\
&= \int_{-\infty}^{\infty} \varphi(t) \int_K e^{(1+it)\rho(H(gk)}\langle A(\pi_\gamma(k(gk))^{-1}v), B(\pi_\gamma(k)^{-1}u)\rangle dk\, dt \\
&= \int_{-\infty}^{\infty} \varphi(t) \left(\int_K e^{-(1+it)\rho(H(gk))}\langle v, \pi_\gamma(k(gk))A^*B\pi_\gamma(k)^{-1}u\rangle dk \right) dt \\
&= \int_{-\infty}^{\infty} \varphi(t)\langle v, E_{\gamma,\gamma}(A^*B\colon it\colon g)u\rangle dt
\end{aligned}$$

in the notation of 8.12.5. Thus

$$\langle B_\gamma^\xi(\varphi)(\exp sH)f_{A,v}, f_{B,u}\rangle$$

$$= \int_{-\infty}^{\infty} \varphi(t)e^{-(q/2)s}\langle v,(e^{-its}A^*BB_\gamma(-it) + e^{its}B_\gamma(it)^*\pi_\gamma(s^*)A^*B\pi_\gamma(s^*)^{-1}u\rangle dt$$

$$+ \int_{-\infty}^{\infty} \varphi(t)\langle v, R(A^*B, it, s)u\rangle dt$$

where $s^* \in s$ and 8.12.6 implies that $\|R(A^*B, it, s)\| \leqslant Ce^{-(q/2+\delta)s}$ for some $\delta > 0$.

Let $A_1, \ldots, A_m$ be an orthonormal basis of $\mathrm{Hom}_M(V_\gamma, H_\xi)$ and let $v_1, \ldots, v_\alpha$ be an orthonormal basis of V_γ. Then

$$\int_G \mathrm{tr}\, B_\gamma^\xi(\varphi)(g)B_\gamma^\xi(\varphi)(g)^*\, dg$$

$$= \int_0^\infty (\sinh s)^q\, \mathrm{tr}(B_\gamma^\xi(\varphi)(\exp sH)B_\gamma^\xi(\varphi)(\exp sH)^*)\, ds$$

$$= \sum_{i,j,k,l} \int_0^\infty (\sinh s)^q e^{-(q)s}\Bigg[\int_{-\infty}^{\infty} \varphi(t)\langle v_i, \{e^{-its}A_k^*A_lB_\gamma(-it)$$

$$+ e^{its}B_\gamma(it)^*\pi_\gamma(s^*)A_k^*A_l\pi_\gamma(s^*)\}v_j\rangle dt \cdot \int_{-\infty}^{\infty} \bar{\varphi}(t)\langle\{e^{its}A_k^*A_lB_\gamma(-it)$$

$$+ e^{-its}B_\gamma(it)^*\pi_\gamma(s^*)A_k^*A_l\pi_\gamma(s^*)\}v_j, v_i\rangle dt\Bigg] ds$$

$$+ \sum_{i,j,k,l} \int_0^\infty (\sinh s)^q\Bigg(\int_{-\infty}^{\infty} \varphi(t)\langle v_i, R(A_k^*A_k, it, s)v_j\rangle dt$$

$$\cdot \int_{-\infty}^{\infty} \varphi(t)\langle R(A_k^*A_l, it, s)v_j, v_i\rangle dt\Bigg) ds$$

$$+ 2\,\mathrm{Re} \sum_{i,j,k,l} \int_0 (\sinh s)^q e^{-(q/2)s}\Bigg[\int_{-\infty}^{\infty} \varphi(t)\langle v_i, \{e^{-its}A_k^*A_lB_l(-it)$$

$$+ e^{its}(B_\gamma(it)^*\pi_\gamma(s^*)A_k^*A_l\pi_\gamma(s^*)\}v_i\rangle dt$$

$$\cdot \int_{-\infty}^{\infty} \bar{\varphi}\langle R(A_k^*A_l, it, s)v_j, v_i\rangle dt\Bigg] ds = A(\varphi) + B(\varphi) + C(\varphi).$$

Let $|t_0| > \varepsilon$. Let $\{\varphi_k\}$ be an element of $C_0^\infty(R - \{0\})$ whose support is contained in $(-1/k + t_0, t_0 + 1/k)$ and is such that $\int_{-\infty}^{\infty} |\varphi_k(t)|^2 dt = 1$. We assert that $\lim_{k\to\infty} B(\varphi_k) = \lim_{k\to\infty} C(\varphi_k) = 0$. Indeed using Schwarz's inequality and the estimates on $R(A, it, s)$ we have

$$|B(\varphi_k)| \leqslant \mathrm{const}\frac{1}{k}\int_0^\infty (\sinh s)^q e^{-qs}e^{-\delta s} ds \leqslant (\mathrm{const})\frac{1}{k} \to 0 \text{ as } k \to \infty.$$

$C(\varphi_k)$ is handled similarly. This implies that

(2) $$\lim_{k\to\infty}\int_G \operatorname{tr}(B_\gamma^\xi(\varphi_k)(g)(B_\gamma^\xi(\varphi_k)(g))^*)dg = \lim_{k\to\infty} A(\varphi_k).$$

Now

$$\int_G \operatorname{tr}(B_\gamma^\xi(\varphi_k)(g))(B_\gamma^\xi(\varphi_k)(g))^*)dg = (\dim H_\gamma)^2 \int_{-\infty}^{\infty} |\varphi_k(t)|^2 P_\xi(t)^{-1}dt.$$

We assert that $\lim_{k\to\infty}\int_{-\infty}^{\infty} |\varphi_k(t)|^2 P_\xi(t)^{-1}dt = P_\xi(t_0)^{-1}$. Indeed this follows directly from the definition of the φ_k and the continuity of $P_\xi(t)$ on $|t| > \varepsilon$. Thus

(3) $$\lim_{k\to\infty} A(\varphi_k) = P_\xi(t_0)^{-1}(\dim H_\gamma^\xi)^2.$$

We now concentrate on $A(\varphi_l)$. Arguing as above

$$\begin{aligned}
&\lim_{m\to\infty} A(\varphi_m)\\
&\quad= \lim_{m\to\infty}\sum_{i,j,k,l}\int_0^\infty (\sinh s)^q\, e^{-(q)s}\Big[\int_{-\infty}^{\infty} \varphi_m(t)e^{-its}dt\langle v_i, A_k^*A_lB_\gamma(-it_0)v_j\rangle\\
&\qquad+\int_{-\infty}^{\infty} \varphi_m(t)e^{-its}dt\langle v_i, B_\gamma(it_0)^*\pi_\gamma(s^*)A_k^*A_r\pi_\gamma(s^*)v_j\rangle\Big]\\
&\qquad\cdot\Big[\int_{-\infty}^{\infty} \overline{\varphi_m(t)}e^{its}dt\langle A_k^*A_lB_\gamma(-it_0)v_j, v_i\rangle\\
&\qquad+\int_{-\infty}^{\infty} \overline{\varphi_m(t)}e^{-its}dt\langle B_\gamma(it_0)^*\pi_\gamma(s^*)A_k^*A_r\pi_\gamma(s^*)v_j, v_i\rangle\Big]ds.
\end{aligned}$$

Now $(\sinh s)^q e^{-(q)s} = (2)^{-(q)}$ + a sum of negative integral powers of e^s. Using the properties of the φ_l we may assume that (setting $\hat{\varphi}(s) = \int_{-\infty}^{\infty} \varphi(t)e^{-its}ds$)

$$\begin{aligned}
A(\varphi_l) &= \sum_{i,j,k,r}\int_0^\infty \hat{\varphi}_l(s)\overline{(\hat{\varphi}_l(s))}ds\\
&\quad\cdot|\langle v_i, A_k^*A_rB_\gamma(-it_0)v_j\rangle|^2\\
&\quad+\sum_{i,j,k,r}\int_0^\infty \hat{\varphi}_l(s)\overline{\hat{\varphi}_l(-s)})ds(\langle v_i, A_k^*A_rB_\gamma(-it_0)v_j\rangle\\
&\quad\cdot\langle B_\gamma(it_0)\pi_\gamma(s^*)A_k^*A_r\pi_\gamma(s^*)v_j, v_i\rangle)\\
&\quad+\sum_{i,j,k,r}\int_0^\infty \hat{\varphi}_l(-s)\overline{(\hat{\varphi}_l(-s)}\\
&\quad\cdot|\langle v_i, B_\gamma(it_0)^*\pi_\gamma(s^*)A_k^*A_r\pi_\gamma(s^*)v_j\rangle|^2ds\\
&\quad+\sum_{i,j,k,r}\int_0^\infty \hat{\varphi}_l(-s)\overline{(\hat{\varphi}_l(s))}\langle v_i, B_\gamma(it_0)^*\pi_\gamma(s^*)A_k^*A_r\pi_\gamma(s^*)v_j\rangle\\
&\quad\cdot\langle A_k^*A_rB_\gamma(-it_0)v_j, v_i\rangle ds.
\end{aligned}$$

Now 8.14.12 implies that

$$\sum_{i,j,k,r} |\langle v_i, A_k^* A_r(B_\gamma(-it_0)v_j)\rangle|^2 = (\dim H_\gamma^\xi)^2 |d(\xi, (1 + it)\rho, s)|^2.$$

Similarly

$$\sum_{i,j,k,r} |\langle v_i, B_\gamma(it_0)^* \pi_\gamma(s^*) A_k^* A_r \pi_\gamma(s^*) v_j\rangle|^2 = (\dim H_\gamma^\xi)^2 |d(\xi, (1 + it)\rho, s)|^2.$$

We assert that $\lim_{l\to\infty} \int_0 \hat{\varphi}_l(s)(\overline{\hat{\varphi}_l(-s)})ds = 0$. Now

$$2\int_0^\infty \hat{\varphi}_l(s)(\overline{\hat{\varphi}_l(-s)})ds$$

$$= \int_{-\infty}^\infty \hat{\varphi}_l(s)(\overline{\hat{\varphi}_l(-s)})ds + \int_{-\infty}^\infty (\operatorname{sign}(s))\ \hat{\varphi}_l(s)(\hat{\varphi}_l(-s))ds$$

$$= 2\pi(\varphi_l * \bar{\varphi}_l)(0) - 2\pi \int_{-\infty}^\infty \frac{(\varphi_l * \bar{\varphi}_l)(x)}{x}\, dx.$$

Now since $t_0 \neq 0$ we see that $\lim_{l\to\infty} (\varphi_l * \bar{\varphi}_l)(0) = 0$.

$$\operatorname{supp}(\varphi_l * \bar{\varphi}_l) \subset (t_0 - 1/l, t_0 + 1/l)$$

and we may assume that l is large enough that $[-\varepsilon, \varepsilon] \cap (t_0 - 1/l, t_0 + 1/l) = \phi$. Then

$$\left|\int_{-\infty}^\infty \frac{(\varphi_l * \varphi_l)(x)}{x}\, dx\right| \leqslant C \int_{-\infty}^\infty |\varphi_l * \varphi_l(x)|\, dx$$

(C can be taken to be sup $1/x$ on $(-1/l + t_0, t_0 + 1/l)) \leqslant C(\int_{-\infty}^\infty |\varphi_l(x)| dx)^2$. Now $(\int_{-\infty}^\infty |\bar{\varphi}_l(x)| dx)^2 \leqslant (\int_{-\infty}^\infty |\bar{\varphi}_l(x)|^2 dx) 2/l = 2/l$ by the Schwarz inequality. We therefore have $\lim_{l\to\infty} \int_0^\infty \hat{\varphi}_l(s)(\hat{\varphi}_l(-s))ds = 0$. We finally have

$$\lim_{l\to\infty} A(\varphi_l) = \lim_{l\to\infty} (\dim H_\gamma^\xi)^2 |d(\xi, (1 + it)\rho, s)|^2$$

$$\cdot\left(\int_0^\infty \hat{\varphi}_l(s)\overline{\hat{\varphi}_l(s)}ds + \int_0^\infty \hat{\varphi}_l(-s)\overline{\hat{\varphi}_l(-s)}ds\right)$$

$$= 2\pi\, (\dim H_\gamma^\xi)^2 |d(\xi, (1 + it)\rho, s)|^2$$

by the Plancherel theorem for R.

This implies that $|d(\xi, (1 + it)\rho, s)|^2 = CP_\xi(t)^{-1}$ for $t \neq 0$ where C is a positive constant. Q.E.D.

8.16 Exercises

8.16.1 Use exercise 7.13.1 to write out the "$\bar{\pi}$ realization" of the principal series of $SL(n, R)$ and $SL(n, C)$. (Hint: If g is in G express $m(g)$ and $a(g)$ in terms of the $\Delta_k(g)$.)

8.16.2 Let (π, H) be an irreducible K-finite representation of G (notation as in 8.2). Let H_F be the space of all K-finite vectors (see 8.6.3) in H. Let $H(N) = \pi(\mathfrak{n})H_F$. Suppose that $1 \leqslant \dim H_F/H(N) < \infty$. Show that (π, H) is infinitesimally equivalent with a subrepresentation of a (nonunitary) principal series representation.

8.16.3 Show that in the notation of 8.16.2, $\tilde{H}_F^{\xi,\nu}/\tilde{H}^{\xi,\nu}(N) \neq 0$. Use the intertwining operators of 8.14 to show that as an M representation $\tilde{H}_F^{\xi,\nu}/\tilde{H}^{\xi,\nu}(N) = \sum_{s\in W(A)} H_{\xi s}$, for most ν.

8.16.4 Let $\mathfrak{a}^+$ be a connected component of $\mathfrak{a}' = \{H \text{ in } \mathfrak{a} | \lambda(H) \neq 0 \text{ for all } \lambda \text{ in } \Lambda\}$. Let $\Lambda^+ = \{\lambda \text{ in } \Lambda | \lambda(H) > 0 \text{ for all } H \text{ in } \mathfrak{a}^+\}$. Let $A^+ = \exp \mathfrak{a}^+$. Let $\psi: K \times A^+ \times K/M \to G$ be defined by $\psi(k_1, a, kM) = k_1 k a k^{-1}$. Use 7.2.5 to show that ψ is injective and that $G - \operatorname{Im}\psi$ is a finite union of submanifolds of lower dimension.

8.16.5 We retain the notation of 8.16.4. Show that if dg is invariant measure on G and if dg is given by the volume form ω then $(\psi^*\omega)_{(e,a,eM)} = \prod_{\lambda\in\Lambda^+} (\sinh \lambda (\log a))^{m_\lambda} dk_1 d a d(kM)$. Here dk_1 and $d(kM)$ are invariant normalized measure on K and K/M, respectively, and da is suitably normalized invariant measure on A. Also $m_\lambda = \dim \mathfrak{g}_\lambda$. (Hint: Let $\mathfrak{n} = \sum_{\lambda\in\Lambda^+} \mathfrak{g}_\lambda$. Let $X_1, \ldots, X_r$ be a basis of $\mathfrak{n}$ so that $B(X_i, \theta X_j) = -\delta_{ij}$ and that $[H, X_i] = \lambda_i(H)X_i$ for H in $\mathfrak{a}$. Let $Z_i = (2)^{-\frac{1}{2}}(X_i + \theta X_i)$, $i = 1, \ldots, r$. Then $B(Z_i, Z_j) = -\delta_{ij}$. Let $U_1, \ldots, U_m$ be a basis of $\mathfrak{m}$ so that $B(U_i, U_j) = -\delta_{ij}$. Let $H_1, \ldots, H_l$ be an orthonormal basis of $\mathfrak{a}$ relative to B. Then we may assume that $\omega(\theta X_1, \ldots, \theta X_r, U_1, \ldots, U_m, H_1, \ldots, H_l, X_1, \ldots, X_r) = 1$. Show that $\psi_{*(e,a,eM)}(Z, H, W) = (\operatorname{Ad}(a)^{-1}Z)_a + H_a + (\operatorname{Ad}(a)^{-1}W - W)_a$. (Here we have identified $T(K/M)_{eM}$ with the linear span of the Z_i.)

8.16.6 (8.9.15 continued) Show that

$$\int_G f(g)dg = \int_{K \times A^+ \times K} f(k_1 a k_2) \prod_{\lambda \in \Lambda^+} (\sinh \lambda(\log a))^{m_\lambda} dk_1 da\, dk_2$$

in the notation of 8.16.5. (Hint: 8.16.4 implies that there is a C^∞ function $h: k \times A^+ \times K/M \to R$ so that

$$\int_G f(g)dg = \int_{K \times A^+ \times K/M} f(k_1 k a k^{-1}) h(k_1, a, kM) dk_1 da\, dkM.$$

Up to a constant factor we may change the integral over K/M to one over K, then using the invariance of dk we have

$$\int_G f(g)dg = \int_{K \times A^+ \times K} f(k_1 a k) h(k_1, a, k^{-1}M) dk_1 da\, dk.$$

Now $\int_G f(k_1 g k_2) dg = \int_G f(g) dg$. Hence $h(k_1, a, k^{-1}M) = h(e, a, eM)$. The result now follows from 8.16.5.)

8.16.7 Let for ξ in $\hat{M}$, ν in $\mathfrak{a}_C^*$, s in W, $\mathscr{A}_\gamma(\xi, \nu, s) = \mathscr{A}(\xi, \nu, s)|_{H_\gamma}$, whenever defined (see 8.14.11). Show that $\mathscr{A}_\gamma(\xi, \nu, s)$ is a rational function of ν. (Hint: Let γ_ξ be as in 8.14.9. Let $\lambda: S(\mathfrak{g}) \to U(\mathfrak{g})$ be defined as in 7.10.3. Let $\eta^{\xi,\nu}: S(\mathfrak{p}) \otimes V_{\gamma_\xi} \to H^\xi$ be defined by $\eta^{\xi,\nu}(u \otimes v) = \tilde{\pi}_{\xi,\nu}(\lambda(u)) f_\nu$ where $f_\nu(k) = A(\pi_{\gamma_\xi}(k)^{-1} v)$ and $A \neq 0$ is a fixed element of $\mathrm{Hom}_M(V_\gamma, H_\xi)$. Show that $U(\mathfrak{g}) = \lambda(S(\mathfrak{p}))U(\mathfrak{k})$. Thus $\mathrm{Im}\, \eta^{\xi,\nu} = \tilde{\pi}_{\xi,\nu}(U(\mathfrak{g}))H_{\gamma_\xi}$. 8.13.12 implies that $\eta^{\xi,\nu}$ is almost always surjective. Fix ν_0 so that η^{ξ,ν_0} is surjective. Let I be the kernel of η^{ξ,ν_0}. Then $S(\mathfrak{p}) \times V_{\gamma_0} = I \oplus W$, $W = \sum_{\gamma \in \hat{K}} (m_\xi(\gamma) V_\gamma = W_\gamma)$ as a representation of K (here we let K act on $S(\mathfrak{p})$ by the extension of $\mathrm{Ad}(k)|_\mathfrak{p}$ and on $S(\mathfrak{g}) \otimes V_{\gamma_0}$ by the tensor product action. We note that $\eta^{\xi,\nu}(k(u \otimes v)) = \tilde{\pi}_{\xi,\nu}(k)\eta^{\xi,\nu}(u \otimes v))$. Show that $\eta^{\xi,\nu}: W_\gamma \to H_\gamma$ is a polynomial mapping. Now

$$\begin{aligned} A(\xi, \nu, s)\eta^{\xi,\nu}(u \otimes v) &= A(\xi, \nu, s)\tilde{\pi}_{\xi,\nu}(\lambda(u)) f_\nu \\ &= \tilde{\pi}_{\xi s, s(\nu-\rho)+\rho}(\lambda(u)) A(\xi, \nu, s) f_\nu \\ &= d(\xi, \nu, s)\eta^{\xi s, s(\nu-\rho)+\rho}(u \otimes v). \end{aligned}$$

This implies that $\mathscr{A}_\gamma(\xi, \nu, s) = \eta^{\xi s, s(\nu-\rho)+\rho} \circ (\eta^{\xi,\nu}|_{W_\gamma})^{-1}$.)

8.16.8 Let M be an orientable manifold with volume form ω. Let f be in $C_0^\infty(M \times M)$. Define for φ in $L^2(M, \omega)$,

$$T\varphi(x) = \int_M f(x, y)\varphi(y)\, dy$$

(here dy means integrate relative to ω). Show that T is of trace class and that $\operatorname{tr} T = \int_M f(x, x)dx$. (Hint: Let $U_1, \ldots, U_m$ be a collection of open subsets of M with $\psi_i: U_i \to \{z \text{ in } R^n \mid |z_i| < \pi\} = W$ and (U_i, ψ_i) a chart for M, and such that $\bigcup_{1 \leqslant j,k \leqslant m} U_j \times U_k \supset \operatorname{supp} f$. Let $\varphi_1, \ldots, \varphi_m$ be a partition of unity for $U_1 \cup U_2 \cup \cdots \cup U_m$ subordinate to the U_j. Let $f_{ij}(x, y) = \varphi_i(x)\varphi_j(y)f(x, y)$. Let $T_{ij}\varphi(x) = \int_M F_{ij}(x, y)\varphi(y)dy$. Clearly it is enough to show that T_{ij} is trace class and that $\operatorname{tr} T_{ij} = \int_M f_{ij}(x, x)dx$. "Pull back" f_{ij} to $W \times W$; extend the pull-back of f_{ij} to be periodic of period 2π in each coordinate. Use the techniques of 8.8 on the n torus to complete the proof.)

8.16.9 Use 8.12.6 and 8.15.12 to prove that if $G = SO(2n + 1, 1)$ or Spin$(2n + 1, 1)$ and if ξ is in $\hat{M}$, ν is in $\mathfrak{a}^*$, $\nu \neq 0$ then $(\pi_{\xi,\nu}, H^{\xi,\nu})$ is irreducible (note that there are no tildes).

8.16.10 Use the technique of the proof of 8.11.9 to give explicit formulas for $B_\gamma(\nu, s)$ if $G = SU(1, 1)$, $SU(2, 1)$, or $SL(2, C)$.

8.16.11 Carry through the details of the proof of 8.12.6 in the case when γ, τ are both the trivial representation of K.

8.16.12 Use 7.13.11 to give the Plancherel formula, explicitly, for $SL(2, C)$ (compare with Gelfand *et al.* [1]).

8.16.13 Let G be a semisimple Lie group. Let $G = KAN$ be an Iwasawa decomposition for G. Let (π_1, V_1), (π_2, V_2) be finite dimensional representations of K. Let Ω be as in the proof of 8.12.6. Let $\varphi: G \mapsto L(V_2, V_1)$ be such that $\pi_1(k_1)\varphi(g)\pi_2(k_2)$. Compute, using the techniques of the proof of 8.12.6, $(\Omega\varphi)(a)$ for a in A.

8.17 Notes

8.17.1 The principal series was first defined for $SL(2, R)$ by Bargman [1] as in 8.4.2 and for $SL(2, C)$ by Gelfand and Naimark [1] (see 8.4.9). To our knowledge, the first place where the principal series is defined as an induced representation is in Bruhat [1]. These representations are special cases of the "cuspidal representations" (cf. Lipsman [1]).

8.17.2 8.5.3 and 8.5.5 are taken from Wallach [1].

8.17.3 8.5.8 is due to Harish-Chandra [8]; our proof is slightly simpler than the original proof.

8.17.4 A slightly weaker form of 8.5.11 is in Wallach [1]. The original proof of this result had an error (pointed out to the author by J. Lepowsky who has an independent proof of this result). When the error of Wallach [1] is corrected the proof yields 8.5.11. See also Lepowsky and Wallach [1].

8.17.5 8.5.14 and 8.5.15 are taken from Lepowsky and Wallach [1]. The proof uses the fact that $1_{2\rho}$ is a cyclic vector for $\tilde{H}^{1,2\rho}$. This is just the statement that the Poisson representation of a harmonic function on a symmetric space is unique. The idea in 8.5.14(1)–(6) is taken from Helgason [2].

8.17.6 8.6.7 is due to Harish-Chandra [2].

8.17.7 8.6.10 is due to Harish-Chandra [2]; the proof in this book using 8.6.9 is new.

8.17.8 8.6.21 and its proof are taken from Harish-Chandra [3].

8.17.9 8.7.2 is taken from Harish-Chandra [2].

8.17.10 8.7.4 is a slight generalization of a theorem of Harish-Chandra [2].

8.17.11 8.8.2 is essentially due to Harish-Chandra [2]. The proof of 8.8.2 is taken directly from Harish-Chandra [3].

8.17.12 8.8.3 is usually attributed to Bruhat [1]. It is also noted in Harish-Chandra [3].

8.17.13 8.9.3 and 8.9.4 use ideas of Satake [1]. However, these results are essentially in Cartan [1].

8.17.14 The proof of 8.9.6(1) is taken from Helgason [1].

8.17.15 8.10.9 is taken from Schiffmann [1]. The idea of defining the intertwining operators as integrals is due to Kunze and Stein [1] (their unnormalized intertwining operators are given by the right-hand side of 8.10.10). Similar results are to be found in Knapp and Stein [1] and Helgason [1].

8.17.16 8.10.12 is taken from Schiffmann [1]. The idea of the proof is the celebrated technique of Gindinkin and Karpelevic [1].

8.17.17 8.10.13 is called the "reduction principal" in Helgason [3]. It was proved essentially simultaneously (and for the same reasons) by Schiffmann [1] and Knapp and Stein [1].

8.17.18 8.10.13 is due to Helgason [3] and Schiffmann [1].

8.17.19 The proof of 8.10.17 follows the broad lines of Gindinkin and Karpelevic [1]. The details of the proof are essentially from Schiffmann [1] and Helgason [3]. It should be observed that the proof of 8.10.17 contains an explicit formula for $c(\nu)$ in terms of gamma functions. The convergence statement in 8.10.17 is due to Harish-Chandra [8].

8.17.20 8.11.3 is a result of Kostant [3]. The proof follows Kostant's original ideas. It is, however, longer and less elegant than the original since it does not use results of Kostant [1].

8.17.21 8.11.5 is due to Schiffmann [1].

8.17.22 The meromorphic continuation statement of 8.11.9 goes back to Kunze and Stein [1]. Our proof of this result is new. We note that our proof of 8.11.3 gives explicit formulas for the matrix entries of the intertwining operators in terms of gamma functions. The fact that the matrix entries of the intertwining operators are expressable in terms of gamma functions is due to Schiffmann [1].

8.17.23   8.12.6 is a special case of a general result of Harish-Chandra on the asymptotics of the Eisenstein integral. The specific form of the statement of 8.12.6 is taken from Knapp and Stein [1]. The proof of 8.12.6 using Appendix 8 is new. The technique of computing the restriction of the Casimir operator to A is due to Harish-Chandra (see G. Warner [1]).

8.17.24   Our proof of 8.13.3 is an outgrowth of the work in Wallach [2] Lepowsky and Wallach [1]. Using the subquotient theorem of Harish-Chandra [3], 8.13.3 gives a new proof of the theorem of Harish-Chandra that asserts that given an infinitesimal character there are only a finite number of irreducible representations with this character. 8.13.3 was originally proved using the result of Harish-Chandra [10] mentioned above. The idea of using the chain conditions to prove 8.13.3 is due to Kostant.

8.17.25   8.13.7 is a real analog of a p-adic result of Howe [1].

8.17.26   8.13.18 is a crude form of the irreducibility theorem of Bruhat [1]. A proof similar to ours for complex semisimple Lie groups can be found in Harish-Chandra [3].

8.17.27   The normalized operators $\mathscr{A}(\xi, \nu, s)$ of 8.14.11 are related to the Kunze–Stein intertwining operators. They first normalization was given in Kunze and Stein [1].

8.17.28   8.15.4 is a generalization of the formula of Harish-Chandra [3] for complex semisimple Lie groups. The Plancherel formula for arbitrary semisimple Lie groups with finite center has been obtained by Harish-Chandra. It is as yet unpublished.

8.17.29   The proof of 8.15.12 is taken from Knapp and Stein [1]. The result is a special case of Harish-Chandra's "Mass–Selberg relations".

APPENDIX 1

Review of Differential Geometry

A.1.1 Manifolds

A.1.1.1 *Definition* Let X be a topological space. An n-dimensional C^∞ structure for X is a collection $\mathfrak{A}$ of ordered pairs (U, ψ) where U is an open subset of X and ψ is a homeomorphism of U onto an open subset of R^n (n-dimensional Euclidean space), such that

(i) $\bigcup_{(U,\psi)\in\mathfrak{A}} U = X$.

(ii) If (U, ψ), $(V, \phi) \in \mathfrak{A}$ then $\psi \circ \phi^{-1}: \phi(U \cap V) \to \psi(U \cap V)$ is a C^∞ mapping in the usual sense in R^n.

(iii) If (W, χ) is a pair of an open subset W of X and a homeomorphisms χ of W onto an open subset of R^n such that for each $(U, \psi) \in \mathfrak{A}$

$$\psi \circ \chi^{-1}: \chi(U \cap W) \to \psi(U \cap W)$$

and

$$\chi \circ \psi^{-1}: \psi(U \cap W) \to \chi(U \cap W)$$

are C^∞ then $(W, \chi) \in \mathfrak{A}$.

A.1.1.2 A collection $\mathfrak{B}$ as above satisfying (i), (ii) is called a C^∞ atlas for X. If $\mathfrak{B}$, in addition, satisfies (iii) then $\mathfrak{B}$ is called complete. A complete C^∞ atlas is clearly a C^∞ structure.

A.1.1.3 A pair $(X, \mathfrak{A})$ of a Hausdorff space X, with a countable basis for its topology and an n-dimensional C^∞ structure $\mathfrak{A}$ for X, is called a C^∞ manifold. We will usually use the symbol M to denote $(X,\mathfrak{A})$ and X. If $M = (X,\mathfrak{A})$ and $(U, \psi) \in \mathfrak{A}$ we will call (U, ψ) a chart for M.

A.1.1.4 In practice it is enough to find C^∞ atlas for a topological space to find a C^∞ structure. Indeed using Zorn's lemma it is easy to show that if $\mathfrak{B}$ is an atlas for X there is a unique C^∞ structure $\mathfrak{A}$ for X containing $\mathfrak{B}$.

A.1.1.5 Let M and N be C^∞ manifolds. Then a continuous map $f: M \to N$ is called C^∞ if for each $p \in M$ there are charts (U, ψ) and (V, ϕ) for M and N, respectively, such that $p \in U$, $f(U) \subset V$ and $\phi \circ f \circ \psi^{-1}: \psi(U) \to \phi(V)$ is C^∞. If $f: M \to N$ is C^∞ and bijective such that $f^{-1}: N \to M$ is C^∞ then f is called a diffeomorphism.

A.1.1.6 *Examples* (1) Let $X = R^n$. Take $U = R^n$, ψ = identity. Then $\{(U, \psi)\}$ is a C^∞ atlas for R^n. The corresponding C structure on R^n is called the standard C^∞ structure on R^n.

(2) Let M be a C^∞ manifold and let $U \subset M$ be an open subset of M. Let $\mathfrak{A}$ be the C^∞ structure of M. Set $\mathfrak{B} = \{(U \cap V, \psi|_{U \cap V}) | (U, \psi) \in \mathfrak{A}\}$. Then $\mathfrak{B}$ is a C^∞ atlas for U. The corresponding C^∞ manifold is called an open submanifold of M.

(3) Let M and N be C^∞ manifolds with C^∞ structures $\mathfrak{A}$ and $\mathfrak{B}$ of dimensions m and n, respectively. We set

$$\mathfrak{C} = \{(U \times V, \psi \times \phi) | (U, \psi) \in \mathfrak{A}, (V, \phi) \in \mathfrak{B}\}$$

where $(\psi \times \phi)(x, y) = (\psi(x), \phi(y))$. Then $\mathfrak{C}$ defines an $(n + m)$-dimensional C^∞ atlas for $M \times N$. By $M \times N$ we will mean the C^∞ manifold with C^∞ structure induced by $\mathfrak{C}$.

(4) Let M be a C^∞ manifold and let $\pi: \tilde{M} \to M$ be a covering space for M. Let for each $p \in M$, (U, ψ) be a chart for M so that U is evenly covered by π. (That is, $\pi^{-1}(U) = \bigcup U_j$, a disjoint union such that $\pi: U_j \to U$ is a homeomorphism.) We will say that (U, ψ) is evenly covered by π. Let $\mathfrak{B}$ be the collection of all charts of M which are evenly covered by π. We define

$$\mathfrak{B} = \{(U_i, \psi \circ \pi) | (U, \psi) \in \mathfrak{B}, \pi^{-1}(U) = \bigcup U_j \text{ disjoint union,}$$
$$\pi: U_j \to U \text{ a homeomorphism}\}.$$

Then $\mathfrak{B}$ defines a C^∞ atlas on $\tilde{M}$. With this C^∞ structure $\pi: \tilde{M} \to M$ is C^∞. Furthermore if $\sigma: \tilde{M} \to \tilde{M}$ is a deck transformation (σ a homeomorphism and $\pi(\sigma(x)) = \pi(x)$), then σ is C^∞.

A.1.1.7 If M is a C^∞ manifold and if $f: M \to R = R^1$ is C^∞ we call f a C^∞ function. We denote by $C^\infty(M)$ the space of all C^∞ functions on M. Identifying R^2 with C (as a real vector space and C^∞ manifold) we call a C^∞ map from M to C a complex valued C^∞ function. We denote by $C^\infty(M; C)$ the space of all complex valued C^∞ functions on M. In general if V is an n-dimensional vector space over R or C we look upon V as R^n or C^n as a C^∞ manifold and denote the space of all C^∞ mappings from M to V by $C^\infty(M; V)$.

A.1.1.8 Let (U, ψ) be a chart for M. Then if

$$p \in U, \psi(p) = (x_1(p), \ldots, x_n(p))$$

with $x_i \in C^\infty(U)$. $\{x_1, \ldots, x_n\}$ is called a system of local coordinates for M on U.

A.1.2 Tangent Vectors

A.1.2.1 Let M be an n-dimensional C^∞ manifold. Let $p \in M$. A C^∞ curve in M through p is a C^∞ map $\sigma: (-\varepsilon, \varepsilon) \to M$ ($\varepsilon > 0$) such that $\sigma(0) = p$.

A.1.2.2 If σ is a C^∞ curve through p we define a linear map $L_\sigma: C^\infty(M) \to R$ as follows:

$$L_\sigma(f) = \frac{d}{dt} f(\sigma(t))\bigg|_{t=0}.$$

A.1.2.3 *Definition* We define $T(M)_p$ to be the set of all L_σ, with σ a C^∞ curve through p.

A.1.2.4 Let U be open in M, $p \in U$, and let $x_1, \ldots, x_n$ be local coordinates for M on U, $\psi = (x_1, \ldots, x_n)$ such that (U, ψ) is the corresponding chart. Let $e_1, \ldots, e_n$ be the standard basis for R^n. Set $\sigma_i(t) = \psi^{-1}(\psi(p) + te_i)$ for t small. We set

$$\frac{\partial}{\partial x_{i_p}} f = L_{\sigma_i} f$$

for $f \in C^\infty(M)$. Now if σ is an arbitrary C^∞ curve through p, then $(\psi \circ \sigma)(t) =$

$(x_1(\sigma(t)), \ldots, x_n(\sigma(t)))$, hence the chain rule implies that

$$L_\sigma(f) = \sum_{i=1}^{n} \frac{d(x_i \circ \sigma)}{dt}(0) \frac{\partial f}{\partial x_{i_p}}.$$

This immediately implies that $T(M)_p$ is a vector space with basis

$$\frac{\partial}{\partial x_{i_p}}, \ldots, \frac{\partial}{\partial x_{n_p}}.$$

A.1.2.5 Let $T(M) = U_{p \in M} T(M)_p$. Let $\mathfrak{A}$ be the C^∞ structure of M. Let $\xi \in \mathfrak{A}$, $\xi = (U_\xi, \psi_\xi)$ and let $(x_1^\xi, \ldots, x_n^\xi)$ be the corresponding local coordinates on U. Define $\pi: T(M) \to M$ by $\pi(T(M)_p) = p$.

We define for each $\xi \in \mathfrak{A}$ a map

$$\Psi_\xi: \pi^{-1}(U_\xi) \to \psi_\xi(U_\xi) \times R^n$$

as follows:

If $v \in T(M)_p$, then $v = \sum_i v_i(\partial/\partial x_{i_p})$. Set $\Psi_\xi(v) = (\psi_\xi(p), (v_1, \ldots, v_n))$.

Let now for each $\xi \in \mathfrak{A}$, τ_ξ be the collection of all subsets of $\pi^{-1}(U_\xi)$ of the form $\Psi_\xi^{-1}(W)$ where W is open in $\psi_\xi(U_\xi) \times R^n$. Let $\tau = U_{\xi \in \mathfrak{A}} \tau_\xi$. We leave it to the reader to show that τ defines a basis for a second countable Hausdorff topology on $T(M)$ such that $\pi: T(M) \to M$ is continuous. Furthermore $\{(\pi^{-1}(U_\xi), \Psi_\xi)\}_{\xi \in \mathfrak{A}}$ is a C^∞ atlas for $(T(M), \tau)$. Relative to this C^∞ structure $\pi: T(M) \to M$ is C^∞.

A.1.2.6 Now let M and N be C^∞ manifolds and let $f: M \to N$ be a C^∞ mapping. Let $p \in M$. We define a map

$$f_{*p}: T(M)_p \to T(N)_{f(p)}$$

by $f_{*p}(L_\sigma) = L_{f \circ \sigma}$ where σ is a C^∞ curve through p.

A.1.2.7 Let $x_1, \ldots, x_n$ be local coordinates around p in M and let $y_1, \ldots, y_m$ be local coordinates around $f(p)$ in N. Then using the composite function theorem we find

$$f_{*p}\left(\sum a_i \frac{\partial}{\partial x_{i_p}}\right) = \sum_{i,j} a_i \frac{\partial(y_j \circ f)}{\partial x_{i_p}} \frac{\partial}{\partial y_{j_{f(p)}}}.$$

This implies that f_{*p} is well defined and linear.

A.1.2.8 Let N be a C^∞ manifold. Then N is called a submanifold of M

if $N \subset M$ and the canonical injection $i: N \to M$ satisfies:

(1) i is C^∞.

(2) $i_{*p}: T(N)_p \to T(M)_p$ is injective for each $p \in N$.

In particular an open submanifold is a submanifold.

A.1.3 Vector Fields

A.1.3.1 *Definition* Let M be a C^∞ manifold. A vector field on M is a C^∞ mapping $X: M \to T(M)$ such that $X_p \in T(M)_p$ (the vector field is denoted $p \mapsto X_p$).

A.1.3.2 Let $\mathfrak{X}(M)$ be the set of all vector fields on M. We make $\mathfrak{X}(M)$ into a vector space by defining $(aX + bY)_p = aX_p + bY_p$ for $a, b \in R$, $X, Y \in \mathfrak{X}(M)$, $p \in M$.

A.1.3.3 If $X \in \mathfrak{X}(M)$ and U is an open subset of M with local coordinates $x_1, \ldots, x_n$ then $X|_U = \sum \xi_i(\partial/\partial x_i)$ with $\xi_i \in C^\infty(U)$.

A.1.3.4 Let $X, Y \in \mathfrak{X}(M)$, let $p \in M$, $x_1, \ldots, x_n$ local coordinates on U an open neighborhood of p. Then $X|_U = \sum \xi_i(\partial/\partial x_i)$, $Y|_U = \sum \eta_i(\partial/\partial x_i)$. Set

$$[X, Y]_p = \sum_i \left\{ \sum_j \left(\xi_j \frac{\partial \eta_i}{\partial x_{j_p}} - \eta_j \frac{\partial \xi_i}{\partial x_{j_p}} \right) \right\} \frac{\partial}{\partial x_{i_p}}. \tag{1}$$

From this we see that on U the map $q \to [X, Y]_q$ is C^∞. To see that $[X, Y]$ is independent of choices we note that if $f \in C^\infty(M)$ then

$$X, Y]_p f = X_p(Yf) - Y_p(Xf)$$

where Xf is the C^∞ function $p \to X_p f$.

A.1.3.5 A direct computation shows that if $X, Y, Z \in \mathfrak{X}(M)$ then

$$[X, [Y, Z]] = [[X, Y], Z] + [Y, [X, Z]]. \tag{2}$$

Formula (2) is called the Jacobi identity.

A.1.4 Partitions of Unity

A.1.4.1 Let X be a topological space and let $\mathfrak{U}$ be a covering of X. Then $\mathfrak{U}$ is called locally finite if for each $p \in X$ there is an open neighborhood U of p such that there are only a finite number of elements of $\mathfrak{U}$ that intersect U.

A.1.4.2 We now suppose that M is a C^∞ manifold. Then every open covering of M has a locally finite refinement (see F. Warner [1]).

A.1.4.3 Let $\mathfrak{U}$ be an open covering of M. Then a partition of unity of M subordinate to $\mathfrak{U}$ is a collection $\{\phi_i\}_{i=1}^{\infty}$ of C^∞ functions such that

(i) Supp ϕ_i $(= \overline{\{x \in M | \phi_i(x) \neq 0\}})$ is compact and there is $U \in \mathfrak{U}$ so that supp $\phi_i \subset U$.

(ii) Let $p \in M$ such that there is a neighborhood U of p such that only a finite number of the ϕ_i are not identically zero on U.

(iii) $\phi_i \geqslant 0$ and $\sum_{i=1} \phi_i(x) = 1$ for each $x \in M$.

A.1.4.4 *Theorem* Let M be a C^∞ manifold and let $\mathfrak{U}$ be an open covering of M. Then there is a partition of unity subordinate to $\mathfrak{U}$.

For a proof of this result see F. Warner [1].

APPENDIX 2

Lie Groups

A.2.1 Basic Notions

A.2.1.1 *Definition* Let G be a C^∞ manifold that has the structure of a group. Then G is called a Lie group if the map

$$G \times G \to G$$

given by $(x, y) \to xy^{-1}$ is C^∞.

A.2.1.2 Basic examples of Lie groups are $GL(n, K)$, $K = R$ or C, the groups of $n \times n$ nonsingular matrices over K. $GL(n, K)$ is a C^∞ manifold since $GL(n, K)$ is an open subset of the $(\dim_R K)n^2$ dimensional real vector space of all $n \times n$ matrices over K. The map $(A, B) \to AB^{-1}$ is C^∞ by Cramer's rule.

A.2.1.3 Let G be a Lie group. If $x \in G$ we define $l_x: G \to G$ and $r_x: G \to G$ by $l_x g = xg$ and $r_x g = gx$. Then l_x and r_x are diffeomorphisms of G.

A.2.1.4 A vector field X on G is said to be left invariant (resp. right invariant) if for each $x, g \in G$, $l_{g*x} X_x = X_{gx}$ (resp. $r_{g*x} X_x = X_{xg}$). Let $\mathfrak{g}$ be the subspace of $\mathfrak{X}(G)$ consisting of the left invariant vector fields on G.

A.2.1.5 *Theorem* (1) The map $\mathfrak{g} \to T(G)_e$ (e the identity of G) given by $X \to X_e$ is an onto isomorphism.

(2) If $X, Y \in \mathfrak{g}$ then $[X, Y] \in \mathfrak{g}$.

PROOF Let $X \in \mathfrak{g}$ and suppose $X_e = 0$. Then $X_g = l_{g*e} X_e = 0$. Thus $X = 0$. Hence the map $\mathfrak{g} \to T(G)_e$ is injective. Let now $v \in T(G)_e$. Let $v = L_\sigma$ for σ a C^∞ curve through e. Define $\varphi_g(t) = g\sigma(t)$. Then for each $g \in G$ we see that $X_g = L_{\varphi_g} \in T(G)_g$. Clearly $l_{g*x} X_x = X_{gx}$ for $g, x \in G$. Since the map $(g, t) \mapsto \varphi_g(t)$ is C^∞ on $Gx(-\varepsilon, \varepsilon)$, X is in $\mathfrak{X}(G)$. Finally $X_e = v$. This proves (1).

(2) follows from $l_{g*}[X, Y] = [l_{g*}X, l_{g*}Y]$. (see, e.g., F. Warner [1]).

A.2.1.6 $\mathfrak{g}$ is called the Lie algebra of G.

A.2.1.7 In general a Lie algebra over a field K is a K-vector space $\mathfrak{a}$ with a bilinear pairing $x, y \mapsto [x, y]$ satisfying:

LA-1 $[x, x] = 0, \quad x \in \mathfrak{a}$.

LA-2 $[x, [y, z]] = [[x, y], z] + [y, [x, z]]$

for $x, y, z \in \mathfrak{a}$.

A.2.1.8 We note that $\mathfrak{g}$ is a Lie algebra over R.

A.2.1.9 If $\mathfrak{a}_1$ and $\mathfrak{a}_2$ are Lie algebras over K then a Lie algebra homomorphism $\varphi: \mathfrak{a}_1 \to \mathfrak{a}_2$ is a linear map such that $\varphi[x, y] = [\varphi x, \varphi y]$.

A.2.1.10 Let G and H be Lie groups. Let $\varphi: G \to H$ be a group homomorphism. Then φ is called a Lie homomorphism if φ is C^∞.

A.2.1.11 Let $\varphi: G \to H$ be a Lie homomorphism. Let $\mathfrak{g}$ and $\mathfrak{h}$ be, respectively, the Lie algebras of G and H. We define a map $\varphi_*: \mathfrak{g} \to \mathfrak{h}$ by: $\varphi_* X$ is the element of $\mathfrak{h}$ corresponding under Theorem A.3.1.1 to $\varphi_{*e} X_e$. It is not hard to show that $\varphi_*[X, Y] = [\varphi_* X, \varphi_* Y]$, (see F. Warner [1]). Thus φ_* is a Lie algebra homomorphism.

A.2.1.12 Let G and H be Lie groups. Suppose furthermore that H is a submanifold of G and that $i: H \to G$ is a Lie homomorphism. Then H is called a Lie subgroup of G. Let $\mathfrak{h}$ be the Lie algebra of H, $\mathfrak{g}$ the Lie algebra of G. Then $i_*: \mathfrak{h} \to \mathfrak{g}$ is a Lie algebra homomorphism. Since i_* is injective we may identify $\mathfrak{h}$ with its image $i_*(\mathfrak{h})$ in $\mathfrak{g}$.

A.2.2 The Exponential Map

A.2.2.1 Let G be a Lie group. A one-parameter subgroup of G is a Lie homomorphism $\varphi: R \to G$ of the additive group of real numbers into G. In particular φ is a C^∞ curve through e in G. Let X_φ be the element of $\mathfrak{g}$ the Lie algebra of G corresponding to $L_\varphi \in T(G)_e$.

A.2.2.2 *Theorem* Let $X \in \mathfrak{g}$. Then there is a unique one-parameter subgroup φ_X of G so that $X_{\varphi_X} = X$. Furthermore the map $(\mathfrak{g}, R) \to G$ given by $(X, t) \to \varphi_X(t)$ is C^∞.

This result depends on the existence and uniqueness theorem for ordinary differential equations (see F. Warner [1]).

A.2.2.3 Define for $X \in \mathfrak{g}$, $\exp X = \varphi_X(1)$. Then $\exp tX = \varphi_X(t)$. exp is called the exponential mapping.

Using the exponential map one proves the following theorem.

A.2.2.4 *Theorem* Let G and H be Lie groups and let $\varphi: G \to H$ be a continuous group homomorphism. Then φ is a Lie homomorphism.

The proof goes by showing that $\varphi(\exp tX) = \exp tY$ for some Y. For a proof see F. Warner [1].

A.2.2.5 *Theorem* Let G be a Lie group. Then G has the structure of an analytic manifold so that the map $G \times G \to G$ given by $x, y \to xy^{-1}$ is analytic.

A.2.2.6 See Chevalley [1] for the definition of analytic manifold and see F. Warner [1], for a proof of A.2.2.5.

A.2.2.7 If V is a finite dimensional vector space and if A is a linear map of V to V then set

$$\frac{1 - e^{-A}}{A} = \sum_{m=0}^{\infty} (-1)^m \frac{A^m}{(m+1)!}, \quad A^\circ = 1.$$

A.2.2.8 *Lemma* Let G be a Lie group with Lie algebra $\mathfrak{g}$. Then (using $\operatorname{ad}X \cdot Y = [X, Y]$)

$$\exp_{*X}(Y) = \left(\left(\frac{1 - e^{-\operatorname{ad}X}}{\operatorname{ad}X}\right) \cdot Y\right)_{\exp X}.$$

PROOF Let Z in $\mathfrak{g}$ be so that $(\exp_{*X} Y) = Z_{\exp X}$. Then

$$L_{(\exp -X)^* \exp X} \exp_{*X} Y = Z_e.$$

Hence

$$Z_e f = Z_{\exp Y} f \circ L_{(\exp -X)} = \frac{d}{dt} f(\exp(-X)\exp(X + tY))\Big|_{t=0},$$

for f in $C^\infty(G)$.

Now if W is in $\mathfrak{g}$ and if f is analytic in a neighborhood U of e then

$$f(g \exp tW) = \sum_{n=0}^{\infty} \frac{t^n}{n!}(W^n f)(g)$$

the sum uniformly and absolutely convergent for all t such that $|t|$ is sufficiently small and g is in U.

If V, W are in $\mathfrak{X}(G)$ define $\operatorname{ad} V \cdot (W) = [V, W]$.

(a) $$(\operatorname{ad} V)^m \cdot W = \sum_{k=0}^{\infty} (-1)^k \binom{m}{k} V^{m-k} W V^k.$$

(a) can be proved by the obvious induction. Using (a) we find:

(b) If f is analytic on U then

$$\left(\frac{I - e^{-\operatorname{ad}X}}{\operatorname{ad}X} \cdot Y\right)_e f = \sum_{m=0}^{\infty} \frac{(-1)^m}{(m+1)!} \sum_{k=0}^{\infty} \binom{m}{k} (-1)^k (X^{m-k} Y X^k \cdot f)(e).$$

On the other hand it is easy to see that

(c) $$\frac{d}{dt} f(\exp(X + tY))\Big|_{t=0} = \sum_{n=0}^{\infty} \frac{1}{(n+1)!} \sum_{j=0}^{\infty} (X^{n-j} Y X^j f)(e)$$

for f analytic in a neighborhood of $\exp X$.

Hence

(d) $$\frac{d}{dt} f(\exp(-X)\exp(X + tY))\Big|_{t=0}$$

$$= \sum_{n,m=0} (-1)^m \frac{1}{m!\,(n+1)!} \sum_{j=0}^{n} (X^{n+m-j} Y X^j \cdot f)(e)$$

for f analytic in a neighborhood of e.

To prove the lemma we need only show that the right-hand side of (b) is equal to the right-hand side of (d).

$$\sum_{n,m=0} (-1)^m \frac{1}{m!\,(n+1)!} \sum_{j=0}^{j} (X^{n+m-j} Y X^j f)(e)$$

$$= \sum_{p=0}^{\infty} \sum_{n=0}^{p} (-1)^{p-n} \frac{1}{(p-n)!\,(n+1)!} \sum_{j=0}^{n} (X^{p-j} Y X^j f)(e)$$

$$= \sum_{p=0} \frac{(-1)^p}{(p+1)!} \sum_{j=0}^{p} \left(\sum_{n=j} (-1)^n \binom{p+1}{n+1} \right) (X^{p-j} Y X^j f)(e).$$

Now $\sum_{n=0}^{p+1} (-1)^n \binom{p+1}{n} = 0$ hence

$$\sum_{n=j}^{n} (-1)^n \binom{p+1}{n} = - \sum_{n=j+1}^{p+1} \binom{p+1}{n} (-1)^n = \sum_{n=0}^{j} \binom{p+1}{n} (-1)^n.$$

We assert that if $j \leqslant p$ then

$$\sum_{n=0}^{j} \binom{p+1}{n} (-1)^n = \binom{p}{n} (-1)^j.$$

Indeed the result is true if $j = 0$. If true for $j - 1$ then

$$\sum_{n=0}^{j} \binom{p+1}{n} (-1)^n = \sum_{n=0}^{i-1} \binom{p+1}{n} (-1)^n + \binom{p+1}{n} (-1)^j$$

$$= \binom{p}{j-1} (-1)^{j-1} + \binom{p+1}{j} (-1)^j$$

$$= (-1)^j \left\{ \binom{p+1}{j} - \binom{p}{j-1} \right\} = (-1)^j \binom{p}{j}$$

since $\binom{p+1}{j} = \binom{p}{j-1} + \binom{p}{j}$. Hence

$$\sum_{p=0}^{\infty} \frac{(-1)^p}{(p+1)!} \sum_{j=0}^{p} \left(\sum_{n=j}^{p} (-1)^p \binom{p+1}{n+1} \right) (X^{p-j} Y X^j f)(e)$$

$$= \sum_{p=0}^{\infty} \frac{(-1)^p}{(p+1)!} \sum_{j=0}^{p} (-1)^j \binom{p}{j} (X^{p-j} Y X^j f)(e),$$

which is the right-hand side of (b).

A.2.3 Lie Subalgebras and Lie Subgroups

A.2.3.1 Let G be a Lie group and let $\mathfrak{g}$ be its Lie algebra. Let $\mathfrak{h}$ be a subalgebra of $\mathfrak{g}$.

A.2.3.2 *Theorem* Let the notation be as above. There is a unique connected Lie subgroup H of G so that $\mathfrak{h}$ is the Lie algebra of H (under the identification of A.2.2.1.)

A.2.3.3 The connected subgroup is gotten by taking $\tilde{H}$ to be the subgroup of G generated by $\exp(\mathfrak{h})$, then showing that $\tilde{H}$ actually is a submanifold of G. This is done by taking U a sufficiently small neighborhood of 0 so that $\exp: U \to G$ is injective. Taking $N_e = \exp(U)$ we have (N_e, ψ), $\psi: N_e \to U$. We take as a neighborhood basis at $e \in \tilde{H}$, $\exp(V)$ for $V \subset U$, V open and a neighborhood basis of $\tilde{H}$ the neighborhoods $g \cdot \exp(V)$ as above. Then $\psi: N_e \to U$ is a chart for $\tilde{H}$ at e. Similarly $\psi_h: hN_e \to U$ defined by $\psi_h(hg) = \psi(g)$, $g \in N_e$ defines a chart at $h \in \tilde{H}$. It is then shown that this defines a C^∞ structure on H and relative to this C^∞ structure $\tilde{H} = H$ is the required Lie group. See Helgason [1], or F. Warner [1] for details.

A.2.3.4 *Theorem* Let G be a Lie group and let H be a subgroup which as a topological subspace is closed in G. Then H has the structure of a Lie subgroup of G.

A.2.3.5 To prove this result one first shows that if $\mathfrak{h} = \{X \in \mathfrak{g} | \exp tX \in H$ for all $t \in R\}$ then $\mathfrak{h}$ is a vector subspace of $\mathfrak{g}$. Then one shows that $\exp \mathfrak{h}$ contains a neighborhood of e in H. The Lie group structure is then given as in the proof of A.2.3.3. See Helgason [1] or F. Warner [1] for details.

A.2.4 Homogeneous Spaces

A.2.4.1 Let G be a Lie group and let H be a closed subgroup of G (hence H is a Lie subgroup of G).

A.2.4.2 *Theorem* Let $M = G/H$ (the space of left cosets of G relative to H given the topology that makes the canonical map $\pi: G \to G/H, \pi(g) = gH$ continuous and open). Then M has a unique C^∞ structure so that

(1) There is a neighborhood U of eH in M and a C^∞ map $\psi: U \to G$ so that $\pi\psi(u) = u$.

(2) The map $G \times M \to M$ given by $(g_1, g_2H) \mapsto g_1g_2H$ is a C^∞ mapping.

For a proof of this result see F. Warner [1].

A.2.4.3 Let G be a Lie group and let M be a C^∞ manifold. A differentiable action of G on M is a C^∞ map

$$F: G \times M \to M$$

such that $F(e, m) = m$ for all m in M and

$$F(g_1, F(g_2, m)) = F(g_1 g_2, m).$$

A.2.4.4 We usually denote in a differentiable action $F(g, m)$ by $g \cdot m$. A differentiable action of G on M is said to be transitive if for any $m_1, m_2 \in M$ there is $g \in G$ so that $g \cdot m_1 = m_2$. G is said to act transitively on M.

A.2.4.5 *Theorem* Let G be a Lie group and let M be a C^∞ manifold. Suppose that G acts transitively on M. Let $p \in M$ and $G_p = \{g \in G | g \cdot p = p\}$. Then M is diffeomorphic with G/G_p.

For a proof of this result see F. Warner [1].

A.2.4.6 *Theorem* Let G be a Lie group and let H be a closed normal subgroup of G. Then the topological group G/H is a Lie group.

See F. Warner [9] for a proof of this result.

A.2.5 Simply Connected Lie Groups

A.2.5.1 Let G be a connected Lie group with Lie algebra $\mathfrak{g}$. Let $(\tilde{G}, p)$ be a covering space for G. If $\tilde{e}$ is in $p^{-1}(e)$ then there is a structure of a topological group on G so that $\tilde{e}$ is the identity and p is a group homomorphism. Furthermore $\tilde{G}$ has a C^∞ manifold structure so that p is a local diffeomorphism. Using this it is not hard to show that $\tilde{G}$ is a Lie group where Lie algebra is isomorphic with $\mathfrak{g}$ ($p_*: \tilde{\mathfrak{g}} \to \mathfrak{g}$ is an isomorphism, here $\tilde{\mathfrak{g}}$ is the Lie algebra of $\tilde{G}$).

Combining these remarks we have the following theorem.

A.2.5.2 *Theorem* Let G be a connected Lie group. Then every covering space $(\tilde{G}, p)$ of G has the structure of a Lie group so that p is a Lie homomorphism and p_* is an isomorphism of Lie algebras.

A.2.5.3 Now let $(\tilde{G}, p)$ be the universal covering group of G. Let σ be a deck transformation of G. Let e be the identity of G. Set $g_\sigma = \sigma(\tilde{e})$. Let $\hat{\sigma}(x) = g_\sigma x$. Then $p(\hat{\sigma}(x)) = p(g_\sigma)p(x) = p(x)$. Thus $\hat{\sigma}$ is a deck transformation and $\sigma(e) = \hat{\sigma}(\tilde{e})$. Thus $\hat{\sigma} = \sigma$. Thus the map $\sigma \to g_\sigma$ of $\pi_1(G)$ (the group of deck transformations) into $p^{-1}(e)$ is an injection. If g is in $p^{-1}(e)$ then $\sigma_g(x) = gx$ defines a deck transformation of G.

A.2.5.4 *Theorem* Let G be a connected Lie group and let $(\tilde{G}, p)$ be the universal covering group of G. Then $\pi_1(G)$ is isomorphic with the subgroup $p^{-1}(e)$ of $\tilde{G}$. Furthermore $p^{-1}(e)$ is a central discrete subgroup of $\tilde{G}$. If $\tilde{G}$ is a connected simply connected Lie group and if Γ is a discrete central subgroup of $\tilde{G}$ then $\tilde{G}/\Gamma$ is a Lie group and the natural map of $\tilde{G}$ onto $\tilde{G}/\Gamma$ is a covering space. Furthermore $\pi_1(\tilde{G}/\Gamma)$ is isomorphic with Γ.

PROOF Since p is a local diffeomorphism it is clear that $p^{-1}(e)$ is discrete. Clearly $p^{-1}(e)$ is normal. We leave it to the reader to show that a discrete normal subgroup of a Lie group is central. The last assertions follow directly from A.4.2.9.

A.2.5.5 *Theorem* Let G be a connected and simply connected Lie group with Lie algebra $\mathfrak{g}$. Let H be a Lie group with Lie algebra $\mathfrak{h}$. Let φ be a Lie algebra homomorphism of $\mathfrak{g}$ into $\mathfrak{h}$. Then there is a unique Lie group homomorphism ψ of G onto H so that $\psi_* = \varphi$.

For a proof of this result see F. Warner [1].

APPENDIX 3

A Review of Multilinear Algebra

A.3.1 The Tensor Product

A.3.1.1 Let K be a field. Let $V_1, \ldots, V_r$ and W be vector spaces over K. Then an r-linear map $f: V_1 \times \cdots \times V_r \to W$ is a function such that $v \mapsto f(v_1, \ldots, v_{i-1}, v, v_{i+1}, \ldots, v_r)$ is a linear map of $V_i \to W$ for each $i = 1, \ldots, r$, v_j in V_j, $j \neq i$, $1 \leqslant j \leqslant r$.

A.3.1.2 Note that if f is r-linear from $V_1 \times \cdots \times V_r$ to W and $A: W \to U$ is linear (U a vector space over K) then $A \circ f: v_1 \times \cdots \times V_r \to U$ is r-linear.

A.3.1.3 A tensor product of $V_1, \ldots, V_r$ is a pair $(\mathscr{V}, i)$ where $\mathscr{V}$ is a vector space over K and $i: V_1 \times \cdots \times V_r \to \mathscr{V}$ is r-linear such that if W is a K-vector spa·:e and if $f: V_1 \times \cdots \times V_r \to W$ is r-linear then there is a unique linear map $\tilde{f}: \mathscr{V} \to W$ such that $\tilde{f} \circ i = f$.

A.3.1.4 *Lemma* Let $(\mathscr{V}, i)$, $(\mathscr{V}', i')$ be tensor products of $V_1, \ldots, V_r$. Then there is a unique surjective isomorphism $A: \mathscr{V} \to \mathscr{V}'$ such that $A \circ i = i'$.

PROOF Let $A = \tilde{\imath}'$ and $B = \tilde{\imath}$ (relative to $(\mathscr{V}', i')$). Then clearly $A \circ i = i'$, $B \circ i' = i$, thus AB and BA are the corresponding identity maps.

A.3.1.5 *Lemma* If $V_1, \ldots, V_r$ are vector spaces over K then there exists a tensor product $\mathscr{V}$ of $V_1, \ldots, V_r$ which we denote $V_1 \otimes \cdots \otimes V_r$.

PROOF If S is a set let $V(S)$ be the vector space of all $\varphi: S \to K$ so that $\{s|\varphi(s) \neq 0\}$ is finite. If s is in S let $\varphi_s(t) = 0$ if $s \neq t$, $\varphi_s(t) = 1$ if $s = t$. We write $\sum a_i \varphi_{s_i}$ (a_i in K, s_i in S) as the formal sum $\sum a_i s_i$. Let $V = V(V_1 \times \cdots \times V_r)$. Let $\mathscr{A}$ be the subspace of V spanned by the elements

$$(v_1, \ldots, v_{i-1}, av_i + bw_i, v_{i+1}, \ldots, v_r) - a(v_1, \ldots, v_r)$$
$$- b(v_1, \ldots, v_{i-1}, w_i, v_{i+1}, \ldots, v_r)$$

for all v_i, w_i in V_i, $i = 1, \ldots, r$, a, b in K. Set $\mathscr{V} = V/\mathscr{A}$. Let

$$i: V_1 \times \cdots \times V_r \to \mathscr{V}$$

be given by $i(v_1, \ldots, v_r) = (v_1, \ldots, v_r) + \mathscr{A}$. Then by definition of $\mathscr{A}$, i is r-linear. Let $f: V_1 \times \cdots \times V_r \to W$ be r-linear. Define for

$$\textstyle\sum a_i s_i \text{ in } V, \hat{f}(\sum a_i s_i) = \sum a_i f(s_i).$$

Then $\hat{f}: V \to W$ is linear. Clearly $\ker \hat{f} \supset \mathscr{A}$. Thus $\hat{f}$ induces a linear map $\tilde{f}$ of $\mathscr{V}$ to W. Clearly $\tilde{f} \circ i = f$. The uniqueness of f follows from the fact that $i(V_1 \times \cdots \times V_r)$ spans $\mathscr{V}$.

A.3.1.6 Set $i(v_1, \ldots, v_r) = v_1 \otimes \cdots \otimes v_r$.

A.3.1.7 Let now $V_1, \ldots, V_r$ be K-vector spaces. Let $L(V_1, \ldots, V_r; K)$ be the space of all r-linear maps $V_1 \times \cdots \times V_r \to K$. Let $V_i^* = L(V_i; K)$, the dual space of V_i.

A.3.1.8 If ϕ_i is in V_i^* let $\phi_1 \hat{\otimes} \cdots \hat{\otimes} \phi_r$ be the element of $L(V_1, \ldots, V_r; K)$ given by $(\phi_1 \hat{\otimes} \cdots \hat{\otimes} \phi_r)(v_1, \ldots, v_r) = \phi_1(v_1) \cdots \phi_r(v_r)$.

A.3.1.9 Clearly the map $V_i^* \times \cdots \times V_r^* \to L(V_1, \ldots, V_r; K)$ given by $(\phi_1, \ldots, \phi_r) \to \phi_1 \hat{\otimes} \cdots \hat{\otimes} \phi_r$ is r-linear and hence induces a linear map $\alpha: V_1^* \otimes \cdots \otimes V_r^* \to L(V_1, \ldots, V_r; K)$.

A.3.1.10 Now suppose that the V_i are all finite dimensional. Then if $e_1^j, \ldots, e_{n_j}^j$ is a basis of V_j and $\phi_1^j, \ldots, \phi_{n_j}^j$ is the dual basis then the elements $\phi_{i_r}^1 \hat{\otimes} \cdots \hat{\otimes} \phi_{i_r}^j$ form a basis of $L(V_1, \ldots, V_r; K)$ and the elements $\phi_{i_r}^1 \otimes \cdots \otimes \phi_{i_r}^r$ span $V_1^* \otimes \cdots \otimes V_1^*$. Hence since α is clearly onto, α is an isomorphism. We use α to identify $V_1^* \otimes \cdots \otimes V_r^*$ and $L(V_1, \ldots, V_r; K)$; we also identify $\phi_1 \hat{\otimes} \cdots \hat{\otimes} \phi_r$ with $\phi_1 \otimes \cdots \otimes \phi_r$.

A.3.1.11 From now on in this appendix we will use the term vector space to mean finite dimensional vector space. We define a map

$$\beta\colon (V_1 \otimes \cdots \otimes V_r)^* \to L(V_1, \ldots, V_r; K)$$

by $\beta(u)(v_1, \ldots, v_r) = u(v_1 \otimes \cdots \otimes v_r)$. Then as above β is an onto isomorphism. We therefore see that $(V_1 \otimes \cdots \otimes V_r)^*$ is canonically isomorphic with $V_1^* \otimes \cdots \otimes V_r^*$ and $L(V_1, \ldots, V_r; K)$. We will take all of these natural maps to be identifications.

A.3.1.12 Let V be a K-vector space. We set $\otimes^r V = V \otimes \cdots \otimes V$ r-times. We use the universal property of the tensor product to identify $(V_1 \otimes V_2) \otimes V_3$ with $V_1 \otimes V_2 \otimes V_3$ and $(v_1 \otimes v_2) \otimes v_3$ with $v_1 \otimes v_2 \otimes v_3$; similarly for $v_1 \otimes (v_2 \otimes v_3)$ with $v_1 \otimes v_2 \otimes v_3$.

A.3.1.13 Thus if we set $T(V) = \sum_{k=0} \otimes^k V$. Then $T(V)$ is an associative algebra over K under the operation $\otimes$. $T(V)$ is called the tensor algebra over V.

A.3.1.14 We now assume that K has characteristic 0 (e.g., $K = R$ or C). We define the process of alternation on $\otimes^r V$. Define $A^r\colon \otimes^r V \to \otimes^r V$ by $(A^r(v_1, \ldots, v_r) = 1/r! \sum_{\sigma \in S_r} \text{sgn}(\sigma) v_{\sigma 1} \otimes \cdots \otimes v_{\sigma r}$ where S_r is the symmetric group in a r-letters and $\text{sgn}(\sigma)$ is the sign of σ.

$$(\text{sgn}(\sigma) = \textstyle\prod_{i<j} (x_{\sigma_i} - x_{\sigma_j}) / \prod_{i<j} x_i - x_j, \text{ the } x_i \text{ distinct}).$$

We note that $(A^r)^2 = A^r$. Thus $\otimes^r V = A^r(\otimes^r V) \oplus (I - A^r)(\otimes^r V)$. We set $\Lambda^r V = A^r(\otimes^r V)$, and we define $v_1 \wedge \cdots \wedge v_r = A^r(v_1 \otimes \cdots \otimes v_r)$.

A.3.1.15 We now find a universal problem to which $\Lambda^r V$ is a solution. A map $f\colon V \times \cdots \times V \to W$ is called r-alternating if it is r-linear and if for each $\sigma \in S_r$, $f(v_{\sigma_1}, \ldots, v_{\sigma_r}) = \text{sgn}(\sigma) f(v_1, \ldots, v_r)$. We note that if $j\colon V \times \cdots \times V \to \Lambda^r V$ is defined by $j(v_1, \ldots, v_r) = A^r(v_1 \otimes \cdots \otimes v_r) = v_1 \wedge \cdots \wedge v_r$, and if $f\colon V \times \cdots \times V \to W$ is r-alternating then $\ker f \subset (I - A^r)(\otimes^r V)$, thus $f\colon \Lambda^r V \to W$ is uniquely defined. Let $\Lambda V = \sum \Lambda^r V$. We make ΛV into an algebra by defining $u \wedge v = A^{r+s}(u \otimes v)$ for u in $\Lambda^r V$, v in $\Lambda^s V$.

A.3.1.16 Let $f: V \times \cdots \times V \to W$ be r-linear and suppose that $f(v_1, \ldots, v_i, \ldots, v_j, \ldots, v_r) = f(v_1, \ldots, v_j, \ldots, v_i, \ldots, v_r)$ for all $i, j, v_1, \ldots, v_r$ in V. Then f is called multilinear.

A.3.1.17 Let $(S^r(V), j)$ be the solution of the universal problem: $j: V \times \cdots \times V \to S^r(V)$ is r-linear and symmetric and if W is a K-vector space, $f: V \times \cdots \times V \to W$, r-linear and symmetric, then there is a unique linear map $\hat{f}: S^r(V) \to W$ such that $\hat{f} \circ j = f$.

A.3.1.18 We construct $S^r(V)$ using the symmetrization operator. That is, let $S_r: \otimes^r V \to \otimes^r V$ be defined by $S_r(v_1 \otimes \cdots \otimes v_r) = 1/r! \sum_{\sigma \in S_r} v_{\sigma_1} \otimes \cdots \otimes v_{\sigma_r}$. Then $S_r^2 = S_r$. As in the case of $\Lambda^r V$ we find that $S_r(\otimes^r V) = S^r(V)$ with j given by $S_r \circ i$, $i: V \times \cdots \times V \to \otimes^r V$.

A.3.1.19 Set $S(V) = \sum_r S^r(V)$. We make $S(V)$ into an algebra by defining $u \cdot v = S_{r+k}(u \otimes v)$ for u in $S^r(V)$, u in $S^k(V)$.

A.3.1.20 Let $V_1, \ldots, V_r, W_1, \ldots, W_r$ be vector spaces. Let $A_i: V_i \to W_i$ be linear. We define an r-linear map $v_1, \ldots, v_r \to A_1 v_1 \otimes \cdots \otimes A_r v_r$ of $V_1 \times \cdots \times V_r \to W_1 \otimes W_2 \otimes \cdots \otimes W_r$. This induces a linear map $A_1 \otimes \cdots \otimes A_r: V_1 \otimes \cdots \otimes V_r \to W_1 \otimes \cdots \otimes W_r$.

A.3.1.21 Clearly if $A: V \to W$ is linear and we set $\otimes^r A = A \otimes \cdots \otimes A$ then $\otimes^r A(\Lambda^r V) \subset \Lambda^r W$ and $\otimes^r A(S^r(V)) \subset S^r(W)$. We denote the corresponding restrictions, respectively, by $\Lambda^r A$ and $S^r(A)$.

A.3.1.22 Clear $(A_1 \otimes \cdots \otimes A_r) \circ (B_1 \otimes \cdots \otimes B_r) = A_1 B_1 \otimes \cdots \otimes A_r B_r$. Hence $\Lambda^r A \circ \Lambda^r B = \Lambda^r(A \circ B)$, $S^r(A) \circ S^r(B) = S^r(A \circ B)$.

APPENDIX 4

Integration on Manifolds

A.4.1 k-forms

A.4.1.1 Let M be a C^∞ manifold and let $T(M)$ be the tangent bundle of M. Let $T(M)^*$ be the dual bundle to $T(M)$; see 1.2. Let $\Lambda^k T(M)^*$ be the kth Grassmann power of $T(M)^*$. Then $\Lambda^k T(M)^*$ is a C^∞ vector bundle over M.

A.4.1.2 *Definition* A C^∞ cross section of $\Lambda^k T(M)^*$ is called k-form on M.

A.4.1.3 We note that by the definition of $\Lambda^k T(M)^*$ if $\omega \in \Lambda^k T(M)^*_p$ then ω is a k-alternating form on $X^k T(M)_p$ (the k-fold Cartesian product of $T(M)_p$ with itself). Furthermore if $X_1, \ldots, X_k \in \mathfrak{X}(M)$ and ω is a k-form on M then the map $p \to \omega_p(X_{1_p}, \ldots, X_{k_p})$ is a C^∞ function on M. Indeed if $p \in M$ let U be a neighborhood of p with local coordinates $x_1, \ldots, x_n$. Let $dx_{i_p}(\partial/\partial x_{j_p}) = \delta_{ij}$. Then if

$$\omega \in \Lambda^k T(M)^*_j, \; \omega = \textstyle\sum_{1 < i_1 < \cdots < ih < n} a_{i_1 \cdots i_k} dx_{i_p} \wedge \cdots \wedge dx_{i_k}.$$

ω is a C^∞ cross section of $\Lambda^k T(M)^*$ if and only if ω restricted to U is given by

$$\omega = \sum a_{i_1 \cdots i_k} dx_{i_1} \wedge \cdots \wedge dx_{i_k}$$

and $a_{i_1 \cdots i_k} \in C^\infty(U)$.

A.4.1.4 The totality of k-forms on M is denoted by $\mathscr{D}^k(M)$.

A.4.1.5 *Definition* An n-form ω on M is said to be a volume form if $\omega_p \neq 0$ for any $p \in M$.

A.4.1.6 *Definition* Let M be a C^∞ manifold. Then an orientation of M is a choice of a subatlas of the C^∞ structure of M, $\{(\psi_\alpha, U_\alpha)\}_{\alpha\in I}$ so that if $U_\alpha \cap U_\beta \neq \phi$ then the Jacobian of $\psi_\alpha \circ \psi_\beta^{-1}: \psi_\alpha(U_\alpha \cap U_\beta) \to \psi_\beta(U_\alpha \cap U_\beta)$ is positive.

A C^∞ manifold is said to be orientable if it has an orientation.

A.4.1.7 *Theorem* Let M be a C^∞ manifold. If M has a volume form then M is orientable. If M is orientable then M has a volume form.

PROOF Suppose M has a volume form ω. Let $\{(U_\alpha, \psi_\alpha)\}_{\alpha\in I}$ be an atlas for M with U_α, $\alpha \in I$, connected. Let $\{x_1^\alpha, \ldots, x_n^\alpha\}$ be the corresponding local coordinates on U. Then since $\Lambda^n T(M)_p^*$ is one-dimensional for each $p \in U$ we see that $\omega|_U = \varphi_\alpha dx_\alpha^1 \wedge \cdots \wedge dx_n^\alpha$. Since ω is never zero we see that either $\varphi_\alpha(p) > 0$ for all $p \in U_\alpha$ or $\varphi_\alpha(p) < 0$ for all $p \in U$. If $\varphi_\alpha < 0$, replace ψ_α by $(x_2^\alpha, x_1^\alpha, \ldots, x_n^\alpha)$. We may therefore assume that $\varphi_\alpha > 0$ for all α. Now by definition of Jacobian of a differentiable map we see that if $J_{\psi_\alpha \circ \psi_\beta^{-1}}$ is the Jacobian of $\psi_\alpha \circ \psi_\beta^{-1}$ then $dx_1^\alpha \wedge \cdots \wedge dx^n = J_{\psi_\alpha \circ \psi_\beta^{-1}} dx_1 \ldots dx_n$. But $dx_1^\alpha \ldots dx_n' = \varphi_\alpha^{-1}\omega$. Thus $J_{\psi_\alpha \circ \psi_\beta^{-1}} = \varphi_\alpha/\varphi_\beta > 0$. Hence $\{(U_\alpha, \psi_\alpha)\}_{\alpha\in I}$ is an orientation for M.

Suppose now that M is orientable. Let $\{(U_\alpha, \psi_\alpha)\}_{\alpha\in I}$ be an orientation of M. Let $\{x_1^\alpha, \ldots, x_n^\alpha\}$ be the corresponding local coordinates on U_α. We may assume that $\{U_\alpha\}_{\alpha\in I}$ is locally finite and that $\{\varphi_\alpha\}$ is a partition of unity subordinate to $\{U_\alpha\}_{\alpha\in I}$. Set

$$\omega_\alpha = \begin{cases} \varphi_\alpha dx_1 \ldots dx_n & \text{on} \quad U_\alpha \\ 0 & \text{on} \quad M - U_\alpha. \end{cases}$$

Then $\omega_\alpha \in \mathscr{D}^n(M)$. Furthermore on $U_\alpha \cap U_\beta$,

$$dx_1^\alpha \ldots dx_n^\alpha = \det(\partial x^n/\partial x^n) dx_1^\beta \ldots dx_n^\beta, \text{ hence } dx_1^\alpha \ldots dx_n^\beta$$

is a positive multiple of $dx_1 \ldots dx_n$ on $U_\alpha \cap U_\beta$. Hence $\sum_{\alpha\in I} \omega_\alpha$ defines a volume element for M.

A.4.1.8 Let M and N be C^∞ manifolds. Let $f: M \to N$ be a C^∞ mapping. We define for each k, $f^*: \mathscr{D}^k(N) \to \mathscr{D}^k(M)$ by $(f^*\omega)_p(v_1, \ldots, v_k) = \omega_{f(p)}(f_{*p}v_1, \ldots, f_{*p}v_k)$.

A.4.2 Integration on Manifolds

A.4.2.1 Let M be an orientable C^∞ manifold and let ω be a volume form on M. Let $C_0(M)$ be the space of all complex valued continuous functions on M with compact support. In this section we define the notion

$$\int_M f\omega$$

for $f \in C_0(M)$.

A.4.2.2 Let $\{(U_\alpha, \psi_\alpha)\}_{\alpha \in I}$ be a locally finite orientation for M and let $\{\varphi\}_{\alpha \in I}$ be a partition of unity subordinate to $\{U_\alpha\}_{\alpha \in I}$. We also may assume that $\psi_\alpha(U_\alpha)$ is a cube in R^n. We assume that if $\psi_\alpha = (x_1^\alpha, \ldots, x_n^\alpha)$ then $dx_1^\alpha \wedge \cdots \wedge dx_n^\alpha = f_\alpha \omega, f > 0$.

A.4.2.3 We fix α and show first of all how one integrates functions $f \in C_0(M)$ so that $\operatorname{supp} f \subset U_\alpha$. $\omega|_{U_\alpha} = g_\alpha dx_1^\alpha \ldots dx_n^\alpha$ $(g_\alpha = f_\alpha^{-1})$. Define

$$\mu_\alpha(f) = \int_{\psi_\alpha(U_\alpha)} (f \circ \psi_\alpha^{-1})(g_\alpha \circ \psi_\alpha^{-1}) dx_1 dx_2 \ldots dx_n.$$

Here the above integral is the standard multiple integral of advanced calculus. The change of variable theorem of advanced calculus shows that if (U_β, ψ_β) is another chart for M so that $\operatorname{supp} f \subset U_\beta$ and $\psi_\beta^* \, dx_1 \wedge \cdots \wedge dx_n$ is a positive multiple of ω then $\mu_\beta(f) = \mu_\alpha(f)$. (A chart (U, ψ) so that $\psi^* dx_1 \wedge \cdots \wedge dx_n$ is a positive multiple of ω is called a positive chart.)

A.4.2.4 Thus whenever there is (U, ψ) a positive chart so that $\operatorname{supp} f \subset U$, we have

$$\int_M f\omega$$

well defined.

A.4.2.5 Let $f \in C_0(M)$. Then only a finite number of the U_α intersect $\operatorname{supp} f$. We set

$$\int_M f\omega = \sum_{a\in I} \int_M (\varphi_\alpha f)\omega.$$

We show that $\int_M f\omega$ is well defined.

Let $\{(V_\gamma, \eta_\gamma)\}_{\gamma\in J}$ be another such locally finite positive atlas with partition of unity $\{\xi_\gamma\}_{\gamma\in J}$ subordinate to $\{(V_\gamma)\}_{\gamma\in J}$. $\int_M \varphi_\alpha f\omega = \sum_\gamma \int_M \xi_\gamma\varphi_\alpha f\omega$. Thus

$$\sum_\alpha \int_M \varphi_\alpha f\omega = \sum_{\alpha,\gamma} \int_M \xi_\gamma\varphi_\alpha f\omega = \sum_\gamma \int_M \xi_\gamma f\omega.$$

(All orders of sums can be interchanged since they are actually finite sums.) Hence $\int_M f\omega$ is well defined.

We note for further reference:

A.4.2.6 *Lemma* Let $f \in C_0(M)$, $f \not\equiv 0$, $f \geqslant 0$. Then $\int_M f\omega > 0$.

PROOF This follows directly from the fact that there is a positive coordinate patch for M on which f is positive.

A.4.2.7 Let M_1, M_2 be orientable paracompact C^∞ manifolds. Let $M = M_1 \times M_2$. Let ω_i be a volume form on M_i for $i = 1, 2$. Let $p_i\colon M \to M_i$ be the projection on the ith factor for $i = 1, 2$. Set $\omega = p_1^*\omega_1 \wedge p_2^*\omega_2$. Then ω is a volume form on $M_1 \times M_2$. Using a partition of unity argument and Fubini's theorem in $R^{n_1+n_2}$ we leave it to the reader to prove.

A.4.2.8 *Lemma*

$$\int_M f\omega = \int_{M_1}\left(\int_{M_2} f(x, y)\omega_2\right)\omega_1 = \int_{M_2}\left(\int_{M_1} f(x, y)\omega_1\right)\omega_2.$$

A.4.2.9 The standard change of variable formula of advanced calculus and a partition unity argument would allow the reader to prove the following lemma.

A.4.2.10 *Lemma* Let $\Phi\colon M \to M$ be a diffeomorphism of M. Then

$$\int_M f\Phi^*\omega = \int_M (f \circ \Phi^{-1})\omega.$$

A.4.2.11 *Lemma* Let M_1, M_2 be connected C^∞ manifolds. Let ω be a volume form on M_2. Let $\varphi\colon M_1 \to M_2$ be a w-fold covering (that is, φ is a covering mapping and $\varphi^{-1}(p)$ has w elements for each p in M_2). Then if f is in $C_0(M_2)$

$$w\int_{M_2} f\omega = \int_{M_1} (f \circ \varphi)\varphi^*\omega.$$

A.4.2.12 A proof of this result can be found, for example, in F. Warner [1]. It is also suggested that the reader who is unfamiliar with integration on manifolds read F. Warner's excellent chapter on integration on manifolds.

APPENDIX 5

Complex Manifolds

A.5.1 Basic Concepts

A.5.1.1 *Definition* Let X be a topological space. A C^∞ atlas for X, $\{(U_\alpha, \psi_\alpha)\}_{\alpha\in I}$, is said to be a complex structure if

(1) $\psi: U_\alpha \to C^n$.

(2) If $U_\alpha \cap U_\beta \neq \phi$ then $\psi_\alpha \circ \psi_\beta^{-1}: \psi_\beta(U_\alpha \cap U_\beta) \to \psi_\alpha(U_\alpha \cap U_\beta)$ is holomorphic. That is, $D(\psi_\alpha \circ \psi_\beta^{-1})(x)$ is complex linear for each $x \in \psi_\beta(U_\alpha \cap U_\beta)$. (Here $Df(x)$ is the differential of f at x.)

(3) $\{(U_\alpha, \psi_\alpha)\}_{\alpha\in I}$ is maximal subject to (1) and (2).

A.5.1.2 A pair of a topological space and a complex structure is called a complex manifold. Clearly a complex manifold is a C^∞ manifold.

A.5.1.3 Let M be a complex manifold. Let $\mathfrak{B}$ be the complex structure on M. $(U, \psi) \in \mathfrak{B}$ is called a complex chart. We note that if $x \in M$, then using $\mathfrak{B}$ we may define the structure of a complex vector space on $T(M)_x$. Indeed, let $x \in U$, $(U, \psi) \in \mathfrak{B}$. Then $\psi_{*x}: T(M)_x \to C^n$ is a (real) linear map. Give $T(M)_x$ the structure of a complex vector space that makes φ_{*x} C-linear. If $x \in (V, \phi) \in \mathfrak{B}$ then $\phi_{*x} = (\phi_{*x} \circ \psi_{*\psi(x)}^{-1}) \circ \psi_{*x}$, thus ϕ_{*x} is complex linear relative to the complex structure on $T(M)_x$ defined by (U, ψ). Hence $T(M)_x$ has a well defined complex structure.

A.5.1.4 For each $x \in M$ let $J_x: T(M)_x \to T(M)_x$ be given by multiplication by i. By its definition we have

(1) $J_x^2 = -I$ for each $x \in M$.

(2) If $X \in (M)$ then $x \to J_x X_x$ is in $\mathfrak{X}(M)$.

A.5.1.5 *Definition* A pair (M, J) with M a C^∞ manifold and J a C^∞ cross section of Hom$(T(M), T(M))$ (see 1.3.7) satisfying (1) and (2) is called an almost complex manifold.

A.5.1.6 Let M and N be complex manifolds and let $f\colon M \to N$ be a C^∞ mapping. Then f is said to be holomorphic if $f_{*x}\colon T(M)_x \to T(N)_{f(x)}$ is complex linear. That is, $f_{*x} \cdot J_x = J_{f(x)} f_{*x}$ where we denote by J the almost complex structures of M and N. It is easy to see that this is equivalent to saying that given $x \in M$ there are complex charts (U, ψ) and (V, ϕ) for M and N, respectively, so that $x \in U$ and $f(U) \subset V$ and

$$\phi \circ f \circ \psi^{-1}\colon \psi(U) \to \phi(V)$$

is holomorphic.

A.5.2 The Holomorphic and Antiholomorphic Tangent Spaces

A.5.2.1 Let M be a complex manifold and let (U, ψ) be a complex chart for M. Then $\psi\colon U \to C^n$, hence $\psi(p) = (z_1(p), \ldots, z_n(p))$, $z_i\colon U \to C$, z_i holomorphic. Set $z_i = x_i + \sqrt{-1}y_i$, $x_i\colon U \to R$, $y_i\colon U \to R$, $i = 1, \ldots, n$. Then $\{x_1, y_1, \ldots, x_n, y_n\}$ is a system of local coordinates on U. Furthermore if $p \in M$ then $J_p(\partial/\partial x_{i_p}) = (\partial/\partial y_{i_p})$ and $J_p(\partial/\partial y_{i_p}) = -(\partial/\partial x_{i_p})$.

A.5.2.2 Let $T^C(M) = T(M) \otimes C$, that is, take $T(M)$ to be a real vector bundle and tensor it with the trivial complex line bundle. Then $T^C(M)$ is a complex vector bundle over M. Now $T^C(M)_p$ splits into a direct sum $\mathscr{T}(M)_p \oplus \bar{\mathscr{T}}(M)_p$ where $J_p|_{\mathscr{T}(M)_p} = \sqrt{-1}I$ and $J_p|_{\bar{\mathscr{T}}(M)_p} = -\sqrt{-1}I$. Furthermore since J_p is the extension of a real operator $\dim \mathscr{T}(M)_p = \dim \bar{\mathscr{T}}(M)_p$.

A.5.2.3 Returning to local coordinates we note that if we set

$$\frac{\partial}{\partial z_i} = \frac{1}{2}\left(\frac{\partial}{\partial x_i} - \sqrt{-1}\,\frac{\partial}{\partial y_i}\right)$$

$$\frac{\partial}{\partial \bar{z}_i} = \frac{1}{2}\left(\frac{\partial}{\partial x_i} + \sqrt{-1}\,\frac{\partial}{\partial y_i}\right)$$

then

$$J_p \frac{\partial}{\partial z_{i_p}} = \sqrt{-1}\,\frac{\partial}{\partial z_{i_p}}, \qquad J_p \frac{\partial}{\partial \bar{z}_{i_p}} = -\sqrt{-1}\,\frac{\partial}{\partial \bar{z}_{i_p}}.$$

Thus $(\partial/\partial z_{1_p}), \ldots, (\partial/\partial z_{n_p})$ is a basis of $\mathscr{T}(M)_p$ and $(\partial/\partial \bar{z}_{1p}), \ldots, (\partial/\partial \bar{z}_{n_p})$ is a basis of $\bar{\mathscr{T}}(M)_p$. Let $\mathscr{T}(M) = \bigcup_{p\in M} \mathscr{T}(M)_p$, $\bar{\mathscr{T}}(M) = \bigcup_{p\in M} \bar{\mathscr{T}}(M)_p$ and give each the subspace topology in $T^C(M)$. Let $\pi: \mathscr{T}(M) \to M$ (resp. $\pi: \bar{\mathscr{T}}(M) \to M$) be the restriction of the projection on $T^C(M)$.

A.5.2.4 Using $z_1, \ldots, z_n$ on U we define a map $\Phi: \mathscr{T}(M)|_U \to U \times C^n$ (resp. $\bar{\Phi}: \bar{\mathscr{T}}(M)|_U \to U \times C^n$). Let $p \in U$, $v \in \mathscr{T}(M)_p$, then $v = \sum v_i(\partial/\partial z_{i_p})$. Set $\Phi(v) = (p, (v_1, \ldots, v_n))$, (resp. $v \in \bar{\mathscr{T}}(M)_p$, $\bar{\Phi}(v) = (p, (v_1, \ldots, v_n))$, $v = \sum v_i(\partial/\partial \bar{z}_i)$. Using the complex structure $\mathfrak{B}$ it is easy to check using this formalism that $\mathscr{T}(M)$ and $\bar{\mathscr{T}}(M)$ are C^∞ vector bundles over M. We show that that the above charts on $\mathscr{T}(M)$ actually give $\mathscr{T}(M)$ the structure of a complex manifold.

A.5.2.5 Let $(U_1, \psi_1), (U_2, \psi_2) \in \mathfrak{B}$ so that $U_1 \cap U_2 \neq \phi$. Let $(z_1^2, \ldots, z_n^1)$, $(z_1^2, \ldots, z_n^2)$ be the corresponding holomorphic local coordinates on U_1 and U_2, respectively. Let $\Phi_i: \mathscr{T}(M)|_{U_i} \to U_i \times C^n$ be the map defined as above for $i = 1, 2$. $\Phi_1 \circ \Phi^{-1}: U_1 \cap U_2 \times C^n \to U_1 \cap U_2 \times C^n$ and $\Phi_1 \circ \Phi^{-1}(p, v) = (p, (\psi_1 \circ \psi_2^{-1})_{*\Phi_2(p)}v)$. Now by the definition of holomorphic chart $(p, v) \to (\psi_1 \circ \psi^{-1})_{*\psi_2(p)}v$ is a holomorphic map in (p, v). Thus the $\{(\mathscr{T}(M)|_U, \bar{\Phi})\}$ defined above define a complex structure on $\mathscr{T}(M)$. Furthermore $\pi: \mathscr{T}(M) \to M$ is a holomorphic map.

A.5.3 Complex Lie Groups

A.5.3.1 *Definition* A complex Lie group is a complex manifold G which is also a Lie group so that the map $G \times G \to G$ given by $(x, y) \mapsto xy^{-1}$ is holomorphic.

A.5.3.2 If $\mathfrak{g}$ is the Lie algebra of G and if J is the almost complex structure on G corresponding to the complex structure we note that $g_{*p}J_p = J_{g\cdot p}g_{*p}$, hence if $X \in \mathfrak{g}$, $g_{*p}J_pX_p = J_{g\cdot p}g_{*p}X_p = J_{g\cdot p}X_{g\cdot p}$. Thus JX defined by $p \to J_pX_p$ is in $\mathfrak{g}$. Since $J^2 = -I$ we see that J makes $\mathfrak{g}$ into a complex vector space.

A.5.3.3 It can be shown that if X, Y are in $\mathfrak{g}$ then $[JX, Y] = J[X, Y]$, thus $\mathfrak{g}$ is actually a complex Lie algebra (see Kobayashi-Nomizu [1]).

A.5.3.4 *Theorem* Let G be a Lie group so that the Lie algebra of G, $\mathfrak{g}$, has a complex structure that makes $\mathfrak{g}$ a Lie algebra over C. Then G has the structure of a complex Lie group.

For a proof of this result see Kobayashi-Nomizu [1], chapter IX.

Just as in the real case we have the following lemma.

A.5.3.5 *Theorem* Let G be a complex Lie group and let H be a closed complex subgroup of G (that is, $i: H \to G$ is holomorphic). Then G/H has a unique complex structure so that:

(i) There is a neighborhood U of a H in G/H and a holomorphic map $\psi: U \to G$ so that $\pi \circ \psi(u) = u$ for $u \in U$. ($\pi: G \to G/H$ is the natural map.)

(ii) The natural map $G \times G/H \to G/H$ is holomorphic.

See for example Kobayashi-Nomizu [1].

APPENDIX 6

Elementary Functional Analysis

A.6.1 Banach Spaces

A.6.1.1 *Definition* Let B be a complex vector space. A norm on B is a function $x \to \|x\|$ of B to R so that

(1) $\|x\| \geqslant 0$ for all x in B.
(2) $\|x\| = 0$ if and only if $x = 0$.
(3) $\|ax\| = |a| \, \|x\|$ for a in C, x in B.
(4) $\|x + y\| \leqslant \|x\| + \|y\|$.

A.6.1.2 A pair $(B, \| \ldots \|)$ of a complex vector space and a norm $\|\ldots\|$ is called a normed space.

A.6.1.3 If $(B, \|\ldots\|)$ is a normed space then we make B into a metric space by defining $d(x, y) = \|x - y\|$. We give B the corresponding topology.

A.6.1.4 $(B, \|\ldots\|)$ is called a Banach space if d is complete. The usual proof of the existence of a completion of a metric space proves the following theorem.

A.6.1.5 *Theorem* Let $(B_0, \|\ldots\|)$ be a normed space. Then there exists a Banach space $(B, \|\ldots\|)$ and a continuous, injective linear map, i: $B_0 \to B$ such that

(i) $i(B_0)$ is dense in B.

(ii) $\|i(x)\| = \|x\|$.

A.6.1.6   *Definition*   Let $(B_1, \|\ldots\|_1)$, $(B_2, \|\ldots\|_2)$ be normed spaces. Let $A: B_1 \to B_2$ be a linear map. Then A is said to be bounded if there is a constant C so that

$$\|Ax\|_2 \leqslant C\|x\|_1.$$

A.6.1.7   *Lemma*   A linear map $A: B_1 \to B_2$ is bounded if and only if it is continuous.

PROOF   If A is bounded then $\|A(x - y)\|_2 \leqslant C\|x - y\|_1$. Thus A is continuous. Suppose A is continuous but not bounded. Then there is a sequence $\{\varphi_i\}$ in B_1 so that $\|\varphi_i\|_1 = 1$ and $\|A\varphi_i\|_2 \geqslant i^2$. Thus $\lim_{n \to \infty} \|A(1/n)\varphi_n\|_2 \geqslant \lim_{n \to \infty} n = \infty$ but $\lim_{n \to \infty} 1/n\varphi_n = 0$, a contradiction.

A.6.1.8   *Lemma*   Let $(B_1, \|\ldots\|_1)$ and $(B_2, \|\ldots\|_2)$ be normed spaces. Let $\bar{B}_1$, $\bar{B}_2$ be the completions of B_1 and B_2, respectively. Let $A: B_1 \to B_2$ be a bounded linear operator. Then A has a unique extension to a bounded linear map of $\bar{B}_1$ to $\bar{B}_2$.

PROOF   Let $\{x_n\}$ be a Cauchy sequence in B_1. Then $\{x_n\}$ converges to a typical element x_0 of $\bar{B}_1$. Let C be such that $\|Ax\|_2 \leqslant C\|x\|_1$ for all x in B_1. Clearly $\{Ax_n\}$ is Cauchy and hence converges to y_0 in B_2. We assert that y_0 depends only on x_0 and $\|y_0\|_2 \leqslant C\|x_0\|_1$. In fact, if $\{z_n\}$ is a sequence in B_1 such that $\lim_{n \to \infty} z_n = x_0$ then $\{x_n - z_n\}$ is a sequence in B_1 converging to zero. Hence $\lim_{n \to \infty}(Ax_n - Az_n) = 0$. But $\lim_{n \to \infty} Ax_n = y_0$, thus $\lim_{n \to \infty} Az_n = y_0$. Also, $\|Ax_n\|_2 \leqslant C\|x_n\|_1$. Hence $\lim_{n \to \infty} \|Ax_n\|_2 \leqslant \lim_{n \to \infty} C\|x_n\|_1 = C\|x_0\|_1$.

Define $\bar{A}x_0 = y_0$. Then A is well defined, linear, and $\|Ax_0\|_2 \leqslant C\|x_0\|_1$. Thus $\bar{A}$ is continuous. Since B_1 is dense in $\bar{B}_1$, $\bar{A}$ is the unique extension.

A.6.1.9   Let $(B, \|\ldots\|)$ be a Banach space. Let $L(B, B)$ be the set of all bounded linear maps $T: B \to B$. Define $\|T\| = \sup\{\|Tv\| \mid \|v\| = 1\}$.

A.6.1.10   *Lemma* $(L(B, B), \|\ldots\|)$ is a Banach space. Furthermore, if T, S are in $L(B, B)$ then $\|TS\| \leqslant \|T\| \|S\|$.

PROOF   Clearly $(L(B, B), \|\ldots\|)$ satisfies A.6.1.1,(1)–(3). Let T and S

be in $L(B, B)$, then $\|Tv\| \leqslant \|T\|\,\|v\|\ \|Sv\| \leqslant \|S\|\,\|v\|$, $\|(S + T)v\| \leqslant \|Sv\| + \|Tv\| \leqslant (\|S\| + \|T\|)\|v\|$. But then if $\|v\| \leqslant 1$, $\|(S + T)v\| \leqslant \|S\| + \|T\|$. Hence $\|S + T\| \leqslant \|S\| + \|T\|$. Similarly $\|STv\| \leqslant \|S\|\,\|Tv\| \leqslant \|S\|\,\|T\|\,\|v\|$. Thus $\|ST\| \leqslant \|S\|\,\|T\|$.

Let $\{A_k\}$ be a Cauchy sequence in $L(B, B)$. Then given $\varepsilon > 0$, there is n so that if k, $1 \geqslant n$, $\|A_k v - A_l v\| \leqslant \varepsilon\|v\|$, for all v in B. Thus for fixed v in B, $\{A_k v\}$ is a Cauchy sequence. Set $Av = \lim_{k\to\infty} A_k v$. Then A is clearly linear. Now let n be so large that $\|A_k - A_l\| \leqslant 1$ for k, $1 \geqslant n$. Now $\|Av\| \leqslant \|Av - A_k v\| + \|A_k v\|$. Now $Av = \lim_{k\to\infty} A_k v$. Thus

$$\|Av\| \leqslant \lim_{k\to\infty} \|A_k v - A_n v\| + \|A_n v\| \leqslant \|v\| + \|A_n\|\,\|v\|.$$

Hence $\|Av\| \leqslant (1 + \|A_n\|)\|v\|$. Hence A is in $L(B, B)$. Let $\varepsilon > 0$ be given. let n be such that if $k, l \geqslant n$ then $\|A_k - A_l\| < \varepsilon/2$. Then if $\|v\| = 1$, $k \geqslant n$, $\|(A - A_k)v\| \leqslant \|Av - A_l v\| + \|A_k v - A_l v\|$. Let l be so large ($l \geqslant n$) that $\|Av - A_l v\| < \varepsilon/2$. Then $\|(A - A_k)v\| < \varepsilon$. Thus if $k \geqslant n$, v in B, $\|v\| = 1$, $\|(A - A_k)v\| < \varepsilon$. Hence, $\|A - A_k\| < \varepsilon$. This implies $\lim_{k\to\infty} A_k = A$.

A.6.1.11 Let $GL(B)$ be the set of all T in $L(B, B)$ such that T is surjective and there exists S in $L(B, B)$ such that $ST = TS = I$ the identity.

A.6.1.12 *Lemma* If A is in $L(B, B)$ and $\|A\| < 1$ then $I - A$ is in $GL(B)$.

PROOF Let $S_n = I + A + \cdots + A^n$. We assert that $\lim_{n\to\infty} S_n$ exists in $L(B, B)$. Indeed, if $k \geqslant l$,

$$\|S_k - S_l\| = \|A^{l+1} + \cdots + A^k\| = \|A^{l+1}(I + \cdots + A^{k-l-1})\|$$
$$\leqslant \|A\|^{l+1}(1 + \|A\| + \cdots + \|A\|^{k-l-1}) \leqslant \|A\|^{l+1}\left(\frac{1}{1 - \|A\|}\right).$$

Hence $\{S_n\}$ is Cauchy. A.7.1.11 implies $\lim_{n\to\infty} S_n = S$ exists and is in $L(B, B)$. We compute

$$\begin{aligned}(I - A)S &= \lim_{n\to\infty} (I - A)S_n = \lim_{n\to\infty} ((I - A)(I + A + \cdots + A^n)) \\ &= \lim_{n\to\infty} (I + A + \cdots + A^n - (A + A^2 + \cdots + A^{n+1})) \\ &= \lim_{n\to\infty} (I - A^{n+1}).\end{aligned}$$

Now $\lim_{n\to\infty} \|A^n\| \leqslant \lim_{n\to\infty} \|A\|^n = 0$. Thus $(I - A)S = I$. Clearly $(I - A)S_n = S_n(I - A)$. Thus $S(I - A) = I$.

A.6.2 Hilbert Spaces

A.6.2.1 *Definition* A pair $(H, \langle\ ,\ \rangle)$ of a complex vector space and a Hermitian inner product on H is called a pre-Hilbert space.

A.6.2.2 *Lemma* Let $(H, \langle\ ,\ \rangle)$ be a pre-Hilbert space. If $x, y \in H$ then

$$|\langle x, y\rangle|^2 \leqslant \langle x, x\rangle \langle y, y\rangle$$

with equality if and only if there are complex scalars a, b not both zero so that $ax + by = 0$.

PROOF Since the above inequality is clear for $y = 0$ we assume $y \neq 0$. Let $\lambda \in C$. Then $0 \leqslant \langle x - \lambda y, x - \lambda y\rangle = \langle x, x\rangle - \bar{\lambda}\langle x, y\rangle - \lambda\langle y, x\rangle + \lambda\bar{\lambda}\langle y, y\rangle$. Hence taking $\lambda = \langle x, y\rangle/\langle y, y\rangle$ we see

$$0 \leqslant \langle x, x\rangle - \frac{\langle y, x\rangle\langle x, y\rangle}{\langle y, y\rangle} - \frac{\langle x, y\rangle\langle y, x\rangle}{\langle y, y\rangle} + \frac{\langle x, y\rangle\langle y, x\rangle}{\langle y, y\rangle}.$$

Hence $0 \leqslant \langle x, x\rangle - (\langle x, y\rangle\langle y, x\rangle)/\langle y, y\rangle$, the desired inequality. If equality holds then $\langle x - (\langle x, y\rangle/\langle y, y\rangle)y, x - (\langle x, y\rangle/\langle y, y\rangle)y\rangle = 0$, hence $x = (\langle x, y\rangle/\langle y, y\rangle)y$. This proves the lemma.

A.6.2.3 Now if $(H, \langle\ ,\ \rangle)$ is a pre-Hilbert space define $\|x\| = \sqrt{\langle x, x\rangle}$. A.6.2.2 implies that $(H, \|\ldots\|)$ is a normed vector space. (Indeed, $\|x + y\|^2 = \langle x + y, x + y\rangle = \langle x, x\rangle + \langle x, y\rangle + \langle y, x\rangle + \langle y, y\rangle \leqslant \|x\|^2 + 2\|x\|\,\|y\| + \|y\|^2 = (\|x\| + \|y\|)^2$. Thus $\|x + y\| \leqslant \|x\| + \|y\|$.)

A.6.2.4 *Definition* A pre-Hilbert space $(H, \langle\ ,\ \rangle)$ is called a Hilbert space if it is complete relative to the norm.

A.6.2.5 *Lemma* Let $(H_0, \langle\ ,\ \rangle)$ be a pre-Hilbert space. Then there is a Hilbert space $(H, \langle\ ,\ \rangle)$ and a linear isometry $i: H_0 \to H$ (that is, i is linear and $\langle ix, iy\rangle = \langle x, y\rangle$) so that $i(H_0)$ is dense in H.

PROOF See A.6.1.5.

A.6.2.6 We may thus complete any pre-Hilbert space to a Hilbert space. We will generally identify $i(H_0)$ with H_0 and use A.7.2.5 to say that if $(H_0, \langle\ \rangle)$ is a pre-Hilbert space then $(H_0, \langle\ \rangle)$ is a dense subspace of its completion $(H, \langle\ \rangle)$.

A.6.2.7 *Definition* Let $(H, \langle\ \rangle)$ be a Hilbert space. H is said to be separable if it contains a countable dense subset. An orthonormal basis of H is a sequence $\{\varphi_n\}_{n=1}$ of elements of H so that

(i) $\langle \varphi_n, \varphi_m \rangle = \delta_{nm}$.
(ii) The linear span of the $\{\varphi_n\}$ is dense in H.

A.6.2.8 We note that if $\{\varphi_n\}_{n=1}$ is an orthonormal basis of H and if $f \in H$ is so that $\langle f, \varphi_n \rangle = 0$, then $\langle f, \sum a_i \varphi_i \rangle = 0$ for all $a_i \in C$. Thus $f = 0$. Conversely suppose that $\{\varphi_n\}_{n=1}$ is an orthonormal subset of H (that is, satisfies (i) above) and is such that if $\langle f, \varphi_n \rangle = 0$ for all n, then $f = 0$. We assert that $\{\varphi_n\}$ is an orthonormal basis of H.

In fact we prove the following.

A.6.2.9 Suppose $\{\varphi_n\}$ is an orthonormal sequence in H so that if $f \in H$ and $\langle f, \varphi_n \rangle = 0$ for all n then $f = 0$. Then the series

$$\sum_{n=1} \langle f, \varphi_n \rangle \varphi_n$$

converges to f.

PROOF Set $c_n = \langle f, \varphi_n \rangle$. Set $S_N = \sum_{n=1}^{N} c_n \varphi_n$. Set $h_N = f - S_N$. Then $\langle h_N, S_N \rangle = 0$ and $f = S_N + h_N$. Thus $\|f\|^2 = \|S_N\|^2 + \|h_N\|^2$. Hence $\|S_N\|^2 \leqslant \|f\|^2$. Since $\|S_N\|^2 = \sum_{i=1}^{N} |c_i|^2$ we see that the series $\sum_{i=1}^{\infty} |c_i|^2$ converges and $\sum_{i=1} |c_i|^2 \leqslant \|f\|^2$. Now if $M > N$ then $\|S_M - S_N\|^2 = \sum_{j=N+1}^{M} |c_j|^2$. This implies that the sequence $\{S_n\}$ is Cauchy. Hence $\lim_{N\to\infty} S_N = f_0$ exists in H since H is complete. Now $\langle f_0, \varphi_n \rangle = \lim_{N\to\infty} \langle S_N, \varphi_n \rangle = c_n = \langle f, \varphi_n \rangle$ for all n. Thus $\langle f - f_0, \varphi_n \rangle = 0$ for all n. The lemma is now proved.

A.6.2.10 *Lemma* Let H be a Hilbert space, then H has an orthonormal basis if and only if H is separable.

PROOF Let H have an orthonormal basis $\{\varphi_n\}$. Let $P \subset H$ be the subset consisting of all linear combinations. $\sum_{n=1}^{N} (a_n + \sqrt{-1}b_n)\varphi_n$, a_n, b_n rational. Then P is dense in H. Hence H is separable. Suppose H is separable. Let $P = \{\xi_n\}$ be a countable dense subset. Define $Q_1 = \{\xi_1\}$ if $\xi_1 \neq 0$, otherwise $Q_1 = \varnothing$. Suppose that Q_N has been defined. Let $Q_{N+1} = Q_N \cup \{\xi_{N+1}\}$ if ξ_{N+1} is independent of Q_N. Let $Q_{N+1} = Q_N$ if ξ_{N+1} is dependent on Q_N. Let $Q = \bigcup_{N=1} Q_N$. Then label the elements of Q as $Q = \{\rho_n\}_{n=1}$. The space of linear combinations of the $\{\rho_n\}$ is dense in H. Furthermore $\{\rho_1, \ldots, \rho_N\}$ is a linearly independent set for each N. We may therefore apply the Gram-Schmidt orthogonalization process to $\{\rho_1, \ldots, \rho_N\}$ for each N. (that is, $\varphi_1 = \rho_1/\|\rho_1\|$, $\varphi_2 = (\rho_2 - \langle\rho_2, \varphi_1\rangle\varphi_1/\|\rho_2 - \langle\rho_2, \varphi_1\rangle\varphi_1\|), \ldots$). We then get an orthonormal sequence $\{\varphi_n\}$ whose linear span is dense in H. This proves the lemma.

A.6.2.11 Let $(H, \langle\ \rangle)$ be a Hilbert space. Then relative to

$$\|x\| = \langle x,x\rangle^{1/2}, (H, \|\ldots\|)$$

is a Banach space. We may thus carry the material of A.6.1 over to Hilbert spaces.

A.6.2.12 *Definition* A in $L(H, H)$ is said to be completely continuous (compact) if for any bounded infinite subset $S \subset H$, AS has an infinite convergent subsequence. Let $K(H) \subset L(H, H)$ be the space of all completely continuous operators in $L(H, H)$.

A.6.2.13 We note that $L(H, H)$ is an algebra under composition ($\|A_1A_2\| \leqslant \|A_1\|\ \|A_2\|$) and $K(H)$ is an ideal in $L(H, H)$.

A.6.2.14 *Lemma* If $\{A_n\}$ is a sequence in $K(H)$ converging to A in $L(H, H)$ then $A \in K(H)$.

PROOF Let $\{f_n\}$ be a bounded sequence in H with bound $\|f_n\| < c$. Using the Cantor diagonal process we can find a subsequence $\{f_{n_j}\}$ so that $\{A_k f_{n_j}\}$ converges for each k. We may assume $\{f_{n_j}\} = \{f_n\}$. We show $\{Af_n\}$ converges. Given $\varepsilon > 0$ there is m so that if $k \geqslant m$ then $\|A - A_k\| < \varepsilon/4c$. Fix such a k. There exists l so that if $p, q \geqslant l$ then $\|A_k f_p - A_k f_q\| < \varepsilon/4$. Now if p, $q \geqslant l$, $\|Af_p - Af_l\| \leqslant \|A_k(f_p - f_l)\| + \|(A - A_k)(f_p - f_l)\| \leqslant \varepsilon/4 + 2c\varepsilon/4c < \varepsilon$. Thus $\{Af_p\}$ is a Cauchy sequence. This proves the lemma.

A.6.2.15 *Lemma* Let $(H, \langle\ ,\ \rangle)$ be a Hilbert space. Let H_0 be a closed subspace of H. Set $H_0^{\perp} = \{v \text{ in } H_0 | \langle v, w\rangle = 0 \text{ for all } w \text{ in } H_0\}$. Then $H_0^{\perp}$ is closed and $H = H_0 \oplus H_0^{\perp}$.

PROOF $H_0^{\perp}$ is closed since $H_0^{\perp} = \bigcap_{w \in H_0} \{v \text{ in } H | \langle v, w\rangle = 0\}$. Let x be in H, x not in H_0. Let $d = \inf\{\|x - w\| \mid w \text{ in } H_0\}$. If $d = 0$ then there is a sequence $\{w_n\}$ in H_0 such that $\lim_{n\to\infty} \|x - w_n\| = 0$. Thus $\lim_{n\to\infty} w_n = x$. This implies that (since H_0 is closed) x is in H_0. Since we have assumed otherwise, $d > 0$.

Let $\{w_n\}$ in H_0 be a sequence such that $\lim_{n\to\infty} \|x - w_n\| = d$. Now

$$\begin{aligned}
&\|x - w_n\|^2 + \|x - w_m\|^2 \\
&\quad = \langle x - w_n, x - w_n\rangle + \langle x - w_m, x - w_m\rangle \\
&\quad = \langle x, x\rangle - \langle w_n, x\rangle - \langle x, w_n\rangle + \langle w_n, w_n\rangle + \langle x, x\rangle - \langle w_m, x\rangle \\
&\qquad - \langle x, w_m\rangle + \langle w_m, w_m\rangle \\
&\quad = 2\left\langle x - \frac{w_n + w_m}{2}, x - \frac{w_n + w_m}{2}\right\rangle + \frac{1}{2}\langle w_m - w_n, w_m - w_n\rangle.
\end{aligned}$$

Thus

$$\|x - w_n\|^2 + \|x - w_m\|^2 = 2\left\|x - \frac{w_n + w_m}{2}\right\|^2 + \frac{1}{2}\|w_m - w_m\|^2.$$

Thus $\frac{1}{2}\|w_m - w_n\|^2 \leqslant \|x - w_n\|^2 + \|x - w_m\|^2 - 2d$. Since $\|x - (w_n + w_m/2)\| \leqslant d$, this clearly implies $\{w_k\}$ is a Cauchy sequence in H_0. Thus $\lim_{k\to\infty} w_k = x_0$ exists and x_0 is in H_0. $\|x - x_0\| = d$. We show that $x - x_0$ is orthogonal to H_0. In fact suppose that w is in H_0; then

$$\begin{aligned}
d^2 \leqslant \|x - (x_0 - \lambda w)\|^2 &= \|x - x_0 - \lambda w\|^2 \\
&= \|x - x_0\|^2 - 2\mathrm{Re}(\lambda\langle x - x_0, w\rangle) + |\lambda|^2\|w\|^2.
\end{aligned}$$

$|\lambda|^2\|w\|^2 \geqslant 2\mathrm{Re}\lambda\langle x - x_0, w\rangle$, for all w in H_0. Let w be a unit vector in H_0. Then if t is in R, $t > 0$, we have $t^2\|w\|^2 \geqslant t|\mathrm{Re}\langle x - x_0, w\rangle|$ (taking $\lambda = -t$ if $\mathrm{Re}\langle x - x_0, w\rangle < 0$). Then

$$t\|w\|^2 \geqslant |\mathrm{Re}\langle x - x_0, w\rangle|.$$

Thus letting $t \to 0$ we see $\mathrm{Re}\langle x - x_0, w\rangle = 0$. Similarly $\mathrm{Im}\langle x - x_0, w\rangle = 0$. Thus $x - x_0$ is in $H_0^{\perp}$. The rest of the lemma is now clear.

A.6.2.16 Let A be in $L(H, H)$. Then A is said to be self-adjoint if $\langle Ax, y\rangle = \langle x, Ay\rangle$ for all x, y in H.

A.6.2.17 *Theorem* Let $(H, \langle\ ,\ \rangle)$ be a separable Hilbert space. Let A be a self-adjoint element of $K(H)$. Then

(1) if $H_\mu = \{z \in H | Az = \mu z\}$, $H_\mu \neq (0)$ implies μ is real. If $\mu \neq 0$, $\dim H_\mu < \infty$.

(2) The set of μ such that $H_\mu \neq (0)$ is either finite or countable. Let for $A \neq 0$, $\{\mu | \mu \neq 0, H_\mu \neq (0)\} = \{\lambda_k\}_{k=1}^N$, $|\lambda_k| \geqslant |\lambda_{k+1}|$, N a positive integer or infinity. If $N = \infty$, $\lim_{k\to\infty} |\lambda_k| = 0$.

(3) $H = H_0 + \sum_{k=1} H_{\lambda_k}$ a Hilbert space direct sum (that is, if f is in H then $f = f_0 + \sum_{k=1} f_k$ with f_k in H_{λ_k} and $\langle f_k, f_l\rangle = 0$ if $k \neq l$).

PROOF If $H_\mu \neq 0$ then there is v in H, $v \neq 0$, so that $\langle Av, v\rangle = \lambda\langle v, v\rangle$. But $\langle Av, v\rangle = \langle v, Av\rangle = \bar{\lambda}\langle v, v\rangle$. Hence $\lambda = \bar{\lambda}$. Let $H_\mu = \{v | Av = \mu v\}$, $\mu \neq 0$. Suppose $\dim H_\mu = \infty$. Then there is a sequence $\{\varphi_k\}_{k=1}^\infty$ of unit vectors in H_μ so that $\langle \varphi_k, \varphi_l\rangle = \delta_{kl}$ (use the Gram-Schmidt process). But then $\{A\varphi_k\} = \{\mu\varphi_k\}$ has a convergent subsequence. Let φ_{k_j} be such that $\lim_{j\to\infty} \mu\varphi_{k_j} = \mu\varphi_0$. Set $\varphi_{k_j} = \psi_j$. Then $\lim_{j\to\infty} \psi_j = \varphi_0$. Now $\langle \psi_j, \psi_{j+1}\rangle = 0$. Thus $\lim_{j\to\infty} \langle \psi_j, \psi_{j+1}\rangle = 0$. Hence $\lim_{j\to\infty} \langle \psi_j, \psi_{j+1}\rangle = \langle \varphi_0, \varphi_0\rangle = 0$. But $1 = \lim_{j\to\infty} \langle \psi_j, \psi_j\rangle = \langle \varphi_0, \varphi_0\rangle$. This contradiction completes the proof of (1).

Let us now assume that $A \neq 0$. Let $\{\psi_n\}$ be a sequence of unit vectors in H so that $\lim_{n\to\infty} \|A\psi_n\| = \|A\|$. Then $\{A\psi_n\}$ has a convergent subsequence $\{A\psi_{n_j}\}$ and if $\lim_{j\to\infty} A\psi_{n_j} = \varphi$ then $\|\varphi\| = \|A\|$. Now if x is in H then $\langle A^2x, A^2x\rangle\langle x, x\rangle \geqslant |\langle A^2x, x\rangle|^2$ by A.6.2.2 and equality holds if and only if A^2x and x are linearly dependent. Set $\xi_1 = \varphi/\|A\|$. Then $\|A\| \geqslant \|A\xi_1\|$ $\lim_{j\to\infty} \|A^2(\|A\|^{-1}\psi_{n_j})\| = \|A\|$. Thus $\|A\xi_1\| = \|A\|$. Now $\|A\|^2 = \|A\xi_1\|^2 \leqslant \|A^2\xi_1\| \leqslant \|A\|\,\|A\xi_1\| \leqslant \|A\|^2$. Thus $\|A\xi_1\|^2 = \|A^2\|$. This implies that $A^2\xi_1$ and ξ_1 are linearly dependent. Hence $A^2\xi_1 = \lambda^2\xi_1$, $\lambda^2 = \|A\|^2$. If $\lambda = \|A\|$ then $(A^2 - \lambda^2)\xi_1 = 0$. Hence $(A - \lambda)(A + \lambda)\xi_1 = 0$. This clearly implies that

(a) there is a unit vector φ_1 in H and λ_1 in R so that $A\varphi_1 = \lambda_1\varphi_1$ and $|\lambda_1| = \|A\|$.

Let now $H_0 = \ker A$. Then H_0 is closed and in the notation of A.6.2.15, $H = H_0 \oplus H_0^\perp$. Since A is self-adjoint, $AH_0^\perp \subset H_0^\perp$ (indeed if v is in $H_0^\perp$ and w is in H_0, $\langle Av, w\rangle = \langle v, Aw\rangle = \langle v, 0\rangle = 0$). Set $H^1 = H_0^\perp$.

Now H_{λ_1} and $H_{-\lambda_1}$ are finite dimensional (hence closed). Thus $H = (H_{\lambda_1} \oplus H_{-\lambda_1}) \oplus (H_{\lambda_1} \oplus H_{-\lambda_1})^\perp$. $H^2 = (H_{\lambda_1} \oplus H_{-\lambda_1})^\perp \cap H^1$, then $AH^2 \subset H^2$. (a) now applies to H^2. Let $|\lambda_2| = \|A|_{H^2}\|$. Then $(H_{\lambda_2} \oplus H_{-\lambda_2}) \subset H^2$ and $H_{\lambda_2} \oplus H_{-\lambda_2} \neq (0)$ define H^3 to be $(H_{\lambda_2} \oplus H_{-\lambda_2})^\perp \cap H^2$. We may thus define λ_k inductively so that $|\lambda_1| > |\lambda_2| > \cdots > |\lambda_k|$ and H^{k+1} to be $(\sum_{i=1}^k H_{\lambda_i} \oplus H_{-\lambda_i}))^\perp \cap H^k$. Then applying (a) to H^k we find λ_{k+1},

(i) If after k stages $H^{k+1} = (0)$. Then $\dim H_0^\perp < \infty$ and (2) and (3) are true.

(ii) $H^k \neq (0)$ for all k. Let $H^0 = \bigcap_{k=1}^{\infty} H^k$. Suppose $|\lambda_k| \geqslant \delta > 0$ for $k = 1, 2, \ldots$. Let φ_k in $(H_{\lambda_k} \oplus H_{-\lambda_k})$ be a unit vector. Set $\xi_k = 1/|\lambda_k|\ \varphi_k$. Then $|\xi_k| < \delta$. Since A is completely continuous $\{A\xi_k\}$ has a convergent subsequence. Arguing as in the proof of (1) we find a contradiction ($\langle \varphi_k, \varphi_l \rangle = \delta_{k,l}$). This implies that $\lim_{k\to\infty} |\lambda_k| = 0$. But by definition of H^k, $\|A|_{H^k}\| = |\lambda_k|$. Thus if v is in H^0, $Av = 0$. This implies that $H^0 \subset H_0 \cap H_0^{\perp}$, a contradiction.

A.6.2.18 The proofs of A.6.2.13 and A.6.2.15 were taken with slight modifications from Shilov [1].

APPENDIX 7

Integral Operators

A.7.1 Measures on Locally Compact Spaces

A.7.1.1 Let X be a locally compact Hausdorff space. Let $C(X)$ and $C_0(X)$ be, respectively, the spaces of continuous complex valued functions on X and the continuous complex valued functions on X with compact support. Let for $f \in C(X)$, $\|f\|_\infty = \sup_{x \in X} |f(x)|$. Set

$$B(X) = \{f \in C(X) | \ \|f\|_\infty < \infty\}.$$

A.7.1.2 $B(X)$ is complete relative to $\|\ldots\|_\infty$.

PROOF Let $\{f_n\}$ be a Cauchy sequence in $B(X)$. Then for each $\varepsilon > 0$ there is an m so that if $k, l \geqslant m$, $\|f_k - f_l\|_\infty < \varepsilon$. But this says that if $x \in X$, $|f_k(x) - f_l(x)| < \varepsilon$, for $k, l \geqslant m$. Thus for each $x \in X$, $\{f_n(x)\}$ is a Cauchy sequence in C. Thus we may define $f(x) = \lim_{n \to \infty} f_n(x)$.

f is continuous. Indeed let $x_0 \in X$ be given and let $\varepsilon > 0$ be given. Let m be so that if $k, l \geqslant m$, $\|f_k - f_l\| < \varepsilon/4$. Let $k > m$ and let U be a neighborhood of x_0 so that $|f_k(x) - f_k(x_0)| < \varepsilon/4$ for $x \in U$. We note that if $x \in X$ then $|f_k(x) - f(x)| = \lim_{m \to \infty} |f_k(x) - f_m(x)| \leqslant \varepsilon/4$. Thus if $x \in U$ then

$$|f(x) - f(x_0)| \leqslant |f(x) - f_k(x)| + |f_k(x) - f_k(x_0)| + |f_k(x_0) - f(x_0)|$$
$$< 3\varepsilon/4 < \varepsilon.$$

Thus $f \in C(X)$. Clearly $\|f\|_\infty < \infty$. Indeed, if m is so that $\|f_l - f_k\|_\infty < 1$ for $l, k \geqslant m$ then $\|f\|_\infty \leqslant 1 + \|f_m\|_\infty < \infty$.

We note that in the case X is compact $C(X) = C_0(X) = B(X)$.

A.7.1.3 *Definition* A measure on X is a linear function $\mu: C_0(X) \to C$ so that if $K \subset X$ is compact there is a constant N_K so that if $f \in C_0(X)$ is such that supp $f \subset K$ then $|\mu(f)| \leqslant N_K \|f\|_\infty$.

A.7.1.4 A measure is said to be positive if whenever $f \in C_0(X)$ is real valued and $f(x) \geqslant 0$ for all x in X then $\mu(f) \geqslant 0$.

A.7.1.5 We say that a positive measure on X is normal if for each $f \in C_0(X)$ such that $f \geqslant 0$ and $f \not\equiv 0$ then $\mu(f) > 0$.

A.7.1.6 Fix a normal positive measure on X. We make $C_0(X)$ into a pre-Hilbert space (see A.6.2) by defining $\langle f_1, f_2 \rangle = \mu(f_1 \bar{f}_2)$. Let $L^2(X, \mu)$ be the completion of $(C_0(X), \langle\ ,\ \rangle)$.

A.7.1.7 *Theorem* If X is a separable metric space (that is, X has a countable dense subset and the topology of X is given by a metric d), then $L^2(X, \mu)$ is separable.

PROOF By the above assumptions X has a countable basis $\{X_j\}$, $j \in N$, for its topology so that $\bar{X}_j$ (the closure of X_j is denoted $\bar{X}_j$) is compact. Now there is a collection of functions $\varphi_j \in C_0(X)$ so that supp $\varphi_j \subset X_j$, $\varphi_j \geqslant 0$, if $U \subset X$ is open then all but a finite number of the φ_j are zero on U, and $\sum_{j=1}^{\infty} \varphi_j(x) = 1$ for each $x \in X$. Suppose that $f \in C_0(X)$ is such that $\langle f, \varphi_j \rangle = 0$ for all j. We show that $f = 0$. By the positivity of μ we find that if f is real valued $\mu(f) \in R$. If $\langle f, \varphi_j \rangle = 0$ for all φ_j then $\langle \operatorname{Re} f, \varphi_j \rangle = \langle \operatorname{Im} f, \varphi_j \rangle = 0$ for all j. We may therefore suppose that f is real. Suppose that $f(x_0) \neq 0$. Then there is a connected neighborhood U of x_0 in X so that $f(x) \neq 0$ for $x \in U$. Thus either $f > 0$ on U or $f < 0$ on U. By possibly substituting $-f$ for f we may assume that $f > 0$ on U. Let φ_j be so that $\varphi_j(x_0) > 0$ and supp $\varphi_j \subset U$. Then $\varphi_j f \in C_0(X)$ and $\mu(\varphi_j f) > 0$, a contradiction. The argument of Lemma A.7.1.1 now completes the proof.

A.7.1.8 If M is an orientable manifold then it is a standard result that M is metrizable (see Kobayashi-Nomizu [1]). Let ω be a volume element on M.

Then we note that if we define for $f \in C_0(M)$, $\mu(f) = \int_M f\omega$, μ is a positive normal measure on M. The results of this section therefore apply to $f \mapsto \int_M f\omega$.

A.7.2 Integral Operators

A.7.2.1 Let X be a locally compact metrizable space. Let μ be a normal measure on X. We set $\mu(f) = \int_X f(x)dx$.

A.7.2.2 Let $k \in C_0(X \times X)$. We define an operator

$$A_k\colon C_0(X) \to C_0(X) \quad \text{by}$$

$$A_k(f)(x) = \int_X k(x, y)f(y)dy.$$

A.7.2.3 *Lemma* For each $k \in C_0(X \times X)$, A_k extends to a completely continuous operator on $L^2(X, \mu)$. A_k is self-adjoint if and only if $k(x, y) = \overline{k(y, x)}$.

PROOF We first show that A_k is bounded, hence extends to $L^2(X, \mu)$. Let $f \in C_0(X)$ and suppose that $\|f\| = 1$. Then

$$\|A_k f\|^2 = \int_X \left| \int_X k(x, y)f(y)dy \right|^2 dx.$$

Lemma A.6.2.2 implies

$$\left| \int_X k(x, y)f(y)dy \right|^2 \leqslant \int_X |k(x, y)|^2 dy \, \|f\|^2.$$

Hence

(1) $$\|A_k\|^2 \leqslant \int_X \left(\int_X |k(x, y)|^2 \, dy \right) dx.$$

Hence A_k is bounded.

Let now $\{\varphi_n\}$ be an orthonormal basis of $L^2(X, \mu)$ with $\varphi_n \in C_0(X)$. Let $\psi_{n,m}(x, y) = \varphi_n(x)\varphi_m(y)$. Defining the product measure on $X \times X$ via $(\mu \times \mu)(h) = \int_X (\int_X h(x, y)dx)dy$ we see that $\psi_{n,m}$ is an orthonormal basis of $L^2(X \times X, \mu \times \mu)$. Let $a_{n,m} = \langle k, \psi_{n,m} \rangle$. Set $k_p = \sum_{n+m<p} a_{n,m}\psi_{n,m}$. Then

A_{k_p} has finite dimensional image and is therefore completely continuous. By (1) above $\|A_k\| \leqslant \|k\|$. Thus since $A_{k_p} - A_k = A_{(k-k_p)}$ we see that $\|A_{k_p} - A_k\| \leqslant \|k_p - k\|$. This implies that $\lim_{p\to\infty} A_{k_p} = A_k$. Hence Lemma A.6.2.12 implies that A_k is completely continuous.

Now consider

$$\langle A_k f_1, f_2\rangle = \int_X \left(\int_X k(x, y) f_1(y) dy \right) \overline{f_2(x)} dx = \int_{X \times X} k(x, y) f_1(y) \overline{f_2(x)} dy\, dx.$$

Thus if $k(x, y) = \overline{k(y, x)}$ then $\langle A_k f_1, f_2\rangle = \langle f_1, A_k f_2\rangle$. Using the $\psi_{n,m}$ it is easy to see that this condition is also necessary.

APPENDIX 8

The Asymptotics for Certain Sturm-Liouville Systems

A.8.1 The Systems

A.8.1.1 Let $Q\colon (0, \infty) \to M_n(C)$ be a C^∞ mapping. Set $R^+ = (0, \infty)$ and define for f in $C^2(R^+; C^n)$, $Tf = -(d^2f/dt^2) + Qf$.

A.8.1.2 *Lemma* Let z, w be in C^n and let $0 < a < \infty$. Then there exists a unique f in $C^2(R^+; C)$ such that

(1) $Tf = 0$

(2) $f(a) = z$, $(df/dt)(a) = w$. In particular the space of all f in $C^2(R^+; C^n)$ such that $Tf = 0$ has complex dimension $2n$.

This result follows directly from the existence and uniqueness theorem for systems of O.D.E. (see, for example, Coddington and Levinson [1]).

A.8.1.3 By replacing Q by $Q = Q - \lambda I$ with λ in C, A.8.1.2 applies to the eigenvalue problem $Tf = \lambda f$.

A.8.2 The Asymptotics

A.8.2.1 Let $Q\colon R^+ \to M_n(C)$ be C^∞ and let $Tf = -(d^2f/dt^2) + Qf$ for f in $C^2(R^+, C^n)$.

A.8.2.2 Let $\|Q(t)\|$ be the operator norm of $Q(t)$ on C^n (we use the usual Hilbert structure on C^n, $\|(z_1, \ldots, z_n)\| = (\sum z_i \bar{z}_i)^{1/2}$). We assume that

$$\int_a^\infty (1 + t) \|Q(t)\| \, dt < \infty \tag{1}$$

for some (hence all) $a > 0$.

A.8.2.3 *Theorem* Suppose that Q satisfies (1). Let $\varepsilon > 0$ be given. Let μ be in R, $|\mu| > \varepsilon$, and let f be a solution to $Tf = \mu^2 f$. Then

(1) there are elements z, w in C^n and a constant $C > 0$ so that

$$\|f(t) - e^{i\mu t}z - e^{-i\mu t}w\| < C_\varepsilon \int_t^\infty (1 + s)\|Q(s)\| \, ds$$

for t sufficiently large, with C_ε depending only on ε.

(2) If $\lim_{t\to\infty} f(t) = 0$ then $f = 0$.

A.8.2.4 A.8.2.3 will be proved after we develop some lemmas.

A.8.2.5 Let $H^+ = \{z \text{ in } C | \operatorname{Im} z \geqslant 0\}$. Let $H_\varepsilon^+ = \{z \text{ in } H^+ | \, |z| \geqslant \varepsilon\}$.

A.8.2.6 *Lemma* Let for z in C, $\varphi(z) = (e^{2iz} - 1)/2iz$ (recall $(e^x - 1)/x \doteq \sum_{n=0} x^n/(n + 1)!$). Then $|\varphi(z)| \leqslant 1$ for z in H^+.

PROOF We leave this result as an exercise to the reader.

A.8.2.7 Let for $a > 0$, $\mathfrak{B}_a$ be the space of all bounded continuous functions from $[a,\infty]$ to C^n. Let $\|f\|_a = \sup\{\|f(t)\| \mid t \geqslant a\}$. A.7.1.2 implies that $(\mathfrak{B}_a, \|\ldots\|_a)$ is a Banach space.

A.8.2.8 Let for h in $\mathfrak{B}_a$, μ in H^+,

$$(L_{a,\mu}h)(t) = \int_t^\infty \varphi(\mu(s - t))(s - t)Q(s)h(s)ds$$

for $t \geqslant a$.

A.8.2.9 *Lemma* $L_{a,\mu}\colon \mathfrak{B}_a \to \mathfrak{B}_a$ and $\|L_{a,\mu}\| \leqslant \int_a^\infty (1 + s)\|Q(s)\|ds$. Furthermore the map $\mu \to L_{a,\mu}$ is continuous from H^+ to $L(\mathfrak{B}_a, \mathfrak{B}_a)$.

PROOF We leave it to the reader to check that $(L_{a,\mu}h)$ is continuous on $[a, \infty)$. Now

$$\|(L_{a,\mu}h)(t)\| \leqslant \int_t^{\infty} (s - t)|\varphi(\mu(s - t))| \, \|Q(s)\| \, ds \, \|h\|_a$$
$$\leqslant \int_a^{\infty} (1 + s)\|Q(s)\| \, ds \, \|h\|_a.$$

This clearly proves the first assertion of the lemma.

Suppose now that $b > a$, μ, ν in H^+, then

$$\|((L_{a,\mu} - L_{a,\nu})h(t)\| = \left\| \int_t^{\infty} (\varphi(\mu(s - t)) - \varphi(\nu(s - t)))(s - t)Q(s)h(s)ds \right.$$
$$\leqslant \sup_{0 \leqslant t \leqslant b - a} |\varphi(\mu t) - \varphi(\nu t)| \int_a^b (1 + s)\|Q(s)\| \, ds \, \|h\|_a$$
$$+ 2\int_t^{\infty} (1 + s)\|Q(s)\| \, ds \, \|h\|_a.$$

This certainly implies continuity.

A.8.2.10 Let now $a_0 > 0$ be so large that $\int_{a_0}^{\infty} (1 + s)\|Q(s)\| \, ds < \frac{1}{2}$. Then fixing $\mathfrak{B} = \mathfrak{B}_{a_0}$, $L_\mu = L_{a_0,\mu}$ for μ in H^+ and $\| \ \| = \| \ \|_{a_0}$, we see that $\|L_\mu\| < \frac{1}{2}$. A.6.1.2 now implies that $(I - L_\mu)^{-1} : \mathfrak{B} \to \mathfrak{B}$ exists. Clearly $\|(I - L_\mu)^{-1}\| < 2$.

A.8.2.11 If z is in C^n we identify z with the function $t \mapsto z$. Then z is in $\mathfrak{B}$. Define $h_{\mu,z} = (I - L_\mu)^{-1}z$. Then $\mu, t \mapsto h_{\mu,z}(t)$ is continuous on $H^+ \times [a_0, \infty)$ and $z \mapsto h_{\mu,z}$ is linear. Furthermore $\|h_{\mu,z}\| < 2\|z\|$.

A.8.2.12 *Lemma* Let $g_{\mu,z}(t) = e^{i\mu t}h_{\mu,z}(t)$ for $t \geqslant a$. Then $Tg_{\mu,z} = \mu^2 g_{\mu,z}$. Let $\sigma_1(t, \mu, z)$ be the solution of $Tf = \mu^2 f$ on $(0, \infty)$ agreeing with $g_{\mu,z}$ on $[a, \infty)$. Then

$$\|\sigma_1(t, \mu, z) - e^{i\mu t}z\| \leqslant 2\int_t^{\infty} (1 + s)\|Q(s)\| \, ds \, \|z\|$$

for $t \geqslant a$, μ in H^+.

PROOF We leave it to the reader to check that if f is in $\mathfrak{B}$ then $L_\mu f$ is in C^2 and

$$(L_\mu f)'(t) = -\int_t^\infty e^{2i\mu(s-t)} Q(s) f(s)\, ds,$$

$$(L_\mu f)''(t) = Q(t)f(t) + 2i\mu \int_t e^{2i\mu(s-t)} Q(s) f(s)\, ds.$$

This implies $(L_\mu f)'' + 2i\mu(L_\mu f)' = Qf$ for f in $\mathfrak{B}$.

Now $h_{\mu,z} = (1 - L_\mu)^{-1} z$. Hence $L_\mu h_{\mu,z} + z = h_{\mu,z}$. This clearly implies $h_{\mu,z} = (L_\mu h_{\mu,z})'$, $h_{\mu,z} = (L_\mu h_{\mu,z})''$. Hence $h_{\mu,z} + 2i\mu h_{\mu,z} = Qh_{\mu,z}$.

Now $g_{\mu,z}(t) = e^{i\mu t} h_{\mu,z}(t)$. Hence

$$g''_{\mu,z}(t) = -\mu^2 e^{i\mu t} h_{\mu,z}(t) + e^{i\mu t} h'_{\mu,z}(t) + 2i\mu e^{i\mu t} h'_{\mu,z}(t).$$

Thus $g''_{\mu,z}(t) - Qg_{\mu,z}(t) = -\mu^2 g_{\mu,z}(t)$. This implies the first assertion of the lemma.

Now

$$h_{\mu,z}(t) = z + \int_t^\infty \varphi(\mu(s - t))(s - t) Q(s) h_{\mu,z}(s)\, ds$$

for $t \geqslant a$. Thus

$$\|h_{\mu,z}(t) - z\| \leqslant \int_t^\infty (s - t)\|Q(s)\|\, ds\, \|h_{\mu,z}\| \leqslant \int_t^\infty (1 + s)\|Q(s)\|\, 2\|z\|\, ds$$

since $\|h_{\mu,z}\| < 2\|z\|$. This proves the last assertion.

A.8.2.13 Let now for μ in H^+, $|\mu| > 0$, h in $\mathfrak{B}_a$,

$$(K_{a,\mu}h)(t) = \frac{1}{2i\mu}\left[\int_a^t e^{-2i\mu(s-t)} Q(s) f(s) ds + \int_t^\infty Q(s) f(s) ds\right]$$

for $t \geqslant a$.

A.8.2.14 *Lemma* $K_{a,\mu}$ defines a bounded operator from $\mathfrak{B}_a$ to $\mathfrak{B}_a$ for each $a > 0$ and each μ in H^+, $|\mu| > 0$. Also,

$$\|K_{a,\mu}\| \leqslant \frac{1}{|\mu|}\int_a^\infty \|Q(s)\|\, ds.$$

Furthermore $\mu \mapsto K_{a,\mu}$ from $H^+ - \{0\}$ to $L(\mathfrak{B}_a, \mathfrak{B}_a)$ is continuous.

PROOF Using the fact that $|e^{-2i\mu(s-t)}| \leqslant 1$ for μ in H^+ and $s \leqslant t$, we see that

$$\|K_{a,\mu} f(t)\| \leqslant \frac{1}{|\mu|}\int_a^\infty \|Q(s)\|\, ds\, \|f\|_a.$$

This proves the first assertion.

Also if μ, ν are in $H^+ - \{0\}$,

$$\|K_{a,\mu}f(t) - K_{a,\nu}f(t)\| \leqslant \frac{1}{2}\left|\frac{1}{\mu} - \frac{1}{\nu}\right| \int_a^\infty \|Q(s)\|\, ds\, \|f\| + \left\| \frac{1}{2i\mu}\int_a^t e^{-2i\mu(s-t)}Q(s)f(s)ds - \frac{1}{2i\nu}\int_a^t e^{-2i\nu(s-t)}Q(s)f(s)ds \right\|.$$

Let $a < b$, $|\mu - \nu| < \delta$, μ, ν in $H^+ - \{0\}$ and b be so large that

$$\frac{1}{2|\mu|}\int_b^\infty \|Q(s)\|\, ds < \varepsilon/2$$

$$\frac{1}{2|\nu|}\int_b^\infty \|Q(s)\|\, ds < \varepsilon/2.$$

Then

$$|K_{a,\mu}f(t) - K_{a,\nu}f(t)| \leqslant \frac{1}{2}\left|\frac{1}{\mu} - \frac{1}{\nu}\right| \int_a^\infty \|Q(s)\|\, ds\, \|f\|_a + \varepsilon\|f\|_a + \sup_{0 \leqslant x \leqslant b-a}\left|\frac{e^{2i\mu x}}{2\mu} - \frac{e^{2i\nu x}}{2\nu}\right| \int_a^b \|Q(s)\|\, ds\, \|f\|_a.$$

Thus $\mu \mapsto K_{a,\mu}$ is indeed continuous on $H^+ - \{0\}$.

A.8.2.15 Let $\varepsilon > 0$ be given and let $a_\varepsilon > 0$ be such that

$$(1) \qquad \frac{1}{\varepsilon}\int_{a_\varepsilon} \|Q(s)\|\, ds < \frac{1}{2}.$$

Then $\|K_{a_\varepsilon,\mu}\| < \frac{1}{2}$ if μ is in H_ε^+. Hence applying A.6.1.2 again we find that $(I - K_{a_\varepsilon,\mu})^{-1}$ exists on $\mathfrak{B}_{a_\varepsilon}$ and $\mu \mapsto (I - K_{a_\varepsilon,\mu})^{-1}$ is continuous on H_ε^+.

A.8.2.16 Let for z in C^n, μ in H_ε^+, $w_{\mu,z}^\varepsilon = (I - K_{a_\varepsilon,\mu})^{-1}z$. Then $w_{\mu,z}^\varepsilon$ is in $\mathfrak{B}_{a_\varepsilon}$ and $\|w_{\mu,z}\| < 2\|z\|$.

A.8.2.17 *Lemma* $w_{\mu,z}^\varepsilon$ is in C^2 on (a_ε, ∞). Set $v_{\mu,z}^\varepsilon(t) = e^{-i\mu t}w_{\mu,z}(t)$ for $t \geqslant a_\varepsilon$. Then $Tv_{\mu,z}^\varepsilon = \mu^2 v_{\varepsilon,z}$ for μ in H_ε^+. Let $\sigma_2^\varepsilon(t, \mu, z)$ be the unique solution of $Tf = \mu^2 f$ agreeing with $v_{\mu,z}^\varepsilon(t)$ on $[a_\varepsilon, \infty)$ for μ in H_ε^+. Then if $C_\varepsilon(\mu) = (1/2i\mu)\int_{a_\varepsilon}^\infty e^{-2i\mu s}Q(s)w_{\mu,z}^\varepsilon(s)ds$ for μ in R we have

$$\|\sigma_2(t, \mu, z) - e^{-i\mu t}z - C_\varepsilon(\mu)e^{i\mu t}z\| \leqslant \frac{1}{|\mu|}\int_t^\infty \|Q(s)\|\, ds\, \|z\|$$

for $t \geqslant a_\varepsilon \notin |\mu| > \varepsilon$.

PROOF Arguing as in A.8.2.12 we see that if f is in $\mathfrak{B}_a$ then

(a) $$(K_{a,\mu}f)'(t) = \int_a^t e^{-2i\mu(s-t)}Q(s)f(s)ds, \ t \geqslant a.$$

(b) $$(K_{a,\mu}f)''(t) = 2i\mu\int_a^t e^{-2i\mu(s-t)}Q(s)f(s)ds + Q(t)f(t)$$

for $t \geqslant a$.

(a), (b) combined with the definition of $w_{\mu,z}^{\varepsilon}$ imply that

(c) $$(w_{\mu,z}^{\varepsilon})''(t) = 2i\mu(w_{\mu,z}^{\varepsilon})'(t) + Q(t)w_{\mu,z}^{\varepsilon}(t).$$

(c) clearly implies that, since $v_{\mu,z}^{\varepsilon}(t) = e^{-i\mu t}w_{\mu,z}^{\varepsilon}(t)$,

(d) $$Tv_{\mu,z}^{\varepsilon} = \mu^2 v_{\mu,z}^{\varepsilon} \text{ for } \mu \text{ in } H_\varepsilon^+ \text{ on } [a_\varepsilon, \infty).$$

Finally if $\mu \in R$, $|\mu| \geqslant t$

$$\begin{aligned}
&\left\| w_{\mu,z}^{\varepsilon}(t) - z - (2i\mu)^{-1}e^{2i\mu t}\int_{\alpha_\varepsilon}^{\infty} e^{-2i\mu s}Q(s)w_{\mu,z}^{\varepsilon}(s)\,ds \right\| \\
&= \left\| w_{\mu,z}^{\varepsilon}(t) - z - (K_{a_\varepsilon,\mu}w_{\mu,z}^{\varepsilon})(t) - (2i\mu)^{-1}\int_t^{\infty} Q(s)w_{\mu,z}^{\varepsilon}(s)\,ds \right\| \\
&= \frac{1}{2|\mu|}\left\| \int_t^{\infty} Q(s)w_{\mu,z}^{\varepsilon}(s)\,ds \right\| \leqslant \frac{1}{|\mu|}\int_t \|Q(s)\|\,ds\,\|z\|.
\end{aligned}$$

This proves the lemma.

A.8.2.18 *The proof of A.8.2.3* Let μ be in R, $\mu \neq 0$. Then μ is in H_ε^+ for some $\varepsilon > 0$. Let $V_\mu = \{f\colon (0, \infty) \to C^n | Tf = \mu^2 f\}$. We define $A_\mu\colon C^n \times C^n \to V$ by $A_\mu(z, w) = \sigma_1(t, \mu, z) + \sigma_2^{\varepsilon}(t, \mu, w)$. Now A.8.2.12 and A.8.2.17 imply that

(1) $$\lim_{t\to\infty} \|\sigma_1(t, \mu, z) + \sigma_2^{\varepsilon}(t, \mu, w) - e^{i\mu t}z - e^{-i\mu t}w - C_\varepsilon(\mu)e^{i\mu t}w\| = 0.$$

Suppose now that $A_\mu(z, w) = 0$. Then (1) implies that $z = w = 0$. Since A is linear and dim $V_\mu = 2n$ (A.8.12), we see that A_μ is surjective. Hence if f is as in A.8.2.2., $f(t) = \sigma_1(t, \mu, z) + \sigma_2(t, \mu, w)$. A.8.2.2.(1), (2) now follow from (1) above and A.8.2.12, A.8.2.17.

A.8.2.19 The construction of the $\sigma_1(t, \mu, z)$ and $\sigma_2(t, \mu, z)$ is taken with modification to systems from Dunford and Schwartz [1].

Bibliography

ATIYAH, M.

[1] *K-theory*, Benjamin, New York, 1967.

BARGMAN, V.

[1] Irreducible unitary representations of the Lorentz group I, *Ann. of Math.*, **48** (1947), 568–640.

BERS, L., F. JOHN and M. SCHECHTER

[1] *Partial Differential Equations*, Wiley-Interscience, New York, 1964.

BOREL A.

[1] *Linear Algebraic Groups*, Benjamin, New York, 1969.

[2] Kahlerian coset spaces of semi-simple Lie groups, *Proc. Nat. Acad. Sci. U.S.A.*, **40** (1954), 1147–1151.

BOREL, A., AND G. D. MOSTOW

[1] On Semi-simple automorphisms of Lie algebras, *Ann. of Math.*, **61** (1955), pp. 389–504.

BOREL, A., AND F. HIRZEBRUCH

[1] Characteristic classes and homogeneous spaces, I, II, III, *Amer. J. Math.*, **80** (1958), 458–538; **81** (1959), 315–382; **82** (1960), 491–504.

BOTT, R.

[1] Homogeneous differential operators, *Differential and Combinatorial Topology* (S. S. Cairns, ed.), Princeton Univ. Press, Princeton, New Jersey, 1965, pp. 167–186.

[2] Homogeneous Vector Bundles, *Ann. of Math.*, **66** (1957), pp. 203–248.

BOURBAKI, N.

[1] *Groupes et Algebres de Lie, Chapitre I, Algebres de Lie*, Act. Sci. Ind. 1285, Hermann, Paris, 1960.

BRUHAT, F.

[1] Sur les representations induites des groupes de Lie, *Bull. Soc. Math. France*, **84** (1956), pp. 97–205.

CARTAN, E.

[1] Sur certains formes riemannienes remarquables des geometries a group fundemental simple, *Ann. Sci. Ecole Norm. Supp.*, **44** (1927), 345–467.

CEREZO, A., AND R. ROVIERRE

[1] Solution elementaire d'un operator differential lineaire invariant a gauche sur un group de Lie real et sur un espace homogene reductif compact, *Am. Sci. Ecole Norm. Sup.* (4), **2** (1969), 561–581.

CHEVALLEY, C.

[1] *Theory of Lie Groups*, Princeton Univ. Press, Princeton, New Jessey, 1946.

CODDINGTON, E., AND N. LEVINSON

[1] *Theory of Ordinary Differential Equations*, McGraw-Hill, New York, 1955.

DUNFORD AND J. SCHWARTZ

[1] *Linear Operators*, Vol. 3, Wiley-Interscience, New York, 1972.

FROBENIUS, G.

[1] *Berl. Ber*, **501** (1898)

GELFAND, I. M., M. I. GREER, AND I. I. PYATETSKII-SHAPIRO

[1] *Representation Theory and Automorphic Functions*, Saunders, Philadelphia, Pennsylvania, 1969.

GELFAND, I. M., M. I. GRAEV, AND N. A. VLENKIN

[1] *Generalized Functions*, Vol. 5, Academic, 1966.

GELFAND, I. M., AND M. A. NAIMARK

[1] Unitary representations of the Lorentz group, *Izvestia Akad. Nauk. U.S.S.R.*, **11** (1947), 411.

[2] *Unitäre Darstellung* der *Klassichen Gruppen*, Akademie Verlag, Berlin, 1957.

GINDINKIN, S., AND F. KARPELEVIC

[1] Plancherel measure of Riemannian symmetric spaces of non-positive curvature, *Soviet Math. Dokl.*, **3** (1962), 962–965.

GREENFIELD, S., AND N. R. WALLACH

[1] Remarks on global hypo-ellipticity, *Trans. Amer. Math. Soc.*, to appear.

HARISH-CHANDRA

[1] On some applications of the universal enveloping algebra of a semi-simple Lie algebra, *Trans. Amer. Math. Soc.*, **70** (1951), 28–96.

[2] Representations of a Semi-Simple Lie group on a Banach space I, II, III, *Trans. Amer. Math. Soc.*, **75** (1953), 185–243; **76** (1954), 26–65, 234–253.

[3] The Plancherel formula for complex semi-simple Lie groups, *Trans. Amer. Math. Soc.*, **76** (1954), 485–528.

[4] The characters of semi-simple Lie groups, *Trans. Amer. Math. Soc.*, **83** (1956), 98–163.

[5] On a lemma of F. Bruhat, *J. Math. Pures Appl.*, **35** (1956), 203–210.

[6] Differential operators on a semi-simple Lie algebra, *Amer. J. Math.*, **79** (1957), 87–120.

[7] Fourier transform on a semi-simple Lie algebra I, II, *Amer. J. Math.*, **79** (1957) 193–257, 653–686.

[8] Spherical functions on a semi-simple Lie group I, II, *Amer. J. Math.*, **80** (1958), 241–310.

[9] Discrete series I, II, *Acta Math.*, **113** (1965), 241–318, **116** (1966), 1–111.

[10] Invariant eigen-distributions on a semi-simple Lie group, *Trans. Amer. Math. Soc.*, **119** (1965) 457–508.

HELGASON, S.

[1] *Differential Geometry and Symmetric Spaces*, Academic, New York, 1962.

[2] Lie groups and symmetric spaces, *Battelle Recontres*, Benjamin, New York, 1968, pp. 1–71.

[3] A duality for symmetric spaces with applications to group representations, *Advan. Math.*, **5** (1970), 1–154.

HERMANN, R.

[1] *Lie Groups for Physicists*, Benjamin, New York, 1966.

HIRZEBRUCH, F. (see also A. Borel and F. Hirzebruch)

[1] *Topological Methods in Algebraic Geometry*, Springer-Verlag, New York,

HOCKING, J., AND G. YOUNG

[1] *Topology*, Addison-Wesley, Reading, Massachusetts, 1961.

HOCHSCHILD, G.

[1] *The Structure of Lie Groups*, Holden-Day, San Francisco, California, 1965.

HOWE, R.

[1] Some qualitative results on the representation theory of GL_n over a p-adic field, Institute for Advanced Study, Lecture Notes.

HUNT, G.

[1] A theorem of Elie Cartan, *Proc. Amer. Math. Soc.*, **7** (1956), 307–308.

IWASAWA, K.

[1] On some types of topological groups, *Ann. of Math.*, **50** (1949), 507–558.

JACOBSON, N.

[1] *Lectures on Abstract Algebra*, I, II, Van Nostrand, Princeton, New Jessey, 1951, 1952.

[2] *Lie Algebras*, Wiley-Interscience, New York, 1962.

KNAPP, A., AND E. STEIN

[1] Intertwining operators for semi-simple groups, *Ann. of Math.*, **93** (1971), 489–578.

KOBAYASHI, S., AND K. NOMIZU

[1] *Foundations of Differential Geometry*, Vol. II, Wiley-Interscience, New York, 1969.

KOSTANT, B.

[1] The principal three dimensional subgroups and the Betti numbers of a complex simple Lie group, *Amer. J. Math.*, **81** (1959), 973–1032.

[2] Lie algebra cohomology and the generalized Borel-Weil theorem, *Ann. of Math.*, **74** (1961), 329–387.

[3] On the existence and irreducibility of certain series of representations, *Bull. Amer. Math. Soc.*, **75** (1969), 627–642.

KUNZE, R., AND E. STEIN

[1] Uniformly bounded representations I, II, III, IV, *Amer. J. Math.*, **82** (1960), 1–62; **83** (1961), 723–786; **89** (1967), 385–442, to appear.

LANG, S.

[1] *Analysis II*, Addison-Wesley, Reading, Massachusetts, 1969.

LEPOWSKI, J., AND N. R. WALLACH

[1] Finite dimensional and infinite dimensional representations of non-compact, linear, semi-simple Lie groups, to appear, *Trans. Amer. Math. Soc.*

LIPSMAN, R.

[1] On the characters and equivalence of continuous series representations, *J. Math. Soc. Japan*, **23** (1971), 452–480.

LOOMIS, L., AND J. STERNBERG

[1] *Advanced Calculus*, Addison-Wesley, Reading, Massachusetts, 1968.

MACKEY, G. W.

[1] Induced representations of locally compact groups I, *Ann. of Math.*, **55** (1952), 101–139.

MOSTOW, G. D. (see also A. Borel and G. Mostow)

[1] A new proof of E. Cartan's theorem on the topology of semi-simple Lie groups, *Bull. Amer. Math. Soc.*, **55** (1949), 969–980.

NARASIMHAN, M., AND K. OKAMOTO

[1] An analogue of the Borel-Weil-Bott theorem for Hermitian symmetric pairs of non-compact type, *Ann. of Math.*, **91** (1970), 486–511.

PALAIS, et al.

[1] *Seminar on the Atiyah-Singer Theorem*, Annals of Mathematics Studies, Princeton, New Jessey, 1965.

PETER, F., AND H. WEYL

[1] Die Vollständigkeit der primitiven Darstellungen einer geschlossenen kontinuielichen Gruppe, *Math. Ann.*, **97** (1927), 737–755.

SAMELSON, H.

[1] On curvature and characteristic of homogeneous spaces, *Mich. J. Math.*, **5** (1958), 13–18.

SATAKE, I.

[1] On representations and compactifications of symmetric Riemannian spaces, *Ann. of Math.*, **71** (1960), 77–110.

SCHIFFMAN, G.

[1] Integrales d'entrelacement et functions de Whittaker, *Bull. Soc. Math. France*, **99** (1971), 3–72.

SCHMID, W.

[1] On a conjecture of Langlands, *Ann. of Math.*, **93** (1971), 1–42.

SCHUR, I.

[1] Neue Begrundung der Theorie der Gruppen Charactere, *Sitz Ber. Preuss. Akad, Berlin* (1905), 406–513.

SHILOV, G.

[1] *An Introduction to the Theory of Linear Spaces*, Prentice-Hall, Englewood Cliffs, New Jersey, 1961.

"SEMINAIRE SOPHUS LIE"

[1] Theorie des algebres de Lie, Topology des groupes de Lie, Paris, 1955.

STEENROD, N.

[1] *The Topology of Fibre Bundles*, Princeton Univ. Press, Princeton, New Jersey, 1951.

STIEFEL, E.

[1] Kristallographische bestimmung der charactere der geschossenen lieshen gruppen, *Comm. Math. Helv.*, **17** (1945), 165–260.

WALLACH, N. R. (see also S. Greenfield and N. R. Wallach and J. Lepowsky and N. R. Wallach)

[1] Cyclic vectors and irreducibility of principal series representations I and II, *Trans. Amer. Math. Soc.*

[2] Induced representations of lie algebras, and a theorem of Borel-Weil, *Trans. Amer. Math. Soc.*, **136** (1969), 181–187.

[3] Induced representations of lie algebras II, *Proc. Amer. Math. Soc.*, **21** (1969), 151–166.

[4] On Harish-Chandra's generalized C-functions, to appear.

WANG, H. C.

[1] Closed manifolds with homogeneous complex structure, *Amer. J. Math.*, **76** (1954), 1–32.

WARNER, F.

[1] *Foundations of Differential Geometry and Lie Groups*, Scott, Foresman and Co., Glenview, Illinois, 1971.

WARNER, G.

[1] *Harmonic Analysis of Semi-Simple Lie Groups*, Vol. I, II, Springer-Verlag, Berlin, 1972.

WEIL, A.

[1] *L'integration Dans Les Groupes Topologiques et Les Applications*, Hermann, Paris, 1940.

WEYL, H.

[1] Theorie der darstellung kontinuerlichen halbenfachen groupen durch lineare transformationen, I, II, III, *Math. Zeit*, **23** (1925), 271–309, **24** (1926), 328–376, **24** (1926), 377–395.

WHITTAKER, F. T., AND G. N. WATSON

[1] *A Course of Modern Analysis*, 4th ed., Cambridge Univ. Press, London and New York, 1927.

WOLF, J.

[1] The action of a real semi-simple Lie group on a complex flag manifold, II: Unitary representations on partially homomorphic cohomology spaces, to appear.

ZELOBENKO, D. P.

[1] The analysis of irreducibility of elementary representations of a complex semi-simple Lie group, *Math. U.S.S.R.-Izvestya*, *Z*, (1968), 105–128.

Index